WINTER COLLEGE ON FUNDAMENTAL NUCLEAR PHYSICS (VOL. 2)

Errata

1. Pages 867 and 868 should be interchanged.
2. Page 871 should be page 869.
3. Page 869 should be page 870.
4. Page 870 should be page 871.

WINTER COLLEGE ON FUNDAMENTAL NUCLEAR PHYSICS

VOLUME 2

Proceedings of the

WINTER COLLEGE ON FUNDAMENTAL NUCLEAR PHYSICS

Volume 2

International Centre for Theoretical Physics, Trieste, Italy
7 Feb-30 Mar 1984

Editors

K Dietrich
M Di Toro
H J Mang

World Scientific

Published by
World Scientific Publishing Co. Pte. Ltd.
P. O. Box 128, Farrer Road, Singapore 9128

WINTER COLLEGE ON FUNDAMENTAL NUCLEAR PHYSICS – VOL. 2

ISBN 9971-978-25-3

Printed in Singapore by Singapore National Printers (Pte) Ltd.

CONTENTS

VOLUME 1

III. STATISTICAL MECHANICS AND RELATION TO NUCLEAR DYNAMICS

IV. THEORY OF STRONG INTERACTIONS

VOLUME 2

SPECIALIZED LECTURES

I. NUCLEAR STRUCTURE

VOLUME 3

I. NUCLEAR STRUCTURE

RAPIDLY SPINNING NUCLEI

Hans Ryde
Division of Cosmic and Subatomic Physics
University of Lund
Sweden

Among the many-body quantumsystems appearing in nature, the atomic nucleus is a unique one in the sense that the number of constituent nucleons is sufficiently large to allow of correlations, while at the same time the number is finite. Thus the strong interaction between the fermions allows the shapes of the nucleonic orbitals to influence the overall shape of the nucleus. Non-spherical nuclei can therefore be generated by a few anisotropic nucleonic configurations in the valence shell. A new dimension of the study of nuclear dynamics has been opened up through the extension of nuclear spectroscopic investigations to large angular momenta, where the centrifugal and Coriolis forces are sufficiently strong to modify the single-particle basis of the nuclear correlations. This development is founded on the exploration of the properties of nuclear energy levels at high spin. The pertinent level schemes are thus established from studies of the gamma radiation from rapidly rotating nuclei produced in heavy-ion fusion reactions or in Coulomb excitation. Several survey articles dealing with progress in this field have recently been published. Some of them are referred to in refs [1-6].

In these lectures we are first going to discuss the experimental information on high-spin states: how to populate these states and the experimental techniques used for the studies of the decay of the levels thus reached in these nuclear reactions. In the succeeding lectures we shall then be considering the interpretation of the data obtained in the light of pertinent theoretical ideas on the properties of rapidly rotating nuclei.

1. Formation and Decay of High-Spin Nuclear States

The availability of heavy-ion accelerators and improved detector systems for gamma rays have in recent years made it possible to reach high-spin nuclear states and to study their decay properties in detail. The experiments can be performed either via the multiple Coulomb excitation process or in fusion-evaporation reactions with the use of beams of heavy ions. The double-magic nucleus ^{208}Pb, for example, is an almost ideal projectile for the population of high-spin states in Coulomb excitation experiments, partly because of its large nuclear charge of Z = 82 and partly due to its very rigid structure, with the first excited state as high in energy as 2.6 MeV. Special care has to be taken in these experiments due to the large Doppler shifts involved. Beautiful experiments have been performed along these lines by groups working at the UNILAC-accelerator [7] operated by GSI in Darmstadt. A technique has been developed to obtain totally Doppler-corrected gamma-ray spectra following multiple Coulomb excitation and in that way levels with spins as high as 30 $\hbar$ have been observed in the actinide region, where the fusion-evaporation reactions would mostly lead to fission.

We shall leave the multiple Coulomb excitation technique here, referring the reader to the survey article by D Schwalm [7], and concentrate on the commonly used fusion-evaporation reactions as a means to reach high-spin states in the compound nuclei. Beams of particles in the mass region 30-50 and with an energy of 4-5 MeV per nucleon are frequently used in these experiments with a cross-section approaching one barn. As an illustration let us follow the reaction between a 147 MeV ^{40}Ar ion and a ^{124}Sn target nucleus , depicted in fig 1. The colliding nuclei fuse and form a composite system with angular momenta that may be as large as 60 $\hbar$, which would indicate a rotational frequency of about 2 x 10^{20} rotations per second. The compound nucleus is highly excited with an energy of 53.8 MeV, and evaporates neutrons over a characteristic time of 10^{-19} s. For the rare-earth system discussed here the threshold for proton emission, that is, the binding energy and the Coulomb barrier, strongly favours the neutron emission. The neutrons are efficient in removing energy from the excited compound system, but very inefficient in removing angular momentum. The

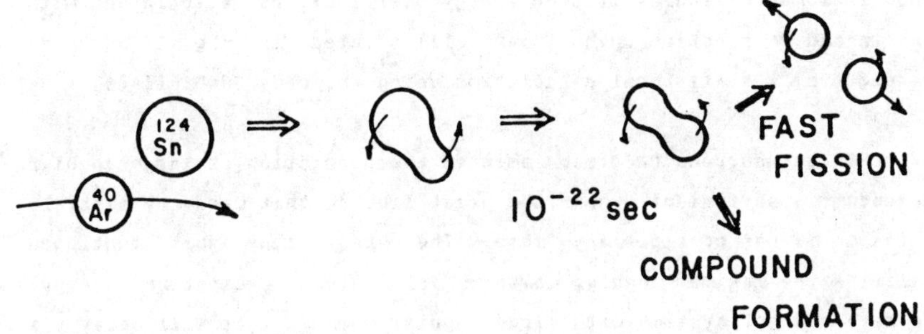

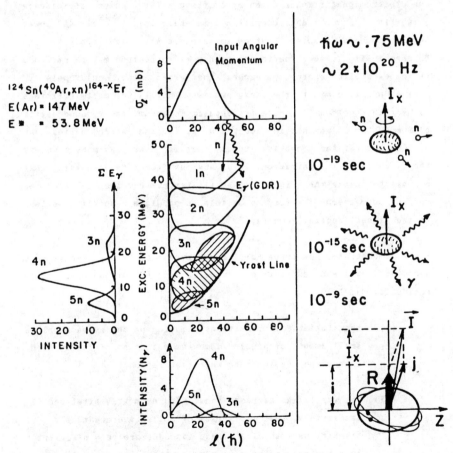

Fig 1. Schematic illustration of compound formation and decay. The left side of the figure shows a statistical calculation by GROGI II.

angular momentum and excitation-energy distributions associated with 1n, 2n and 3n reaction channels are illustrated in fig 1 as they result from a statistical calculation using the code GROGI II [8].

After three neutrons have been emitted the population at large angular momentum is sufficiently near the yrast line so that decay by particle emission is not possible any more. The yrast line just mentioned delimits the maximum angular momentum for a given excitation energy. We have now got systems with large angular momenta that will decay via the electromagnetic interaction by emitting a large number of cascade γ-rays in ^{161}Er on a characteristic time scale of 10^{-15} s. At lower angular momenta, however, neutron emission is still favoured, as can be seen in fig 1, so a fourth and even a fifth neutron may be emitted. As the excitation energy approaches the yrast line, quadrupole γ-radiation will dominate the decay process. This is the most efficient way for the nucleus to reduce its angular momentum. The ground state of the system is reached in about 10^{-9} s, during which period the nucleus has rotated about 10^{11} times, that is, nearly as many times as the Earth since its formation. It is obvious from this discussion that one has to choose the right projectile-target combination and the correct beam energy in order to be able to populate a specific nucleus at the highest angular momenta.

Fig 2 demonstrates how three different types of γ-ray decays are expected from the de-excitation processes for deformed evaporation residuals, namely

(i) the statistical transitions removing on the average only a small amount of angular momentum but being of fairly high energy and leading into the yrast region

(ii) the yrast-like cascades consisting of mostly stretched E2 transitions, that is, series of collective quadrupole transitions with $\Delta I=2$. These cascades are very efficient in removing angular momentum. As there are many alternative paths for these cascades within the yrast region, the discrete γ-rays can in most cases not be identified.

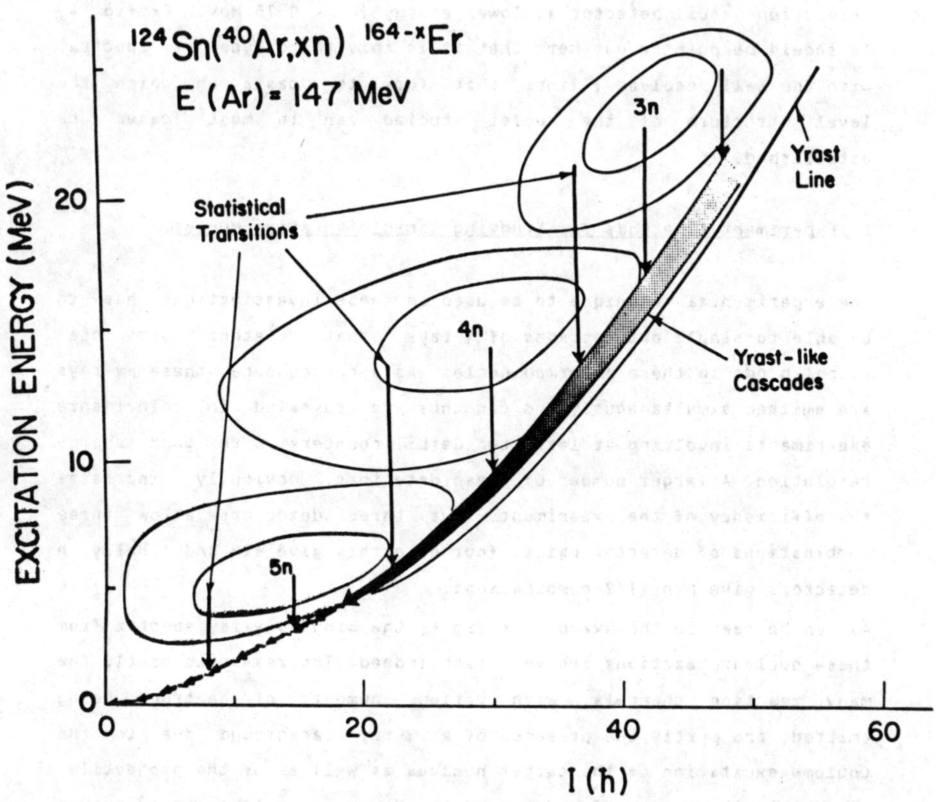

Fig 2. The γ-ray spectrum emitted in a heavy-ion reaction consists of statistical transitions, yrast-like cascades and discrete transitions.

(iii) the discrete transitions forming a few rotational bands leading to the ground state of the daughter nucleus.

The first two classes are clearly distinguished in the γ-ray spectrum (E_γ = 0.5 MeV to 4.0 MeV) recorded with a low-resolution NaI(Tl) detector and shown in fig 3. The third class of γ-rays, due to the discrete transitions, appears in a spectrum obtained with a high-resolution Ge(Li) detector at lower energy (E_γ < 0.75 MeV). Cf fig 4. It should be pointed out here that it is this third type of spectra, with the well-resolved γ-lines, that form the basis on which the level structure of the nuclei studied can in most cases be established.

2. Experimental Methods for Studying Rapidly Rotating Nuclei

The experimental technique to be used in these investigations has to be able to single out cascades of γ-rays, that together form rotational bands in these deformed nuclei. As a consequence these γ-rays are emitted simultaneously and can thus be revealed in coincidence experiments involving at least two Ge(Li) counters with good energy resolution. A larger number of these detectors, obviously, increases the efficiency of the experiment. Thus, three detectors allow three combinations of detector pairs, four detectors give six and finally n detectors give n(n-1)/2 combinations.

As can be seen in the example of fig 4, the singles γ-ray spectra from these nuclear reactions are very rich indeed. The reason is partly the many reaction channels, with various numbers of neutrons being emitted, and partly the presence of a γ-ray background due to the Coulomb excitation of the target nucleus as well as of the projectile, and due to the radioactive decay of the daughter nuclei, whose ground states are often unstable. Fortunately, the latter processes normally give rise to fewer transitions each, that is, the pertinant γ-ray spectra have a lower multiplicity. This fact means that already the coincidence requirement as such removes some of the γ-rays corresponding to low-multiplicity cascades. This is clearly demonstrated in fig 4: the intensity of a number of γ-lines in the single spectrum is considerably reduced in the one-fold spectrum, which is recorded on condition that a second γ-ray detector is also being triggered. With this experience it is then natural to introduce

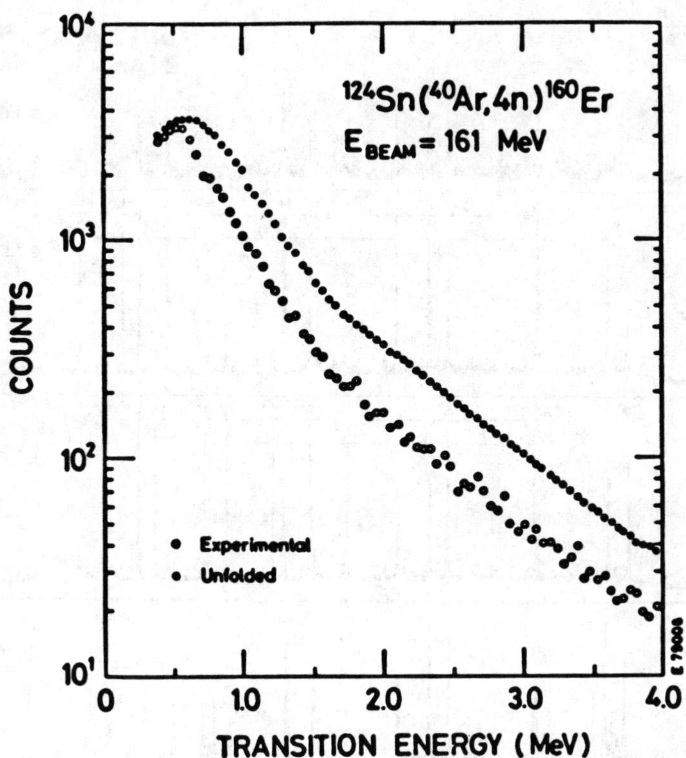

Fig 3. The γ-ray spectrum from a heavy-ion reaction recorded with 0.5 MeV < E_γ < 4.0 MeV with a low-resolution NaI (Tl) detector.

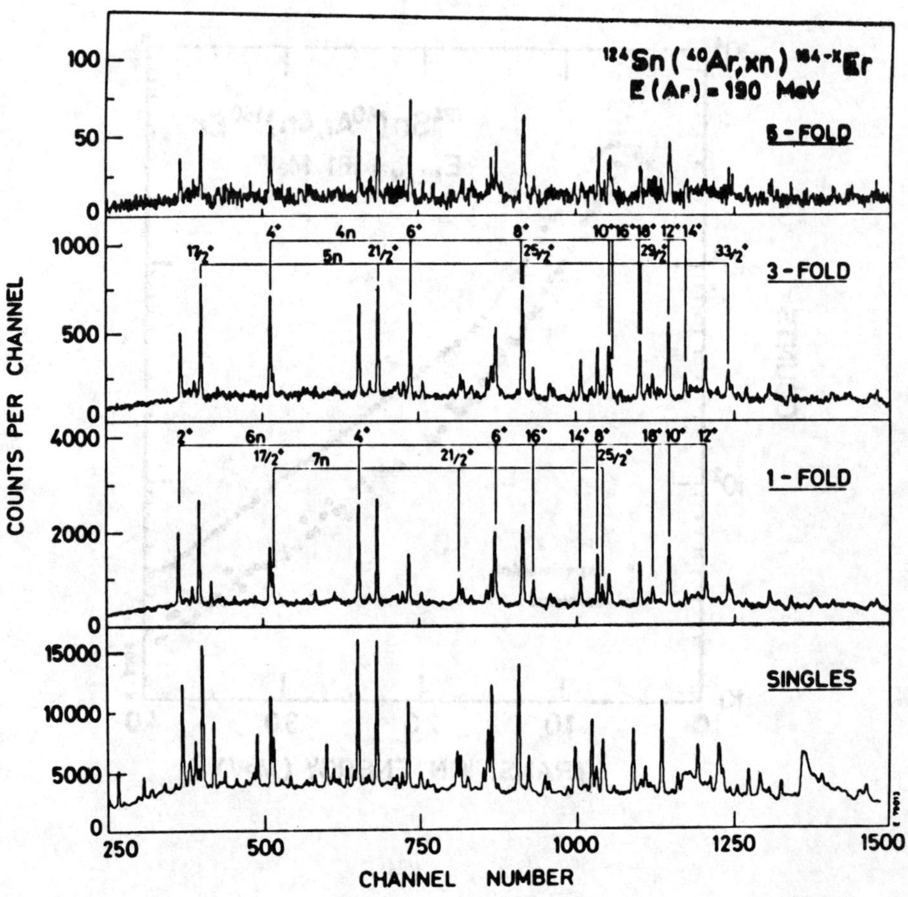

Fig 4. The γ-ray spectrum from a heavy-ion reaction with E_γ<0.75 MeV, recorded with a high-resolution Ge(Li) detector.

a greater number of detectors in order to further reduce the low-multiplicity events and to be able to select the reaction channels with higher multiplicities. The results of such experiments are shown in fig 4, where the three-fold and five-fold spectra become successively simpler. Furthermore, it should be noted that the intensity ratios between transitions due to the 4n and the 6n reaction channels increase with increasing fold-ness. This is a consequence of the higher multiplicity in the 4n channel and is in agreement with the expectations from the statistical model as described in fig 1. Fig 5 shows the principle for a simple experimental arrangement with four NaI(Tl) detectors functioning as a multiplicity filter and four Ge(Li) detectors for recording high-resolution, coincident γ-ray spectra.

The methods described so far have relied on the multiplicity for selecting the proper reaction channel, that is, the selection has been made through the angular momentum of the residual nucleus. (Cf fig 1). Another possible method is to use the excitation energy associated with a particular reaction channel. It can thus be seen in fig 1 how a larger total excitation energy is related to channels with fewer emitted particles. In order to make use of this fact for selecting a particular reaction channel it is important to be able to absorb the total energy of the emitted γ-radiation, that is, we need to completely surround the target with detectors. Such a device is often called a sum spectrometer. A simple sum spectrometer is shown in fig 6, where the target, in which the beam is stopped, is surrounded by a large NaI(Tl) crystal and is at the same time watched by two Ge(Li) detectors, in which coincident γ-rays are measured. In fig 7 the total sum spectrum of the γ-radiation is shown together with the parts emanating from the 4n and the 3n reaction channels, respectively. This figure also demonstrates the power of the method: by selecting a proper gate on the total sum spectrum, almost clean spectra belonging to the appropriate reaction channel can be recorded.

A combination of the two methods discussed here would be a great advantage, as it would allow us to select events coming exclusively from a certain region in the energy-angular momentum plane, that is, preferably from specified parts of the yrast region. The sum spectrometer would then have to be divided into a sufficient number of individual detectors to allow for simultaneous measurements of the total excitation energy and the multiplicity. These instruments have

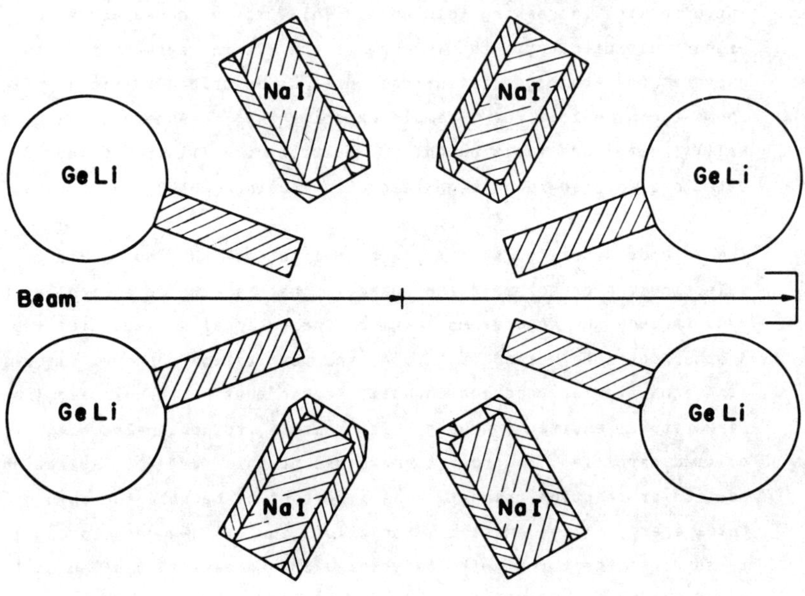

Fig 5. The principle for a multi-detector arrangement for studies of the decays of high-spin nuclear states.

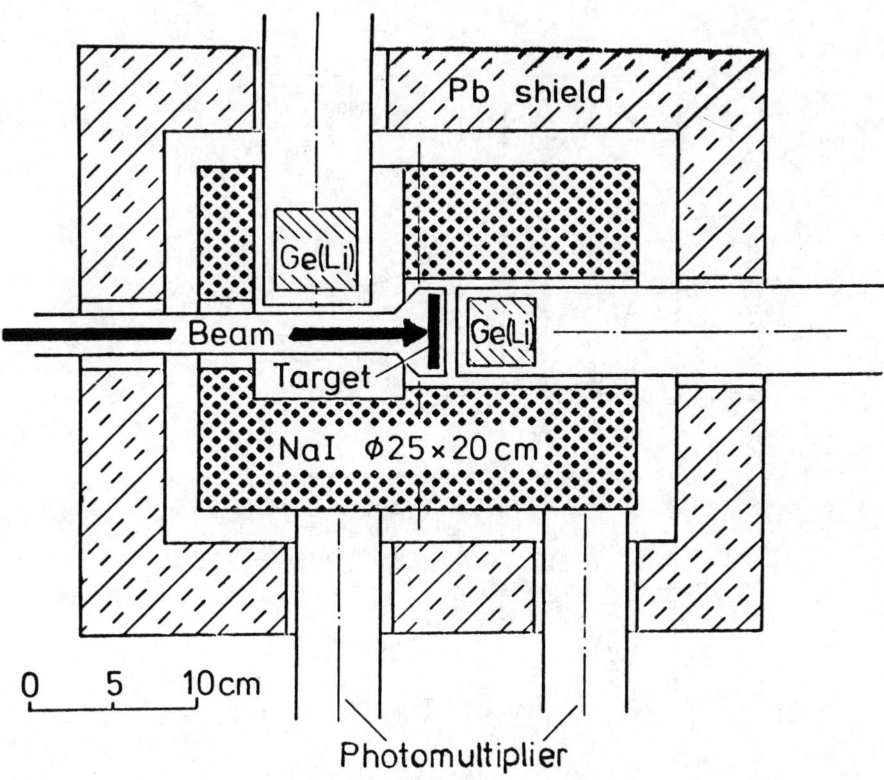

Fig 6. Sum-spectrometer for total absorption of the γ-radiation.

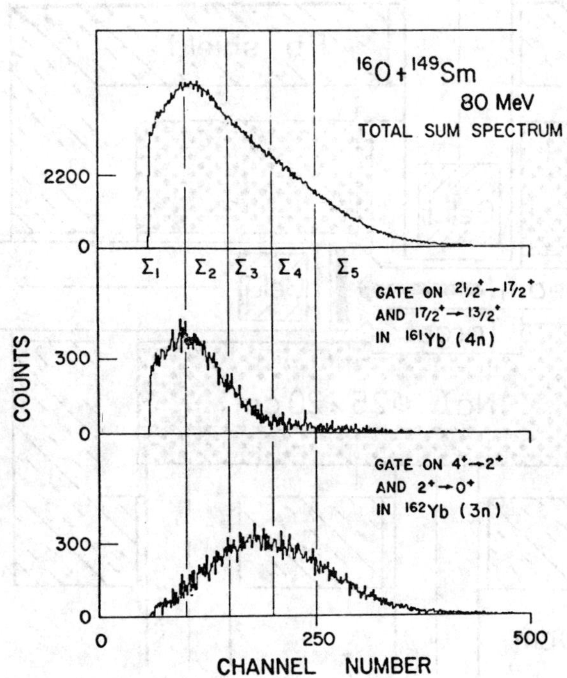

Fig 7. Total and gated γ-ray spectra from a sum spectrometer.

been called crystal balls. They are very complicated and expensive to build. Such instruments have been constructed at the Oak Ridge National Laboratory [9] and at the Max-Planck Institute in Heidelberg: the Darmstadt-Heidelberg Crystal Ball [10].

By the methods just discussed, fairly clean reaction channels can thus be selected for the study of a specific nuclear system. However, the high-resolution γ-ray spectra from the Ge(Li) detectors still often show a considerable background upon which the photo-peaks are superimposed. This background is due to Compton-scattered γ-rays and these events can be suppressed if an anti-coincidence requirement is introduced between the pulses from the Ge(Li) counter and those from a large detector surrounding the Ge-detector. Such an anti-Compton spectrometer is shown in fig 8 and the improved peak-to-background ratio obtained with such a device is demonstrated in the lower part of fig 9.

An ideal γ-ray detecting system would then combine the concept of the crystal ball and the advantages of an anti-Compton spectrometer. Many such systems are now being built and the introduction of bismuth germanate (BGO), replacing NaI(Tl), will make the construction much more compact. This depends on the much shorter radiation length for γ-rays in BGO as compared to NaI(Tl). Fig 10 shows such a system which is in operation at the tandem accelerator laboratory at Daresbury, England [11]. Finally, fig 11 displays a drawing for NORDBALL, a very compact system planned within a Scandinavian collaboration. In this system BaF_2 is also introduced as a scintillator material, having very good time resolution properties.

3. Interpretation of the Experimental Data

With the refined techniques just described, it has in the last few years been possible to extend the limit of discrete line spectroscopy towards very high spins for many nuclear systems. Rotational structures built on low-lying quasiparticle configurations can thereby be followed up to very high spin values. Level schemes with spin values up to $I \sim 40\,\hbar$ have thus been constructed from discrete line coinci-

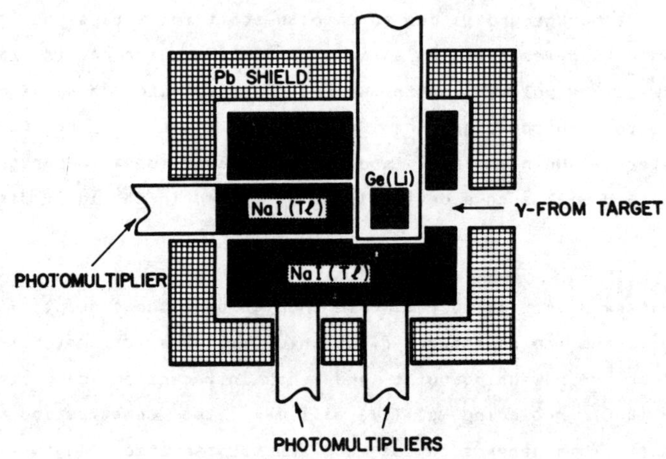

Fig 8. Anti-Compton spectrometer for suppression of the Compton background in the γ-ray spectra.

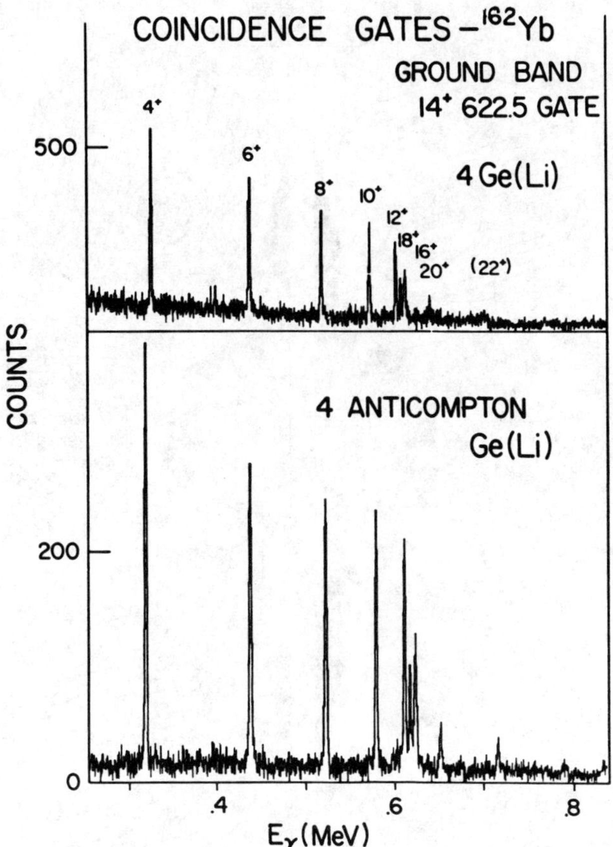

Fig 9. Demonstration of the improvement of the peak-to-background ratio in a γ-ray spectrum recorded by an anti-Compton spectrometer.

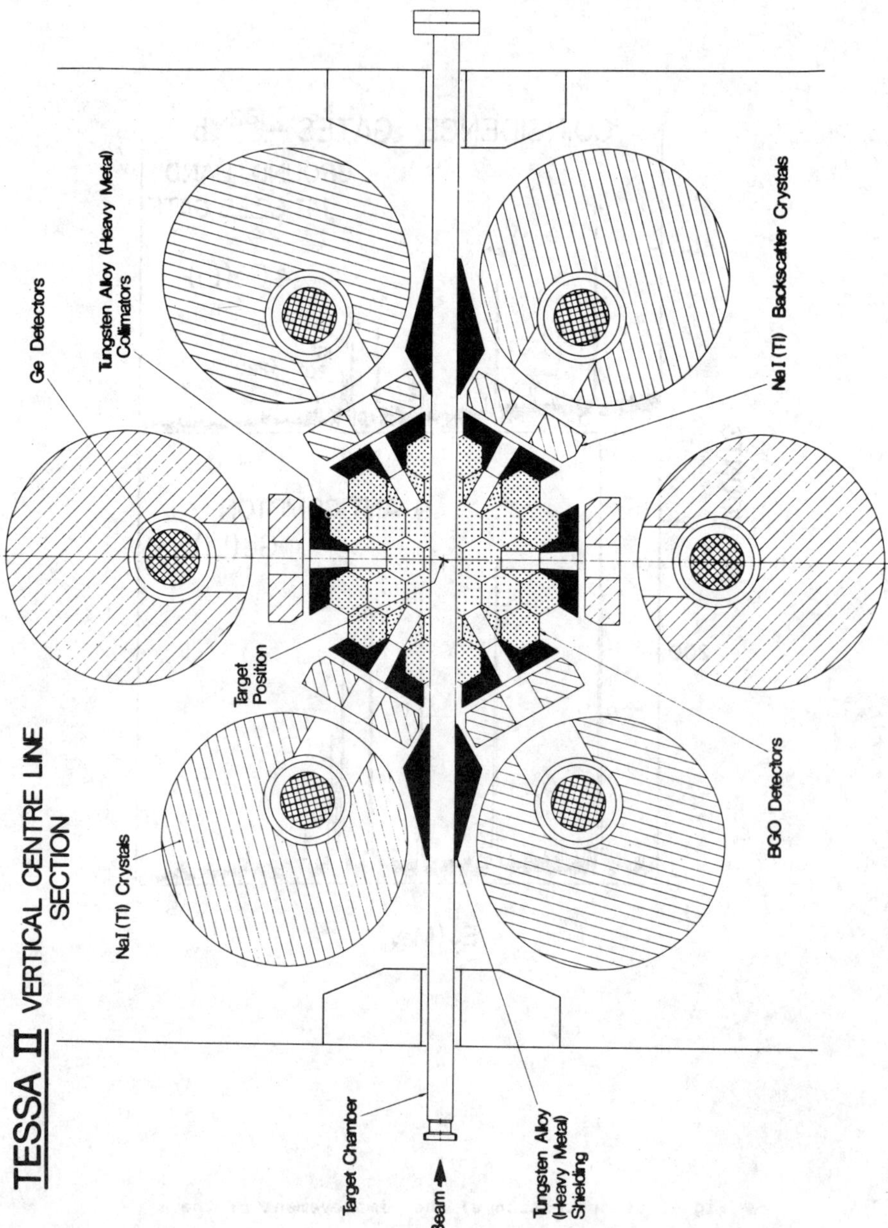

Fig 10. TESSA - the Total Energy Suppression Shield Array - consisting of six anti-Compton spectrometers and 62 BGO crystals operated as an effective crystal ball. At NSF, Daresbury, England

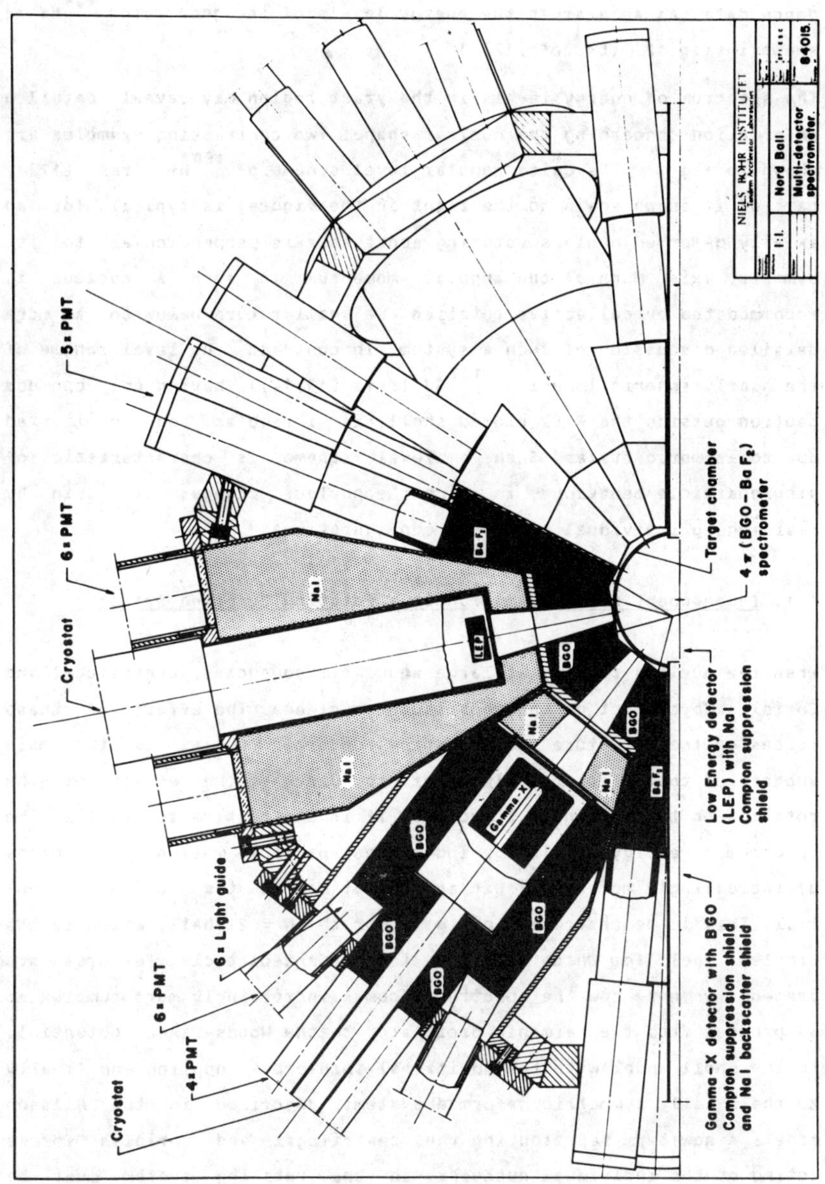

Fig 11. NORDBALL - a compact detector system consisting of 10 gamma-X [Ge(Li)] detectors with BGO Compton suppression shields, 10 Low Energy Detectors (LEP-Ge) with NaI (Tl) Compton suppression shields and a central 4π BGO+BaF_2 spectrometer.

dence data. As an example the energy levels of the good rotor ^{168}Hf is shown in fig 12. (Cf ref [12].)

The spectrum of energy levels in the yrast region may reveal detailed information concerning the nuclear shape. Two contrasting examples are shown in fig 13. The quite regular level scheme of ^{168}Hf (ref [12]), part of it shown again to the right in the figure, is typical for an axially deformed nucleus rotating about an axis perpendicular to its symmetry axis. Much of the angular momentum in such a nucleus is accommodated by collective rotation. We shall return below to a more detailed discussion of such a system. In contrast, the level scheme of the nearly spherical nucleus ^{147}Gd (refs [13,14]), having only one odd neutron outside the N=82 closed shell, is irregular and complicated due to isomeric states. Such a level scheme is characteristic of singleparticle behaviour, in which the nucleus changes its spin by rearranging individual nucleonic configurations.

3.1. Independent-Particle Motion in a Rotating Deformed System

When the nucleus rotates at large angular frequencies, centrifugal and Coriolis forces act on the constituent nucleons. The effect of these forces on the structure of rotating deformed nuclei is the main subject of these lectures. In order to discuss the effect of the rotation on the intrinsic structure, it is instructive to recall the spectrum of energy levels for independent-particle motion in a series of increasingly more realistic approximations to the nuclear potential. This is demonstrated in fig 14 for the N = 2 shell, which is the simplest shell for which all the independent-particle features are present. We note how the spectrum becomes increasingly more complex as we proceed from the harmonic oscillator to the Woods-Saxon potential, to the shell model with the additional spin-orbit coupling and finally to the axially symmetric deformed systems described in the Nilsson model. A new term representing the centrifugal and Coriolis forces acting on the individual nucleons in the rotating system must be included in the single-particle Hamiltonian for a rotating system

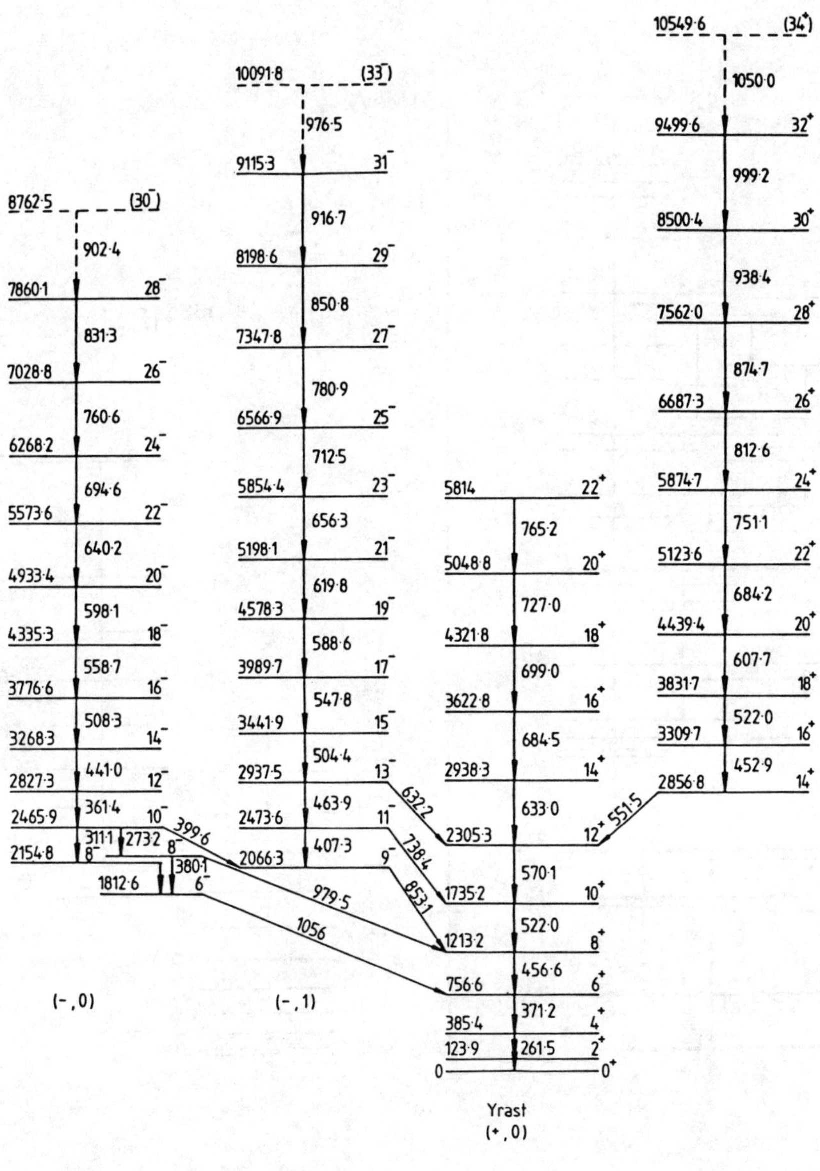

Fig 12. The level scheme for ^{168}Hf

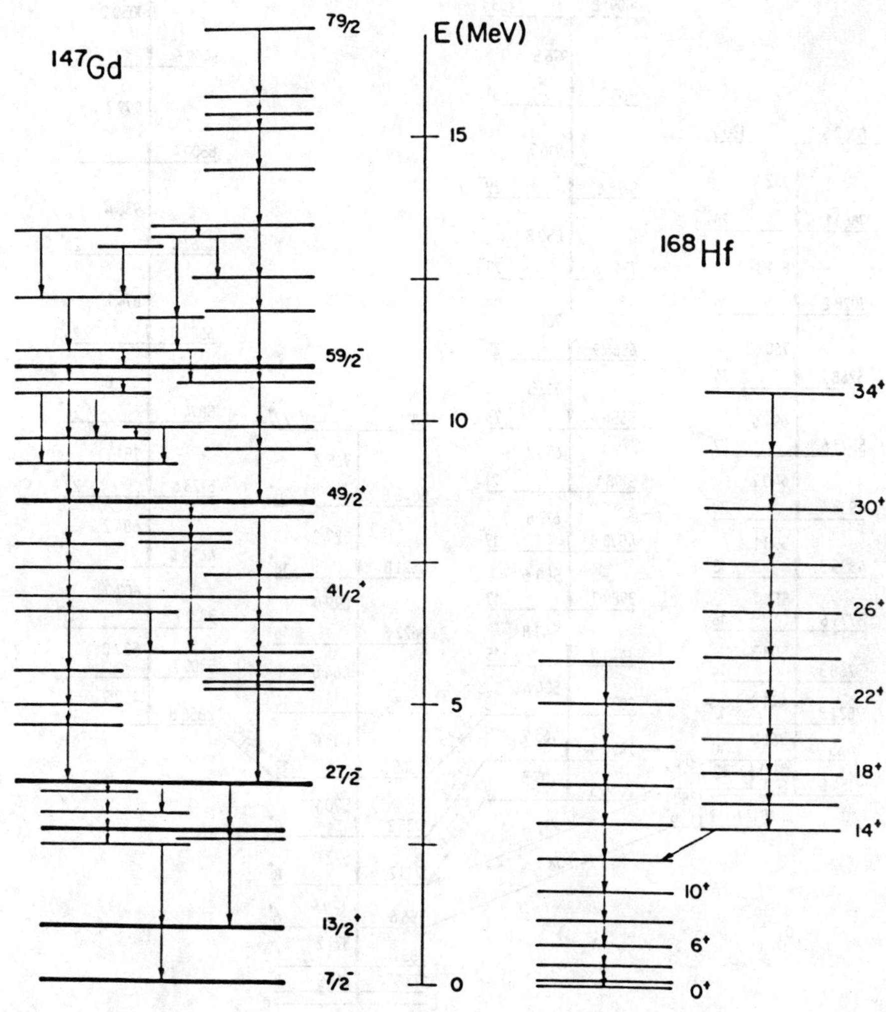

Fig 13. Spectra of energy levels in the yrast region for nuclei, whose angular momentum is dominated by single-particle alignment - ^{147}Gd - and by collective rotation - ^{168}Hf.

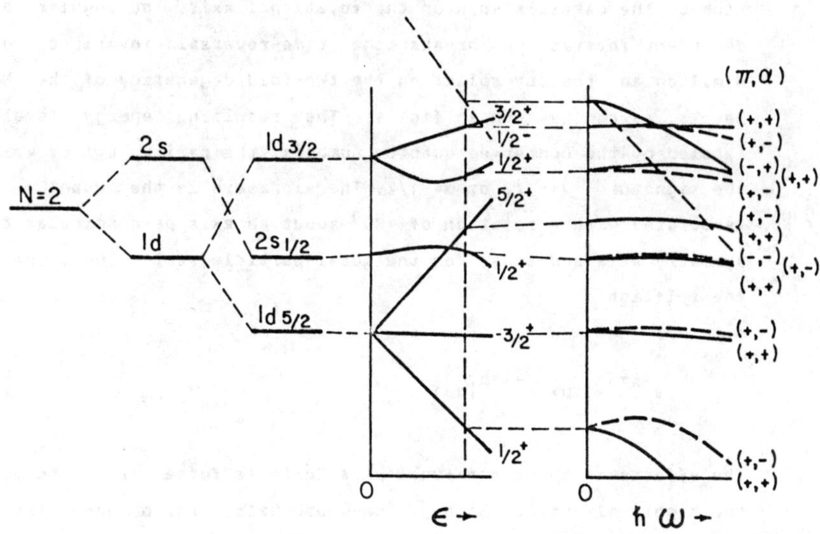

	Harm. Oscil.	Woods Saxon	Shell Model $\ell \cdot s$	axially deformed	Coriolis + Centrifugal $-\hbar\omega j_1$
Deg.	$2(2\ell+1)$ $2n+\ell$	$2(2\ell+1)$	$2j+1$	2	1
q.n labeling	N, π	N, ℓ, π	N, ℓ, j, π	$[N n_3 \Lambda]\ \Omega^{\pi}$	$\alpha \quad \pi$

Fig 14. Energy levels for the N=2 shell, corresponding to independent-particle motion in a series of increasingly more realistic approximations to the nuclear potential.

$$h = h_{def} - \omega j_x . \qquad (1)$$

Here ω is the angular frequency of the rotation and j_x is the projection of the particle spin on the rotational axis. The angular-velocity dependent interaction breaks the time-reversal invariance of the Hamiltonian, thereby splitting the two-fold degeneracy of the Nilsson levels, as can be seen in fig 14. The resulting energy levels are labeled by the conserved quantum numbers, the parity, π=+ or π=-, and the signature, α=+1/2 or α=-1/2. The signature is the quantum number associated with a rotation of 180° about an axis perpendicular to the symmetry axis and thus, for the quasi-particle states $|\alpha\mu\rangle$, defined by the relation

$$e^{-i\pi j_x}|\alpha\mu\rangle = e^{-i\pi\alpha}|\alpha\mu\rangle \qquad (2)$$

The effects of the centrifugal plus Coriolis forces are strongest on the highly-alignable, high-j, low-Ω orbitals, which have large j_x-values, as can be seen from eq (1). In the s-d shell, as shown in fig 14, the Ω = 1/2 orbital derived from the $d_{5/2}$ subshell, that is, the $1/2^+$[220] Nilsson orbital, is the most alignable orbital and therefore is most influenced by the rotation.

Since, for the rare-earth nuclei, the level spacing, d, near the Fermi surface is small compared to the pairing correlation energy, Δ, - e.g. for neutron d ~ 300 keV and Δ ~ 1.0 MeV - the pairing force, given by h_{pair}, has to be added as a collective two-body interaction scattering pairs of particles between the eigenstates of the average field

$$h = h_{def} - h_{pair} - \omega j_x \qquad (3)$$

If a particle-number, non-conserving, pairing theory - e.g. the BSC approximation - is employed, an additional term must be included in eq (3) to conserve particle number on the average.

These features are illustrated in fig 15, where a few specific neutron orbitals lying close to the Fermi surface in a rare-earth nucleus have been calculated. The energy levels calculated for a rotating deformed potential with the centrifugal and Coriolis forces acting on the individual particle configurations (fig 15b) are often referred to as the Routhians. The term Routhian is applied to the excitation energy in the rotating system, as the transformation from the laboratory system to the rotating intrinsic frame is equivalent to Routh's procedure [15] for a change of variables in classical mechanics. It should be noted how the introduction of the pairing correlations in fig 15c changes the spectrum. The quasi-particle Routhians shown in fig 15c are calculated by the Cranked Shell Model (SCM) code of Bengtsson and Frauendorf [16].

As mentioned above, we can now study very accurately the energy systematics of the quasi-particle configurations as a function of rotational frequency, and thus test the premises of the independent-particle motion in a rotating deformed potential. We shall now see how this can be done.

3.2 The Analysis of the Experimental Spectra

The total signature α_t of a state is given by the spin I

$$\alpha_t = I + \text{even number} \tag{4}$$

The sequence of states with a given α_t belonging to a rotational band is interpreted as corresponding to a certain configuration of quasi-particles in the trajectories of a diagram such as fig 15c. For such a band the experimental frequency, ω, is obtained from the measured energies, E(I), by means of

$$\hbar\omega(I) = \frac{E(I+1)-E(I-1)}{I_x(I+1)-I_x(I-1)}, \tag{5}$$

where I_x represents the component of the angular momentum perpendicular to the symmetry axis. It is related to I by the relation

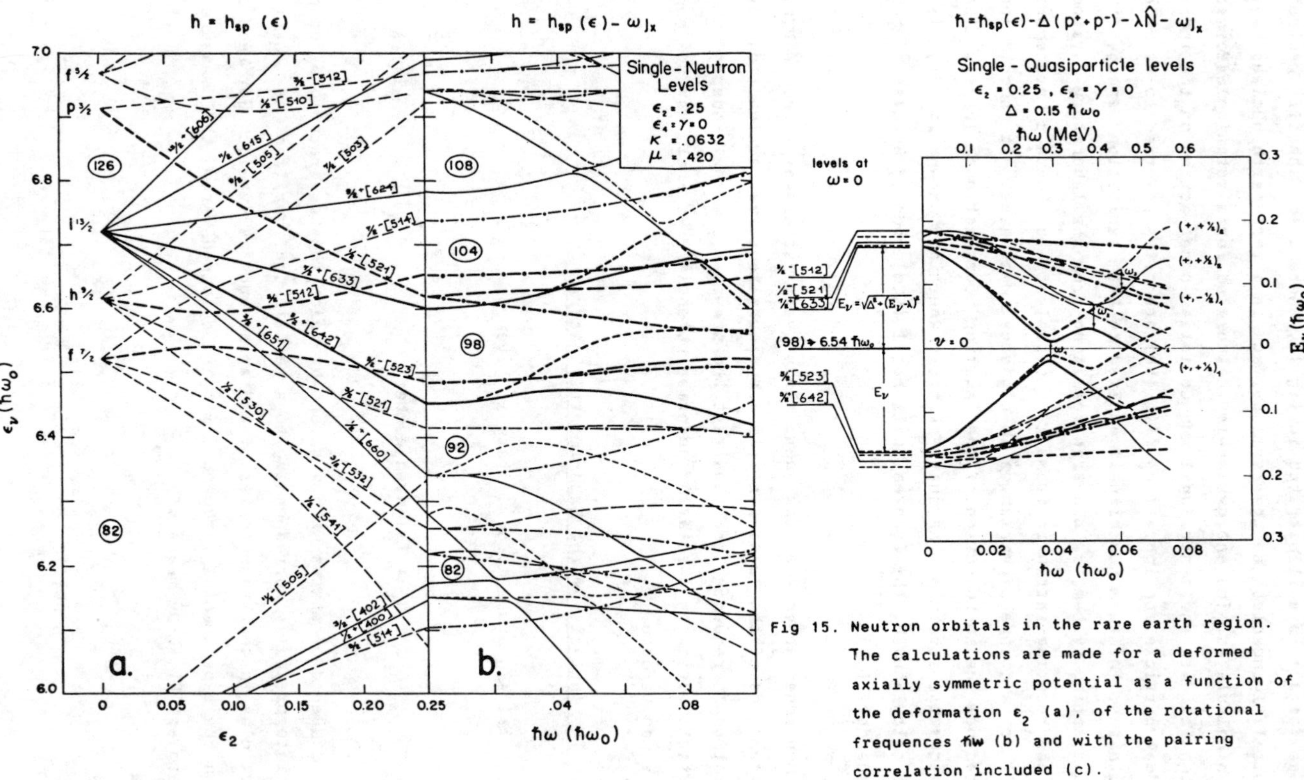

Fig 15. Neutron orbitals in the rare earth region. The calculations are made for a deformed axially symmetric potential as a function of the deformation ϵ_2 (a), of the rotational frequences $\hbar\omega$ (b) and with the pairing correlation included (c).

$$I_x(I) = \sqrt{(I+1/2)^2 - K^2}. \tag{6}$$

The projection, K, of the angular momentum onto the symmetry axis may normally be obtained from the spin of the band head if K is sufficiently well conserved.

The transition I+1 → I-1 defines thus, through eq (5), the frequency, ω, ascribed to the spin I. Eq (6) provides the corresponding value $I_x(I)$. The sequence of discrete points $I_x(\omega(I))$ obtained in this way is connected by interpolation to the continuous function $I_x(\omega)$. The energy in the rotating frame, E'(I), which we have called the Routhian, is defined for the transition I+1 → I-1 as follows

$$E'(I) = \frac{E(I+1)+E(I-1)}{2} - \omega(I)\cdot I_x(I). \tag{7}$$

The Routhian E'(ω), can then be obtained by changing the variable I to the frequency ω and interpolating.

The independent particle Hamiltonian given by eq (3) is valid in the intrinsic frame. Thus, in order to preserve the independent particle picture we have chosen to express the experimental data in the intrinsic frame. The calculations result in relative excitation energies e'(ω) and aligned angular momenta i(ω) with respect to a reference configuration. For the comparison, these relative quantities are obtained from the experimental functions $I_x(\omega)$ and E'(ω) by subtracting the experimental functions $I_{xr}(\omega)$ and $E'_r(\omega)$ for the reference configuration:

$$\begin{aligned} i(\omega) &= I_x(\omega) - I_{xr}(\omega) \\ e'(\omega) &= E'(\omega) - E'_r(\omega) \end{aligned} \tag{8}$$

In the case of odd mass nuclei, the odd-even mass-difference must be added to the excitation energies in order to obtain the proper value of e'(ω).

The choice of reference is very important for the determination of the functions $e'(\omega)$ and $i(\omega)$. A convenient and very frequently used choice of reference is the parametrization according to the Harris formula [17] for doubly-even nuclei:

$$I_{xr}(\omega) = \omega \mathcal{J}_0 + \omega^3 \mathcal{J}_1$$
$$E'_r(\omega) = -(\omega^2/2)\cdot \mathcal{J}_0 - (\omega^4/4)\cdot \mathcal{J}_1 + \hbar^2/8\mathcal{J}_0 \qquad (9)$$

Experimental energies corresponding to the ground state sequence in ^{168}Hf are shown in fig 16a, where the experimentally measured energies are plotted as a function of the spin I. (Cf also fig 12). In fig 16b the measured energies are converted to excitation energies in the rotating intrinsic frame and presented as a function of the angular frequency of rotation, $\hbar\omega$.

Calculations of the energies of the completely paired ground state and the lowest two-quasiparticle excitation of $i_{13/2}$ neutrons are shown as full lines in fig 16b. The $i_{13/2}$ neutrons with l=6 and j=13/2 have the largest particle angular momentum available for rare-earth nuclei. Therefore these are the particles most strongly influenced by the rotational induced forces, ωj_x. For no rotation, that is, for $\hbar\omega$ = 0, the completely paired ground state is lowered relatively to the two-quasineutron states by two times the pairing gap energy, Δ. During rotation, the energy of the highly alignable two-quasineutron configuration, having a large j_x-value, is reduced as a result of the rotational induced forces. As can be seen in fig 16b the energy associated with these forces on the $i_{13/2}$ two-quasineutron configurations is, at a frequency hω=0.26 MeV, equal in magnitude, but opposite in sign to the pairing gap energy. Thus the energy of this two-quasineutron state becomes degenerate with the paired ground-state configuration. Since the gamma-ray cascade tends to follow the configuration which is lowest in energy, the cascade will at this band crossing or "backbending" change from a rotational band based on the aligned two-quasineutron configuration to a band based on the paired ground state configuration. This is equivalent to the change from a gapless to a normal superconductor.

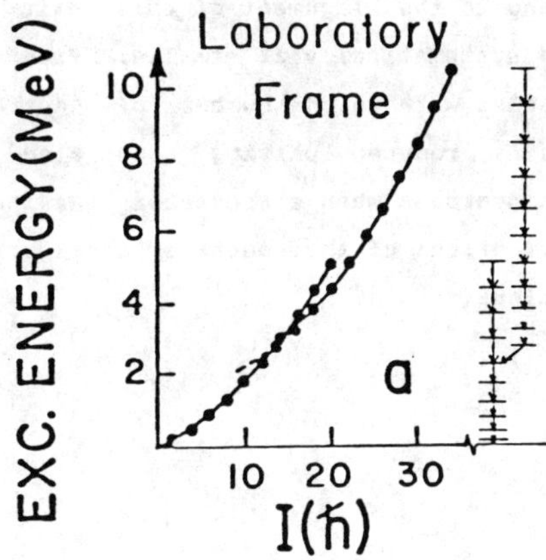

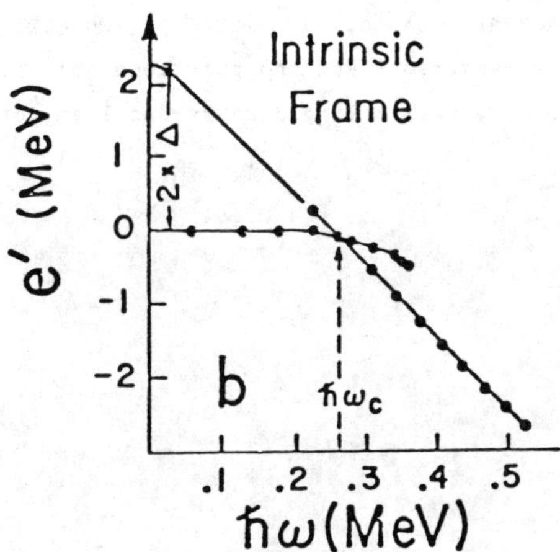

Fig 16. a) The excitation energies for levels in ^{168}Hf

b) The Routhians e'(ω) for levels in ^{168}Hf

Such band crossing frequencies, $\hbar\omega_c$, based on the alignment of a pair of $i_{13/2}$ quasineutrons can be established from the experimental data as is demonstrated in fig 16b. It is therefore possible to determine the magnitude of the pairing gap energy, Δ, on a relative scale from such values of $\hbar\omega_c$, which reflect how fast the nucleus must rotate for the rotational induced forces to compensate for the pairing forces. Band crossings corresponding to the alignment of this pair of $i_{13/2}$ quasineutrons are observed systematically at smaller frequencies in most decay sequences of nuclei with an odd number of neutrons. This decrease in $\hbar\omega_c$ indicates a reduced pairing associated with the aligned pair of $i_{13/2}$ quasineutrons when a spectator quasineutron is present. The pairing contributions of the spectator configuration are "blocked" in the odd-N nucleus.

ACKNOWLEDGEMENTS

The author of these lecture notes has drawn heavily on the many good survey articles in this field written by J D Garrett, G B Hagemann and B Herskind of the Niels Bohr Institute as well as on articles and manuscripts written by R Bengtsson. For the privilege of using this material as well as for many stimulating discussions I am very grateful.

REFERENCES

1. J D Garrett, B G Hagemann, and B Herskind, Nuclear Physics A400 (1983) 113c.

2. B Herskind, Proc of the International Conference on Nuclear Physics, Florence, Italy 1984.

3. J D Garrett, G B Hagemann and B Herskind, Comments on Nuclear and Particle Physics, 1984

4. Z Szymański, Fast Nuclear Rotation, Oxford University Press, Oxford 1983.

5. M J A de Voigt, J Dudek and Z Szymański, Revs Modern Physics 55 (1983) 949.

6. J C Lisle, Proc of the Fifth Nordic Meeting on Nuclear Physics, Jyväskylä, Finland 1984.

7. D Schwalm, Nuclear Physics A396 (1983) 339c.

8. J R Grover and J Gilat, Phys Rev 157 (1967) 802,814.

9. M Jääskeläinen, O G Sarantites, R Woodward, F A Dilmanian, J T Hood, R Jääskeläinen, D C Hensley, M L Halbert and J H Barker, Nuclear Instrument and Methods 204 (1983) 385.

10. V Metag, R D Fischer, W Kühn, R Mühlhans, R Novotny, D Habs, U von Helmolt, H W Heyng, R Kroth, D Pelte, D Schwalm, W Hennerici, H J Hennrich, G Himmele, E Jaeschke, R Repnow, W Wahl, E Adelberger, A Lazzarini, R S Simon, R Albrecht and B Kolb, Nuclear Physics A409 (1983) 331c.

11. P J Twin, P J Nolan, R Araeinejad, D J G Love, A H Nelson and A Kirwan, Nuclear Physics A409 (1983) 343c.

12. R Chapman, J C Lisle, J N Mo, E Paul, A Simcock, J C Willmott, J R Leslie, H G Price, P M Walker, J C Bacelar, J D Garrett, G B Hagemann, B Herskind, A Holm and P J Nolan, Phys Rev Letters 51 (1983) 2265.

13. O Bakander, C Baktash, J Borggreen, J B Jensen, J Kownacki, J Pedersen, G Sletten, D Ward, H R Andrews, O Häusser, P Skensved and P Taras, Nuclear Physics A389 (1982) 93.

14. T L Khoo, P Chowdhury, H Emling, D Frekers, R V F Janssens, W Kuhn, A Pakkanen, Y H Chung, P J Daly, Z W Grabowski, H Helppi, M Kortelahti, S Bjørnholm,J Borggreen, J Pedersen and G Sletten, Physica Scripta T5 (1983) 16.

15. J J Routh, in The Advanced Part of a Treatise on the Dynamics of a System of Rigid Bodies, 6th ed. (Macmillan & Co, London, 1905)

16. R Bengtsson and S Frauendorf, Nuclear Physics A327 (1979) 139.

17. S M Harris, Phys Rev 138 (1965) B509.

NUCLEAR-STRUCTURE STUDIES OF ROTATING NUCLEI

Ikuko HAMAMOTO

Department of Mathematical Physics,
Lund Institute of Technology, University of Lund,
Lund, SWEDEN

Table of contents

1. Introduction

The yrast spectroscopy at high-spin in nuclei has been extensively developed for the last ten years, both experimentally and theoretically. Progress has been achieved in the understanding of the nuclear structure at high spin, especially in terms of quasiparticle motion in rotating potentials.[1] There are at present two ways of populating nuclear high-spin states, namely Coulomb excitations and (HI,xn) reactions. By means of Coulomb excitations one can reach, for the moment, states with spin $I \lesssim 32$. Thus, the quantitative information on the yrast spectroscopy of high-spin states comes from the analysis of discrete γ-rays emitted from evaporation residues in (HI,xn) reactions. The highest spin (the highest rotational frequency) obtained in this kind of analysis is at present $I \lesssim 45$ ($\hbar\omega_{rot} \lesssim 550$ keV). This is certainly the region of spin at which one expects a drastic change of pair-correlations as well as the deviation of the nuclear shape from axial-symmetry.

Examples of measured quantities (and extracted physical properties) are ; energy levels with spin-assignments (spin-alignments, moments of inertia, size of pair-correlation, band-crossings etc.), collective E2 transitions (band-structure, shape,etc.), E1 or M1 moments (proton- or neutron-configurations, fingerprint of one-particle motion in rotating potentials, etc.), E2 transitions with $\Delta I = 1$ (deviation of nuclear shape from axial-symmetry), and so on.

My basic policy in the present lecture is to keep our model as simple as possible and try to draw some definite conclusions which would be obtained from any reasonable model. The point on which I would especially like to concentrate my discussions is the problems in connection with the possible deviation of nuclear shape from axial-symmetry. In sect. 2 I explain the models and the notations used in the study of nuclear high-spin states. In sect. 3 the analysis based on energy-levels is presented both for the axially-symmetric intrinsic-shape and for the triaxial intrinsic-shape. In sect. 4 the rotational effect on electromagnetic properties of high-spin states in nuclear yrast spectroscopy is analyzed in connection with recent experimental data.

Before proceeding to sect. 2, I explain briefly some of the quantities and the terminology, which I shall use later and which may not be familiar to the people who are not the specialists in the high-spin field.

(i) By "a band" I mean a fixed (or very smoothly varying with spin I) intrinsic structure. The states within a band are connected by large collective E2 transitions, since the main part of nuclear deformation is of quadrupole (λ=2) type.

(ii) A rotational frequency, ω_{rot}, is defined by using the canonical relation

$$\hbar\omega_{rot} \equiv \frac{dE_I}{dI} \approx \tfrac{1}{2}\,(E_{I+1} - E_{I-1}) \tag{1}$$

where E_I expresses the energy as a function of spin I. It has a proper physical meaning (namely, collective rotational frequency), when it is defined between the two states belonging to the same band.

(iii) It is useful to distinguish between the two definitions of moments of inertia[2)]

$$\mathcal{J}^{(1)} \equiv \frac{\hbar I}{\omega_{rot}} = \hbar^2 I\left(\frac{dE}{dI}\right)^{-1} \tag{2.1}$$

and

$$\mathcal{J}^{(2)} \equiv \hbar\left(\frac{d\omega_{rot}}{dI}\right)^{-1} = \hbar^2\left(\frac{d^2E}{dI^2}\right)^{-1} \tag{2.2}$$

When the yrast spectrum contains band-crossings as is drawn in fig. 1, the two moments of inertia, $\mathcal{J}^{(1)}$ and $\mathcal{J}^{(2)}$, can be defined by using either the energy expression of the yrast line (namely, the envelope of individual yrast bands) expressed by dashed-line in fig. 1 or that of a given band denoted by a solid-line in fig. 1. The former is expressed by $\mathcal{J}^{(1)}_{yr}$ and $\mathcal{J}^{(2)}_{yr}$, while the latter is denoted by $\mathcal{J}^{(1)}_{band}$ and $\mathcal{J}^{(2)}_{band}$. Then, from fig. 1 we see, for example, that $\mathcal{J}^{(2)}_{yr} \geq \mathcal{J}^{(2)}_{band}$. Then, which moment of inertia is obtained from an experiment depends on the kind of experimental data and the way of extracting the moment of inertia from the data.

(iv) Spin-alignment of an excitation, α, is defined by the relation [3)]

$$i_\alpha(\omega_{rot}) = I_\alpha(\omega_{rot}) - I_o(\omega_{rot}) \tag{3}$$

for a given ω_{rot} where I_α expresses the spin of the state, α, of interest while I_o denotes that of a reference state. The reference state is usually taken to be either the yrast state or the state of the ground band of the neighbouring even-even nucleus.

(v) The signature quantum-number, α, is defined as;

$$\text{odd-A nuclei} \quad \begin{cases} \alpha = 1/2 \text{ for } I = 1/2,\ 5/2,\ 9/2,\ldots. \\ \alpha = -1/2 \text{ for } I = 3/2,\ 7/2,\ 11/2,\ldots \end{cases} \tag{4.1}$$

$$\text{even-even nuclei} \quad \begin{cases} \alpha = 0 \text{ for } I = 0,\ 2,\ 4,\ldots \\ \alpha = 1 \text{ for } I = 1,\ 3,\ 5,\ldots \end{cases} \tag{4.2}$$

This quantum number [4)] comes originally from the invariance property of the cranking Hamiltonian with respect to a rotation of 180° about the cranking-axis, in which the intrinsic shape is assumed to have R-symmetry.

2. Models and Notations

When we investigate the effect of rotational perturbation on the one-particle motion in nuclei, it is very convenient to examine particles in so-called "high-j" orbits, such as the $i_{13/2}$-neutron-orbit or the $h_{11/2}$-proton orbit in rare-earth nuclei. This is partly because there is much less ambiguity in the wave functions of "high-j" orbits than in those of other orbits ("normal-parity orbits"), and partly because the rotational perturbation is so strong in high-j orbits that even a small rotation has a drastic effect on the wave functions of the orbits. Therefore, in the numerical examples which I take in the following I use mostly the phenomena related to "high-j" orbits. Experimental information on the rotational perturbation on "normal-parity orbits" is less detailed. Though it was difficult to find a good quantum number appropriate for the interesting regions of rotational frequency, say, a few hundred KeV, it has been found [5] that the pseudospin formalism [6], which was developed for both spherical and deformed nuclei, is also very useful for the description of "normal-parity orbits" at the relevant rotational-frequency region.

For simplicity we consider the quadrupole (λ=2)-deformation only. Thus, the nuclear shape is described by the deformation parameters, β and γ (see, for example, ref. 7). We express the intrinsic coordinate-system by the suffix (1,2,3), while the laboratory coordinate-system is denoted by (x,y,z). By including only the monopole pairing-interaction treated in the BCS approximation, our intrinsic Hamiltonian is written as

$$H_{intr} = \sum_{\nu} (\varepsilon_{\nu}-\lambda)a^{+}_{\nu}a_{\nu} + \frac{\Delta}{2} \sum_{\mu,\nu} \delta(\bar{\mu},\nu)\,(a^{+}_{\mu}a^{+}_{\nu} + a_{\nu}a_{\mu}) \tag{5}$$

where ε_{ν} expresses the one-particle energies in the potential

$$V = k(r)\,\beta\left[Y_{20}(\theta,\phi)\cos\gamma + \sqrt{\tfrac{1}{2}}(Y_{22}(\theta,\phi) + Y_{2-2}(\theta,\phi))\sin\gamma\right] \tag{6}$$

and λ is determined so that the expectation value of nucleon-number operator is equal to the nucleon-number of the system. It is often convenient to know that for a "single-j-shell" the potential V in (6) is equivalently written as

$$V = \frac{\kappa}{j(j+1)}\{(3j_3^2 - j(j+1))\cos\gamma + \sqrt{3}\,(j_2^2-j_1^2)\sin\gamma\} \tag{7}$$

where κ is used as an energy unit [8] and its appropriate value is something between 2 and 2.5 MeV, depending on nuclei.

When we want to study the nuclear rotational motion from a microscopic point of view, for the moment we have, in essence, only two kinds of models available. Namely, cranking model and particle-rotor model.

2.1 Cranking Model

The idea of the cranking model in nuclei, introduced by D.R.Inglis[9], was to describe the nuclear collective rotation as a rotation of the deformed single-particle potential [10]. By considering a uniform rotation around the 1-axis with cranking frequency ω, the cranking Hamiltonian is written as

$$H_{CR} = H_{intr} - \omega J_1 \tag{8}$$

where

$$J_1 = \sum_{\mu\nu} \langle\mu|j_1|\nu\rangle\, a^+_\mu a_\nu \tag{9}$$

and ω is determined so that

$$\langle J_1\rangle = I \tag{10}$$

The Hamiltonian H_{CR} in (8) together with H_{intr} in (5) can be diagonalized by Bogoliubov-Valatin transformation

$$a_\alpha = \sum_i (A^i_\alpha\, b_i + B^i_\alpha b^+_i) \tag{11}$$

which is reduced to the well-known BCS transformation at $\omega = 0$.

The cranking Hamiltonian, H_{CR}, is invariant [4] with respect to a rotation of π about the 1-axis,

$$R_1(\pi) = \exp(-i\pi J_1) \tag{12}$$

Thus, for example, one-particle orbits can be labelled by eigenvalues of $R_1(\pi)$,

$$r = \exp(-i\pi j_1) = \begin{cases} -i \text{ for } j_1 = \frac{1}{2}, \frac{5}{2}, \ldots \\ +i \text{ for } j_1 = \frac{3}{2}, \frac{7}{2}, \ldots \end{cases} \tag{13}$$

Since in the cranking model the total angular-momentum I is identified with the component J_1 of the total angular-momentum of the particles, one has the relation[11)]

$$r = \exp(-i\pi I) \quad (14)$$

As "signature quantum-number" one uses either r defined in (14) or, equivalently, α defined by the relation

$$\exp(-i\pi\alpha) = \exp(-i\pi I) \quad (15)$$

which leads to the relations (4.1) and (4.2).

It is well-known that the range of γ between 0° and 60° is sufficient for the description of intrinsic shape[12)]. However, in order to specify a cranked system, one needs three times larger range of γ-variables (namely, $-120^o \leq \gamma \leq 60^o$), corresponding to the three axis about which the system with a given intrinsic shape can be cranked. See fig. 2[13)]. In the three sectors (namely $-120^o \leq \gamma \leq -60^o$, $-60^o \leq \gamma \leq 0^o$, and $0^o \leq \gamma \leq 60^o$) the intrinsic shape is the same, and only the relation between the symmetry-axis and the cranking-axis is different. For example, the rotation with $\gamma=60^o$ (-120°) expresses that of oblate shape (prolate shape) around the symmetry-axis, while the rotation with $\gamma=0^o$ (-60°) means that of prolate-shape (oblate-shape) around the axis perpendicular to the symmetry-axis.

The attractive feature of cranking model is: 1) The calculation is simple for high spin as well as low spin. 2) The collective and the single-particle aspect of nuclear structure at high spin can be treated on the same footing.

However, the rotational frequency is not an exact constant of motion in nuclei, and there will be a spread of rotational frequency for a given I. If many degrees of freedom contribute to the rotation, one expects relatively small fluctuations about the average. In contrast, if a single degree of freedom contributes significantly to the rotation, one expects departures from uniform rotation. Then, it becomes necessary to treat explicitly the coupling of the single degree of freedom to the rotational motion. This leads to a fluctuation in the collective rotational frequency. A typical example of this situation[8)] is the crossing between the ground-band and the S-band, which will be illustrated at the end of sect.3.1.

In the analysis of experimental data at least for $\omega_{rot} \lesssim 400$ keV a so-called "cranked shell-model[3, 14)](CSM)" has been surprisingly successful since 1977. The basic points of CSM are: 1) The cranking frequency plays a role as a fundamental quantity in the analysis. 2) Potential-parameters (namely, deformation parameters and pair-correlation parameter) are kept constant independent of both cranking frequency and configurations. More elaborate cranking-model calculations than CSM, namely Hartree-Fock-Bogoliubov cranking calculations (HFBC), were performed by several groups[15] already in the early seventies. In HFBC the effect of rotational motion on the parameters of the one-body field is taken into account self-consistently. However, HFBC has not been very practical and successful in the understanding of experimental information, since due to its complicated character it was not clear what (namely, a deficiency of the cranking model or an improper choice of two-body interaction or....?) was the origin of the discrepancy between the model prediction and experimental data.

In my present talk I take the spirit of CSM. Namely, I keep always both potential-parameters and the λ-value independent of rotational frequency (or angular momentum) and configurations.

2.2 Particle-Rotor Model

The model[7] consisting of a particle (or particles) coupled to a rotor has been very useful in illustrating various general features of rotating systems. The total angular-momentum I is treated quantum-mechanically as a good quantum-number, and thus the exchange of angular momentum between individual particles and the potential produced by the rest of the system is treated more reasonably than in cranking model. The particle-rotor Hamiltonian is written as

$$H_{PR} = H_{intr} + \sum_{h=1}^{3} \frac{\hbar^2}{2\mathcal{J}_h} (I_h - j_h)^2 \tag{16}$$

where the angular momentum of particle(s) is denoted by j_h and the inertia parameter is expressed by $\mathcal{J}_h$ which is a function of β, γ and Δ. For example, in the case of an axially-symmetric (about the 3-axis) intrinsic shape the sum in (16) is taken only over h=1 and 2, and H_{PR} is diagonalized by using the basis wave-functions (with $\Omega = K$)

$$\Psi_{KIM} = \sqrt{\frac{2I+1}{16\pi^2}} \left(\phi_\Omega D^I_{MK} + (-1)^{I+K} \phi_{\bar{\Omega}} D^I_{M-K}\right) \tag{17}$$

where the quantum-numbers corresponding to the operators j_3, I_z and I_3 are denoted by Ω, M and K, respectively, and

$$H_{intr}\phi_\Omega = E_\Omega \phi_\Omega \tag{18}$$

$$\left(H_{intr} + \frac{\hbar^2}{2J}(I_1^2 + I_2^2)\right)\Psi_{KIM} = E_{KI}\Psi_{KIM} \tag{19}$$

The quantal state (17) is not an eigenstate of a rotation of π about the 1-axis, $R_1(\pi)$. However, for example, if we take the $K = \Omega = \frac{1}{2}$ case and a particle with angular-momentum j, it is easy to show the relation

$$\Psi_{KIM} = \sqrt{\frac{2I+1}{16\pi^2}}\left(\Phi_{j,\Omega=\frac{1}{2}}\, D^I_{M\frac{1}{2}} + (-1)^{I+\frac{1}{2}}\, \Phi_{\overline{j,\Omega=\frac{1}{2}}}\, D^I_{M-\frac{1}{2}}\right) =$$

$$= \sum_{\Omega_1}\left[1+(-1)^{I+\Omega_1+1}\right] d^j_{\frac{1}{2}\Omega_1}\left(\frac{\pi}{2}\right)\Phi_{j\Omega_1} D^I_{M\Omega_1} \tag{20}$$

where the eigenvalue of j_1 is expressed by Ω_1. From the expression (20) it is seen that the states with $I = \frac{1}{2}$, 5/2, 9/2,....(I = 3/2, 7/2, 11/2,....) consist of the states with Ω_1 = 1/2, 5/2, 9/2, ...(Ω_1 = 3/2, 7/2, 11/2,....). Therefore, the definition (15) (or (14)) of the signature quantum-number α (or r) in terms of I which is a good quantum-number in the particle-rotor model is seen not to be inconsistent with the one defined in the way of cranking model. (cf.(13)).

Though there are various sophisticated versions[16)] of particle-rotor model in the literature, in the present lecture I take the simplest version. Namely, I keep all the parameters in H_{PR} independent of I and configurations.

3. Analysis based on Energy-Levels

3.1 Axially-Symmetric Intrinsic Shape

In the present subsection we consider the cranking model with $\gamma=0^{o}$. Namely, a prolate shape with the 3-axis as a symmetry axis is cranked around the 1-axis. Since there are no collective rotations around the symmetry-axis in a quantum-mechanical system, the rotational axis must be perpendicular to the 3-axis. Rotations about different axes perpendicular to the symmetry axis are equivalent. Thus, in the cranking model we take the 1-axis as a cranking axis, while in the particle-rotor model (and presumably in real nuclei) the rotational axis is not fixed in the space though it is perpendicular to the 3-axis. A comparison between the cranking model and the particle-rotor model in the case of axially-symmetric intrinsic shape can be found, for example, in ref. 17.

In figs. 3 and 4 the one-particle ($\Delta=0$) and the one-quasiparticle spectra ($\Delta\neq0$) obtained by solving H_{CR} with the $j=\frac{13}{2}$ shell are shown. In fig. 4 the Fermi-level, λ, is placed between the $\Omega=3/2$ and $\Omega=5/2$ levels at $\omega=0$. We note: a) There is no coupling between the $\alpha=\frac{1}{2}$ and $\alpha=-\frac{1}{2}$ orbits. b) At $\omega=0$ each orbit is doubly degenerate, corresponding to the time-reversal symmetry of the system. c) The matrix-element $\langle j,\Omega=\pm\frac{1}{2}|j_1|j,\Omega=\mp\frac{1}{2}\rangle \neq 0$ is responsible for all the signature splitting. Thus, only the orbits with $|\Omega| = \frac{1}{2}$ show the splitting proportional to ω for an infinitesimally small ω. d) The spin-alignment, i, of the particle (or the quasiparticle) is obtained as an expectation value of j_1, which in the present case is equal to the negative of the slope of the one-particle (or the one-quasiparticle) eigenvalues plotted in figs. 3 and 4. e) The lowest-lying orbit has the signature

$$\alpha = (-1)^{j-\frac{1}{2}}\frac{1}{2} \tag{21}$$

for both $\Delta=0$ and $\Delta\neq0$. f) For $\Delta=0$ the lower-lying orbits get the asymptotic value of i(=13/2, 11/2, 9/2,....) already for smaller values of ω. g) For $\Delta\neq0$ the lowest-lying quasiparticle orbits (with $\alpha=\pm\frac{1}{2}$) get an appreciable amount of spin-alignment already at a small value of ω for any value of λ. h) For $\Delta\neq0$ the quasiparticle orbits, which are the continuation of the quasiparticle orbits with positive quasiparticle-energies at $\omega=0$, are shown. This set of orbits form a complete set of quasiparticles, which are the physical quasi-particles based on the quasiparticle vacuum for a given value of ω.

By using the quasiparticle spectra in fig. 4 I explain briefly some examples of what one can learn about nuclear properties. i) The quasiparticle vacuum which has $\alpha=0$ and a positive parity expresses (usually) the yrast states of even-even nuclei and is reduced to the BCS quasiparticle-vacuum at $\omega=0$. ii) At ω_{bb} a band-crossing takes place along the yrast-line of the even-particle-number system. This can be seen from fig. 5 in which the spin-alignment of the quasiparticle vacuum is plotted as a function of ω. It is noted that if there is no interaction between the quasiparticle vacuum and the lowest two-quasiparticle state (AB, in which the orbits A and B are occupied) the sum of the two quasiparticle-energies, E_A+E_B, vanishes at ω_{bb}. iii) All the quasiparticle states both in the even- and the odd-particle-number system experience the band-crossing at ω_{bb}, except the quasiparticle states in which A or B is occupied (Pauli principle). A similar statement can be given on the interaction between two quasiparticle-orbits, such as the interaction between A and C at ω_{AC}. iv) The spin-alignment of a many-quasiparticle state for a given ω is the sum of both the spin-alignemnt of individual quasiparticle and that of the quasiparticle-vacuum for the same ω.

In fig. 6 we show an example of the quasiparticle spectra obtained by using a cranked modified-oscillator model in which both a "high-j" orbit (namely, the $i_{13/2}$-orbit in this case) and "normal-parity" orbits are included. It is seen that the spin-alignments of the lowest-lying "high-j" orbits (A and B) are much larger than those of the lowest-lying "normal-parity" orbits (E and F) and, thus, the first band-crossing along the yrast line is almost always caused by the quasiparticles in "high-j" orbits.

In order to compare the results of the cranking-model calculations with experimental information, one has to convert the observed energies from the laboratory system to the rotating frame of reference. The practical procedure employed usually is described, for example, in ref.18. One can compare, for example, energies in the rotating frame of reference (called often as Routhian) of each band, band-crossing frequencies, spin-alignment of quasiparticle excitations, change of spin-alignments at band-crossings, and so on. Here we do not go into details of the comparison, since there have been many articles about it. Though there are some observed minor features that may not be easily explained in the simple shell-model with cranked potential, it is pretty amazing that the simple cranking model, assuming ω-independent values of potential parameters and thereby ignoring possible effects of rotational motion on those parameters, works pretty well in explaining the energies and the alignments of bands, as well as the characteristics of band-crossings. It is, indeed, an important goal of the analysis to identify the effects of rotation on various nuclear fields and thereby elucidate new aspects of the dynamics of rotating nuclei, by examining observed properties of rotating nuclei in terms of quasiparticle spectra.

It may be worth mentioning the difficulty[8)] inherent in cranking model in the description of crossing of the bands with a large difference of spin-alignments, irrespective of whether the parameters in the model are chosen self-consistently or not. A typical situation involving the crossing of the ground-band and the S-band observed in rare-earth nuclei is illustrated in fig. 7, where the interaction between the two bands which is pretty small in any case is neglected. In the cranking model, the yrast-line for $I=0 \rightarrow I_1$ is obtained from the ground band for rotational frequency $\omega_{rot} = 0 \rightarrow \omega_{bb}$, and at $\omega_{rot} = \omega_{bb}$ the configuration of the yrast-line changes suddenly (in the absence of the interaction between the two bands) from the configuration of the ground band to that of the S- band. Namely, the cranking model leads to the yrast-line for $I_1 < I < I_2$ which is expressed by the (almost) straight dashed-line in fig. 7, while the actual yrast line follows the ground band for $I_1 < I < I_o$ and the S-band for $I_o < I < I_2$. Other way to say it is that since in the cranking model the total angular-momentum I is identified with the expectation value of the component J_1 of the total angular-momentum the yrast states for $I_1 < I < I_2$ are obtained from superpositions of the yrast states with $I = I_1$ and I_2 and do not contain the component with the relevant value of I. A proper description of the yrast-line between I_1 and I_2 is outside of the applicability of the cranking model in its simplest form, since the two bands should interact for a given angular-momentum I while the cranking model considers a mixing of the two bands for a given rotational frequency. The angular momenta which corresponds to a given rotational frequency are very different in the two bands which have very different spin-alignments from each other. Since the angular-momentum shift $\Delta I \equiv I_2 - I_1$ is equal to the difference between the spin-alignments of the crossing two bands, it may be as large as 10 in the present example. The unreliability of cranking model description of the yrast-line for $I_1 < I < I_2$ may be seen from the estimated large variation in the angular momentum fluctuations, $\langle J_1^2 \rangle - \langle J_1 \rangle^2$, around ω_{bb}, as is shown in fig. 8. It is a condition for the validity of cranking model to have an almost constant fluctuation, so that the width of the wave packet in the J_1 space should not strongly depend on rotational frequencies.

3.2 Triaxial Intrinsic Shape

Though around the ground-state of medium or heavy nuclei there has been no clear-cut evidence for the deviation of nuclear shape from axial symmetry, the deviation is generally expected for high-spin states. In fig. 9 the situation is illustrated, in which the presence of spin-aligned particles in rotating nuclei leads the system to a triaxial shape ($\gamma \neq 0$) even if the shape of the ground state is axially symmetric. When a calculated energy-minimum occurs at $\gamma \neq 0$, the calculated potential-energy becomes usually very flat in the γ-direction, as is schematically shown in fig. 10. In such a case calculations for a given fixed value of γ may not have a meaning, since there is always a fluctuation in a quantum-mechanical system. In the present lecture I shall not touch any further this important and difficult problem of fluctuation. In the present subsection I first examine the triaxial shape preferred by aligned particles and the effect of triaxial shape on the signature-splitting, by using a simple cranking approximation. Then, I discuss the problem of the cranking approximation in the description of the rotational motion of triaxial shape. The region of rotational frequency which I am thinking of is less than a few hundred keV in actual nuclei, and not an extremely-high-frequency limit.

First, we investigate which value of γ is preferred[19] by a spin-aligned single quasiparticle in a j=13/2 shell (assuming λ, Δ and κ to be independent of γ). By using the BCS approximation in treating the pair correlation, we write the quasiparticle energy of the Hamiltonian (8) with (5) and (7)

$$h = \{ [\frac{\kappa}{j(j+1)} \{ (3j_3^2 - j(j+1))\cos\gamma + \sqrt{3}(j_2^2 - j_1^2)\sin\gamma \} - \lambda]^2 + \\ + \Delta^2 \}^{\frac{1}{2}} - \omega j_1 \tag{22}$$

(Note that in this approximation we ignore $\Delta\nu = 2$ matrix-elements of the Coriolis interaction. See, for example, ref. 20.). From the expression (22) it is seen that for a given value of (λ, Δ, ω) the quasiparticle energy is a minimum for j_1=j and the γ-value determined by the equation

$$-2\cos(\gamma - 60^0) = \frac{\lambda}{\kappa} \qquad \text{for} \qquad -2 \leq \frac{\lambda}{\kappa} \leq 2 \tag{23}$$

For example, γ=0 for λ/κ=-1.0 (namely, for the Fermi-level lying at the bottom of the shell), while $\gamma=-83^0$ for λ/κ=1.6 (namely for the Fermi-level at the top of the shell). And by increasing the value of λ/κ from -1.0 to +1.6 (namely, by filling the shell gradually from the bottom to the top), the values of γ decreases monotonically from 0° to -83°. This simple (classical) relation between the γ-value of the energy-minimum and the degree of the shell-filling is in agreement with a more detailed numerical solution of the quantal equations in ref. 21.

Next, we examine the effect of $\gamma \neq 0$ on the signature-splitting by using the $j=\frac{13}{2}$ shell model. In fig. 11 the one-particle ($\Delta=0$) spectra obtained by solving H_{CR} are shown for $\gamma=+20^{o}$ and $\gamma=-20^{o}$. For $\gamma \neq 0$ both the matrix-elements $\langle j, \Omega=\pm\frac{1}{2}|j_1|j,\Omega=\mp\frac{1}{2}\rangle$ and $\langle j,\Omega=\pm\frac{1}{2}|j_2^2 - j_1^2|j,\Omega=\mp\frac{3}{2}\rangle$ are responsible for the signature-splitting. The general pattern of the spectra with $\gamma=-20^{o}$ is similar to the one with $\gamma=0$ shown in fig. 3, while the spectra with $\gamma=+20^{o}$ look very different. The complicated pattern of the latter spectra comes from the fact that for small values of $\omega \neq 0$ the sign of the signature-splitting of the orbits which are degenerate at $\omega=0$ is alternatively changing from the bottom of the shell to the top. The alternative structure is understood[19] in terms of the expectation values of j_1, in the absence of both pair-correlation and rotation, which are plotted in fig. 12 for the $\alpha=\frac{1}{2}$ orbits. The corresponding expectation values of j_1 for $\alpha=-\frac{1}{2}$ orbits are the same magnitudes as those for $\alpha=\frac{1}{2}$ orbits but have an opposite sign. (Energy-eigenvalues are the same for the $\alpha=\frac{1}{2}$ and the $\alpha=-\frac{1}{2}$ orbits in the absence of rotation). The alternative sign occurs for $0^{o}<\gamma\leq 60^{o}$ and $-120^{o}\leq \gamma\leq -60^{o}$.

In fig. 13 we show examples of quasiparticle spectra obtained by solving H_{CR} for the $j=\frac{13}{2}$ shell. As can be seen from figs. 11 and 12, when in the case of positive γ-values we place the Fermi-level around the second (or fourth or sixth or.....) lowest doubly-degenerate particle-orbit at $\omega=0$, we expect the anomalous signature for the lowest-lying quasiparticle-orbit at small values of ω. ("Anomalous signature" means that the $\alpha=(-1)^{j+\frac{1}{2}}\frac{1}{2}$ quasiparticle orbit lies lower than the $\alpha=(-1)^{j-\frac{1}{2}}\frac{1}{2}$ orbit.) The position of the Fermi-level in fig. 13 is chosen to be one of such particular positions, which corresponds approximately to the neutron-number N=93 of the $i_{13/2}$ orbits in rare-earth nuclei. From the calculations such as those shown in fig. 13, it is seen that for particular values of λ the anomalous signature for the lowest quasiparticle state appears for $0^{o}<\gamma\leq 60^{o}$ and $-120^{o}\leq \gamma<-60^{o}$ at relatively small ω-values and that the magnitude of the signature-splitting depends very much on γ-values.

Experimentally, the anomalous signature for the lowest "high-j" quasiparticle state has never been observed in odd-A nuclei. There was a publication[22] in which the inversion of signature-splitting of the Routhians (namely, the phenomenon in which the "anomalous signature" becomes the "normal signature" at a certain ω-value) was discussed in odd-odd nuclei with N $\approx$ 90. The analysis was based on CSM, and the configuration considered consisted of the aligned $i_{13/2}$ neutron and the aligned $h_{11/2}$ proton. By fitting the CSM Routhians to observed signature-splittings, the authors in ref. 22 determined the γ-values for each configuration and each nuclei. Though the idea of ref. 22 looks interesting, one may raise many questions about their way of determining the γ-values, such as: 1) Are the parameters in the nuclear potential so stable for $\omega \lesssim$ 300 keV? 2) How about the effect of neutron-proton interaction in odd-odd nuclei on the signature splitting? 3) Fluctuation in the γ-variable has presumably to be taken into account in the analysis of such an accuracy. 4) Has the cranking approximation in the description of the rotational motion of triaxial shape been carried out in a reasonable way? And so on. In the rest part of the present subsection I would like to discuss[23] particularly about the question 4).

In the cranking model as usually formulated (or as described above) triaxiality is taken into account in the one-body field, but not in the way of rotation. - This is an inconsistent theory. We note: (a) In general, the principal axes of inertia tensor for rotational motion need not coincide with the principal axes of the density distribution, or the one-body potential. Further, the rotation around one of the principal axes of the triaxial shape in the density distribution may not in general correspond to the local energy-minimum, even if we restrict ourselves within the cranking approximation. (b) All three inertia parameters should play a role in determining the rotational motion, in order to construct a rotational state with good angular-momentum. It is also noted that especially for $\gamma \neq 0$ the quantity on the right-hand-side of eq. (1) extracted from experimental data may not really correspond to the cranking frequency as usually formulated, since the rotational axis would not be fixed to a particular intrinsic axis (such as the 1-axis). I do not have a definite answer to the question of how serious effect the inclusion of the full consequences of triaxial shape has on the nuclear spectroscopy. The seriousness depends naturally on the magnitude of I (or rotational frequency) and the physical quantities discussed. In the following I would like to show[23] three examples which may illustrate some part of the question described above.

The first example is the relative magnitude of the moments of inertia $\mathcal{J}^{(1)}$ and $\mathcal{J}^{(2)}$ defined in eqs. (2.1) and (2.2). First, in the cranking model with fixed potential-parameters and a given spin-alignment i, one obtains

$$\mathcal{J}^{(2)}_{\text{band}} \approx \mathcal{J}^{(1)}_{\text{band}} \quad \text{for } i = 0 \tag{24.1}$$

$$\mathcal{J}^{(2)}_{\text{band}} < \mathcal{J}^{(1)}_{\text{band}} \quad \text{for } i > 0 \tag{24.2}$$

Since in the yrast bands one generally expects a sizable positive value of i, one has the relation $\mathcal{J}^{(2)}_{\text{band}} < \mathcal{J}^{(1)}_{\text{band}}$. This relation does not fit in with some of recent experimental data, since in the A = 170 region values of $\mathcal{J}^{(2)}_{\text{band}}$ which are systematically larger than $\mathcal{J}^{(1)}_{\text{band}}$ have been observed of an extensive region of $(E_{I+1} - E_{I-1})/2$ values. Thus, next, we solve the Hamiltonian of triaxial rotor

$$H_{rot} = \sum_{k=1}^{3} A_k I_k^2 \tag{25}$$

keeping I as a good quantum-number. In the case of axially-symmetric intrinsic shape the eigenvalue of H_{rot} in (25) is proportional to I(I+1), and thus one obtains $\mathcal{J}^{(2)}_{\text{band}} \approx \mathcal{J}^{(1)}_{\text{band}}$. In the case of triaxial intrinsic shape one can show the relation $\mathcal{J}^{(2)}_{\text{band}} > \mathcal{J}^{(1)}_{\text{band}}$ in the following way. For a small I (see p. 186 of ref. 7) one has the expression

$$E_{rot} = AI^2 + BI^4 \tag{26}$$

Thus, one obtains

$$\mathcal{J}^{(1)}_{\text{band}} = \frac{1}{2(A + 2BI^2)}$$
$$\mathcal{J}^{(2)}_{\text{band}} = \frac{1}{2(A + 6BI^2)} \tag{27}$$

The fact that γ is non-zero and small means the negative sign for the B coefficient, which leads to the relation $\mathcal{J}^{(2)}_{\text{band}} > \mathcal{J}^{(1)}_{\text{band}}$. For a very large I (see p. 190 of ref. 7) one has

$$I_1 \approx I \gg 1 \tag{28}$$

for the yrast states, labeling the principal axes such that $A_1<A_2<A_3$. Then, one obtains

$$E_{rot} = A_1 I(I+1) + (\tilde{n}+\tfrac{1}{2})\hbar\omega \tag{29}$$

where

$$\hbar\omega = 2I[(A_2-A_1)(A_3-A_1)]^{\frac{1}{2}} \equiv bI \tag{30}$$

Therefore, one obtains again the relation $\mathcal{J}^{(2)}_{band} > \mathcal{J}^{(1)}_{band}$ from the expressions

$$\mathcal{J}^{(1)}_{band} = \frac{1}{2A_1 + \frac{b}{I}}$$
$$\mathcal{J}^{(2)}_{band} = \frac{1}{2A_1} \tag{31}$$

For an intermediate values of I the Hamiltonian (25) is numerically diagonalized. The result is shown in fig. 14 in terms of the parameters $(\hbar^2/2J_\kappa)\equiv A_\kappa$. It is seen that in the region of $20\lesssim I\lesssim 40$ all plotted values become more or less straight-lines and the $\mathcal{J}^{(2)}_{band}$-values are appreciably larger than the $\mathcal{J}^{(1)}_{band}$-values except the case of the rigid-body J_κ-values. Now, no alignemnt of intrinsic spin is included in the present triaxial-rotor model, and in actual nuclei one expects an appreciable amount of spin-alignment in the yrast region. Thus, the $\mathcal{J}^{(2)}_{band}$-values calculated by including spin-alignments could become smaller than the values given by the present triaxial-rotor model. However, it is seen that the treatment of rotational motion in the way of triaxial rotor has a possibility of producing the relation $\mathcal{J}^{(2)}_{band} > \mathcal{J}^{(1)}_{band}$, which is impossible to be obtained in the simple cranking model with a fixed intrinsic structure.

The second example is the "signature inversion" problem as an illustration of the role of all three moments of inertia in the rotational motion. We solve the particle-rotor model Hamiltonian (16) with (5) and (7), in which a quasiparticle in a single $j=\frac{3}{2}$ shell is coupled to a triaxial rotor. The choice of $j=\frac{3}{2}$ is not very typical of "high-j" in rotating nuclei, but the simplicity of the $j=\frac{3}{2}$ shell makes our understanding of the result easier. For the $j=\frac{3}{2}$ shell we have

$$\varepsilon_\nu/\kappa = -0.80 \text{ and } +0.80 \tag{32}$$

irrespective of the γ-value. In fig. 15 we show a numerical example of the difference between the spin-alignments of the yrast quasiparticle states with two signatures ($\alpha=\pm\frac{1}{2}$). The abscissa expresses the energy-difference $(E_{I+1} - E_{I-1})/2$, while the ordinate denotes the angular-momentum difference $I(\alpha=-\frac{1}{2})-I(\alpha=+\frac{1}{2})$. Note that the Fermi-level used in fig. 15 is such that "anomalous" signature for the lowest quasiparticle state is expected for relatively small ω-values in the cranking model. The result of the cranking model obtained by using the same intrinsic Hamiltonian is also shown in fig. 15. One understands the general trend of (rigid-body $\rightarrow$ hydrodynamical $\rightarrow$ cranking), since the cranking calculation (with $-\omega J_1$) would correspond to the values $J_1 \gg J_2$, J_3. From the example in fig. 15 it is seen that the "rotational frequency", at which the inversion of the relative magnitudes of the two spin-alignments occurs, depends significantly on the structure of rotation. Similarly, it is easy to show that the "rotational frequency", at which the inversion of the relative magnitudes of the two energies (with $\alpha=\pm\frac{1}{2}$) referring to the rotating coordinate-system occurs, depends very much on the way of rotation. One may expect that the moments of inertia of nuclei are something between the hydrodynamical values and the rigid-body values. Therefore, it does not seem to be reasonable to use the cranking model as usually formulated in the determination of the γ-values, by comparing the calculated quantities such as the signature-inversion frequency with the corresponding observed quantities.

The third example is to illustrate that even in the cranking model the rotation around one of the principal axes of triaxial shape in the one-body potential may not correspond to the local energy-minimum. We consider the classical orbits for configurations with a single-j-quasiparticle in the presence of the pair-correlation treated in the BCS approximation. The quasiparticle energy of the cranking Hamiltonian is written as (22). The points (j_1, j_2, j_3) on the sphere with a radius of j, in which h is stationary, are shown in fig. 2b of ref. 19. Now, it is straightforward to show, for example, that for a given small $\gamma < 0$ with a small ω the simultaneous inclusion of rotation around the 2-axis is favourable in the sense that the Hamiltonian

$$h' = h - \omega_2 j_2 \tag{33}$$

gives a local energy-minimum (for $j_2>0$ and not for $j_2<0$) for $\omega_2 \neq 0$ and $\omega_2 << \omega$, of which the energy is lower than that of the local minimum of h. This example suggests that we need to crank the triaxial system around at least two principal axes of the one-body potential, in order to get a local energy-minimum in the cranking model.

4. Rotational Effect on Electromagnetic Properties

Experimental information on the electromagnetic moments is, generally speaking, still in a preliminary stage. However, very exciting data on magnetic-dipole and electric-quadropole transition-moments are being obtained, from which one can study exclusively the fingerprint of the one-particle motion in rotating potentials. The study is the basis for the analysis of the shell-structure effects on nuclear high-spin states. Energies of states are, of course, one of the physical quantities from which one gets the information of the rotational perturbation on the system. However, it is seen that electromagnetic moments can be used as a richer and more stringent source for studying the rotational perturbation.

Since the rotational perturbation on the electric-dipole (E1) moments has been discussed in ref. 24 and there has been no clear-cut E1 data on high-spin states, in the following I do not take the examples of E1 transitions.

The rotational perturbation on the magnetic-dipole (M1) or the electric-quadropole (E2) transitions has been discussed [25] already in the early seventies, in connection with experimental data on the "attenuation of Coriolis-coupling". The numerical result presented in ref. 25 is on the same line as that in the present talk, but the present talk contains a more general and systematic discussion including the possibility of the triaxial shape in nuclei.

The magnitudes of static magnetic-dipole moments are expected to be sensitive to spin-alignment, the size of pair-correlation, and the relevant particle-configurations (e.g. proton or neutron). However, unfortunately the measurement of static magnetic-dipole moments of individual high-spin states has not yet been successful because of the low-statistical accuracy of the observed precessions, although a lot of experimental efforts have been made[26] in the measurement by using very large transient magnetic fields. Therefore, in the limited time of the present talk I do not include the static magnetic-dipole moments of high-spin states, though there have been a number of theoretical calculations since the sixties.

4.1 Experimental Information

We start by looking at some examples of experimental data, which show a strong rotational perturbation.

In fig. 16, we show the decay scheme of yrast states with positive parity in Yb-isotopes. (The predominant component of the wave functions of the neutron one-quasiparticle (1qp) states comes from the $i_{13/2}$-orbit.) For convenience, we call[27] the transitions as type B (A), if the angular momentum increases (decreases) by one unit going from the $\alpha = -1/2$ state to the $\alpha = +1/2$ state. These transitions are in principle mixtures of M1 and E2. Taking into account the measured (usually small) E2 admixture, one obtains the ratio of the B(M1)-value for type B to the one for type A. The observed dependence of the ratio on neutron number is very unique. Especially, the M1 components are negligible in the case of ^{165}Yb. Furthermore, for example, the ratio 10 in the case of ^{161}Yb is a remarkably large number, if we notice that in the absence of rotational perturbation on the intrinsic wave functions the ratio is about 1.1 for K=3/2, as is estimated from the well-known formula[7],

$$B(M1;K, I_1 \rightarrow K, I_2=I_1\pm1) = \frac{3}{4\pi}\left(\frac{e\hbar}{2mc}\right)^2 (g_K-g_R)^2 \; K^2 \langle I_1 K 1 o | I_2 K\rangle^2 \qquad (34)$$

for $K > 1/2$.

In fig. 17 the measured electric-quadrupole transition-moments (with $\Delta I=2$) of $^{154}_{66}Dy_{88}$ are shown. A number of this kind of data have been reported. The measured variation of Q_0(eff) is often used for obtaining the variation of the γ-value (namely, triaxiality) as a function of I, by using the classical expression such as $\hat{Q}_{\pm2}$ in (45). A comment on this way of determining the γ-value will be given later.

In fig. 18 the ratio of $q_1(K=7/2)_{eff}$ with $\Delta I=1$ to $q_0(K=7/2)_{eff}$ with $\Delta I=2$ where

$$q_n(K=\tfrac{7}{2})_{eff} \equiv \frac{\left(\frac{16\pi}{5} B(E2:I\rightarrow I-\Delta I)\right)^{\frac{1}{2}}}{\langle I \; \frac{7}{2} \; 2 \; o \mid I-\Delta I, \frac{7}{2}\rangle} \qquad (35)$$

is plotted, which is obtained by using measured B(E2)-values in the yrast negative-parity band of $^{157}_{67}Ho_{90}$. The strong signature-dependence of the ratio $(q_1/$ $/q_0)_{eff}$, seen in fig. 18 must mean that either $q_0(K=7/2)_{eff}$ or $q_1(K=7/2)_{eff}$, or both $q_0(K=7/2)_{eff}$ and $q_1(K=7/2)_{eff}$ must have a strong signature-dependence.

In my knowledge there has been no measurements of static electric-quadrupole moments of high-spin states.

4.2 M1 Transitions estimated in Cranking Model [27, 28]

First, we evaluate B(M1)-values by using a simple version of the cranking model and see what is the definite conclusion from the evalution. We take M1 operator as

$$\theta^{\nu}_{M1} \equiv \sqrt{\frac{3}{4\pi}}\,[(g^{eff}_{\ell} - g_R)\ell_{\nu} + (g^{eff}_{s} - g_R)s_{\nu}] \tag{36}$$

The B(M1)-values defined by

$$B(M1; I_1 \to I_2) = \sum_{\mu M_2} |\langle I_2 M_2|\theta^{\nu}_{M1}|I_1 M_1\rangle|^2 \tag{37}$$

can be written as

$$B(M1; I_1 \to I_2) \approx |\langle I_2, M_2{=}I_2|\theta^{\mu=I_2-I_1}_{M1}|I_1, M_1{=}I_1\rangle|^2 \tag{38}$$

for $I_1 \approx I_2 \gg \lambda = 1$. Now, in the cranking model the angular momentum I is identified with the component J_1 of the total angular momentum of the paritcles. Namely, the 1-axis is the classical direction of I. Thus, a state $|I, M{=}I\rangle$, appearing in (38), is interpreted as a cranked state by choosing the 1-axis as a quantization axis.In fig. 19 an example of quasiparticle spectra in a rotating system is shown as a function of rotational frequency, which is obtained from the diagonalization of a cranked Nilsson Hamiltonian, while in fig. 20 the calculated B(M1)-values of the transitions between the corresponding 1qp states are shown. Defining the energy difference between the 1 qp states with $\alpha = \pm 1/2$ as

$$\Delta E_{qp} = E'(\alpha{=}-1/2) - E'(\alpha{=}+1/2) \tag{39}$$

We write M1 transition energies as

$$\Delta E_{M1} = |\Delta E_{qp} \pm \hbar\omega| \quad \text{for type } {A \atop B} \tag{40}$$

in a cranking approximation.

By examining the numerical results such as figs. 19 and 20 in the case of axially-symmetric shape, one can draw the following two definite conclusions: For a given quasiparticle pair (with α=+1/2 and -1/2);1) The transition with a smaller transition energy has a larger B(M1)-value. 2) The B(M1)-value of the transition with a larger transition-energy vanishes, when the energy of the partner-transition vanishes. These two conclusions are schematically illustrated in fig. 21. It is seen that these conclusions are in agreement with the experimental observation shown in fig. 16.

The above conclusion 1) may be understood as a tendency toward the limit of the complete spin-alignment. Namely, it is noticed that in the limit the B(M1)-value with a smaller transition-energy is of the order of the B(M1)-values for single-particle transitions, while the one with a larger transition-energy vanishes. Although the spin-alignments of each quasiparticle in the example of fig. 19 are still far from the values for the complete spin-alignment, one sees already the feature of the selection rule for the B(M1)-values which appears in the limit of complete spin-alignment. The conclusion 2) is understood in the following way: If the intrinsic Hamiltonian is axially symmetric around the 3-axis, one obtains

$$[H_{cranking}, J_3] \equiv [H_{intr} - \hbar\omega J_1, J_3] = i\,\hbar\omega\, J_2 \tag{41}$$

By taking the matrix element of the operator in (41) between the 1 qp states with $\alpha = \pm 1/2$, we obtain

$$(E'_{-\frac{1}{2}} - E'_{+\frac{1}{2}})\langle\alpha=-\tfrac{1}{2}|J_3|\alpha=+\tfrac{1}{2}\rangle = \hbar\omega\,\langle\alpha=-\tfrac{1}{2}|iJ_2|\alpha=+\tfrac{1}{2}\rangle \tag{42}$$

where $E'_{\pm\frac{1}{2}}$ expresses the eigenvalues of the cranking Hamiltonian

$$(H_{intr} - \hbar\omega\; J_1)|\;\alpha=\pm\tfrac{1}{2}\rangle = E'_{\pm\frac{1}{2}}|\alpha=\pm\tfrac{1}{2}\rangle \tag{43}$$

From (42) we see that if the quantity $(E'_{-\frac{1}{2}}-E'_{+\frac{1}{2}} - \hbar\omega)$ vanishes which is the transition energy of type B, the matrix element $\langle\alpha=-\frac{1}{2}|iJ_2-J_3|\alpha=+\frac{1}{2}\rangle$ vanishes which is proportional to the M1 matrix-element of the transition of type A.

There may be some arguments to say that the estimate based on the cranking model is a too crude approximation. However, by using the same intrinsic Hamiltonian in the particle-rotor model, and thus by treating the angular momentum as a good quantum-number, one can calculate the corresponding M1 transition matrix-elements. We show such numerical examples in sect. 4.4. Here we briefly summarize the calculated results as follows: The two conclusions, which were drawn above from the simple cranking model, remain semiquantitatively unchanged when angular momentum is treated as a good quantumnumber in particle-rotor model calculation. Especially, the behaviour of the measured ratio of $B(M1)_B$-values to $B(M1)_A$-values as a function of neutron number, which is shown in fig. 16, can be nicely reproduced in the particle-rotor model, without introducing a deviation from axially-symmetric shape. In particular, the measured almost-vanishing $B(M1)_A$-values in ^{165}Yb are in a nice agreement with the energy-degeneracy criterion described above.

4.3 E2 Moments estimated in Triaxial Rotor[29)]

By writing the quadrupole-operator qunatized along the rotation-axis (the intrinsic 3-axis) as $\hat{Q}_\mu(Q_\nu)$, one has the relation

$$\hat{Q}_\mu = D^2_{\mu o} Q_o + (D^2_{\mu 2} + D^2_{\mu -2}) Q_2 \tag{44}$$

where the deformation of the rotor is expressed by the parameters Q_o and Q_2. One then gets the quadrupole operators

$$\hat{Q}_o = -\frac{1}{2} Q_o + \sqrt{\frac{3}{2}} Q_2 \qquad \text{for static quadrupole moment}$$

$$\hat{Q}_{\pm 1} = (\mp \sqrt{\frac{3}{2}} Q_o \pm Q_2)(\frac{I_3}{I}) \text{ for } \Delta I=\pm 1 \text{ transitions} \tag{45}$$

$$\hat{Q}_{\pm 2} = \sqrt{\frac{3}{8}} Q_o + \frac{1}{2} Q_2 \qquad \text{for } \Delta I=\pm 2 \text{ transitions}$$

From the expression (45) we note the following points: (a) In the case of axially-symmetric shape all the quadrupole moments (with $\Delta I=0$, ± 1, ± 2) are determined by one parameter Q_o, while for triaxial shape the ratio of static moments to collective E2 transition (with $\Delta I=\pm 2$) depends significantly on the ratio of Q_o to Q_2. (b) The effect of $Q_2 \neq 0$ (i.e. $\gamma \neq 0$) is such that the quantity $\hat{Q}_{\pm 1}$ is reduced (enhanced) if the quantity $\hat{Q}_{\pm 2}$ is enhanced (reduced). Therefore, the ratio of the $B(E2;I\rightarrow I-1)$-value to the $B(E2;I\rightarrow I-2)$-value can be used as a sensitive measure of the γ-value. (c) The expression (45) gives the values of the E2 matrix-elements averaged over the two signatures, and a more detailed analysis of the tunneling between $\pm$ values of I_3 is necessary in order to describe the signature-dependent effects. (d) By using the definition of Q_2 and Q_o in terms of β and γ, the expression for $\hat{Q}_{\pm 2}$ in (45) can be seen to be proportional to $\cos(\gamma+30^o)$.

4.4 Analysis based on Particle-Rotor Model

Now, we take into account the intrinsic state describing a single quasi-particle with angular-momentum j. We take the Hamiltonian in (16) together with the expressions (5) and (7). We diagonalize the Hamiltonian by using the BCS 1qp states for the particle wave-functions. The calculation is sufficient for drawing the conclusion about the problems treated in the following.

We do not discuss here the details of the particle-rotor model and its connection with the cranking model, but would emphasize that the cranking model as usually formulated involves neglect of correlations that appear to be crucial for the description of the E2 matrix-elements considered in the present section. In particular, we mean the detailed correlation of the vectors $\vec{I}$ and $\vec{j}$ as seen in the intrinsic coordinate-system.

The basic formulae for the particle-rotor model with triaxial shape can be found in refs. 7 and 30. In the numerical calculations such as ref. 30 the moments of inertia of hydrodynamical type

$$J_K = \frac{4}{3} J_o \sin^2(\gamma + k \frac{2}{3}) \tag{46}$$

are conveniently used. If one takes the expressions (7) and (46), it is sufficient to investigate the region of $0^o \leq \gamma \leq 60^o$. However, the expression (46) means that J_2 is largest for $0^o \leq \gamma \leq 60^o$, while the rigid moments of inertia calculated from the ellipsoidal shape of the core, which gives rise to the one-body potential (7), has the largest component around the 1-axis. That means, in available numerical calculations such as those in ref. 30 only the region of so-called "rotation around an intermediate axis" (namely, the "$-60^o \leq \gamma \leq 0^o$ region" in the sense of the cranking model[13]) has been investigated. In order to investigate the "$0^o \leq \gamma \leq 60^o$ region" in the sense of the cranking model, we invert the sign of γ in $J_k(\gamma)$ in (46) while keeping $V(\gamma)$ in (7). Thus, in the following numerical evaluation we employ the sign of γ corresponding to the convention[13] of the cranking model, namely

$$\begin{aligned} \gamma > 0: &\quad V(\gamma) \leftrightarrow J_k(-\gamma) \\ \gamma = 0: &\quad V(\gamma=0) \leftrightarrow J_3 = 0,\ J_1 = J_2 \\ \gamma < 0: &\quad V(\gamma) \leftrightarrow J_k(\gamma) \end{aligned} \tag{47}$$

We stress here that the moments of inertia of the type (46) is an assumption and, in fact, numerical results have often a crucial dependence on the form of $J_k(\gamma)$ assumed.

In fig. 22 we show examples of calculated B(E2)-values for the $h_{11/2}$-shell. Chosen Fermi-levels are such that, roughly speaking, the upper graph (with $\lambda = (-0.70)\kappa$) corresponds to the positive-parity yrast states of the $N \approx 91$ nucleus (when the Fermi-level is translated into the $i_{13/2}$-model), while the lower graph (with $\lambda = (-0.1.0)\kappa$) corresponds to the negative-parity yrast states of odd-A Ho-isotopes. For convenience, B(E2)-values are expressed in the unit of the unperturbed axially-symmetric B(E2)-values with K=3/2 and K=7/2, respectively, which are smoothly-varying functions of I. Examples of calculated static quadrupole-moments are shown in fig. 23, which are expressed in the unit of $(-\frac{1}{2})Q_0$ that is equal to the classical value for axially-symmetric shape as is seen from the expression for $\hat{Q}_0$ in (45).

Conclusions[29], which can be drawn from numerical examples such as those in figs. 22 and 23 and are not obtained from the expressions (45) are: 1) For $\gamma=0$, B(E2)-values have a negligible amount of signature-dependence both for $\Delta I=1$ and $\Delta I=2$ transitions, while for $\gamma \neq 0$ a considerable amount of signature-dependence ("zigzag") appears especially in B(E2; $\Delta I=1$)-values. The lower the Fermi-level is placed in the $h_{11/2}$-shell, the more conspicuous the "zigzag" becomes in the presently investigated region of I. The conspicuous signature-dependence of the calculated B(E2; $\Delta I=1$)-values comes from the constructive or destructive phase between the core E2 matrix-elements with $\Delta K=0$ and $\Delta K=2$. - This structure of the wave-functions does not exist in the simplest version of cranking model. 2) The B(E2; $\Delta I=2$)-values may approach to the classical value of the γ-dependence, which is proportional to $(\cos(\gamma+30^o))^2$, for larger values of I. By comparing the ratio $(\cos(15^o)/\cos(45^o))^2 = 1.87$ with the ratio of B(E2; $I \to I-2$, $\gamma=-15^o$)/B(E2; $I \to I-2$, $\gamma=+15^o$) estimated from fig, 22, we see that the B(E2; $\Delta I=2$)-values approach the classical values very slowly with increasing I. 3) Calculated values of static quadrupole moments approach to the expression $\hat{Q}_0$ in (45), also very slowly with increasing I. The calculated values show a small signature-dependence.

The signature-dependence ("zigzag") which appears in the calculated B(E2; $\Delta I=1$)-values in the case of $\gamma \neq 0$ may be illustrated by a simple formula, when we consider the model in which a $j=\frac{1}{2}$ particle is coupled to an even-even core. As is well-known, a $j=\frac{1}{2}$ particle is totally decoupled from the rotation of the core. (Or, in other words, the spin-alignment of a $j=\frac{1}{2}$ particle is all the time complete.) For example, expressing the angular momentum of the core by R_o, the two states with $(R_o\ \frac{1}{2})I$ where $I=R_o+\frac{1}{2}$ and $R_o-\frac{1}{2}$ are energetically degenerate. The $B(M1:(R_o+2\ \frac{1}{2})R_o+\frac{3}{2}\rightarrow(R_o\ \frac{1}{2})R_o+\frac{1}{2})$-value vanishes. In the limit of large I the E2 moments between the yrast states are written as (cf. expressions in (45))

$$\hat{Q}_o = -\frac{1}{2}Q_o + \sqrt{\frac{3}{2}}Q_2$$

$$\hat{Q}_{\pm 1} = \left(\mp\sqrt{\frac{3}{2}}Q_o \pm (1\mp(-1)^{I-\frac{1}{2}}2)Q_2\right)\left(\frac{I_3}{I}\right) \tag{48}$$

$$\hat{Q}_{\pm 2} = \sqrt{\frac{3}{8}}Q_o + \frac{1}{2}Q_2$$

Namely, in the case of $Q_2 \neq 0$ a strong signature-dependence of the $B(E2:I\rightarrow I\pm 1)$-values remains in the limit of large I, while such a signature-dependence may be seen in neither $B(E2:I\rightarrow I\pm 2)$-values nor static quadrupole moments. It is noted that the condition $Q_2 \neq 0$ in the present formula does not mean that the energy minimum occurs at $\gamma \neq 0$, but it means that the expectation value of Q_2 is non-zero. That means, a large fluctuation of nuclear shape in the γ-variable may lead to $Q_2 \neq 0$, even if the energy-minimum occurs at $\gamma = 0$.

Coming back to the experimental information: A considerable amount of possible signature-dependence in the $B(E2:\Delta I=1)$-values was reported[31)] already two years ago in the analysis of the negative-parity band of ^{157}Ho, as is shown in fig. 18. As can be seen from fig. 22, such a strong signature-dependence is impossible to be obtained theoretically, if one assumes an axially-symmetric shape for the nucleus. The observed signature-dependence in ^{157}Ho shown in fig. 18 is far more conspicuous than the calculated one for $\gamma=\pm 15^o$ shown in the lower part of fig. 22. The calculated magnitude of the signature dependence ("zigzag") depends on the three moments of inertia used, and actual moments of inertia can be pretty different from those of hydrodynamical type. However, as is shown in the following, for the proton Fermi-level of ^{157}Ho it does not seem to be possible to obtain the calculated result, in which the ratio $(q_1/q_o)_{eff}$ is much larger than unity in agreement with experimental data. Namely, by expressing the average value of $q_1(K)_{eff}$ with $\Delta I=+1$ and -1 by $\bar{q}_1(K)_{eff}$, in the limit of $I>>K$ one obtains

$$\left|\frac{\bar{q}_1(K)_{eff}}{q_o(K)_{eff}}\right| = \left|\frac{1-\sqrt{\frac{2}{3}}\frac{Q_2}{Q_o}}{1+\sqrt{\frac{2}{3}}\frac{Q_2}{Q_o}}\right| \quad \frac{K_{eff}}{K} \tag{49}$$

For the yrast quasiparticle orbits we have

$$\frac{K_{eff}}{K} < 1 \tag{50}$$

for K > 5/2. Then, for $\gamma < 0$, namely for the positive value of (Q_2/Q_0), the ratio (49) is much smaller than unity. For K=7/2 as in the case of ^{157}Ho, the ratio (49) can hardly become larger than 1.2 for $\gamma < 30^o$. The value $\gamma > 30^o$ seems to be rather unlikely in the low-energy yrast states of ^{157}Ho, considering experimental information other than B(E2:ΔI=±1)-values. Therefore, if we accept the measured magnitudes of the ratio shown in fig. 18, it seems that the data on ^{157}Ho may be pretty difficult to be explained quantitatively by introducing a triaxial shape in the present form, though the deviation of the nuclear shape from axially-symmetry seems to be absolutely necessary for obtaining the "zig-zag".

On the left-hand-side of fig. 24 we show calculated B(M1)-values for λ= (0.10)κ. On the right-hand-side the difference between the energies E'(α) with $\alpha = \frac{1}{2}$ and $-\frac{1}{2}$, which refer to the rotating coordinate system, is plotted as a function of ω defined as

$$\hbar\omega \equiv \frac{dE}{dI} \tag{51}$$

Though in the case of $\gamma \neq 0$ the definition of ω in (51) does not correspond really to the cranking frequency in the simple cranking model[23], the plotting is one way of showing the signature-splitting of the yrast spectra. The value of ω, for which the difference $E'(\alpha=\frac{1}{2})-E'(\alpha=-\frac{1}{2})$ is plotted on the r.h.s., corresponds approximately to the value of I on the l.h.s. figure. It can be seen that the information extracted from B(M1)-values is quite independent of that from B(E2)-values. While B(E2)-values exhibit directly the information on the shape of the system, B(M1)-values are more closely related to the signature-splitting of the yrast spectra. However, if the shape of the system deviates considerably from axial-symmetry, the effective g-factor may have an appreciable dependence on the direction of the rotation axis as seen in the intrinsic space. Since we know very little about this dependence, the usefulness of B(M1)-values for studying the nuclear structure of triaxial systems may be limited.

References

1) See, for example ; J.D.Garrett and J.J.Gaardhøje, XIV Mazurian School Lectures, 1981, to be published in Nucleonika : I.Hamamoto, NORDITA preprint 81/28, to appear in Heavy Ion Sciences, ed. A.Bromley, Plenum Publishing Company : Lecture notes in Proc.Nuclear Physics Workshop, Trieste, Italy, 1981 (ed. C.H.Dasso) North-Holland

2) A.Bohr and B.R.Mottelson, Physica Scripta, 24(1981)71

3) A.Bohr and B.R.Mottelson, Proc.Int.Conf.nuclear Structure, Tokyo, 1977

4) A.L.Goodman, Nucl.Phys.A230(1974)466

5) A.Bohr, I.Hamamoto and B.R.Mottelson, Physica Scripta 26(1982)267

6) K.T.Hecht and A.Adler, Nucl.Phys.A137(1969)129 : R.D.Ratna Raju, J.P.Draayer and K.T.Hecht, Nucl.Phys.A202(1973)433 : A.Arima, M.Harvey and K.Shimizu, Phys.Lett.30B(1969)517

7) A.Bohr and B.R.Mottelson, Nuclear Structure, vol.2 (Benjamin, New York, 1975)

8) I.Hamamoto, Nucl.Phys.A271(1976)15

9) D.R.Inglis, Phys.Rev.96(1954)1059

10) S.G.Nilsson, Mat.Fys.Medd.Dan.Vid.Selsk.29, no.16(1955)

11) A.Bohr, Proc.Int.School of Physics "Enrico Fermi", course LXIX (ed. A.Bohr and R.A.Broglia), North-Holland, 1977

12) A.Bohr, Mat.Fys.Medd.Dan.Vid.Selsk, 26, no.14(1952)

13) In the present lecture I employ so-called "Lund convention", which was used in the article : G.Andersson, S.E.Larsson, G.Leander, P.Möller, S.G.Nilsson, I.Ragnarsson, S.Åberg, R.Bengtsson, J.Dudek, B.Nerlo-Pomorska, K.Pomorski and Z.Szymanski, Nucl.Phys.A268(1976)205. Note that somewhat different conventions in the definition of γ-values are sometimes used in the literature, though the physics discussed is the same.

14) R.Bengtsson and S.Frauendorf, Nucl.Phys.A327(1979)139

15) For example ; B.Banerjee, H.J.Mang and P.Ring, Nucl.Phys.A215(1973)366 : P.C.Bhargava and D.J.Thouless, Nucl.Phys.A215(1973)515 : A.L.Goodman, Nucl. Phys.A230(1974)466 : A.Faessler, K.R.Sandhya, F.Grummer, K.W.Schmid and R.R. Hilton, Nucl.Phys.A256(1976)106 : H.J.Mang, Phys.Reports 18C(1975)327 : K.Neergaard, V.V.Pashkevich and S.Frauendorf, Nucl.Phys.A262(2976)61

16) For example ; R.R.Hilton, H.J.Mang, P.Ring, J.L.Egido, H.Herold, M.Reinecke, H.Ruder and G.Wunner, Nucl.Phys.A366(1981)365 : J.Rekstad and T.Engeland, Phys.Lett.89B(1980)316

17) J.Almberger, I.Hamamoto and G.Leander, Nucl.Phys.A333(1980)184 : A.Bohr and B.R.Mottelson, Physica Scripta 22(1980)461
18) For example ; R.Bengtsson, J. de Phys.(Paris) Colloq.41(1980)C10-84
19) I.Hamamoto and B.R.Mottelson, Phys.Lett.127B(1983)281
20) B.R.Mottelson, Proc.Symp.on High-Spin Phenomena in Nuclei (1979), ANL/PHY-79-4 : A.Bohr and B.R.Mottelson, Physica Scripta 22(1980)461
21) For example ; S.Frauendorf and F.R.May, Phys.Lett.125B(1983)245
22) R.Bengtsson, H.Frisk, F.R.May and J.A.Pinston, Nucl.Phys.A415(1984)189
23) I.Hamamoto, Phys.Lett., in press
24) I.Hamamoto and H.Sagawa, Nucl.Phys.A327(1979)99
25) S.A.Hjorth, H.Ryde, K.A.Hagemann, G.Løvhøiden and J.C.Waddington, Nucl. Phys.A144(1970)513 : S.A.Hjorth, A.Johnson and G.Ehrling, Nucl.Phys.A184 (1972)113
26) For example ; H.R.Andrews et al, Phys.Rev.Lett. 45(1980)1835 : E.Grosse, Proc. of 1982 INS international Symposium on Dynamics of Nuclear Collective Motion, p.134 (Tokyo, 1982) : O.Häusser et al, Phys.Rev.Lett.48(1982)383
27) I.Hamamoto, Phys.Lett.102B(1981)225
28) I.Hamamoto, Phys.Lett.106B(1981)281
29) I.Hamamoto and B.R.Mottelson, Phys.Lett.132B(1983)7
30) J.Meyer-ter-Vehn, Nucl.Phys.A249(1975)111 and 114 : S.E.Larsson, G.Leander and I.Ragnarsson, Nucl.Phys.A307(1978)189
31) G.B.Hagemann, J.D.Garrett, B.Herskind, G.Sletten, P.O.Tjøm, A.Henriquez, F.Ingebretsen, J.Rekstad, G.Løvhøiden and T.F.Thorsteinsen, Phys.Rev.25C (1982)3224
32) J.J. Gaardhøje, Speciale, University of Copenhagen, 1980: L.L. Riedinger, Nucl. Phys. A347 (1980) 141.
33) J. Kownacki, J.D. Garrett, J.J. Gaardhøje, G.B. Hagemann, B. Herskind, S. Jönsson, N. Roy, H. Ryde and W. Walús, Nucl. Phys. A394 (1983) 269.
34) N. Roy, S. Jönsson, H. Ryde, W. Walús, J.J. Gaardhøje, J.D. Garrett, G.B. Hagemann, and B. Herskind, Nucl. Phys. A382 (1982) 125.
35) A. Pakkanen et al, Phys. Rev. Lett. 48 (1982) 1530.
36) G.B. Hagemann, J.D. Garrett, B. Herskind, G. Sletten, P.O. Tjøm, A. Henriquez, F. Ingebretsen, J. Rekstad, G. Løvhøiden and T.F. Thorsteinsen, Phys. Rev. 25C (1982) 3224.

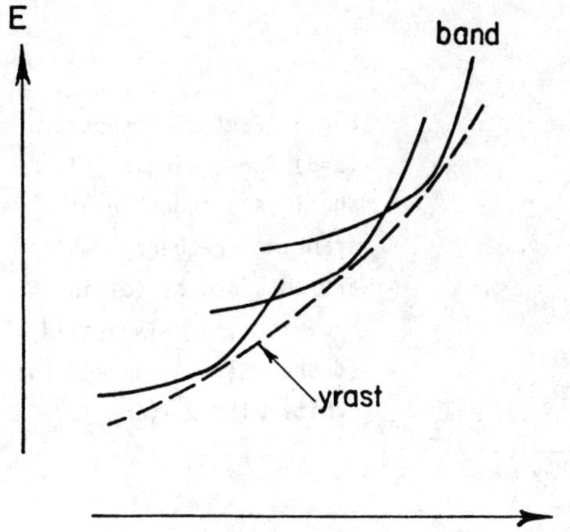

Fig. 1. A segment of an yrast spectrum involving a series of band crossings. The individual rotational bands are drawn by solid lines, while the envelope of individual yrast bands is indicated by a dashed line which characterizes the overall structure.

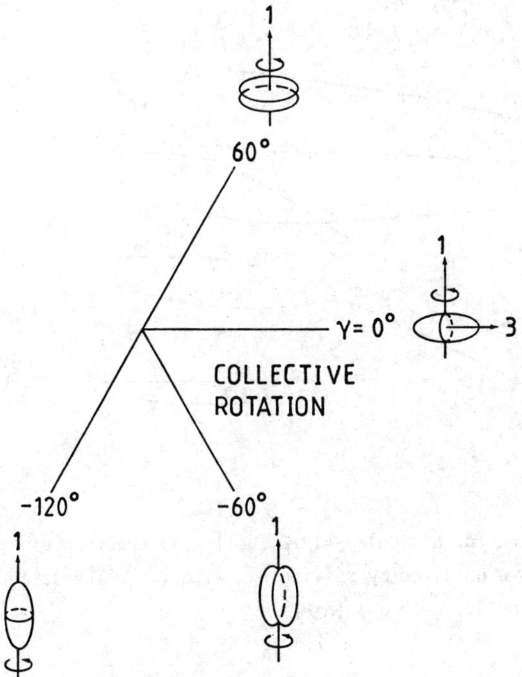

Fig.2. Definition of γ-variable in the relation between the intrinsic shape and the cranking axis, according to the "Lund convention" (ref. 13).

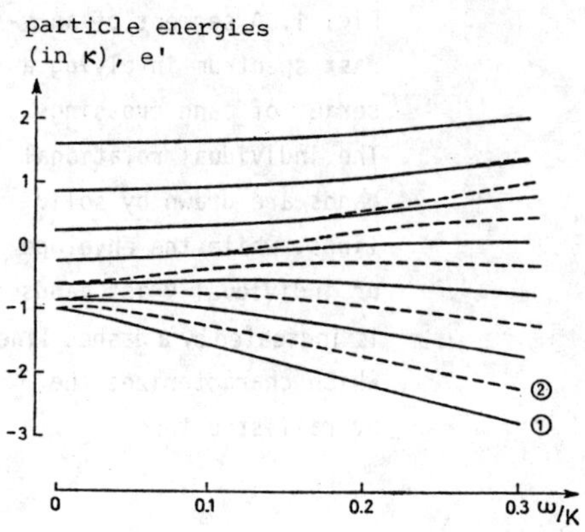

Fig.3. Particle spectra (Δ=0) for a single j=13/2 shell as a function of cranking frequency, which are obtained by solving H_{CR} with $\gamma=0^{0}$. The solid (dashed) lines express the orbits with $\alpha=\frac{1}{2}(\alpha=-\frac{1}{2})$.

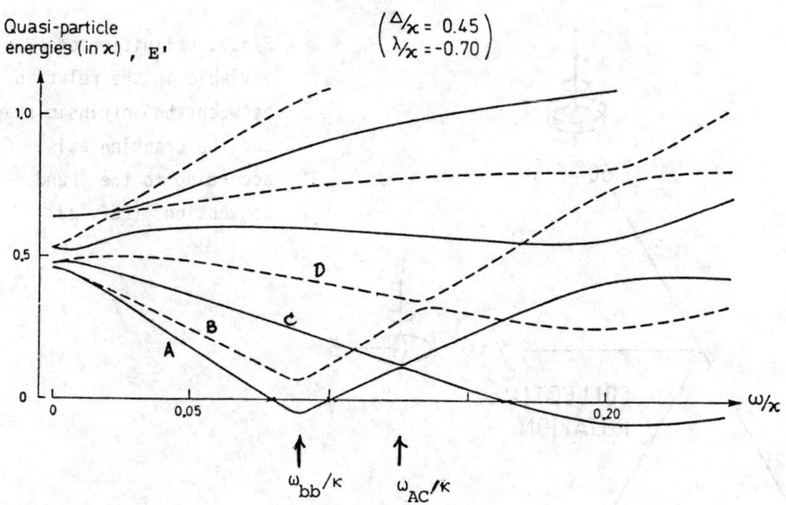

Fig. 4. Quasiparticle spectra for a single j=13/2 shell as a function of cranking frequency, which are obtained by solving H_{CR} with $\gamma=0^{0}$. The solid (dashed) lines express the orbits with $\alpha=\frac{1}{2}$ ($\alpha=-\frac{1}{2}$).

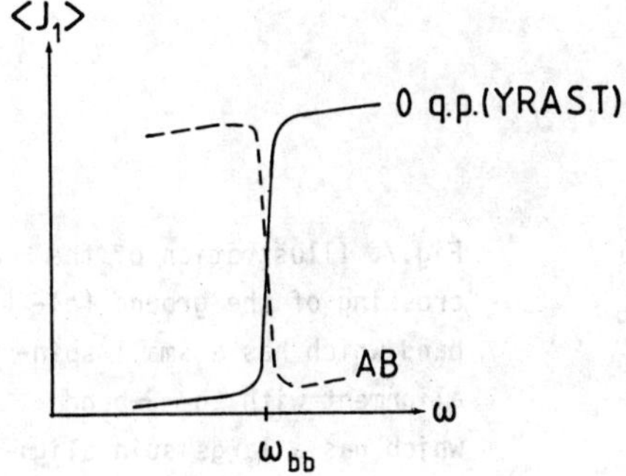

Fig. 5. Spin-alignments of the quasiparticle vacuum and the two-quasiparticle state (AB) as a function of cranking frequency.

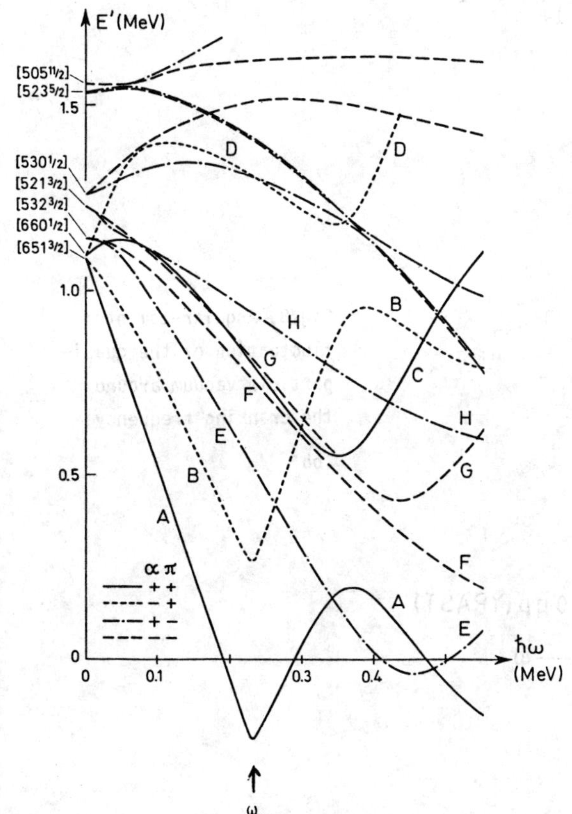

Fig.6. Quasiparticle spectra as a function of cranking frequency, which are calculated from a cranked modified-oscillator model for the fixed parameters suitable for neutrons in the low-lying states of ^{160}Yb. The asymptotic quantum-numbers [N n_3 Λ Ω] of Nilsson orbits at ω=0 are denoted on the left hand side of the ordinate.

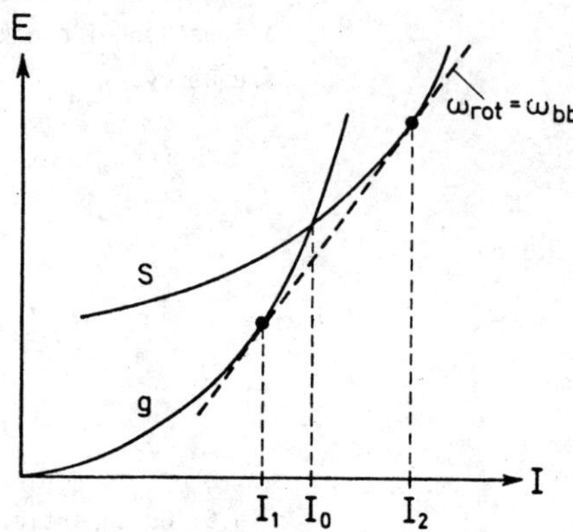

Fig.7. Illustration of the crossing of the ground (g)-band which has a small spin-alignment with the S-band which has a large spin-alignment. The figure is taken from ref. 11. In the illustration the interaction between the two bands is neglected.

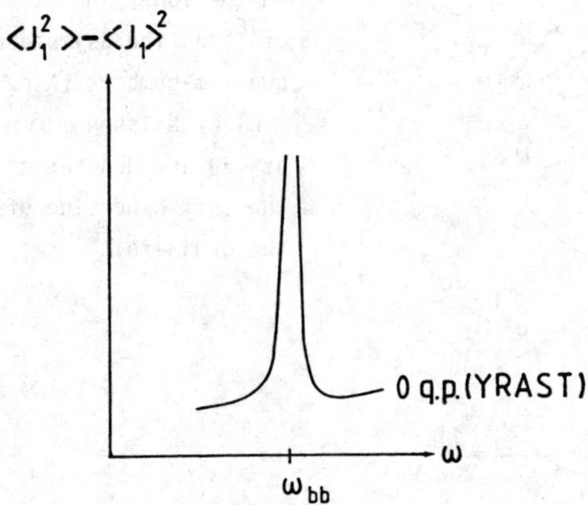

Fig.8. Angular-momentum fluctuation of the quasi-particle vacuum around the cranking frequency ω_{bb}.

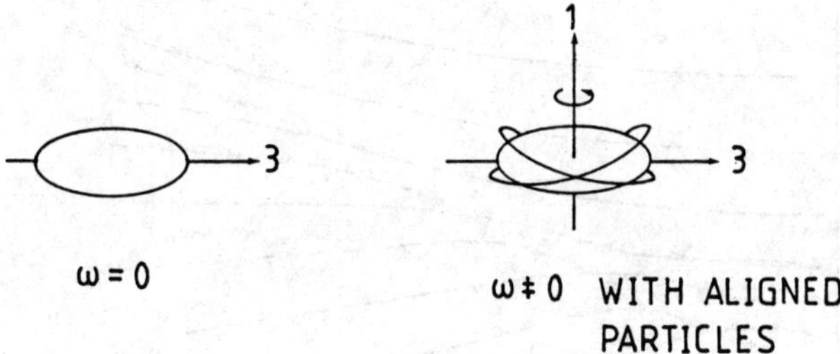

Fig.9. Illustration of the situation, in which the presence of spin-aligned particles in rotating nuclei leads the system to a triaxial shape ($\gamma \neq 0$).

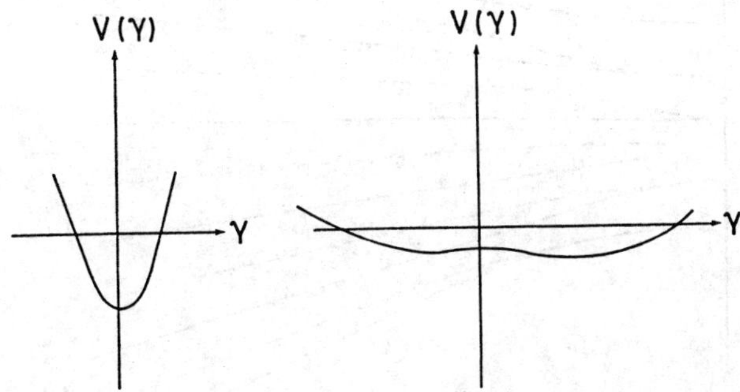

Fig.10. Schematic illustration of the potential energy as a function of the γ-variable.

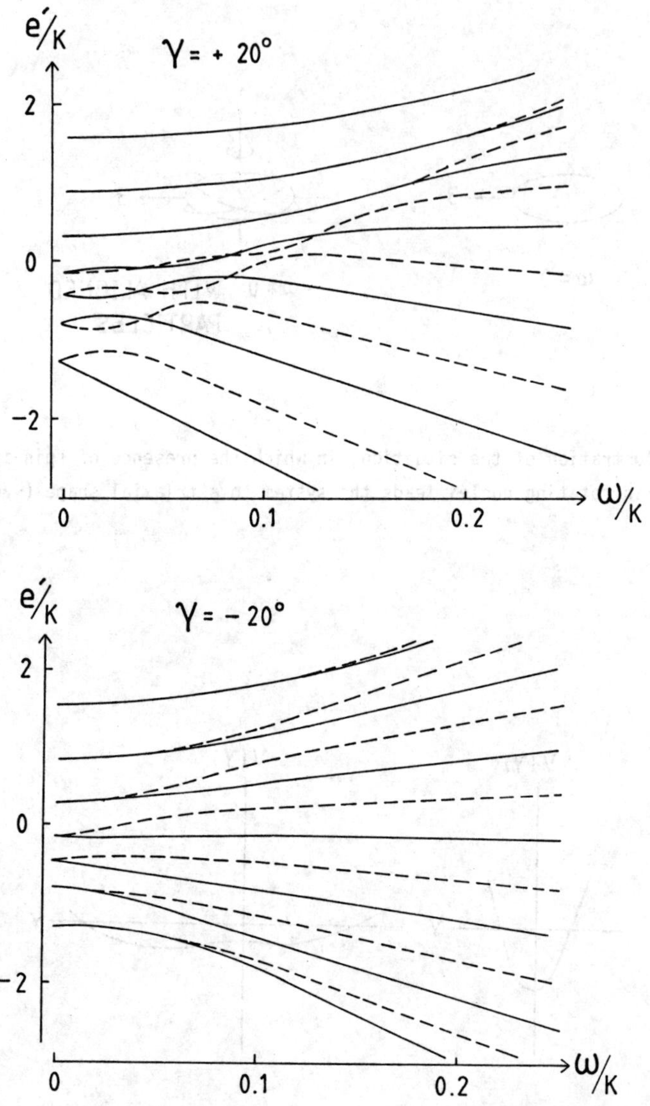

Fig. 11. Particle spectra (Δ=0) for a single j=13/2 shell as a function of cranking frequency, which are obtained by solving H_{CR} with $\gamma \neq 0$. The solid (dashed) lines express the orbits with $\alpha=\frac{1}{2}$ ($\alpha=-\frac{1}{2}$).

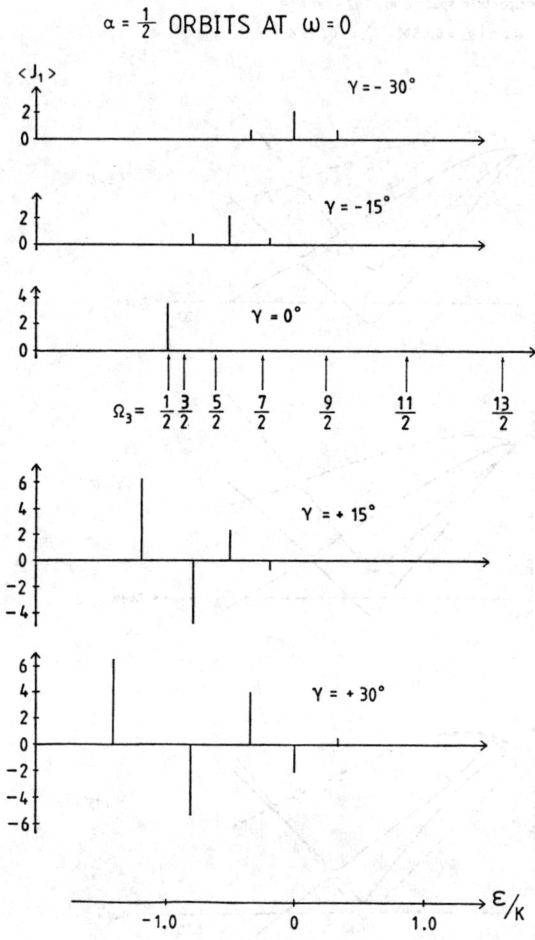

Fig. 12. The expectation values of j_1 for the $\alpha=\frac{1}{2}$ orbits as a function of eigenvalues of the one-particle intrinsic Hamiltonian with the potential (7) for a single j=13/2 shell.

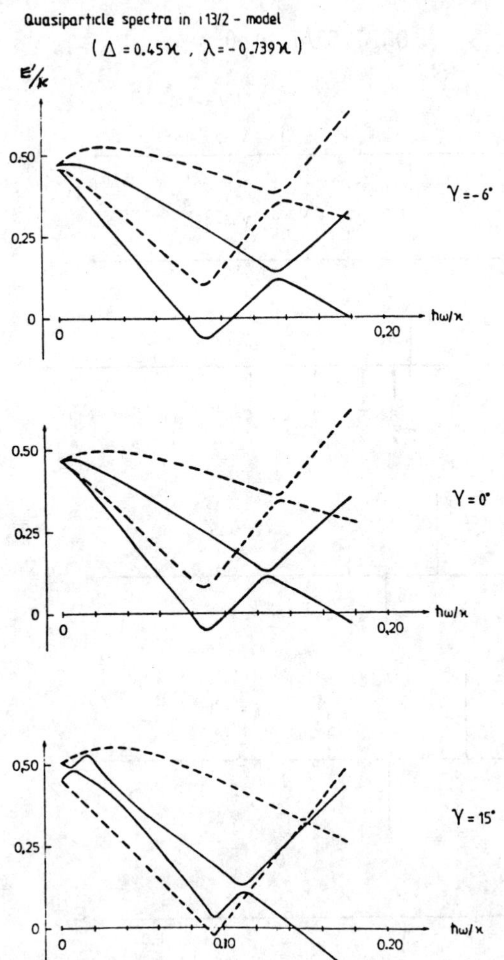

Fig. 13. Examples of quasiparticle spectra for a single j=13/2 shell as a function of cranking frequency. The solid (dashed) lines express the orbits with $\alpha=\frac{1}{2}$ ($\alpha=-\frac{1}{2}$).

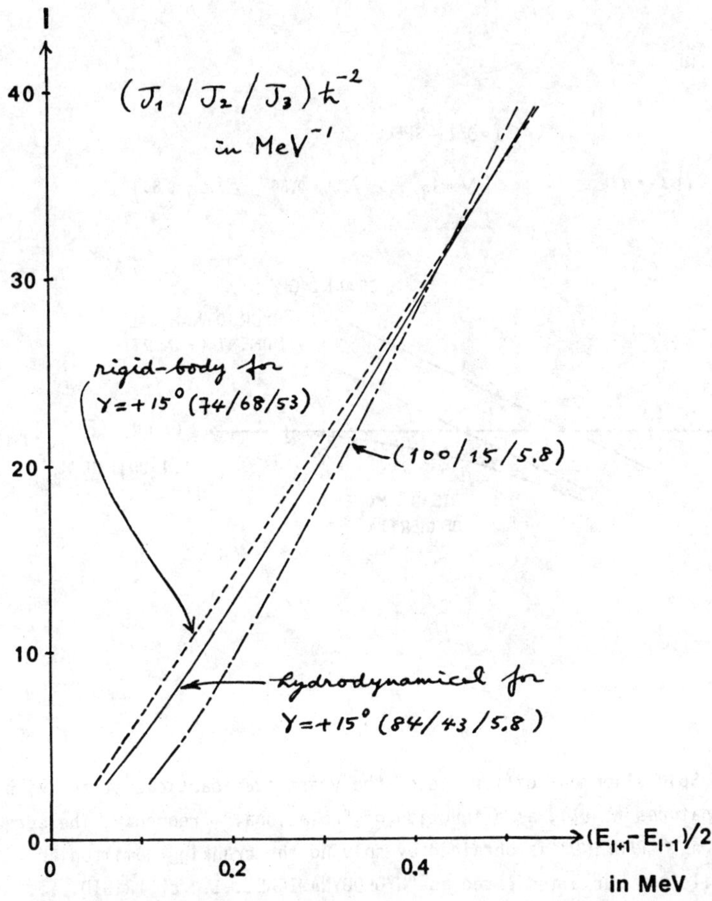

Fig.14. Angular-momentum I versus "rotational frequency" $(E_{I+1}-E_{I-1})/2$, obtained from the diagonalization of the triaxial-rotor Hamiltonian (25) with $(\hbar^2/2\mathcal{J}_\kappa)\equiv A_\kappa$. The solid-line is obtained by using the hydrodynamical moment of inertia for $\gamma=+15^0$, the dashed-line by using the rigid-body moment of inertia for $\gamma=+15^0$, and the dash-dotted line is chosen somewhat arbitrary for comparison. The two moments of inertia, $\mathcal{J}^{(1)}_{band}$ and $\mathcal{J}^{(2)}_{band}$ defined in eqs. (2.1) and (2.2), can be obtained from this figure straightforwardly.

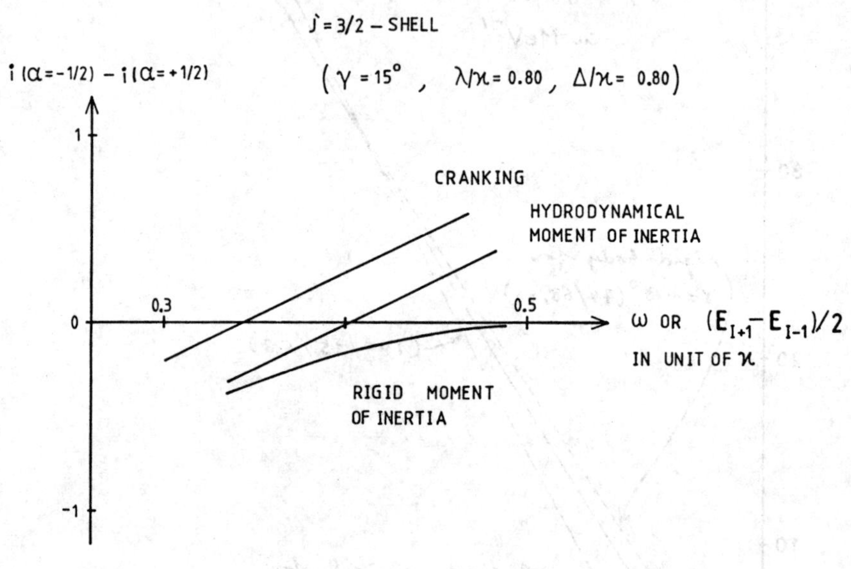

Fig.15. Spin-alignment difference of the yrast quasiparticle states with two signatures ($\alpha=\pm\frac{1}{2}$), as a function of "rotational" frequency. The curve denoted as "CRANKING" is obtained by solving the cranking Hamiltonian (8), while the curve indicated as "HYDRODYNAMICAL....." or "RIGID....." is obtained from the particle-rotor model in which the respective moments of inertia for $\gamma=+15^0$ are used. The intrinsic Hamiltonian (5) used is the same in the two models. See text for the details.

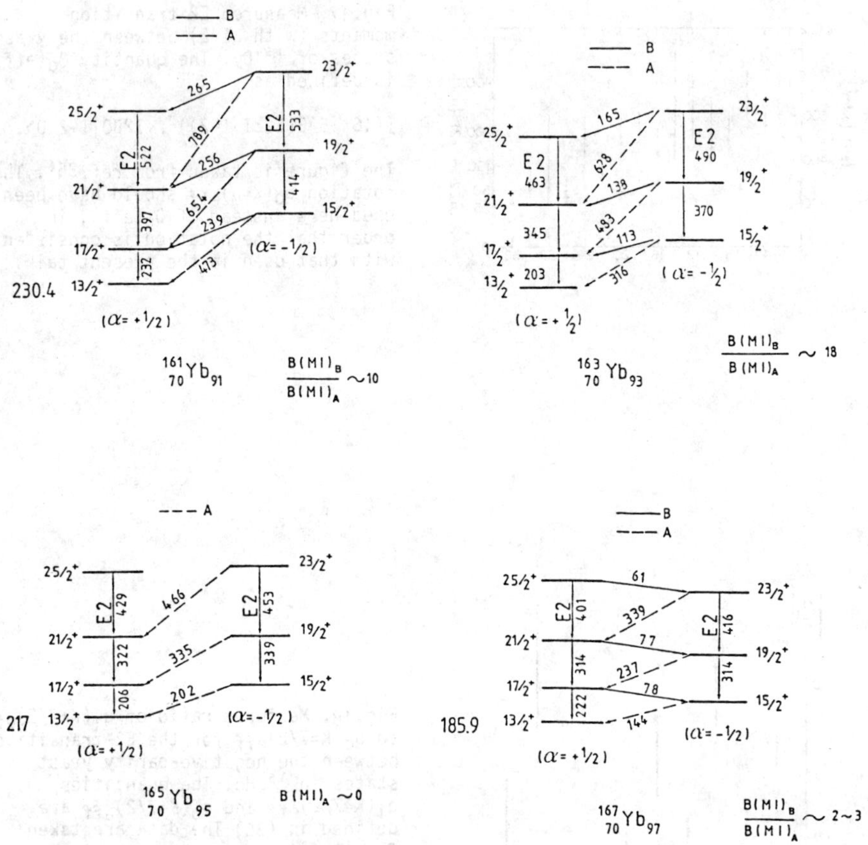

Fig. 16. Measured decay-scheme and ratio of B(M1)-values of yrast states with positive parity in Yb-isotopes. Both the excitation energy of 13/2+ state and the transition energies are given in keV. In ^{165}Yb the transitions of type B are not detected, because of very small energies of the gamma-rays. The data on ^{161}Yb, ^{163}Yb, ^{165}Yb and ^{167}Yb are taken from refs. 32, 33, and 34, respectively.

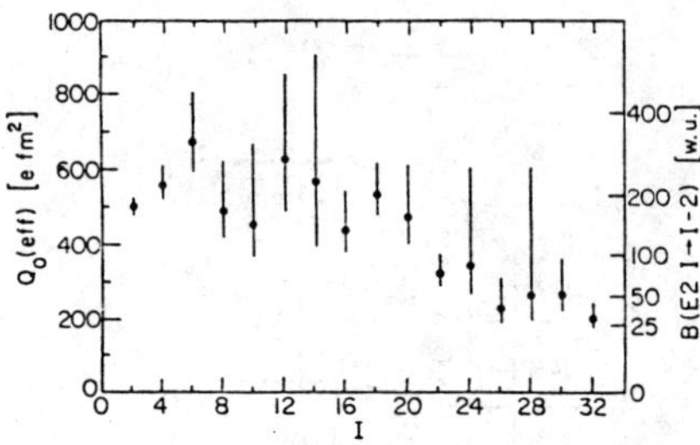

Fig.17. Measured E2 transition moments (with $\Delta I=2$) between the yrast states of ^{154}Dy. The quantity Q_0(eff) is defined as

$$((16\pi/5)B(E2:I\rightarrow I-2))^{\frac{1}{2}}/\langle I200|I-2\ 0\rangle.$$

The figure is taken from ref.35 . The notation $q_0(K=0)_{eff}$ should have been used here instead of Q_0(eff), in order that the notation is consistent with that used in the present talk.

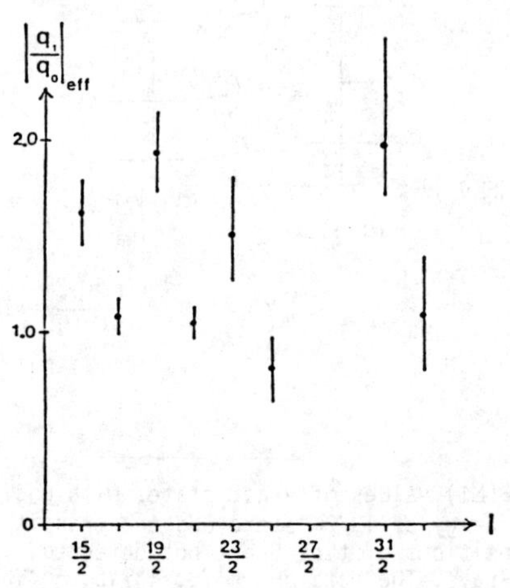

Fig.18. Measured ratio of $q_1(K=7/2)_{eff}$ to $q_0(K=7/2)_{eff}$ for the E2-transitions between the negative-parity yrast states in ^{157}Ho. The quantities $q_1(K=7/2)_{eff}$ and $q_0(K=7/2)_{eff}$ are defined in (35) The data are taken from ref.36 .

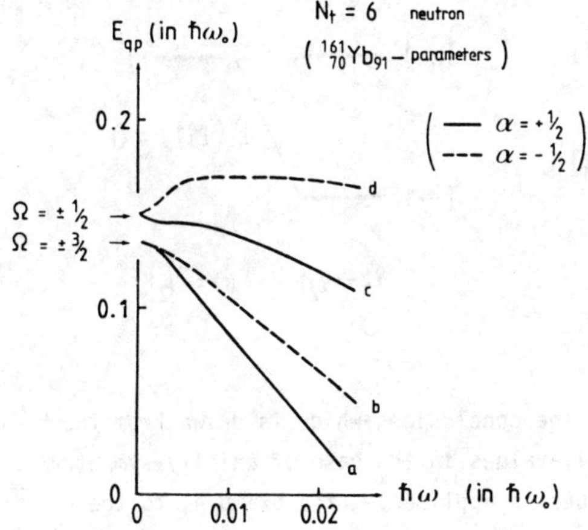

Fig.19. Low-lying quasiparticle spectra for N_t=6 neutrons as a function of cranking frequency, which are calculated from a cranked Nilsson potential for the fixed parameters suitable for the low-lying states of ^{161}Yb. Used parameters are: Δ=0.135 $\hbar\omega_0$, λ=6.398 $\hbar\omega_0$, ε_2= 0.21, ε_4=-0.015, and γ=0.

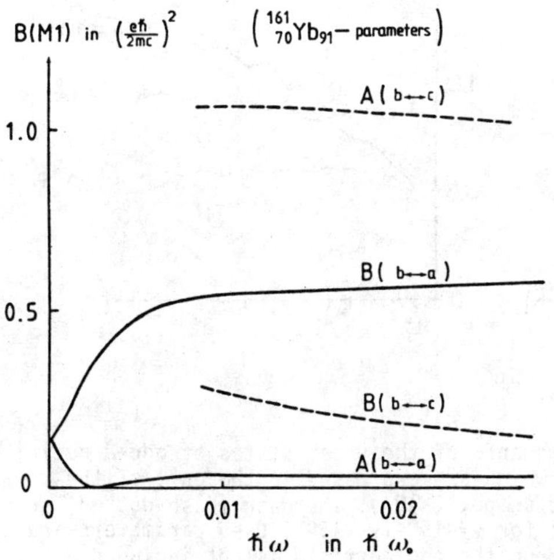

Fig.20. The B(M1:ΔI=1)-values of the transitions between the 1qp states shown in fig.19, which are calculated by using the cranking formula (38). The type of transitions is described by A or B defined in sect.4.1, while the relevant 1qp states are expressed by a, b and c indicated in fig.19.

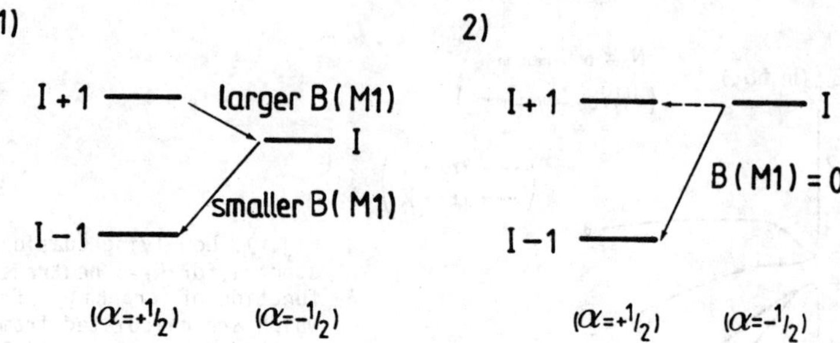

Fig. 21. Schematic illustration of the conclusion, which is drawn from the cranking-model calculation of B(M1)-values in the case of axially-symmetric shape. The assigned α-quantum-number is applicable, for example, to the yrast states of the $i_{13/2}$-particle model.

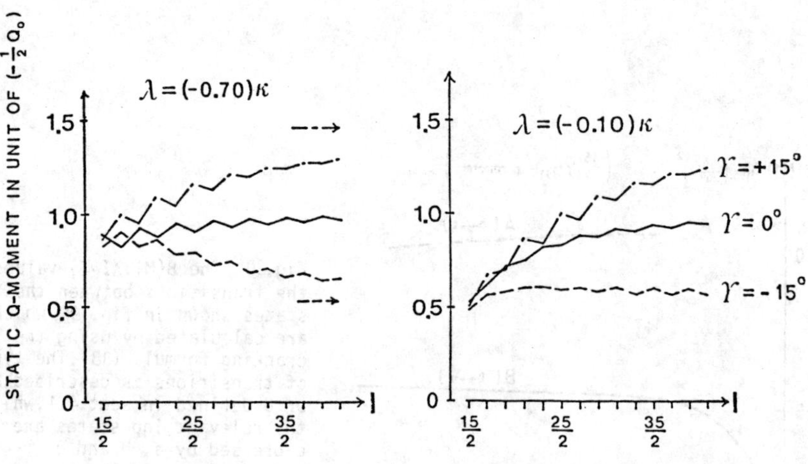

Fig.23. Calculated static quadrupole-moments of the yrast states of odd-A nuclei in the h11/2-shell particle-rotor model, which are expressed in the unit of the asymptotic value for axially-symmetric shape, $(-\frac{1}{2}Q_0)$. The dash (dash-dotted) arrow shows the asymptotic value $\hat{Q}_0$ in (45) for $\gamma=-15^0$ ($\gamma=+15^0$). Used parameters are $J_0\kappa=72$ and $\Delta=0.45\kappa$. The contribution from the odd-particle is not included.

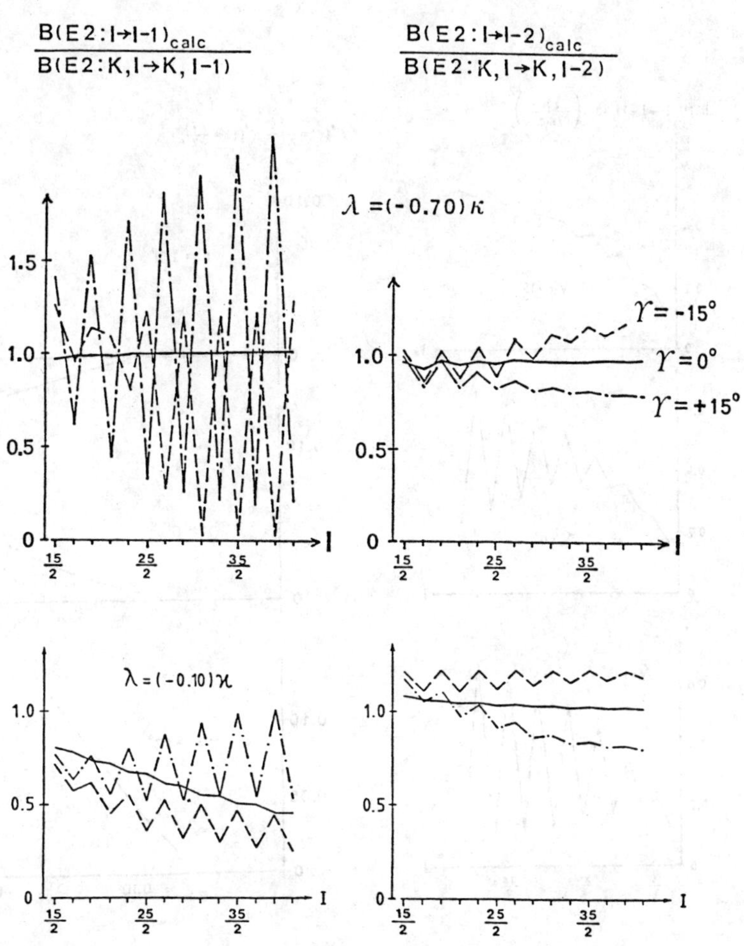

Fig.22. Calculated B(E2)-values between the yrast states of odd-A nuclei in the h11/2-shell particle-rotor model, which are expressed in the unit of respective axially-symmetric B(E2)-values. The Fermi-level $\lambda=(-0.70)\kappa$ ($\lambda=(-0.10)\kappa$) and $K=3/2$ ($K=7/2$) are used for the upper (lower) part, while B(E2)-values with $\Delta I=1$ ($\Delta I=2$) are plotted in the left (right) part. Used parameters are: $J_0\kappa=72$, $\Delta=0.45\kappa$ and $e_{eff}=2e$. The contribution from the odd-particle (with e_{eff}) to the E2 matrix-elements is very small compared with that from the core.

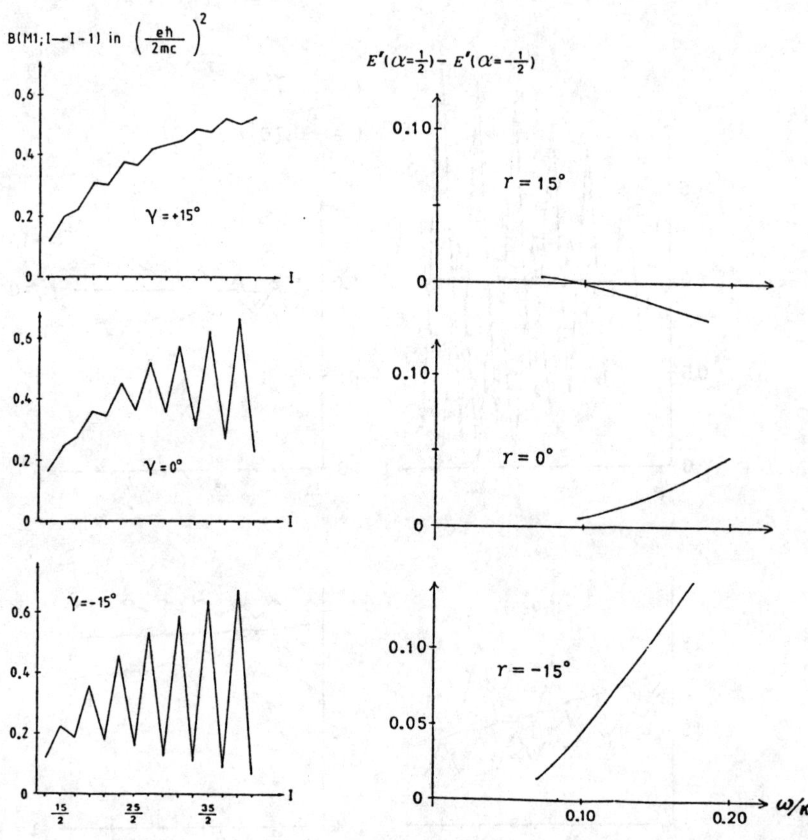

Fig. 24. Calculated B(M1)-values between the proton yrast states in the h11/2-shell particle-rotor model, and the difference between the energies with $\alpha=+\frac{1}{2}$ and $-\frac{1}{2}$ referring to the rotating coordinate system. See the text. Used parameters are: $\lambda=(0.10)\kappa$, $J_o\kappa=72$, $\Delta=0.45\kappa$, $g_s=3.91$, $g_\ell=1.0$, $g_R=0.42$.

Hartree-Fock-Bogoliubov Theory with Cranking Contraints, Projection on Angular Momentum Eigenstates

P. Ring

Physik-Department der Technischen Universität München
D 8046 Garching

Abstract

Mean field theory with conservation of angular momentum and number symmetry is discussed in the context of a microscopic description of the high spin region of heavy deformed nuclei. Symmetry conservation on the average before the variation corresponds to the selfconsistent Cranking model. Various versions with and without number projection, with and without temperature are presented. Excited rotational bands and rotating giant resonances are described in the framework of cranked random phase approximation. In the second part projection techniques are discussed: Systematic approximations for large deformations and exact numerical projections in heavy nuclei.

<u>Content</u>

1. Introduction

In a microscopic description of strong interacting finite Fermi systems, as nuclei, the mean field approximation plays an important role. Strong correlations with a long range in a system of many particles can often be taken into account by symmetry violations. In such cases one ends up with wavefunctions of a rather simple structure: The groundstate can then be described by product states of independently moving particles or quasiparticles. Excited states are characterized by time dependent product states or linear combinations of states with a simple structure, as ph states.

The best-known example for such a scheme is the spherical shell model, where translational symmetry is broken. It is very successful in the region of closed shells, but it also provides a basis of nearly all microscopic calculations in nuclei. For a small number of valence particles residual interactions not yet taken into account by the symmetry violating mean field can be treated by a full diagonalization in a limited configuration space.

For many valence particles in heavy nuclei additional approximations have to be introduced. Again mean field theory turns out to be very successful in the limit of strong correlations, where one observes phase transitions to deformed and to superfluid mean fields connected with breaking of rotational invariance and number conservation. Hartree-Fock-Bogoliubov (HFB) theory with effective forces is indeed an extremely powerful method to describe the ground state of a very large number of nuclei[1-3].

The wavefunctions obtained in this way, however, have only a very small overlap with the exact eigenstates of the many-body Hamiltonian. They are "intrinsic" wavefunctions, which can be used to calculate observable quantities, such as energies or electromagnetic matrix elements by simple rules[4,5]. Wavefunctions in the laboratory frame, i.e. symmetry conserving functions are obtained by projection onto good particle numbers, angular momentum and eventually linear momentum[6).

For the ground state the intrinsic wavefunction will be in general a product state, for excited states one has time-dependent products and linear combinations of many-quasiparticle states. The projected wavefunctions obtained in this way are very general. In many models they are exact eigenstates to the Hamiltonian, in other cases they have a 99 per cent overlap with exact eigenstates and one ends up with a very good approximation to full shell model diagonalization, which is actually impossible for heavy nuclei because of the astronomical number of configurations in such cases.

Besides this in a sense "technical" advantage of the mean field approximation, it gives us a deeper understanding of the physics going on in these many-body systems: It would be probably extremely hard to disentangle a deformed shape of the nucleus from the millions of components in a shell model wavefunction. From the symmetry violating "intrinsic" functions we can recognize immediately the occurrence of phase transitions.

So far we did not discuss how to calculate the "intrinsic" wavefunctions. The nonlinear equations, which determine the static mean field, are usually derived from a variational principle. To be consistent the variation should be carried out after projection[7-9]. This is in principle a very complicated task. Only in special cases as for number projection[10] and very recently also for angular momentum projection from axially symmetric functions[11] the projected variational equations could be solved numerically. Several approximation schemes have been investigated. For strong symmetry violations the Kamlah expansion[12,13] provides a systematic tool: Zeroth order gives unprojected variation, i.e. simple HFB, first order gives the Selfconsistent Cranking Model (SCC)[14], which includes symmetry conservation on the average. It is widely used for the microscopic description of high spin states[15,16]. In this case a rotating mean field is introduced with an additional symmetry violation, namely time reversal.

In the first of this series of lectures the SCC method is discussed and applications for a microscopic description of the yrast line in rapidly rotating heavy nuclei are presented.

Within the same scheme also excited many-quasiparticle bands can be described. For the investigation of collective rotational bands additional correlations have to be taken into account. We thus introduce in the second lecture the Cranked Random Phase Approximation (CRPA). On the basis of a statistical description with temperature it can be used to study the behaviour of giant resonances in hot rotating nuclei. In the third lecture we discuss approximate projection techniques. In particular we give a derivation of the cranking model and show how to calculate transition probabilities from intrinsic wavefunctions. The fourth lecture finally is devoted to exact projection techniques and recent applications.

2. The Selfconsistent Cranking Model (SCC)

The cranking model has turned out to be a very powerful tool for a microscopic description of high spin states in nuclei. It is basically a classical model and has been indroduced by Inglis[17] in a semiclassical way. Over the years there have been many attempts to give a microscopic derivation from the many body problem. We will present in section 4 of these lectures a derivation from a symmetry conserving variational principle. In this section we simply discuss its properties and some of its applications.

The simplest way to end up with the cranking model is just to see it as a variational principle with constraint on the angular momentum. According to Lagrange it is equivalent to a unrestricted variation of a new functional

$$\delta \langle \Phi | H - \omega J_x | \Phi \rangle = 0 \qquad (2.1)$$

where ω is a simple parameter determined from the subsidiary condition

$$\langle \Phi | J_x | \Phi \rangle = \sqrt{I(I+1)} \qquad (2.2)$$

From the textbooks of classical mechanics[18] we know that $H - \omega J_x$ is just the Hamiltonian in a frame which rotates with the angular velocity ω around a fixed axis in space. In that sense we call $|\Phi\rangle$ in the following an intrinsic wavefunction. It has to be used with care. Since the variation in eq. (2.1) is carried out usually only in a restricted set of functions of the Hilbert space, as for instance symmetry violating Slater determinants, the wavefunctions $|\Phi\rangle$ have only little to do with exact eigenstates of the Hamiltonian H in the laboratory frame, which are eigenstates of the angular momentum operator J^2.

2.1 The Cranked HFB-Equations

The Hartree-Fock-Bogoliubov equations for a general twobody Hamiltonian have been derived in one of the first lectures of this school[19]. Since J_x is a single particle operator they have in the rotating frame the simple form

$$\begin{pmatrix} h-\omega j_x & \Delta \\ -\Delta^* & -h^*+\omega j_x^* \end{pmatrix} \begin{pmatrix} U \\ V \end{pmatrix}_k = E_k \begin{pmatrix} U \\ V \end{pmatrix}_k \tag{2.3}$$

where the single particle field h and the pairing potential Δ are given by

$$h = \varepsilon - \lambda + \Gamma \tag{2.4}$$

$$\Gamma_{kk'} = \sum_{ll'} \bar{v}_{kl'k'l}\, \rho_{ll'} \qquad \Delta_{kk'} = \sum_{l<l'} \bar{v}_{kk'll'}\, \kappa_{ll'} \tag{2.5}$$

and ρ and κ are generalized densities

$$\rho_{kk'} = \langle \Phi |\, c_{k'}^{+} c_k \,| \Phi \rangle \qquad \kappa_{kk'} = \langle \Phi |\, c_{k'} c_k \,| \Phi \rangle \tag{2.6}$$

c_k^+, c_k are a set of basis operators and $|\Phi\rangle$ is the HFB wavefunction, which is a vacuum for the quasiparticles

$$\alpha_k^+ = \sum_l U_{lk} c_l^+ + V_{lk} c_l \tag{2.7}$$

$$\alpha_k |\Phi\rangle = 0 \tag{2.8}$$

The cranked HFB-equations (2.3) is a nonlinear eigenvalue problem. Its eigenvectors characterize quasiparticles α_k^+ As compared to the diagonalization of a simple single-particle potential in the shell model it has two complications

i) it has twice the dimension, because it contains two fields Γ and Δ

ii) it is nonlinear, i.e. the fields Γ and Δ depend on the solution.

Because of i) we have always pairs of eigenvalues $E_k, -E_k$ with opposite sign. One corresponds to α_k^+ the other to α_k An exchange of α_k^+ and α_k corresponds to a change in the sign of the eigenvalue E_k and in a change of $(U, V)_k \rightarrow (V^* U^*)_k$ Since the HFB-wavefunction $|\Phi\rangle$ is defined by a complete set of annihilation operators α_k (eq.2.8) we have to go through all the pairs and to decide if α_k^+ belongs to E_k or the $-E_k$

This decision in the quasiparticle scheme fixes the groundstate $|\Phi\rangle$. It corresponds to the choice of the occupation numbers 0 and 1 in a simple shell model potential. Usually the potential is occupied from the bottom. In the quasiparticle scheme this corresponds to the choice of only positive energies E_k for the quasiparticle operators $\alpha_k^\dagger$. However sometimes this choice is not consistent with the required symmetries for $|\Phi\rangle$ such as number parity or signature. In such cases symmetries can eventually require negative quasiparticle energies[20].

From these considerations we see that the HFB equations determine not only the groundstate (or the yrast state at each angular momentum) but also excited states (or excited bands). In this case we have to choose a different configuration, i.e. we have to exchange one or several quasiparticle creation operators with the corresponding annihilation operators and define a new vacuum

$$|\tilde{\Phi}\rangle = \alpha_1^\dagger \alpha_2^\dagger |\Phi\rangle \tag{2.9}$$

for the new quasiparticles

$$\tilde{\alpha}_1^\dagger = \alpha_1 , \quad \tilde{\alpha}_2^\dagger = \alpha_2 , \quad \tilde{\alpha}_3^\dagger = \alpha_3^\dagger , \ldots \tag{2.10}$$

This is a formal trick to use the same program for the calculation of the groundstate band in even nuclei and one quasiparticle configurations in odd nuclei or two quasiparticle bands in even nuclei and so on.

For the discussion of physical properties of these states is is, however, more convenient to use as vacuum $|\Phi\rangle$ only the groundstate band.

2.2 A Model with Seperable Forces

The solution of the cranked HFB equations (2.1) involves a threefold nonlinearity problem. This can be seen very clearly on a model with seperable forces, which is widely used in the literature for actual calculations[21].

$$H = \varepsilon - \tfrac{1}{2}\chi Q^\dagger Q - G P^\dagger P + \frac{1}{2\mathcal{J}_c}\left(\sqrt{I(I+1)} - \langle J_x\rangle\right)^2 \tag{2.11}$$

ε are spherical single energies

Q_μ are multipole operators (quadrupole, hexadecupole...) in the ph channel (c^+c)

P_μ^+ are multipole operators (monopole, quadrupole...) in the pp channel (c^+c^+). In particular P_0^+ creates a Cooper pair and describes monopole pairing

$\mathcal{J}_c$ is a small core which introduces cranking by the variation of the rotational energy.

As usual in seperable models we neglect exchange terms and obtain for the HFB matrix (2.3) in the rotating frame

$$\begin{pmatrix} \varepsilon - \lambda - \beta Q - \omega j_x & \Delta P \\ -\Delta P & -\varepsilon + \lambda + \beta Q + \omega j_x \end{pmatrix} \tag{2.12}$$

where

$$\omega = \left(\sqrt{I(I+1)} - \langle J_x \rangle\right) / \mathcal{J}_c \tag{2.13}$$

$$\beta_\mu = \chi_\mu \langle Q_\mu \rangle \tag{2.14}$$

$$\Delta_\tau = G_\tau \langle P_\tau^+ \rangle$$

The deformation parameters β_μ characterize the shape of the nucleus, the pairing parameters Δ_τ the form of the pairing potential and the cranking frequency determines the size of the Coriol is field β characterize the degree of violation of rotational invariance, Δ the violation of gauche invariance and ω the violation of time reversal.

2.3 The Canonical Basis and Aligned Configurations

From the Bloch-Messiah theorem[22] we know that an arbitrary HFB wavefunction can be written in the so-called canonical basis as a BCS function. For the ground state without blocking we therefore have

$$|\Phi\rangle = \prod_\mu (u_\mu + v_\mu a_\mu^+ a_{\bar\mu}^+)|0\rangle \tag{2.16}$$

This basis is determined by a diagonalization of the density matrix ρ. For systems with time reversal symmetry the conjugate states can be chosen as time reversed pairs. For non vansishing angular velocity ω this is no longer the case: The canonical basis depends on ω and cannot be determined a priori. In such cases neither the field h nor the field Δ are diagonal, but writing down eq.(2.3) in the canonical basis we find for the occupation numbers

$$\left.\begin{matrix} v_\mu^2 \\ u_\mu^2 \end{matrix}\right\} = \frac{1}{2}\left(1 \mp \frac{\tilde{\varepsilon}_\mu - \lambda}{\sqrt{(\tilde{\varepsilon}_\mu-\lambda)^2 + \tilde{\Delta}_\mu^2}}\right) \tag{2.17}$$

with the effective single particle energies

$$\tilde{\varepsilon}_\mu = \frac{1}{2}\left(h_{\mu\mu} + h_{\bar{\mu}\bar{\mu}}\right) - \frac{1}{2}\omega\left(j^x_{\mu\mu} + j^x_{\bar{\mu}\bar{\mu}}\right) \tag{2.18}$$

and the effective gap parameters

$$\tilde{\Delta}_\mu = \Delta_{\mu\bar{\mu}}(\omega) \tag{2.19}$$

It is clear that aligned levels from the intruder orbits with large matrix elements $(j^x_{\mu\mu} + j^x_{\bar{\mu}\bar{\mu}})_{\mu=\mu_0}$ will be occupied first for large angular velocities.

Because of the strong K-mixing in these levels μ_0 the pairing matrix elements $\tilde{\Delta}_{\mu_0}$ of a monopole pairing force become very small and we end up with a pair of two blocked particles aligned more or lees along the rotational axis.

$$|\Phi\rangle_{al} = a^+_{\mu_0} a^+_{\bar{\mu}_0} \prod_{\mu \neq \mu_0} \left(u_\mu + v_\mu a^+_\mu a^+_{\bar{\mu}}\right)|0\rangle \tag{2.20}$$

In the case of full alignment we can neglect the splitting caused by the deformation and find

$$j^x_{\mu_0\mu_0} = j \;; \qquad j^x_{\bar{\mu}_0\bar{\mu}_0} = j-1 \tag{2.21}$$

where j is the angular momentum of the intruder shell (e.g. $j = {}^{13}/2$ for neutrons in the rare earth region). In realistic deformed nuclei this limit is never achieved and the aligned angular momentum is smaller.

From this microscopic theory it is therefore very clear how to define the aligned angular momentum: if one can distinguish one pair ($\mu_0 \bar{\mu}_0$)

$$I_{al} = j^x_{\mu_0\mu_0} + j^x_{\bar{\mu}_0\bar{\mu}_0} \tag{2.22}$$

The total angular momentum is then decomposed into a single particle part and a collective part

$$\langle J_x \rangle = I_{al} + \sum_{\mu \neq \mu_0} v_\mu^2 \left(j^x_{\mu\mu} + j^x_{\bar{\mu}\bar{\mu}} \right) \tag{2.23}$$

where the collective part is distributed over many levels.

The experimentalist cannot use this definition, he has to rely on experimentally determined cranking frequencies[23] and he has to parametrize the collective part of the angular momentum by an extrapolation of the ground state band in the framework of the VMI-model[24]

$$I_{al} = I - I_{VMI} \tag{2.24}$$

The cranked HFB matrix (2.3) depends on the chemical potential λ which has to be determined in an iterative way by the constraint on the particle number. Even for fixed angular velocity and frozen mean fields a numerical solution requires a large effort. Marshalek[25] therefore proposed an approximation to determine the canonical basis by diagonalizing the matrix $h - \omega j_x$. In this basis the pairing field Δ is no longer diagonal. Neglecting the offdiagonal matrix elements one ends up with simple BCS theory.

2.4 The Signature

For the discussion of spectra in the cranking model the signature symmetry[15,26,27]

$$R_x = e^{i\pi J_x} \tag{2.25}$$

plays an important role. Since R_x commutes with the cranking Hamiltonian it can be chosen as a selfconsistent symmetry[4].

Then each of the eigenvalues E_k corresponds to a definite eigenvalue

$$r_x = \pm i \tag{2.26}$$

and $E_{k'}$ $-E_k$ correspond to different eigenvalues because the change $\alpha_k \leftrightarrow \alpha_k^+$ corresponds to an exchange of the sign.

In order to find a total wavefunction $|\Phi\rangle$ with positive signature

$$R_x |\Phi\rangle = |\Phi\rangle \tag{2.27}$$

we therefore have an additional restriction on the choice of the α_k, α_k^+ which might eventually lead to negative quasiparticle energies on the yrast line[20].

The signature symmetry can be used in order to reduce the dimension of the HFB matrix (2.3). It turns out ot be useful to introduce a basis symmetrized with respect to R_x

$$\begin{aligned} |K\rangle &= \tfrac{1}{\sqrt{2}}(1 - i R_x)\, |nljm\rangle \\ |\bar{K}\rangle &= T\,|K\rangle \end{aligned} \qquad (m>0) \tag{2.28}$$

Assuming R_x as a selfconsistent symmetry and arranging the levels in the order $(K_1, K_2, \dots \bar{K}_1, \bar{K}_2 \dots)$ we find for the densities ρ and κ and for the potentials Γ and Δ

$$\rho = \begin{pmatrix} \rho_+ & \\ & \rho_- \end{pmatrix} \qquad \kappa = \begin{pmatrix} & -\kappa^T \\ \kappa & \end{pmatrix} \tag{2.29}$$

$$\Gamma = \begin{pmatrix} \Gamma_+ & \\ & \Gamma_- \end{pmatrix} \qquad \Delta = \begin{pmatrix} & -\Delta^T \\ \Delta & \end{pmatrix} \tag{2.30}$$

and for the matrices U and V

$$U = \begin{pmatrix} U_+ & \\ & U_- \end{pmatrix} \qquad V = \begin{pmatrix} & V_- \\ -V_+ & \end{pmatrix} \tag{2.31}$$

where the submatrices $U_\pm V_\pm$ are eventually rectangular matrices in cases of blocking[28]. In cases of time reversal invariance we have

$$\rho_+ = \rho_- \qquad \kappa^T = \kappa \tag{2.32}$$

$$\Gamma_+ = \Gamma_- \qquad \Delta^T = \Delta \tag{2.33}$$

$$U_+ = U_- \qquad V_+ = V_- \tag{2.34}$$

In the case of separable forces we have the same mean field $\Gamma_+ = \Gamma_-$ for both signatures, i.e. signature splitting is only caused by the Coriolis field. This might be different in calculations with realistic forces.

The angular momentum operators J_x has in this basis the form

$$\begin{pmatrix} j_x & \\ & -j_x \end{pmatrix}$$

The HFB-matrix can thus be reduced to

$$\begin{pmatrix} h_+ - \omega j_x & -\Delta^T \\ -\Delta^* & -h_-^* - \omega j_x \end{pmatrix} \tag{2.35}$$

It has only eigenvalues of positive signature. By changing the sign and replacing $U \rightarrow V^*, V \rightarrow U^*$ one obtains the other eigenstates with negative signature.

2.5 Quasiparticle Spectra in the Rotating Frame

In this section we study the behaviour of the quasiparticle energies E_k as a function of the angular velocity. They describe the motion of the independent quasiparticles in the rotating potential defined by the yrast wavefunctions, i.e. we neglect in this discussion polarization effects and blocking effects. We also neglect the change of the potentials Γ and Δ with increasing angular velocity. We call this approximation the Rotating Shell Model (RSM), other groups call it Cranked Shell Model (CSM) or the frozen field approximation.

In fig. 1 we show as an example the quasiparticle energies $E_k(\omega)$ in the rotating frame for the nucleus ^{168}Hf. The configuration space and the Hamiltonian is taken from ref. 21.

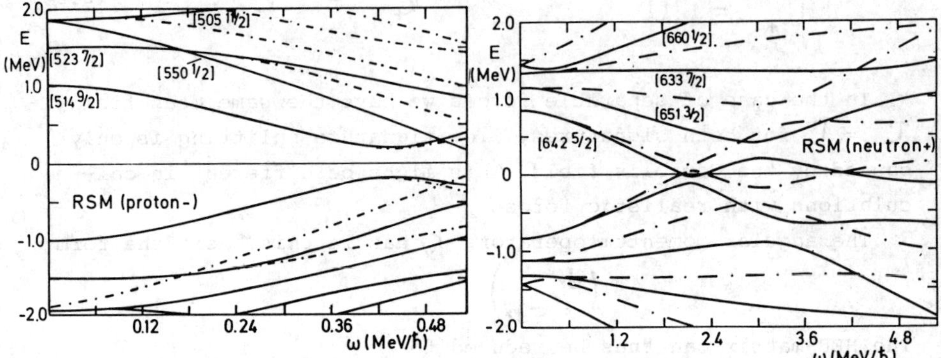

Fig. 1 Quasiparticle energies E_k for ^{168}Hf in the rotating frame for (a) proton levels with negative parity and (b) neutron levels with positive parity as a function of the angular velocity. Full curves corresponds to signature $+i$ and chain curves to signature $-i$

For vanishing angular velocity we see the pairing gap. For increasing frequency we observe a strong ω-dependence for some of the levels and even a vanishing quasiparticle energy. Since the gap parameter Δ is frozen in this calculation we call this phenomenon "gapless superconductivity". This is only possible because the ralation

$$E_k = \sqrt{(\varepsilon_k - \lambda)^2 + \Delta^2} \tag{2.36}$$

which holds for vanishing ω, is no longer true in the rotational frame. Neglecting the splitting of the quasiparticle energies at $\omega = 0$, which is perhaps a reasonable approximation, we can derive an approximate expression

$$E_k(\omega) \approx \sqrt{(\varepsilon_k - \lambda)^2 + \Delta^2} - \omega J_{x\,kk}^{11} \tag{2.37}$$

where J_x'' is the one-quasiparticle part of J_x in the rotating frame. Levels with large alignment, i.e. large values of $J_{x\,kk}''$ are shifted downwards. In fact we can obtain $J_{x\,kk}''$ from the slope of the level k:

$$J_{x\,kk}'' = -\frac{d}{d\omega} E_k(\omega) \tag{2.38}$$

This is an exact relation which can be derived from the variational principle (ref. 4, p. 142).
For ω-values larger than the point of gapless superconductivity we have a negative quasiparticle energy. Because of signature we have to stay at the same level. For even larger ω-values we see a pseudocrossing with two levels coming steeply up from the negative quasiparticle energies. If we stay on the same levels, as we have to do by energy reasons, we follow the yrast line, but the character of our wavefunction changes: The vacuum corresponds now to a twoquasiparticle state, i.e. the level-crossing corresponds to a transition to a aligned two quasiparticle configurations, the so-called s-band. Labelling the quasiparticle energies for small ω-values by the letters A,B,C,..., this would be the AB configuration.

It is clear that for higher frequencies we have additional crossings or additional alignment processes. Some of them are very sudden (sharp crossing), others occur slowly over a large ω-region. Sharp crossings are connected with backbending, smooth crossings only with upbending.

The crossing is characterized by the crossing frequency ω_c and the interaction between the bands. The crossing frequency depends on the size of the gap parameters and stays rather constant over a large region of nuclei. The interaction strength shows a oscillating behaviour as a function of the chemical potential, i.e. as a function of the number of particles in the intruder shell[30,31].

In the crossing region the cranking model provides usually a unphysically large mixing. In fact the cranking wavefunction is one Slaterdeterminant. A mixing phenomenon of the ground state band $|\Phi_g\rangle$ and a aligned two quasiparticle band $\alpha_1^+\alpha_2^+|\Phi_g\rangle$

can in principle be described again by one Slaterdeterminat

$$|\Phi\rangle = \cos(\varphi)\,|\Phi_g\rangle + \sin(\varphi)\,\alpha_1^+\alpha_2^+|\Phi_g\rangle \propto \exp\{tg\varphi\,\alpha_1^+\alpha_2^+\}|\Phi_g\rangle \quad (2.39)$$

The mixing, however, takes place at constant angular velocity and not, as it should be, at constant angular momentum. Since in the aligned band only a part of the angular momentum is produced by collective rotation, the corresponding frequencies might be different. The cranking model is however a variational theory and determines the optimal average frequency.

2.6 Selfconsistent Calculations

So far we neglected in our calculations the selfconsistency in the fields Γ and Δ. In fact the rotation has an influence on these quantities.

Changes in the nuclear shape are usually called "Stretching-Effect". It turns out to be rather small for many well deformed axially symmetric nuclei. Even at high spin we find only small changes in the deformation parameters ß and alignment processes cause a small triaxiality.

In some transitional nuclei, however, we observe large changes in the deformation. In fig. 2 we show as an example the nucleus ^{158}Er which has a shape change $\gamma \approx 0°$ to $\gamma \approx -60°$ at angular momenta $I > 40\,\hbar$

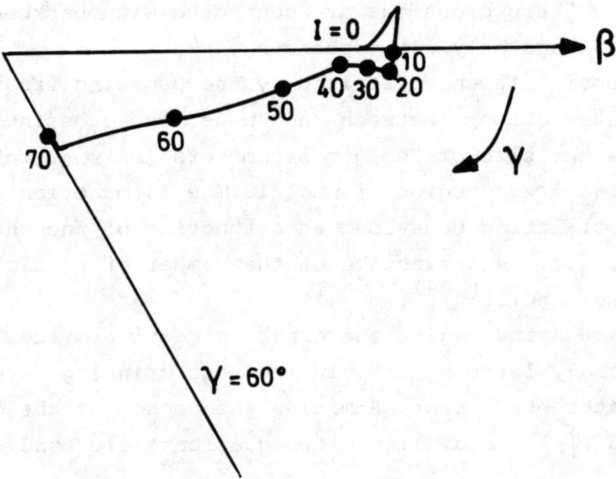

Fig. 2 An example for changes of the nuclear shape at high angular momenta in the nucleus ^{158}Er.

Changes in the pairing field are called Coriolis-anti-Pairing (CAP) effect. They show up in the deviations of rotational spectra from the $I(I+1)$-law. Their microscopic origin is given by the fact that with increasing spin the pairs cannot move in completely time reversed orbits. The conjugate orbits have no longer maximal spacial overlap, which causes for short range forces a reduction of the pairing correlations. To study this effect in detail one has to take into account two aspects

i) The degree of alignment is not the same for all orbits. Intruder shells, in particular those with low K-values at prolate deformations show large mixing. The pairing matrix elements of aligning levels are strongly reduced.

ii) Finite range force as quadrupole pairing cause also level dependent pairing correlations. They also allow pairing in partially aligned configurations (Δ_{21} and Δ_{22}). In particular the Δ_{21} -component plays an important role, because it has the same selection rule $\Delta K \pm 1$ as the Coriolis field.

For very high spins and large angular velocities a complete pairing collaps has been predicted by Mottelson and Valatin[32]. It has been found in simple models[33] and in many realistic calculations based on the selfconsistent cranking (SCC) model[15]. More sophisticated theories, which take more correlations into accaunt, such as full number projection before the variation (PNP) show only a reduction of pairing, but no collaps. Even at such high spins as $I \simeq 60\hbar$ matrix elements $\Delta \simeq$ 400 keV are found.

Fig. 3 shows as an example calculations[34] with the Kumar-Baranger Hamiltonian[21] for the nucleus ^{168}Hf, which has been studied recently in great detail by (HI, xn) reactions[35]. Above spin 20 $\hbar$ a rather constant value for the moment of inertia

$$\mathcal{J}^{(1)} = I/\omega \tag{2.40}$$

has been deduced to the rigid body value (fig.4).

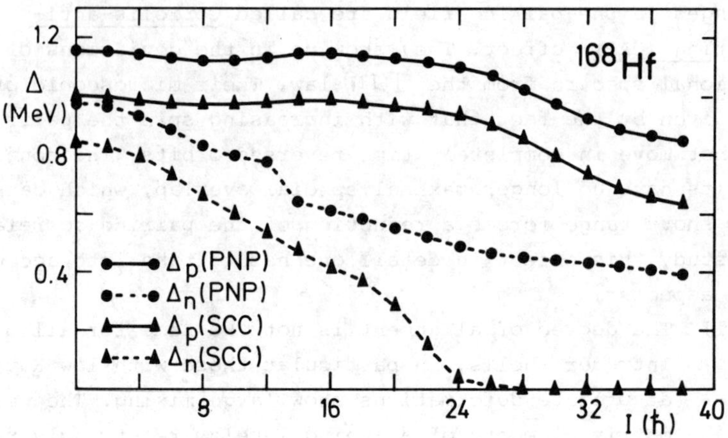

Fig. 3 Theoretical gap parameters for protons (———) and neutrons (-----) for ^{168}Hf as a function of the angular momentum.

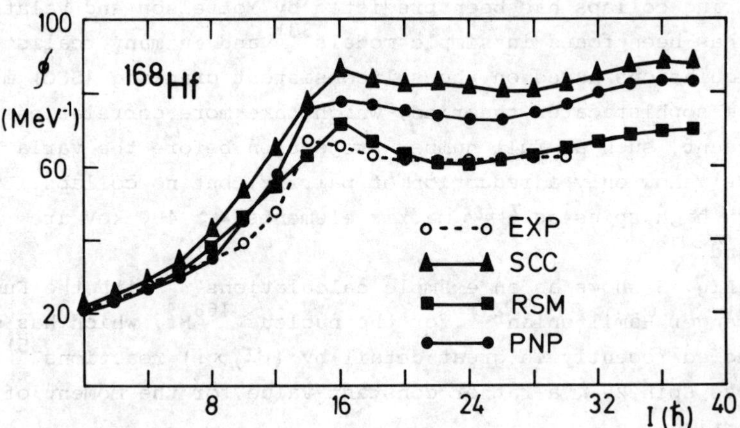

Fig. 4 The moment of inertia $\mathcal{J}^{(1)}$ of ^{168}Hf as a function of the angular momentum. The experimental values (dashed) are from ref. 35

It is clearly seen that this constant value can be reproduced as well by the Rotating Shell Model (RSM) as by the selfconsistent calculations. The corresponding gap parameters have a very different behaviour. For RSM they are kept constant, for SCC they collaps at spin 24 $\hbar$ and for the best theory PNP they are somewhat reduced, but they do not collaps.

The conclusion is that energy levels give us little information on the pairing properties in regions, where alignment processes occur, because in these regions the quasiparticle energies (2.37) deviate strongly from equation (2.36). Fig. 5 shows the moment of inertia

$$\mathcal{J}^{(2)} = \frac{dI}{d\omega} \tag{2.41}$$

which measures the slope of the curve $I(\omega)$ and which is given approximately by the Inglis formula

$$\mathcal{J}^{(2)} \simeq \sum_{k,k'} \frac{|J_{x\,kk'}^{20}|^2}{E_k + E_{k'}} \tag{2.42}$$

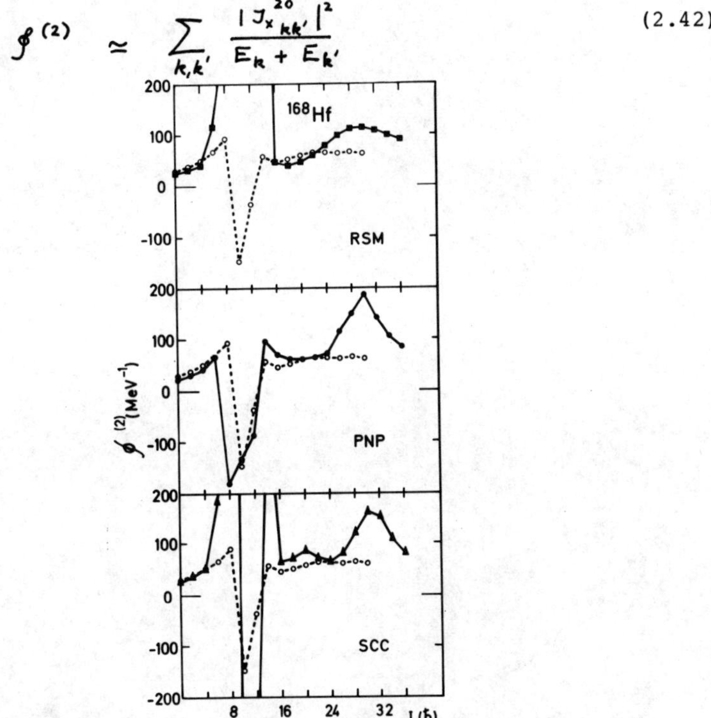

Fig.5 The moment of inertia $\mathcal{J}^{(2)}$ of ^{168}Hf as a function of the angular momentum: a) RSM b) PNP c) SCC

Again we find good agreement with the experiment for all three methods in the critical region for $I \geqslant 18\hbar$. The lower part is described best by number projection.

As a summary we can say that the SCC model is in principle able to account for changes in the mean field in a selfconsistent way. We have to be careful, however, in the region of phase transitions, where the correlations are too small in order to be accounted for by a symmetry broken wavefunction. In this region symmetry conservation on the average is not good enough. In the case of number projection we see that the frozen gap approximation gives usually better results than selfconsistent calculations, a fact which has been observed in many calculations[36).

3. Cranked Random Phase Approximation (CRPA)

So far we have discussed only the yrast line, where the wavefunction was determined as the lowest state for a given angular momentum Quasiparticle bands above the yrastline can by described in a similar way as long as one neglects correlations between the quasiparticles and stays with simple Slater determinants[73).

In this lecture we go a step further and take correlations into account in the framework of time department mean field theory for small amplitudes. We thus can describe collective bands above the yrast line, such as ß- and γ-bands etc., but also giant resonances based on high spin states. Recent experiments have observed such resonances in hot rotating nuclei[37-41).
We therefore discuss in the following linear response theory for rotating systems at finite temperatures.

3.1 Temperature Dependent Linear Response Theory

In temperature dependent mean field theory the expectation value of an arbitrary operator $\hat{O}$ is given by

$$\langle \hat{O} \rangle_T = \sum_n p_n \langle n | \hat{O} | n \rangle \tag{3.1}$$

where $|n\rangle$ are a complete set of multiquasiparticle configurations

$$|n\rangle = \alpha^+_{m_1} \dots \alpha^+_{m_n} | \Phi \rangle \tag{3.2}$$

and the probabilities are given by

$$p_n \sim \exp\left(-(E_{m_1} + \dots E_{m_n})/kT\right) \tag{3.3}$$

with the quasiparticle energies E_m in the intrinsic frame. Using the generalized Wick theorem[42) any expectation value of the form (3.1) can be expressed by the generalized density matrix

$$\mathcal{R} = \begin{pmatrix} \langle c^+_{k'} c_k \rangle_T & \langle c_{k'} c_k \rangle_T \\ \langle c^+_{k'} c^+_k \rangle_T & \langle c_{k'} c^+_k \rangle_T \end{pmatrix} \tag{3.4}$$

where c_k, c^+_k is an arbitrary basis.

Any single particle operator $\hat{F}$ can be expressed in the quasiparticle representation

$$\hat{F} = F^\circ + \sum_{mm'} F^{11}_{mm'} \alpha^+_m \alpha_{m'} + \sum_{m<m'} \left(F^{20}_{mm'} \alpha^+_m \alpha^+_{m'} + h.c. \right) \tag{3.5}$$

Indroducing the notation

$$\left.\begin{aligned} a_\mu &= \alpha_m \\ a_{\tilde{\mu}} &= \alpha^+_m \end{aligned}\right\} \quad \begin{aligned} m &= 1,2,\dots M \\ \mu &= 1,2,\dots M, \tilde{1}, \tilde{2}, \dots \tilde{M} \end{aligned} \tag{3.6}$$

with

$$\{ a_\mu, a_{\mu'} \} = \delta_{\mu\tilde{\mu}'} \tag{3.7}$$

we can write $\hat{F}$ in the form

$$\hat{F} = \mathcal{F}^\circ + \frac{1}{2} \sum_{\mu\mu'} \mathcal{F}_{\mu\mu'} a^+_\mu a_{\mu'} \tag{3.8}$$

with

$$\mathcal{F} = \begin{pmatrix} F^{11} & F^{20} \\ -F^{20*} & -F^{11*} \end{pmatrix}, \quad \mathcal{F}^\circ = F^\circ + \frac{1}{2} \mathrm{tr}(F^{11}) \tag{3.9}$$

The expectation value of $\hat{F}$ can be witten as

$$\langle \hat{F} \rangle_T = \mathcal{F}^\circ + \mathrm{Tr}(\mathcal{F}\mathcal{R}) \tag{3.10}$$

A general Bogoliubov transformation (2.7) can be expressed by the unitary operator

$$e^{i\hat{Z}} \tag{3.11}$$

whose matrix elements are defined by the coefficients U and V:

$$\mathcal{W} = \begin{pmatrix} U & V^* \\ V & U^* \end{pmatrix} = \exp(i\mathcal{Z}) \tag{3.12}$$

The density matrix $\mathcal{R}$ is diagonal in the representation

$$\mathcal{W}\mathcal{R}\mathcal{W}^{+} = \begin{pmatrix} \langle \alpha_{m'}^{+}\alpha_m \rangle_T & \langle \alpha_{m'}\alpha_m \rangle \\ \langle \alpha_{m'}^{+}\alpha_m^{+} \rangle_T & \langle \alpha_{m'}\alpha_m^{+} \rangle \end{pmatrix} = \begin{pmatrix} f_m & \\ & 1-f_m \end{pmatrix} \tag{3.13}$$

with the Fermi occupation numbers

$$f_m = \frac{1}{e^{E_m/kT}+1} \tag{3.14}$$

Assuming the operators a_μ to be time dependent

$$a_\mu(t) = e^{iHt}\, a_\mu(0)\, e^{-iHt} \tag{3.15}$$

and using the Wick theorem again we derive the equation of motion

$$i\dot{\mathcal{R}} = [\mathcal{H}(\mathcal{R}), \mathcal{R}] \tag{3.16}$$

with

$$\mathcal{H}_{\mu\mu'} = \langle \{ [a_\mu, H-\lambda N-\omega J_x], a_{\mu'}^{+} \} \rangle_T \tag{3.17}$$

which corresponds in the original basis to the HFB - matrix (2.3).

The rotating compound state is described as an equilibrium, as a static solution of (3.16)

$$[\mathcal{H}_0, \mathcal{R}_0] = 0 \tag{3.18}$$

These are temperature depentend HFB - equations in the rotating frame, which can also be obtained from the minimization of the free energy[43-45]. Their solution determines the selfconsistent HFB basis (not only $|\Phi\rangle$, as at temperature $T=0$). In this basis $\mathcal{R}_0$ as well as $\mathcal{H}_0$ is diagonal. The eigenvalues are

$$f_\mu = f_m \qquad E_\mu = E_m \tag{3.20}$$

$$f_{\tilde{\mu}} = 1-f_m \qquad E_{\tilde{\mu}} = -E_m \tag{3.21}$$

In the presence of a time dependent external field

$$\hat{F}(t) = \hat{F}\, e^{-iEt} + h.c. \tag{3.21}$$

the equation of motion is

$$i\,\dot{\mathcal{R}} = [\,\mathcal{H}(\mathcal{R}) + \mathcal{F},\, \mathcal{R}] \tag{3.22}$$

Linearizing this equation in the external field

$$\mathcal{R}(t) = \mathcal{R}_0 + \delta\mathcal{R}\, e^{-iEt} + h.c. \tag{3.23}$$

we obtain

$$(E-E_\mu+E_{\mu'})\,\delta\mathcal{R}_{\mu\mu'} = (f_{\mu'}-f_\mu)\,\{\mathcal{F}_{\mu\mu'} + \tfrac{1}{2}\sum_{\nu\nu'} \mathbb{W}_{\mu\mu'\nu\nu'}\,\delta\mathcal{R}_{\nu\nu'}\} \tag{3.24}$$

with the effective interaction

$$\mathbb{W}_{\mu\mu'\nu\nu'}(f_\nu - f_{\nu'}) = \langle [a_\nu^+ a_{\nu'}, \{[a_\mu, H], a_{\mu'}^+\}] \rangle_T \tag{3.25}$$

We now define the response function $\mathbb{R}$ by

$$\delta\mathcal{R} = \mathbb{R}\,\mathcal{F} \equiv \tfrac{1}{2}\sum_{\nu\nu'} \mathbb{R}_{\mu\mu'\nu\nu'}\,\mathcal{F}_{\nu\nu'} \tag{3.26}$$

and find that the response without interaction is given by

$$\mathbb{R}^0_{\mu\mu'\nu\nu'} = \frac{f_{\mu'} - f_\mu}{E - E_\mu + E_{\mu'} + i\eta}\,(\delta_{\mu\nu}\,\delta_{\mu'\nu'} - \delta_{\mu\bar{\nu}'}\,\delta_{\mu'\bar{\nu}}) \tag{3.27}$$

The full response function is determined by the linearized Bethe Salpether equation

$$\mathbb{R} = \mathbb{R}^0 + \mathbb{R}^0\,\mathbb{W}\,\mathbb{R} \tag{3.28}$$

The matrix elements of $\mathbb{W}$ can easily be derived from (3.25). All the indices $\mu,\mu';\nu,\nu'$ run over creation and annihilation operators for quasiparticles ($\mu,\nu \gtrless 0$). There are 16 combinations. In contrast to the quasiparticle RPA matrix, which contains only the H^{22} and the

H^{40} part of the Hamiltonian in the quasiparticle representation we have in the temperature dependent RPA also H^{31} - matrix elements.

In the limit of vanishing external field the system remains in its eigenmodes. They are determined by an eigenvalue problem, the RPA equations, in our case the Cranked RPA equations

$$(E - E_{\mu} + E_{\mu'})\, \delta\mathcal{R}_{\mu\mu'} = (f_{\mu'} - f_{\mu})\, \frac{1}{2} \sum_{\nu\nu'} \mathbb{W}_{\mu\mu'\nu\nu'}\, \delta\mathcal{R}_{\nu\nu'} \tag{3.29}$$

For separable forces the linear response theory becomes very simple. We therefore study a Hamiltonian of the form

$$H = H_o + \frac{1}{2} \sum_{\rho} \chi_{\rho}\, D_{\rho}^{+} D_{\rho} \tag{3.30}$$

H_o is a single particle Hamiltonian and ρ runs over a set of single particle operators D_{ρ} which are either hermitian or antihermitian, as for instance multipole operators in the ph channel $Q_{\mu}^{+} \pm Q_{\mu}$ or in the pp channel $P_{\tau}^{+} \pm P_{\tau}$. χ_{ρ} are strength parameters.

Neglecting exchange terms we thus find

$$\mathbb{W}_{\mu\mu'\nu\nu'} = \chi\, \mathcal{D}_{\mu\mu'}\, \mathcal{D}_{\nu\nu'}^{*} \tag{3.31}$$

Indroducing the matrix

$$R_{\rho\rho'}(E) = \frac{1}{4} \sum_{\mu\mu'\nu\nu'} \mathcal{D}_{\rho\mu\mu'}^{*}\, \mathbb{R}_{\mu\mu'\nu\nu'}(E)\, \mathcal{D}_{\rho'\nu\nu'}$$

which is the Fourier transform of

$$R_{\rho\rho'}(t) = i\, \Theta(t) \langle [D_{\rho}^{+}(t), D_{\rho'}] \rangle_T \tag{3.32}$$

we obtain

$$R(E) = R^{o}(E) + \chi\, R^{o}(E)\, R(E) \tag{3.33}$$

which is solved by a simple matrix inversion

$$R(E) = \frac{R^{o}(E)}{1 - \chi\, R^{o}(E)} \tag{3.34}$$

where the response without interaction is given by

$$R^{\circ}_{\rho\rho'}(E) = \frac{1}{2}\sum_{\mu\mu'} \frac{\mathcal{D}^{*}_{\rho\mu\mu'}\,\mathcal{D}_{\rho'\mu\mu'}}{E - E_{\mu} + E_{\mu'} + i\eta}\,(f_{\mu'} - f_{\mu}) \qquad (3.35)$$

Cranked RPA and linear response theory shall now be applied to the description of lowlying collective bands and of giant resonances at high spins.

3.2 Lowlying Collective Bands

For the description of lowlying collective bands we use the pairing plus quadrupole Hamiltonian and apply linear response theory at T = 0.

The first step is the calculation of the rotating self-consistent field,which corresponds to a solution of the cranked HFB equation. Working in the corresponding rotating quasiparticle basis we calculate the matrix R^{O} of eq. (3.35)

$$R^{\circ}_{\rho\rho'}(E) = \frac{1}{2}\sum_{mm'} \left\{ \frac{D^{20*}_{\rho mm'}\,D^{20}_{\rho' mm'}}{E - E_{m} - E_{m'} + i\eta} - \frac{D^{20}_{\rho mm'}\,D^{20*}_{\rho' mm'}}{E + E_{m} + E_{m'} + i\eta} \right\} \qquad (3.36)$$

Using operators Q symmetrized with respect to signature we obtain for positive signature a 7 x 7 matrix for the seven operators

$$Q_{0},\ Q_{+1},\ Q_{+2},\ (P^{+}_{p} \pm P_{p}),\ (P^{+}_{n} \pm P_{n})$$

and for negative signature a 2 x 2 matrix with the operators

$$Q_{-1},\ Q_{+2}$$

The RPA-eigenfrequencies are found as poles of the response function (3.34)

$$\det\left(\chi\, R^{\circ}(E) - 1\right) = 0 \qquad (3.37)$$

Since we work in a selfconsistent basis we obtain spurious solutions for positive signature

$$[H', J_{x}] = 0\ ;\quad [H', N_{p}] = 0\ ;\quad [H', N_{n}] = 0 \qquad (3.38)$$

at zero energy and for negative signature

$$[H', J_-] = \omega J_- \tag{3.39}$$

at the energy of the cranking frequency ω.
They correspond to rotations in coordinate space and in gauge space.

For even I-values we have postive signature and for odd I-values we have negative signature. In the following we study the excitation energies Ω_μ of the eigenmodes above the yrast line. In fig.6 we show the unperturbed twoquasiparticle energies and the collective energies Ω_μ for ^{164}Er

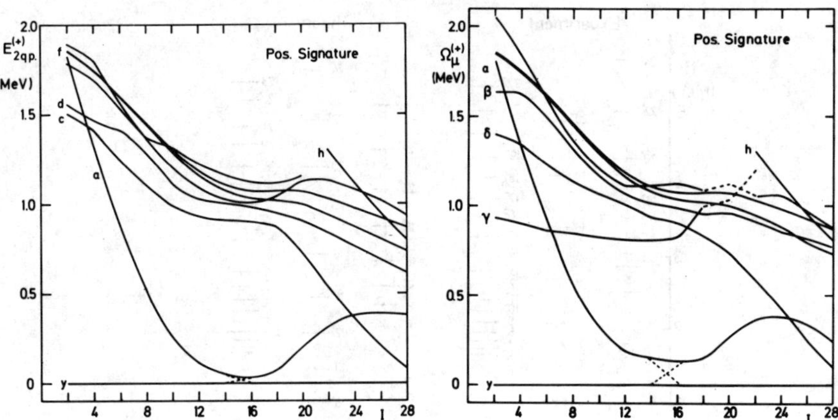

Fig. 6 (a) The spectrum of twoquasiparticle energies with positive signature in the nucleus ^{164}Er as a function of the angular momentum I.
(b) The corresponding spectrum of RPA frequencies

At spin zero we see in the first part the gap $2\Delta \simeq$ 1.5 MeV in the spectrum. Taking into account correlations a few levels are lowered, essentially only the γ-band. There is also a ß- and a δ-band, both are little collective and are a mixture of pairing vibration and ß-vibration.

For increasing frequencies we observe many twoquasiparticle bands coming down. This is caused by alignment. The most pronounced aligned band is a 2qp band in the $\nu i_{13/2}$ intruder orbit (labeled by a).

It comes sharply down and crosses at spin 14 ħ with the ground state band. For higher spin it forms the yrast configuration, which is used as reference in fig. 5.

The γ-band shows nearly no mixing with this aligned band. It is more or less parallel to the yrast line and vanishes after the backbending region, because there it could only be described as a 4-quasiparticle excitation on the s-band which forms the reference. More recent applications of CRPA theory therefore use the groundstate band as reference[47].

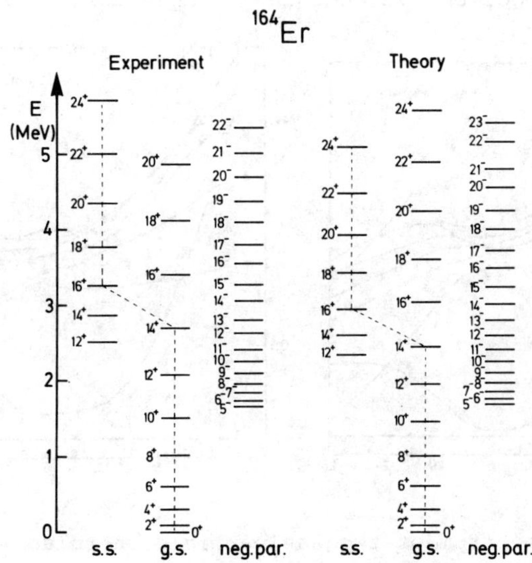

Fig. 7 A comparision of experimental and theoretical spectra for several rotational bands in ^{164}Er. The dashed line indicates the yrast levels.

In fig. 7 we show a comparison with experimental data. We also show a negative parity band, which is essentially a twoquasiparticle band since we have no interaction in the negative parity channel for the pairing plus quadrupole model.

3.3 Ground State Correlations

Having calculated the RPA excitation energies and the corresponding wavefunctions, we can in principle go beyond the mean field approach and calculate groundstate correlations in the yrast line[48]. For the yrast energy we find three types of contributions

$$\Delta E = E_{NM} + E_{GM} + E_{EX} \tag{3.40}$$

E_{NM} is the contribution coming from the virtual admixture of normal modes with the backward runing components Y^{μ}

$$E_{NM} = - \sum_{\mu} \Omega_{\mu} \sum_{m'<m} |Y^{\mu}_{mm'}|^2 \tag{3.41}$$

E_{GM} is the contribution coming from the spurious modes. It removes the spurious rotational energy etc. In the symmetry violating intrinsic wavefunction in RPA order. It is of the form

$$E_{GM} = - \frac{1}{2} \frac{\langle \Delta J^2 \rangle}{\mathcal{J}_{TV}} \tag{3.42}$$

and similar terms containing the fluctuations $\langle \Delta N^2 \rangle, \langle \Delta J_x \Delta N \rangle \ldots$

E_{EX} is finally a contribution from exchange terms neglected in the HFB-energy, which is of the same order as RPA energies and should be taken into account. It is of the form

$$E_{EX} = - \frac{1}{2} \chi \langle \Delta D^{\dagger} \cdot \Delta D \rangle \tag{3.43}$$

and shows a rather smooth behavior.

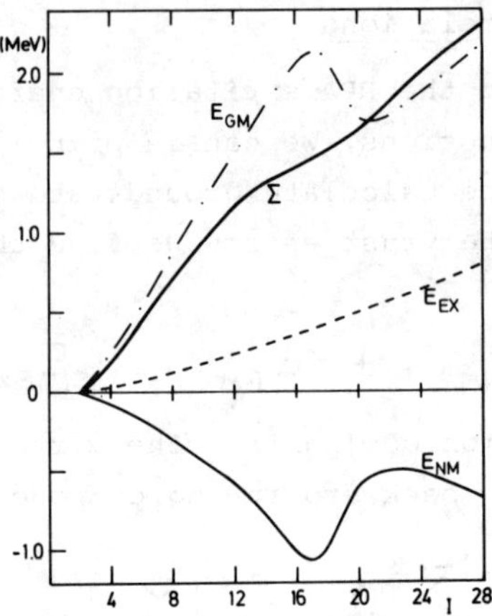

Fig. 8 The different contributions to the correlation energy in eq. 3.40. E_{NM} contains contributions from 80 normal modes Σ is the sum of these contributions. All energies are normalized to I=2, where E_{EX} = -6.95 MeV, E_{NM}= -0.899 MeV and E_{GM} = 2.742 MeV

The two other terms behave rather strange. In the level crossing region the fluctuations $\langle \Delta J_x^2 \rangle$ have a peak and $\mathcal{J}_{TV}$ has singularities and we end up with a cleary reduced correlation energy. On the other side there is a lowlying normal mode which causes an increased contribution of the normal modes. Both effects cancel each other and the sum Σ ,which can also be written as

$$\Sigma = \frac{1}{2}\Big(\sum_{\mu} \Omega_{\mu} - \sum_{m<m'} (E_m + E_{m'})\Big) \tag{3.44}$$

is a smooth function, which can simply be absorbed in a proper adjustment of the effective interaction.

We thus understand at least qualitatively why a simple calculation in the cranking model based on HFB functions with adjustable parameters of the effective interaction is able to reproduce the experimental yrast line rather well.

3.4 Giant Resonances in Hot Rotating Nuclei

In order to investigate the giant dipole resonance in a rotating system with finite temperature we calculate the absorption cross section for a dipole photon of energy E. It is given as a thermal average over all initial states and a sum over all final states

$$\sigma(E) = -4\pi^2\alpha \sum_{if} p_i\,(E_f - E_i)\,\frac{1}{2I_i+1}\,|\langle f\|D\|i\rangle|^2\,\delta(E-E_f+E_i) \qquad (3.45)$$

Assuming that the hot nucleus carries the angular momentum I, we have to restrict the sum in principle only over initial states with angular momentum I. The reduced matrix element $\langle f\|D\|i\rangle$ has to be calculated with projected wavefunctions.

In the cranking approximation discussed in the next lecture we can express these projected matrix elements by intrinsic matrix elements:

$$\frac{1}{2I_i+1}\langle I_f\|D\|I_i\rangle \simeq \langle \Phi_f|\,\tilde{D}_{I_f-I_i}\,|\Phi_i\rangle \qquad (3.46)$$

where $\tilde{D}_\mu$ is the dipole operator quantized along the x-axis

$$\tilde{D}_0 = D_x \;; \qquad \tilde{D}_{\pm 1} = \pm\frac{1}{\sqrt{2}}\,(D_z \mp i\,D_z) \qquad (3.47)$$

In the same spirit the angular momentum is only conserved on the average in the statistical ensemble, i.e. the probabilities p_i are calculated with energies in the rotating frame.

We also have to transform the energy difference $E_f - E_i$ from the laboratory frame to the intrinsic system. We do this in cranking approximation and find

$$\sigma(E) \simeq \sigma_x(E) + \tfrac{1}{2}\sigma_y(E-\hbar\omega) + \tfrac{1}{2}\sigma_y(E+\hbar\omega) + \tfrac{1}{2}\sigma_z(E-\hbar\omega) + \tfrac{1}{2}\sigma_z(E+\hbar\omega) \qquad (3.48)$$

with

$$\sigma_m(E) = -4\pi^2\alpha \sum_{if} p_i\,|\langle\phi_f|D_m|\phi_i\rangle|^2\,\delta(E-E_f+E_i) = -4\pi\alpha E\,\mathrm{Im}\,R_{mm}(E) \qquad (3.49)$$

where m runs over the operators D_x, D_y, D_z. We see that the absorption cross section $\sigma(E)$ in the laboratory frame can be expressed by the absorption cross section σ_m in the intrinsic system

Summarizing these considerations we find that the investigation of giant resonances in a hot rotating system has to be done in three steps:

i) For each angular velocity and each temperature we have to solve the cranked temperature dependent HFB-equations (3.19) This solution defines a rotating mean field, which describes the compound state and which is the basis for the next step.

ii) Linear response theory is used in the intrinsic system to calculate $\sigma_x(E)$, $\sigma_y(E)$, $\sigma_z(E)$

iii) The third step the transformation back to the laboratory scheme in eq. 3.48

Again we use seperable force. In order to discribe correlations in the dipole channel we now add a dipole-dipole term[49]
In the following we study first the influence of the angular momentum and the temperature on the mean field, which is characterized by shape parameters and pairing properties. We replace a full HFB calculation with realistic forces by a Strutinski method at finite temperatures and finite angular velocities[49].

As long as we stay in the mean field approximation pairing seems to collaps as a function of temperature and angular velocity (fig.9) For T>0.5 MeV all

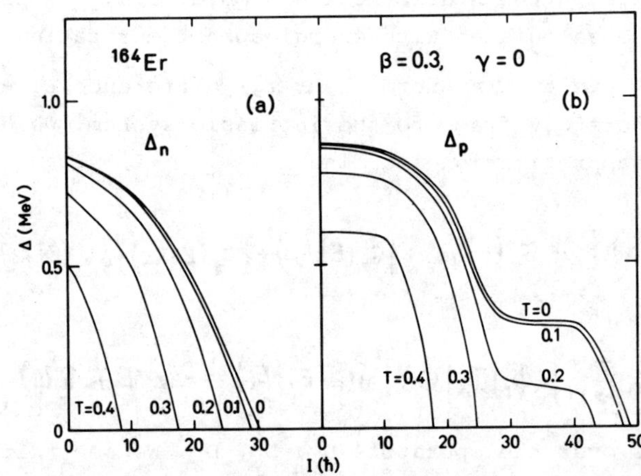

Fig. 9 Gap parameters for protons and neutrons for ^{164}Er as a function of the angular momentum at various temperatures. The deformation parameters are kept fixed at $\beta=0.3$ and $\gamma=0^{\circ}$.

pairing correlations have vanished, We have to emphasize however that both phase transitions with increasing spin and with increasing temperature are only so sharp in calculations of the mean field type.

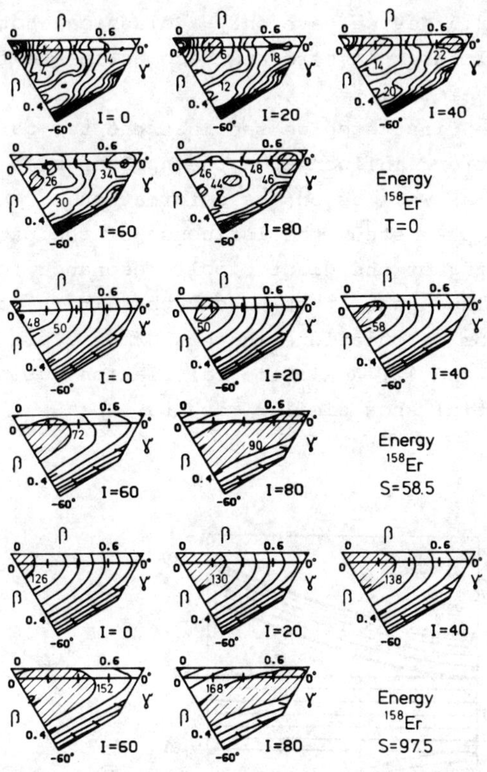

Fig.10 Energy surface for ^{158}Er in the shape parameters ß and γ . They correspond to constant entropy S and constant angular momenta. Contour lines describe an energy difference of 2 MeV. Pairing correlations are neglected. The entropy values S= 58.5 and 97.5 corresponds on the average to temperatures T=1.5 and 2.5 MeV

In fig. 10 we show the influence of angular momentum and temperature on the nuclear shapes in the nucleus ^{158}Er. At low temperatures the nucleus is prolate at spin zero and undergoes a shape transition for spin larger then 40 $\hbar$. For higher temperatures the minima become very flat and shell effects disappear. For T=2.5 MeV we have only a classical hot droplet, which is spherical for spin zero and becomes oblate for increasing angular velocity.

Summarizing the first step we see that the two parameters T and ω have a strong influence on the nuclear mean field, on the pairing field as well as on the deformation.

In the next step we study the influence of the parameters T and ω on the splitting of the giant dipole resonance for fixed deformation and fixed gap. It turns out that it is rather small as can be seen from calculations in the harmonic oscillator model from fig.11 and 12. Realistic calculations using a rotating Woods-Saxon potential show similar results.

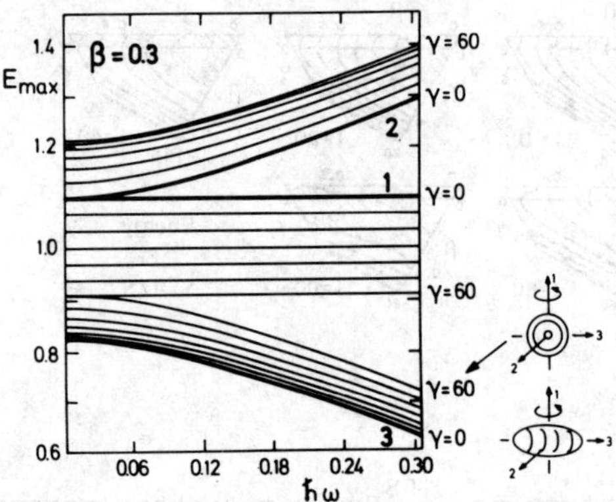

Fig. 11 The eigenfrequencies E_1, E_2, E_3 of the rotating harmonic oscillator for ß=0.3 at various γ-values. The units of the peak energies E_i and for the angular velocity $\hbar\omega$ are $78 \cdot A^{-1/3}$ MeV. i.e. the energy of the GDR at vanishing deformation. In heavy nuclei $\hbar\omega$ never exceeds 1 MeV, which is 0.06 in these units.

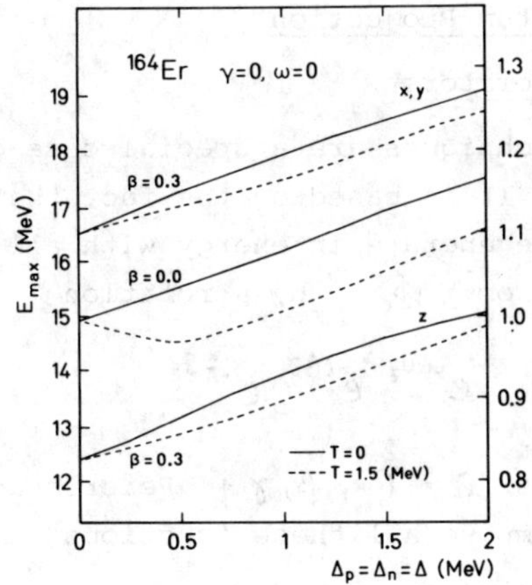

Fig.12 The dependence of the giant dipole resonance peaks on the gap parameters for vanishing angular velocity in the harmonic oscillator model with pairing (from ref.49)

In the harmonic oscillator model without pairing. The resonance peaks do not depend on temperature at all. This can be seen most easily from the expression (3.32). Since the operators D depend only in a linear way on the coordinates and momenta, the commutator gives a temperature independent constant. With pairing the temperature dependence is small.

We also see from fig.11 and 12 that the shape - and the gap parameters have an influence of the splitting of the GDR. This is a well know fact for the shape parameters[50)]. For the pairing it can be understood by the following reasons: i) The unperturbed twoquasiparticle energies are increased by pairing and ii) the number of possible configurations becomes larger, because there are not only ph, but also pp and hh configurations. This fact increases the collectivity and causes for a repulsive force a shift of the resonance peak to higher energies.

A detailed study of the splitting of the giant dipole resonance should us give information on the shape and on the pairing properties for high spins and for finite temperatures. So far such a splitting has not been observed. Severel effects contribute to an increasing width. Therefore it is not yet clear, if and in which regions such effects can be studied.

4. Angular Momentum Projection

4.1 Projection operators

Projection techniques are a special case of the generator coordinate method. It is based on the fact that a deformed wavefunction $|\Phi\rangle$ is degenerate in energy with all wavefunctions $\hat{R}(\Omega)|\Phi\rangle$ obtained from $|\Phi\rangle$ by a rotation

$$\hat{R}(\Omega) = e^{i\alpha J_z} e^{i\beta J_y} e^{i\gamma J_z} \tag{4.1}$$

in the Euler angles $\Omega = (\alpha, \beta, \gamma)$. Peierls and Yoccoz[6]) therefore proposed to superimpose all these functions

$$|\Psi\rangle = \int d\Omega\, f(\Omega)\, \hat{R}(\Omega)\, |\Phi\rangle \tag{4.2}$$

The variation of the energy with respect to f (Ω) corresponds to a diagonalization of H is the space spanned by the functions $\hat{R}(\Omega)|\Phi\rangle$. This subspace is invariant under the elements $\hat{R}$ (Ω)of the rotational group[7]). This means we can find a function f (Ω) which minimizes the energy and causes $|\Psi\rangle$ to be eigenfunction of the angular momentum operators.

To find such a function, we expand f (Ω) in terms of a complete set of eigenfunctions of the symmetry operators expressed in the variables Ω , i.e. by the Wigner functions:

$$f(\Omega) = \frac{2I+1}{8\pi^2} \sum_K g_K\, D^{I\,*}_{MK}(\Omega) \tag{4.3}$$

and obtain the operators[51])

$$\hat{P}^I_{MK} = \frac{2I+1}{8\pi^2} \int d\Omega\, D^{I^*}_{MK}(\Omega)\, \hat{R}(\Omega) \tag{4.4}$$

They can also be written as

$$\hat{P}^I_{MK} = \sum_\alpha |IM\alpha\rangle\langle IK\alpha| \tag{4.5}$$

where $|IM\alpha\rangle$ is a complete orthonormalized set in the Hilbert space and α characterizes all quantum numbers besides the rotational quantum numbers I and M

From eq. (4.5) we see that $\hat{P}^I_{MK}$ is not a projector in the mathematical sense. Insteed of $P^2 = P$ and $P^+ = P$ it obeys the relations

$$\hat{P}^I_{MK}\,\hat{P}^{I'}_{M'K'} = \delta_{II'}\,\delta_{KM'}\,\hat{P}^I_{MK'}\ ; \qquad (\hat{P}^I_{MK})^\dagger = \hat{P}^I_{KM} \tag{4.6}$$

In order to study, how $\hat{P}^I_{MK}$ acts on a "intrinsic" deformed function $|\Phi\rangle$, we decompose it into eigenstates with respect to J_z with eigenvalues

$$|\Phi\rangle = \sum_K |\Phi_K\rangle \tag{4.7}$$

$\hat{P}^I_{MK}$ transforms $|\Phi_K\rangle$ into an eigenfunction of $\hat{J}^2$ and J_z with the quantum numbers I and M annihilates all other components

By the "projection" technique described so far the function in (4.2) is not uniquely defined. A set of parameters g_K is still open.

The projected energy is given by

$$E^I_{proj} = \frac{\langle\Psi^{IM}|H|\Psi^{IM}\rangle}{\langle\Psi^{IM}|\Psi^{IM}\rangle} = \frac{\sum_{KK'} g_K^* g_{K'}\, h^I_{KK'}}{\sum_{KK'} g_K^* g_{K'}\, n^I_{KK'}} \tag{4.8}$$

with

$$\left.\begin{matrix} h^I_{KK'} \\ n^I_{KK'} \end{matrix}\right\} = \langle\Phi|\begin{matrix} H \\ 1 \end{matrix} P^I_{KK'}|\Phi\rangle \tag{4.9}$$

The parameters g_K can then be obtained by the solution of a variational problem

$$\sum_{K'} h^I_{KK'}\, g_{K'} = E^I \sum_{K'} n^I_{KK'}\, g_{K'} \tag{4.10}$$

Using the relation

$$\hat{P}^I_{MK}\,\hat{R}(\Omega) = \sum_{K'} D^I_{MK'}(\Omega)\,\hat{P}^I_{MK'} \tag{4.11}$$

we find for the signature operation

$$\hat{P}^I e^{i\pi J_x} = (-)^I\,\hat{P}^I_{M-K} \tag{4.12}$$

Intrinsic wavefunctions with good signature r_x have K-components (4.7)

$$e^{i\pi J_x} |\Phi_K\rangle = r_x |\Phi_{-K}\rangle \tag{4.13}$$

which shows that for such wavefunctions

$$P^I_{M-K} |\Phi\rangle = r_x^{-1} (-)^I P^I_{MK} |\Phi\rangle \tag{4.14}$$

For wavefunctions $|\Phi\rangle$ with good signature we therefore can restrict the sum over K to $K \geq 0$

$$|\Psi^{IM}\rangle = \sum_{K\geq 0} \frac{g_K}{1+\delta_{K0}} \left(P^I_{MK} + (-)^I r_x^{-1} P^I_{M-K}\right)|\Phi\rangle \tag{4.15}$$

For K = 0 we therefore have

$$r_x = (-)^I \tag{4.16}$$

4.2 Approximate Projection for Large Deformation

The crucial quantities in the calculation of projected matrix elements are the overlap functions

$$N(\Omega) = \langle\Phi|\hat{R}(\Omega)|\Phi\rangle \; ; \quad H(\Omega) = \langle\Phi|H\hat{R}(\Omega)|\Phi\rangle \tag{4.16}$$

Before discussing how to calculate these quantities exactly we discuss an approximation, which is valid for large deformations, i.e. in cases where functions N (Ω) and H (Ω) are strongly peaked at $\Omega=0$. In this limit N (Ω) can be approximated by[4])

$$N(\Omega) \simeq \exp\left\{-\frac{1}{2}\langle J_y^2\rangle\beta^2 - 2\langle J_z^2\rangle \sin^2\left(\frac{\alpha+\gamma}{2}\right) + i\langle J_x\rangle \sin\left(\frac{\alpha+\gamma}{2}\right)\sin\left(\frac{\alpha-\gamma}{2}\right)\right\} \tag{4.17}$$

The fluctuation $\langle J_y^2\rangle^{-1/2}$ therefore measures the width of this function. It decreases in general with the particle number. The following approximation is therefore valid for heavy nuclei. It is a classical approximation, because it assumes that the manybody system has a well defined orientation.

Since H is a twobody Hamiltonian the function H (Ω) is expected to have a very similar behavior as N (Ω). Kamlah[12]) therefore proposed the expansion

$$H(\Omega) = \left(a_0 + a_1 \frac{\partial}{\partial \Omega} + a_2 \frac{\partial^2}{\partial \Omega^2} + \ldots \right) N(\Omega) \qquad (4.18)$$

where $\frac{\partial}{\partial \Omega}, \frac{\partial^2}{\partial \Omega^2}$ are formal expressions for linear, quadratic,.... combinations of first and second order differential operators in the Euler angels. Since the angular momentum operators L_i can be represented in terms of differential operators with respect to these angels, we make the following ansatz

$$H(\Omega) = \left(a_0 + a_1 \Delta L_1 + a_2 \Delta \vec{L}^2 \right) N(\Omega) \qquad (4.19)$$

where $\Delta L_i = L_i - \langle J_i \rangle \qquad (i = x, y, z)$

The coefficients a_o, a_1 and a_2 are determined from the lowest derivatives of H (Ω) at the origin:

$$\begin{aligned} \langle H \rangle &= a_0 + a_2 \langle \Delta \vec{J}^2 \rangle \\ \langle H \Delta J_x \rangle &= a_1 \langle \Delta J_x^2 \rangle + a_2 \langle \Delta \vec{J}^2 \Delta J_x \rangle \\ \langle H \Delta \vec{J}^2 \rangle &= a_0 \langle \Delta \vec{J}^2 \rangle + a_1 \langle \Delta J_x \Delta \vec{J}^2 \rangle + a_2 \langle \Delta \vec{J}^4 \rangle \end{aligned} \qquad (4.20)$$

The solution, which takes into account the third order term $\langle \Delta \vec{J}^2 \Delta J_x \rangle$ in lowest approximation is given by

$$a_0 \equiv \langle H \rangle - \frac{1}{2\mathcal{J}_Y} \langle \Delta \vec{J}^2 \rangle \qquad (4.21)$$

$$a_1 \equiv \omega = \left\langle \left(H - \frac{\Delta \vec{J}^2}{2\mathcal{J}_Y} \right) \Delta J_x \right\rangle \Big/ \langle \Delta J_x^2 \rangle \qquad (4.22)$$

$$a_2 \equiv \frac{1}{2\mathcal{J}_Y} = \langle \Delta (H - \omega J_x) \Delta \vec{J}^2 \rangle / \langle \Delta \vec{J}^4 \rangle \qquad (4.23)$$

We now discuss the different orders of the Kamlah expansion. In lowest order (only $a_0 \neq 0$) we find the unprojected theory.

In first order we find for $K \ll I$ (i.e. $\langle J_z^2\rangle \ll I(I+1)$)

$$E^I = \langle H\rangle + \omega\left(\sqrt{I(I+1)} - \langle J_x\rangle\right) \tag{4.24}$$

A variation of this expression gives the selfconsistent cranking model discussed in section 2.

In second order we find for $K \ll I$

$$E^I = \langle H\rangle - \frac{\langle \Delta J^2\rangle}{2\mathcal{J}_Y} + \omega\left(\sqrt{I(I+1)} - \langle J_x\rangle\right) + \frac{1}{2\mathcal{J}_Y}\left(\sqrt{I(I+1)} - \langle J_x\rangle\right)^2 \tag{4.25}$$

Neclecting the Yoccoz-term $\langle\Delta\vec{J}^2\rangle/2\mathcal{J}_Y$ in the variation we find again the cranking model. This term shows, however, a rather irregular behavior in the band crossing region. For a quadrupole interaction it can be evaluated analytically[52])

$$\frac{\Delta\vec{J}^2}{2\mathcal{J}_Y} = \frac{3}{2}\chi\frac{\langle Q^\dagger\rangle\langle Q\rangle}{\langle\Delta\vec{J}^2\rangle} \tag{4.26}$$

In the band crossing region the fluctuations $\langle\Delta\vec{J}^2\rangle$ show a peak, which reduces the lowering as discussed in section 3. Taking this term into account within a cranking calculation produces a very sharp band crossing, as can be seen from fig. 13, and removes obviously the unphysically strong mixing between groundstate and s-band in the cranking theory. Quantitavitely,

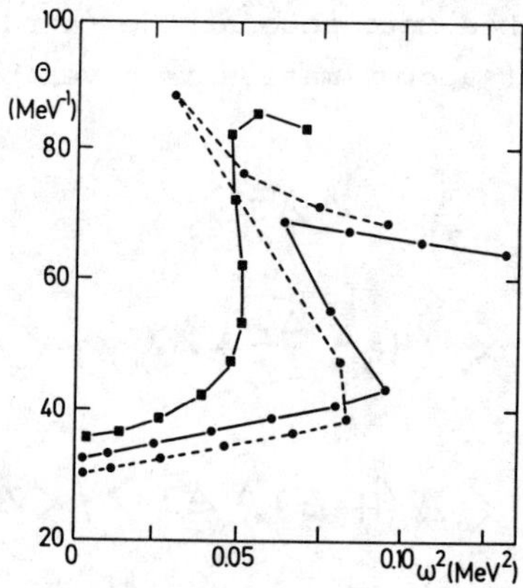

Fig. 13 Backbending plotts für ^{164}Er with (●--●-●) and without (■——■——■) the Yoccoz term (4.26) are compared with the experiment (●—●—●—●)

however, the effect is too large. We also have in such a theory not taken care of fluctuations having their origin in the virtual admixture of normal modes (RPA correlations, which are discussed in section 3)

4.3 Electromagnetic Matrix Elements

Electromagnetic matrix elements, as BEλ-values or spectroscopic multipole moments are in principle reduced matrix elements of spherical tensor operators. They are defined only for eigenstates of the angular momentum, i.e. for wavefunctions in the laboratory frame. Therefore we have to use projected wavefunctions

$$|\Psi_\nu^{I_\nu M_\nu}\rangle = \mathcal{N}_\nu \sum_K g_K \, P^{I_\nu}_{M_\nu K} \, |\Phi_\nu\rangle \tag{4.27}$$

and obtain for the reduced matrix element of a spherical tensor operator $Q_{\lambda\mu}$

$$\langle \Psi_1^{I_1} \| Q_\lambda \| \Psi_2^{I_2}\rangle = \mathcal{N}_1 \mathcal{N}_2 \sum_{\substack{K_1 K_2\\ \mu, \bar{K}}} g^*_{K_1} g_{K_2} (2I_1+1)(-)^{I_1-K_1} \cdot \begin{pmatrix} I_1 & \lambda & I_2 \\ -K_1 & \mu & \bar{K} \end{pmatrix} \times \langle \Phi_1 | Q_{\lambda\mu} P^{I_2}_{\bar{K} K_2} | \Phi_2 \rangle \tag{4.28}$$

This expression looks somewhat asymmetric in the angular momenta I_1 and I_2. In fact it is not and we could have written it in a similar way by the quantities $\langle \Phi_1 | P^I_{KK'} Q_{\lambda\mu} | \Phi_2 \rangle$. For the exact calculation both formulars are identical, however for the following approximations it is reasonable to use the arithmetic average.

In the following we study only intraband matrix elements. We use four approximations[5]):

i) We neglect the difference between the intrinsic states Φ_1 and Φ_2

ii) We use the Kamlah expansion in zeroth order

$$\langle \Phi_1 | Q_{\lambda\mu} P^I_{KK'} | \Phi_2 \rangle = \langle \Phi_1 | Q_{\lambda\mu} | \Phi_2 \rangle \langle \Phi_2 | \hat{R}(\Omega) | \Phi_2 \rangle \tag{4.29}$$

iii) We neglect the K-dependence of the parameters g_K^* and approximate the norm ratios

$$\mathcal{N}_1 / \mathcal{N}_2 \simeq \sqrt{2I_2+1} \;/\; \sqrt{2I_1+1} \qquad (4.30)$$

iv) We use the semiclassical approximation for the 3j-symbol, valid for K << I

$$\begin{pmatrix} I_1 & \lambda & I_2 \\ -K_1 & \mu & K_1-\mu \end{pmatrix} \simeq \frac{(-)^{I_1-K}}{\sqrt{2I_1+1}} \, d^{\lambda}_{I_1-I_2 \; \mu}(\pi/2) \qquad (4.31)$$

Thus we are able to express the complicated projected elements (4.28) in terms of simple intrinsic matrix elements

$$\langle \psi_1^{I_1} \| Q_\lambda \| \psi_2^{I_2} \rangle = \langle \Phi_1 | \tilde{Q}_{\lambda \, I_1-I_2} | \Phi_2 \rangle \, (2I_2+1) \qquad (4.32)$$

where $\tilde{Q}_{\lambda\mu}$ are the multipole operators $Q_{\lambda\mu}$ quantized along the rotational x-axis.

In particular we find for the g-factor

$$g = \mu / I = \langle \Phi | \mu_x | \Phi \rangle / \sqrt{I(I+1)} \qquad (4.33)$$

for the spectroscopic quadrupole moment in units of the axially symmetric rotor with deformation β_0.

$$Q(I) / Q_{SR}(I) = 2 \, (\beta/\beta_0) \, \sin(30° - \gamma) \qquad (4.34)$$

and for the BE2 - values

$$\sqrt{B(E2, I \to I-2)} \Big/ \sqrt{B(E2, I \to I-2)_{SR}} = 2\sqrt{3} \, (\beta/\beta_0) \cos(30° - \gamma) \qquad (4.35)$$

The intrinsic deformation parameters and are defined as

$$\langle \Phi | Q_{20} | \Phi \rangle \propto \beta \cos\gamma \qquad (4.36)$$

$$\langle \Phi | Q_{22} | \Phi \rangle \propto \beta \sin\gamma$$

Fig. 14 shows the validity of this approximation: Spectroscopic quadrupole moments calculated from the formula (4.34) are compared with results from an exact projection.

---

* The choice g_K = const is a resonable choice, since these parameters are highly redundant (see ref. 4 p. 477 and ref. 53)

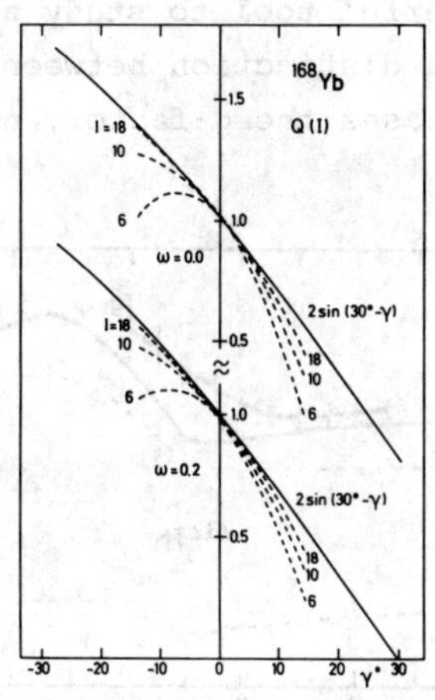

Fig.14 The reduced matrix elements $\langle I \| Q_2 \| I \rangle$ normalized to a symmetric rotor. The dashed lines correspond to different I-values projected exactly from the same intrinsic function for each γ -value. The full line corresponds to formula (4.34). Two cranking frequencies are shown

In recent years magnetic moments and BE2-values along the yrast line have been measured in order to study the behaviour of nuclei at high spins. In well deformed nuclei the BE2-values stay in general close to the rotor values (see fig.15) which can be nicely reproduced by cranking calculations[54,55]. In transitional regions one has found sometime deviations from the rotor values. They could not be explained so far within cranking theory

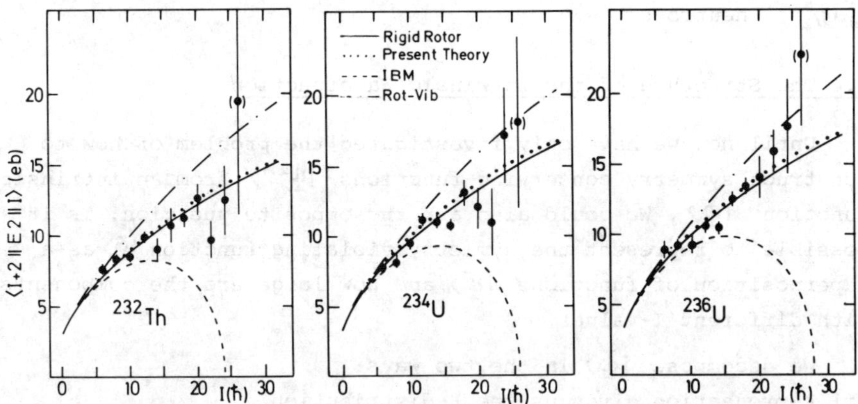

Fig.15 E2-Matrix elements for three actinide nuclei. The experimental values of ref. 56 are compared with cranking calculations (present theory[57]), the SU(3)-limit of the Interacting Boson Model[58]) and the Rotation Vibration Model[59])

The magnetic moments are a very powerful tool to study alignment processes[60,61]) because they allow a distinction between protons and neutrons: Proton alignment increases the g-factor, neutron alignment reduces it (Fig.16)

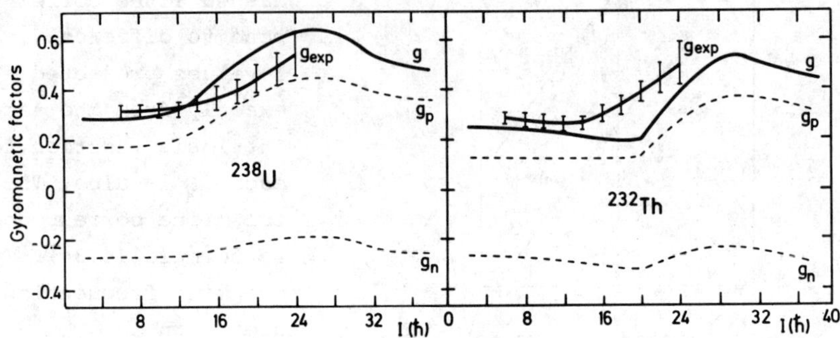

Fig. 16 Gyromagnetic factors of ^{238}U and ^{232}Th as a function of the angular momentum. Experimental values [62]) are compared with theoretical calculations within the Rotating Shell Model[57])

This is in particular important in the actinides, where one has competition between alignment of $i_{13/2}$-protons and alignment of $j_{15/2}$ - neutrons

4.4 The Structure of the Intrinsic Wavefunction

Until now we have only investigated the problem of how to construct symmetry conserving functions $|\Psi^{IM}\rangle$ from an intrinsic function $|\Phi\rangle$. We could also ask the opposite question: Is it possible to represent the symmetry violating function $|\Phi\rangle$ as a superposition of functions $|\Psi^I\rangle$ and how large are the components with different I-values.

We decompose $|\Phi\rangle$ in the two ways:

i) K-projection gives us the K-distribution

$$|\Phi\rangle = \sum_K \sqrt{n_K}\, |\Phi_K\rangle \tag{4.37}$$

where $|\Phi_K\rangle$ is normalized. From eq. (4.5) we find

$$n_K = \sum_I n^I_{KK} \tag{4.38}$$

ii) I-projection gives us the I- distribution

$$|\Phi\rangle = \sum_I \sqrt{n^I}\,|\Phi_I\rangle \tag{4.39}$$

with normalized functions $|\Phi_I\rangle$ we obtain

$$n^I = \sum_K n^I_{KK} \tag{4.40}$$

Fig. 17 shows us the I- distributions[63]) derived from various cranked HFB-functions labeled by the average angular momentum J For J = 0 we see that the component having sharp angular momentum I = 0 is very small (only a few percent). All other components are cut off by the projection.
For higher angular momenta we find in the band crossing region (J = 14ħ) a very broad distribution connected with an increasing fluctuation

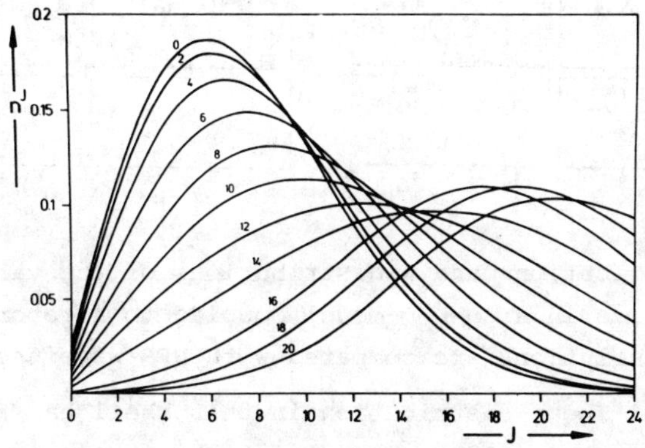

Fig. 17 Probability components with angular momentum I in a deformed HFB wavefunction for different values of $J = \langle J_x \rangle$

For high spins the maximum of the I-distribution coincides with I=J i.e. in the very high spin limit we have the classical picture that the major component of the wavefunction $|\Phi\rangle$ has the proper spin and that $\langle J_x\rangle^2 \gg \langle \Delta J_x^2\rangle$

The K-distribution is dominated by the contributions of the aligning particles. We therefore show in Fig.18 a model calculation of several $i_{13/2}$ -valence particles coupled to a rotor. At I=14 backbending occurs. After that we have two aligned particles in the

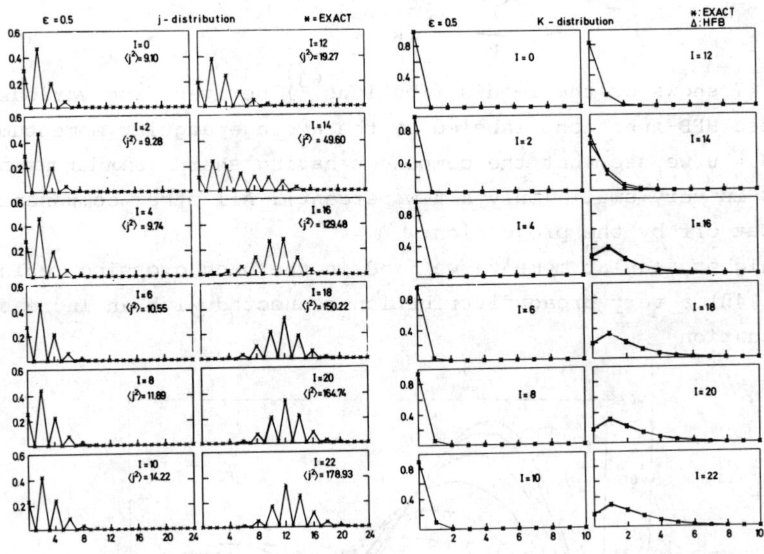

Fig.18 j-distributions and K-distributions of the valence particles in a $i_{13/2}$ -model coupled to a rotor. Exact solutions are compared with HFB wavefunctions[64])

valence shell: The j-distribution in this shell has a rather stable maximum at j = 12 $\hbar$. The K-distribution is concentreted at K = 0 before the backbending. After backbending we have a flat maximum at K = 1 $\hbar$ and decreasing K-contributions up to K = 5 $\hbar$

5. Numerical Applications of Exact Projections

The idea of angular momentum projection is old. It goes back to the fifties. Several methods have been proposed (for the relevant literature see ref.4 chapter 11) since a few years new progress has been made in numerical applications for heavy nuclei, namly

i) Full three-dimensional angular momentum projection from triaxial, time reversal breaking cranked HFB-wavefunctions after the variation in heavy nuclei[53,65])

ii) Full three-dimensional angular momentum projection from cranked HF-wavefunctions in coordinate space after the variation[66])

iii) Unrestricted variation of cranked HFB-wavefunctions after exact number projection[9])

iv) Unrestricted variation of axially symmetric HFB- wave-functions after a one-dimensional projection[11])

v) Configuration mixing of angular momentum projected axially symmetric twoquasiparticle configurations, i.e. projected TDA[67,68]), a program which is sometimes called MONSTER

In this lecture we discuss some of the basic techniques and a few applications

5.2 Projection Techniques

In nearly all cases of numerical projections for heavy nuclei an integral representation (4.4) of the type (4.4) has been used. It has the advantage that in such cases most of the overlap integrals (4.16) are smooth functions of the angels and the integration can be replaced by a sum over a finite sumber of suitably chosen points ($\Omega = 1, \dots, L$).

$$\langle H P^I_{MK} \rangle = \sum_{\Omega=1}^{L} w_\Omega^{IMK} \langle 0|H|\Omega\rangle \langle \hat{R}(\Omega)\rangle \qquad (5.1)$$

where $|\Omega\rangle$ is the rotated Slater determinant

$$|\Omega\rangle = \hat{R}(\Omega) \, |\Phi\rangle \cdot \langle \hat{R}(\Omega)\rangle^{-1} \qquad (5.2)$$

normalized to $\langle 0|\Omega\rangle = 1$. The numbers w_Ω^{IMK} contain weightfactors from the integration and the irreducible representations $\mathcal{D}^{I*}_{MK}(\Omega)$. The sum over Ω can be a single sum for axially symmetric cases, it can be a triple sum, as in the case of treedimensional angular momentum projection, or even more complicated, if in addition to the angular momentum a projection on particle number is carried out.

The most time consuming part in the evaluation of the overlap integrals $\langle \hat{R}(\Omega) \rangle$ and $\langle 0|H|\Omega\rangle$. Since $|\Omega\rangle$ as well as $|0\rangle$ are product states we can use a generalized Wick theorem[69,71], which tells us that we can express any matrix element $\langle 0|\dots|\Omega\rangle$ by the contractions:

$$\begin{pmatrix} \langle 0|c^{\dagger}c|\Omega\rangle & \langle 0|cc|\Omega\rangle \\ \langle 0|c^{\dagger}c^{\dagger}|\Omega\rangle & \langle 0|cc^{\dagger}|\Omega\rangle \end{pmatrix} \tag{5.3}$$

and the overlap $\langle \hat{R}(\Omega) \rangle$ is given by[4])

$$\langle \hat{R}(\Omega) \rangle = \pm \sqrt{\det(\mathcal{U}_{\Omega})} \tag{5.4}$$

where

$$\begin{aligned} \mathcal{U}_{\Omega} &= U^{\dagger} D(\Omega) U + V^{\dagger} D^{*}(\Omega) V \\ \mathcal{V}_{\Omega} &= V^{T} D(\Omega) U + U^{T} D^{*}(\Omega) V \end{aligned} \tag{5.5}$$

are the generalized HFB-coefficients which transform the quasiparticle basis of $|\Phi\rangle$ (characterized by the HFB-coefficients U,V) into the rotated quasiparticle basis $DU, D^{*}V$ and $D(\Omega)$ is the representation of the rotation in the particle basis.

Unfortunately the overlap (5.4) is given only up to phase, a problem which can eventually cause trouble in the integration and which can in principle by avoided by the diagonalization of a non-hermitian matrix[72]).

The contractions (5.3) can be either evaluated in the particle basis

$$\begin{aligned} \langle 0|c^{\dagger}c|\Omega\rangle &= D(\Omega) V^{*} (V \mathcal{U}_{\Omega}^{-1})^{T} \\ \langle 0|cc|\Omega\rangle &= D(\Omega) V^{*} (U \mathcal{U}_{\Omega}^{-1})^{T} \\ \langle 0|c^{\dagger}c^{\dagger}|\Omega\rangle &= D^{*}(\Omega) U^{*} (U \mathcal{U}_{\Omega}^{-1})^{T} \end{aligned} \tag{5.6}$$

or in the quasiparticle basis with respect to $|\Phi\rangle$. In this case the only nontrivial contraction is

$$\langle 0|\alpha\alpha|\Omega\rangle = Z_\Omega = v_\Omega^* u_\Omega^{*-1} \tag{5.7}$$

which can easily be seen from the Thouless theorem[4])

$$|\Omega\rangle = \exp\left(\tfrac{1}{2} Z_{\Omega\, kk'}\, \alpha_k^+ \alpha_{k'}^+\right)|0\rangle \tag{5.8}$$

It is evident that this basis is the most suitable basis for practical calculations.

5.2 Variation after Projection (PHFB)

To be consistent with the variational principle the intrinsic wavefunction should be calculated after a projection onto good quantum numbers

$$\delta E^I = \delta \frac{\langle\Phi|HP^I|\Phi\rangle}{\langle\Phi|P^I|\Phi\rangle} = 0 \tag{5.9}$$

Since an arbitrary variation $|\Phi\rangle$ can be expressed by the Thouless theorem

$$|\Phi + \delta\Phi\rangle = \exp\left(\tfrac{1}{2} C_{kk'}\, \alpha_k^+ \alpha_{k'}^+\right)|\Phi\rangle \tag{5.10}$$

we can parametrize the projected energy surface locally by the parameters $C_{kk'}$. The variational equation (5.9) is equivalent to a vanishing gradient in the multidimensional surface:

$$\frac{\partial}{\partial C_{kk'}^*} E^I = \langle\Phi|\alpha_{k'}\alpha_k (H - E^I) P^I|\Phi\rangle \tag{5.11}$$

These are the Euler- Lagrange equations for variation after projection, the projected HFB equations (PHFB)[10)].

The matrix elements (5.11) of the projected gradient can be calculated using the techniques of section (5.1). Having the gradient at an arbitrary point $|\Phi\rangle$ in the energy surface the variational equations (5.11) can be solved for instance by the gradient method[10,73] or by more sophisticated numerical techniques based on the gradient (5.11).

5.3 Projection from Cranked HFB Functions

Cranked HFB-functions violate axial symmetry. They therefore require a threedimensional angular momentum projection. In section 4.3 we have shown results of an approximate projection up to second order in the Kamlah expansion. We also found that in the band crossing region the second order term has a very strange behaviour. It seems to be therefore worthwhile to investigate, if such a behaviour shows up in the exact projected energy, too, or if it is just an artifact of the Kamlah expansion.

We therefore used cranked HFB wavefunctions for the Baranger-Kumar Hamiltonian. They are labeled by their average angular momentum $\langle J_x \rangle$. From each of these wavefunctions we projected onto good angular momentum. We compared two versions of the projector, namely a projector with constant values of g_K and a projector, where the coefficients g_K were obtained from a diagonalization of eq. (4.10).

It turns out that the g_K-factors of both methods are rather different. Nonetheless the projected wavefunctions are very similar in both cases. In fact their overlap is more than 99%, which shows the high degree of redundancy in the coefficients

Fig. 19a shows the lowering in the yrast energy produced by exact angular momentum projection as a function of the angular momentum together with the correlation energy obtained in second order Kamlah expansion. In fact both values are rather similar and both have a strange kink in the bandcrossing region. This shows us that - at least for cranking functions the Kamlah expansion is very rapidly converging: The zeroth order term $\langle H \rangle$ is several hundreds MeV, the first order term $\omega \langle J_x \rangle$, the rotational energy is in the order of up to 20 MeV. Rge socond order term $\langle \Delta \vec{J}^2 \rangle / 2 \mathcal{J}_Y$ is several MeV large and all the rest is smaller than 100 keV.

Of course we should vary after projection. So far a general variational is in a threedimensional case not possible. The only parameters we can vary is the average angular momentum $\langle J_x \rangle$. In fig.19b we show energy surface obtained in this way for different angular momenta I=0,....,I=20. If the cranking model would be a solution of this generalized variation the minima would correspond to $\langle J_x \rangle = \sqrt{I(I+1)}$. In fig. 20 we see that this

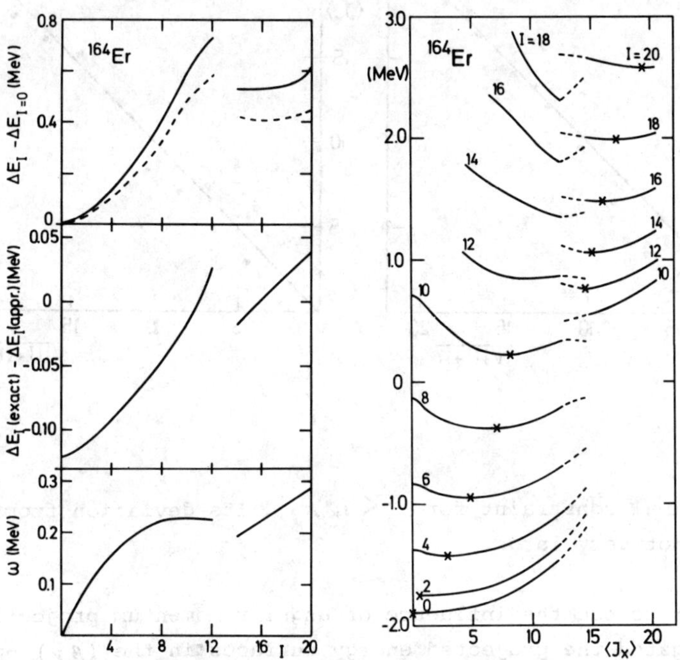

Fig. 19a) Correlation energy as a function of spin I for ^{164}Er. Solid line for the exact and dotted line for the approximate projection.

b) Rotational spectra as a function of I

is indeed in general the case.

5.4 Projected energy surface in Triaxial Nuclei

It is well-known that the experimental data of some transitional nuclei can be rather well described on a phenomenological basis by a triaxial rotor[75-77]. On the microscopic side, however, one has found only very shallow minima in γ-direction in constraint HFB calculations[21,78], which could not explain stable γ-deformations.

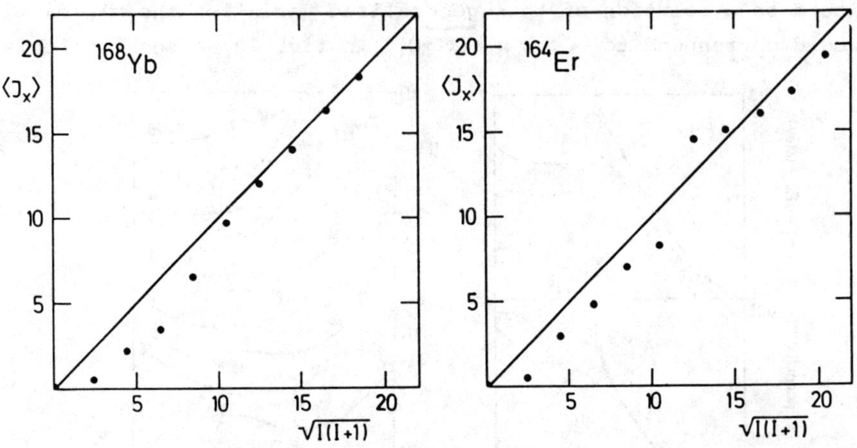

Fig.20 Optimal constraint for $\langle J_x \rangle$. Its deviation from $\sqrt{I(I+1)}$ is not very large.

In order to see the influence of angular momentum projection we investigated the projected energy surfaces in the (β,γ) -plane for well deformed and for transitional nuclei[79]. We find that for well deformed axially symmetric nuclei angular momentum projection has little influence. In the transitional nuclei, however, projection produces a well pronounced minimum at $\gamma \approx 30^{\circ}$. This gives us some indication, why triaxial models in such nuclei are sometimes rather successful. The deeper reason for this behaviour seems to be, that we have in nuclei never a strong violation of axial symmetry. We are either in a case of no symmetry violation, as in ^{168}Er, where projection is not necessary, or in a case of weak symmetry violations, where the arguments of the Kamlah expansion do not apply: A full projection seems to be necessary.

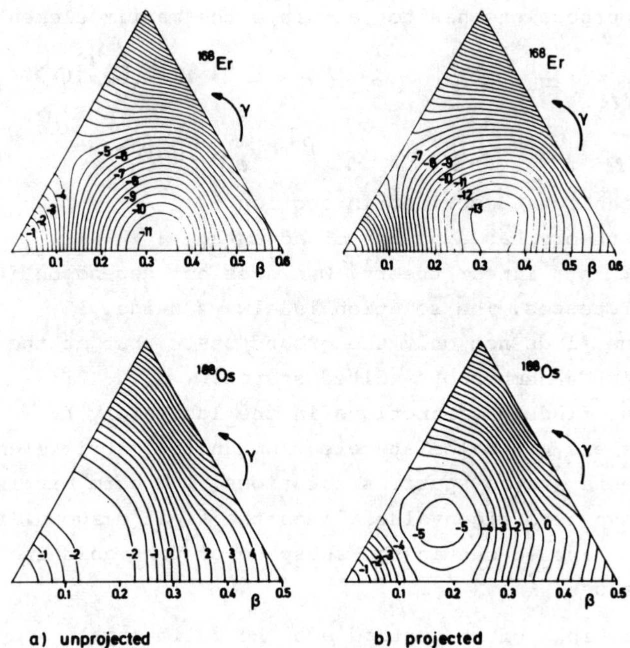

Fig. 21 Energy surfaces in the (β,γ) -plane for the nuclei ^{168}Er and ^{188}Os a) without angular momentum projection b) with exact threedimensional angular momentum projection. The units on equipotential lines are MeV.

5.5 Projected Tamm Dancoff Approximation

So far we have concentrated on a mean field theory in its most simplest version, namely on one single product state, from which eigenfunctions of the symmetry operators were projected. The next step would obviously be a TDA approximation, where many two quasiparticle components are superposed:

$$|\Psi^I\rangle \quad = \quad c_0 \; P^I|\Phi\rangle + \sum_{k<k'} c_{kk'} \, P^I \alpha_k^+ \alpha_{k'}^+ |\Phi\rangle \tag{5.12}$$

The variation of the energy corresponds to a diagonalization of the Hamiltonian in the space spanned by the non-orthogonal basis states

$$P^I|\Phi\rangle \qquad P^I\alpha_k^+\alpha_{k'}^+|\Phi\rangle \tag{5.13}$$

For that purpose one has to calculate the matrix elements

$$\mathcal{A}_{kk'll'} = \langle\Phi|\, \alpha_{k'}\alpha_k \,(H - E^I)\, P^I \,\alpha_l^+\alpha_{l'}^+ |\Phi\rangle \qquad (5.14)$$

$$\mathcal{N}_{kk'll'} = \langle\Phi|\, \alpha_{k'}\alpha_k \; P^I \;\alpha_l^+\alpha_{l'}^+ |\Phi\rangle \qquad (5.15)$$

by the techniques described in section 5.1.

Such a theory has a number of advantages:

i) It is a linear theory. One does not depend on iteration processes. The solution is always unique.

ii) One finds not only the ground state, but at the same time a number of excited states.

iii) One finds wavefunctions in the laboratory frame and it is easy to calculate electromagnetic matrix elements.

iv) There are no spurious solutions. The norm matrix may have zero eigenvalues. Then the final diagonalization is carried out in the subspace of non-vanishing eigenvalues.

The shortcomings of the method are the following:

i) The evaluation of the matrix elements is in general complicated. So far one is restricted to axial symmetric basis functions $|\Phi\rangle$ and to a restricted number of two quasiparticle components.

ii) As usual in TDA the ground state is treated in a better approximation than the excited states. In principle one should include four-quasiparticle components.

iii) The method suffers from shortcomings of a shell model code: It is hard to disentangle the underlying physics from the nearly exact eigenstates.

The method is in principle rather old[80-82]. In the context of violation of particle number it is known under the name Broken Pair Approximation[83-85] or Generalized Seniority[86-89] and gives in these cases a very accurate description of spherical nuclei far from closed shells. Recently it has been applied to the yrast line of deformed nuclei by two groups[67,68] (MONSTER).

In fig. 22 we show the results for ^{164}Er. The agreement with experimental data is very good. In particular there is no smoothing of the level crossing as in most Cranking calculations.

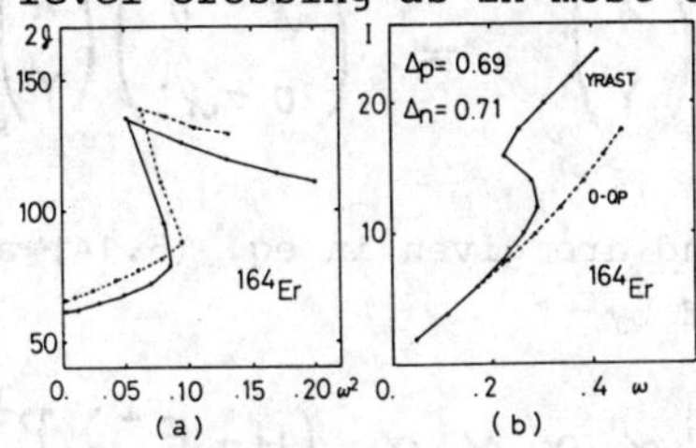

Fig. 22 a) Backbending curve for ^{164}Er. Solid line shows the result of a MONSTER calculation[67] and the dashed line the experiment.

b) Alignment diagram. Both the yrast and the Oqp. bands are given to show the magnitude of the angular momentum alignment.

5.6 Projected Random Phase Approximation (PRPA)

We have seen in section 3.3 that ground state correlations can play an important role in the yrast configurations. These can be included by a rather simple extension of the projected TDA method. In order to derive the equations for the projected RPA we use the time-dependent variational principle with symmetry conserving mean field wavefunctions[90]

$$\delta \int dt \, \langle \Phi | (H - i\partial_t) P^I | \Phi \rangle = 0 \tag{5.16}$$

We start with a static solution $|\Phi_0\rangle$ of the PHFB equations (5.11) and write $|\Phi\rangle$ in the form

$$|\Phi\rangle = \exp(i\hat{z}(t))\,|\Phi_0\rangle \tag{5.17}$$

where

$$\hat{z}(t) = \varepsilon(B^+ e^{-i\Omega t} + h.c.) \tag{5.18}$$

and

$$B^+ = \sum_{k<k'} X_{kk'}\alpha_k^+\alpha_{k'}^+ - Y_{kk'}\alpha_{k'}\alpha_k \tag{5.19}$$

and obtain, after linearizing for small amplitudes the PRPA equations

$$\begin{pmatrix} \mathcal{A} & \mathcal{B} \\ \mathcal{B}^* & \mathcal{A}^* \end{pmatrix} \begin{pmatrix} X \\ Y \end{pmatrix}_\nu = \begin{pmatrix} \mathcal{N} & 0 \\ 0 & -\mathcal{N} \end{pmatrix} \begin{pmatrix} X \\ Y \end{pmatrix}_\nu \Omega_\nu \quad (5.20)$$

where the matrices and are given in eq. (5.14) and (5.15). The matrix is defined by

$$\mathcal{B}_{kk'll'} = \langle \Phi_0 | \alpha_{k'} \alpha_k \alpha_{l'} \alpha_l \; (H - E^I) \, P^I | \Phi_0 \rangle \quad (5.21)$$

So far these equations have been applied only to exactly soluble models[90]. They have many nice properties of usual RPA equations and are in a sence the logical extension of symmetry conserving mean field theory to time-dependent mean fields.

Concluding Remarks

Symmetry conserving mean field theory is a very powerful method to deal with long range correlations in finite Fermi systems an nuclei. It allows the incorporation of a large number of correlations in terms of rather simple eventually time-dependent wavefunctions by violation of symmetries. Additional correlations are taken care of by the projection techniques. Although the projected wavefunctions are rather complicated, in many cases of strong symmetry violations in systems with large particle numbers one can neglect projection. In such cases all important physical quantities can be expressed by intrinsic matrix elements of effective operators. This fact is the basis for the succes of HF and HFB theory in nuclei.

Since nuclei are finite systems, however, we sometime have to deal with symmetry violations, which are not strong, as for instance in regions of phase transitions and in transitional nuclei. In these cases simple-minded mean field theory breaks down and one has to apply full projection before the variation. Examples are triaxial nuclei in transitional regions or the pairing collaps at high angular momenta.

Acknowledgements

I would like to express my gratitude to H.J.Mang and to all the members of the Munich group and to all the guests from other institutes who contributed to obtain the various results discussed in these lectures, in particular to J.L.Egido, M. Faber, C. Federschmid, Y.K. Gambhir, K. Hara, A. Hayashi, S. Iwasaki, U. Mutz and M.L. Robledo

References

1) D. Vautherin, Phys. Rev.C7 (1973)269.
2) D. Gogny Proceedings of the International Conference on Nuclear Selfconsistent Fields, Triest, 1975. G. Ripka and M. Porneuf, Eds. North Holland, Amsterdam,1975.176,209, 265,266
3) J. Dechargé and D. Gogny, Phys.Rev.C21 (1980)1568,265
4) P. Ring and P. Schuck; The Nuclear Mànybody Problem, Springer Verlag, New York 1980
5) P. Ring, A. Hayashi, K. Hara, H. Emling and E. Grosse; Phys.Lett.110b (1982)
6) R.E. Peierls and J.Yoccoz, Proc.Phys.Soc.(London)A70 (1957)
7) H.D. Zeh, Z.Phys.188(1965)
8) N. Onishi, Prog.Theor.Phys.40 (1968)
9) J.L. Egido and P. Ring; Nucl.Phys.A385 (1982)169
10) J.L. Egido and P. Ring; Nucl.Phys.A388 (1982) 19
11) F. Kummar, K.W.Schmidt, privat communication
12) A. Kamlah, Z.Phys.216 (1968) 52.
13) H.J. Mang; Phys.Rep.18C (1975)325.
14) B. Banerjee, H.J. Mang and P. Ring; Nucl.Phys.A215 (1973) 366.
15) P. Ring, R. Beck and H.J.Mang; Z.Phys.231 (1970)10.
16) A.L. Goodmann, Adv.Nucl.Phys.11 (1979)263.
17) D.R. Inglis, Phys.Rev.96 (1954) 1059.
18) L.D. Landau and E.M. Lifshitz, Cours of Theoretical Physics. Pergamon,Oxford, 1959.
19) H.J. Mang, lectures given at this conference
20) B. Banerjee, H.J. Mang and P. Ring; Nucl.Phys.A215 (1973) 366.
21) K. Kumar and M. Baranger; Nucl.Phys.110 (1968)529 ;A122 (1968)273.
22) C. Bloch and A. Messiah; Nucl.Phys.39 (1962)95.
23) T. Bengtsson and S. Frauendorf; Nucl.Phys.A314 (1979)27
24) M.A.J.Mariscotti, G. Scharff-Goldhaber and B. Buck; Phys.Rev.178 (1969)1864.104
25) E.R. Marshalek; Phys.Rev.C15 (1977) 1574

26) A.L. Goodmann; Nucl.Phys.A230 (1974) 466.
27) A. Bohr, Varenna Lectures 69 (1976) 1.
28) J.L. Egido, H.J. Mang and P. Ring; Nucl.Phys.A344(1980)1.
29) A. Goswani, L. Lin and G.L. Struble, Phys.Lett.25B (1967) 451.
30) R. Bengtsson, I. Hamamoto and B.R. Mottelson; Phys.Lett. 73B (1978) 259
31) F. Grummer, K.W. Schmid and A. Faessler; Nucl.Phys.A236 (1979) 1
32) B.R. Mottelson and J.G. Valatin, Phys.Rev.Lett.5 (1960) 511.
33) K.Y. Chan and J.G. Valatin; Nucl.Phys. 82 (1966) 222.
34) U. Mutz and P. Ring; J.Phys.G10 (1984) L39
35) R. Chapman et.al; Manchester University, Annual Report 1983
36) J.L. Egido and P. Ring; Nucl. Physics in print
37) J.O. Newton et. al.;Phys.Rev.Lett.46 (1981) 1383
38) I.I. Gaardhøje, XX. Int. Winter Meeting on nuclear physics, Bormio 1982
39) W. Hennerici et. al.; Nucl.Phys.A396 (1983) 329 C
40) B. Haas et. al.; Phys.Lett.120B (1983) 79
41) A.M. Sandorfi et. al.; Phys.Lett.130B (1983) 19
42) M. Gaudin, Nucl.Phys.15 (1960) 89
43) M. Sano and W. Wakai; Prog.Theor.Phys.48 (1972) 160
44) A.L. Goodmann, Nucl.Phys.A352 (1981) 30
45) K. Tanabe, K. Sugawara-Tanabe and H.J. Mang; Nucl.Phys. A357 (1981) 20
46) E.R. Marshalek; Nucl.Phys.A275 (1977) 416.
47) Y.R. Shimizu and K. Matsuyanagi; Prog.Theor.Phys.70 (1983) 144
48) J.L. Egido, H.J. Mang and P. Ring; Nucl.Phys.A341 (1980) 229
49) P. Ring, L.M. Robledo, J.L. Egido and M. Faber; Nucl. Phys. A419 (1984) 261
50) M. Danos; Nucl.Phys.5 (1958) 23.
51) M.G. Redlich and E.P. Wigner; Phys.Rev.95 (1954) 122.
52) J.L. Egido and P. Ring, Phys.Lett.95B (1980) 331
53) K. Hara, A. Hayashi and P. Ring; Nucl.Phys.A385 (1982)14
54) S. Islam, H.J. Mang and P. Ring; Nucl.Phys.A326 (1979)161

55) Y.K. Gambhir, P. Ring, H.J. Mang; Z.Phys.A306 (1982)155
56) H. Ower et. al.; Nucl.Phys.A388 1982) 421
57) J.L. Egido and P. Ring; Nucl.Phys. (1984) in print
58) A. Arima and F. Iachello, Ann.Phys.99 (1976) 253
59) M. Seiwert, P.O. Hess, J.A. Maruhn and W. Greiner; Phys.Rev.C23 (1981) 2335
60) M. Diebel, A.N. Mantri and U. Mosel; Nucl.Phys.A345 (1980) 72
61) P. Ring, Proc. Int. Symp. on Electromagnetic properties of Atomic Nuclei, Tokyo, Nov. 1983
62) I. Häusser et. al.; Phys. Rev. Lett.48 (1982) 383
63) F. Grümmer, K.W. Schmid and A. Faessler; Nucl.Phys.A306 (1978) 134
64) R.R. Hilton, H.J. Mang, P. Ring, J.L. Egido, H. Herold, M. Reinecke, H. Ruder and G. Wunner; Nucl.Phys.A366 (1981) 365
65) M. Wüst, K. Ansari, U. Mosel; private communication
66) D. Baye and P.-H. Heenen, Phys.Rev. 1984 in print
67) S. Iwasaki and K. Hara; Progr.Theor.Phys.68 (1982) 1782
68) K.W. Schmid, F. Grümmer and A. Faessler; Phys.Rev. in print
69) N. Onishi and S. Yoshida; Nucl.Phys.80 (1966) 367
70) R. Balian and E. Brézin; Nuovo Cim, 64B (1969) 37
71) K. Hara and S. Iwasaki; Nucl.Phys.A332 (1979) 61
72) K. Neegård and W. Wüst; Nucl.Phys.A402 (1983) 311
73) H.J. Mang, B. Samadi and P. Ring; Z.Phys. A279 (1976) 325
74) J.L. Egido, H.J. Mang and P. Ring ; Nucl.Phys.A339 (1980) 390
75) A.S. Davydov and B.F. Filoppov; Nucl.Phys.8 (1958) 237
76) J. Meyer-ter-Vehn, F.S. Stephens and R.M. Daimond; Phys. Lett.32 (1974) 1383
77) H. Toki and A. Faessler; Nucl.Phys.A253 (1975) 231.
78) M. Girod and B. Grammaticos, Phys.Rev.Lett.40 (1978)361
79) A. Hayashi, K. Hara and P. Ring, preprint TU München
80) P.L. Ottaviani and M. Savoia, Phys.Rev.187 (1969)1306
81) M. Kleber, Z.Phys.231 (1970) 421
82) K. Allaart and W.F. Van Gunsteren, Nucl.Phys.A234 (1974)53
83) Y.K. Gambhir, A. Rimini and T. Weber, Phys.Rev.188 (1969) 1573; C3 (1971) 1965

84) S. Haq and Y.K. Gambhir; Phys.Rev.C16 (1977) 2455

85) K. Allaart and E. Boecker, Nucl.Phys.A168 (1972) 33

86) I. Talmi; Nucl.Phys.A172 (1971) 1

87) I. Talmi, Varenna Lectures 69 (1976) 352

88) L.F.F. Lathouwers; Nucl.Phys.A228 (1974) 125

89) B. Lorazo and R. Arvieu, Phys.Lett.49B (1974) 231

90) C. Federschmidt and P. Ring, to be published

The Interacting Boson Model and the Shell Model

Igal Talmi

The Weizmann Institute of Science, Rehovot, Israel

The original version of the interacting bosn model (IBA-1)[1] gives a good description of collective states of even-even nuclei. Such states are described as an assembly of s- and d-bosns. A Hamiltonian with single boson terms and boson boson interactions which commutes with the total number of bosons $N=n_s+n_d$, can reproduce energy levels and transitions in agreement with experiment[2]. In particular, for certain values of the parameters the Hamiltonian can be expressed in terms of scalar products of generators of subgroups of U(6) ((U(5), SU(3) or O(6)). In such cases, eigenvalues are simple algebraic expressions of quantum numbers[1,3]. The calculated low lying levels in those limits of IBA-1 agree with the low lying levels obtained from the approximate solutions of the differential equation corresponding to the Bohr Hamiltonian. These limits are idealized cases of spectra in vibrational, rotational and γ-unstable nuclei respectively.

There are several attractive features of IBA-1:

1) It gives a successful and unified description of a variety of collective spectra including those in transition regions.

2) It is much simpler to diagonalize the IBA-1 Hamiltonian than to solve the differential equation in the geometrical approach. In fact, a representation in terms of d-bosons has been used for the approximate solution of the differential equation.

3) It suggests a new approach for the shell model description of collective states. Various Nilsson-type schemes are simple only in the intrinsic frame, the projection into states with definite J may be approximately valid for strong deformations. In IBA-1 the bosons have definite angular momenta (in a fixed frame) and hence can be probably described by the (spherical) shell model.

The possible connection of the Interacting Boson Model with the shell model and the resulting implications for the boson model[4] will be the main topic of these lectures.

To find a shell model description of s- and d-bosons we must look for basic building blocks with J=0 and J=2. The simplest candidates are pairs of fermions which could be either nucleon-nucleon or nucleon-hole pairs. To obtain nucleon-hole states with J=0 and positive parity the nucleon must be raised into the second major shell. Also there is no argument for having a fixed number of N of such pairs in a given nucleus. Turning to nucleon-nucleon pairs we see that no J=0 positive parity states can be constructed from one valence proton and one valence neutron in heavier nuclei. Thus, we take as building blocks J=0 and J=2 pairs of identical nucleons[4]. Such pairs coupled to J=0 have been used not only after the Bardeen-Cooper Schrieffer (BCS) theory of superconductivity. They were introduced into (atomic) spectroscopy by Racah in 1943 in his theory of the seniority scheme[5]. The concept of pairing and seniority found more direct applications in nuclear spectroscopy. Let us now briefly discuss the main features of the seniority scheme using a much more recent formalism[6].

To create a pair in the j-orbit we operate on the vacuum state |0> (state of a nucleus with closed shells only) by the operator

$$S_j^+ = \frac{1}{2}\sum(-1)^{j-m}a_{jm}^+a_{j-m}^+ \tag{1}$$

The corresponding pair annihilation operator is $(S_j^+)^+ = S_j^-$ and the pairing operator $S_j^+S_j^-$ roughly measures the number of J=0 pairs. It has the eigenvalue (2j+1)/2 for the j^2 J=0 state and 0 for all other j^2 states. The seniority scheme is the set of eigenstates of this operator.

To find eigenstates and calculate the eigenvalues of $S_j^+S_j^-$ we first obtain

$$[S_j^+,S_j^-] = \sum_m a_{jm}^+a_{jm} - \frac{2j+1}{2} = 2S_j^o \tag{2}$$

The definition of S_j^o is due to its commutation relations with S_j^+ and S_j^-, namely

$$[S_j^o,S_j^+] = S_j^+ \qquad [S_j^o,S_j^-] = -S_j^- \tag{3}$$

The commutators (2) and (3) demonstrate that S_j^+, S_j^- and S_j^o are generators of the SU(2) Lie algebra. In other words, their commutation relations are the same as those of the m = +1, m = -1 and m=0 components of an angular momentum vector. Properties of such quasi-spin operators are well known and we readily obtain

$$S_j^+S_j^- = \vec{\underset{\sim}{S}}_j^2 - (S_j^o)^2 + S_j^o = s(s+1) - S_j^o(S_j^o-1) = s(s+1) - \frac{1}{4}\left(\frac{2j+1}{2} - n\right)\left(\frac{2j+5}{2} - n\right) \tag{4}$$

The quantum number s is an integer or half integer which determines the eigenvalues of the square of the quasi-spin vector $\vec{\underset{\sim}{S}}_j$.

Seniority of a given state is loosely speaking the number of unpaired nucleons. More precisely, a state in the j^v configuration has seniority v if it has no J=0 pairs, i.e.

$$S_j^-|j^v vJM> = 0$$

The relation of the quantum number v to s in eq. (4) can be obtained by applying the operator (4) to such a state obtaining

$$s = \frac{1}{2}\left(\frac{2j+1}{2} - v\right) \tag{5}$$

There are two solutions for s in terms of v to the quadratic equation but due to $s \geq 0$ we obtain

$$s = \frac{1}{2}\left(\frac{2j+1}{2} - v\right)$$

The eigenvalues of (4) can be expressed as

$$\frac{1}{4}(n-v)(2j+3-n-v)$$

Operating on a state with given seniority by S_j^+ increases n by 2 but leaves the seniority unchanged. The vacuum state (n=0) has the maximum value of $s=(2j+1)/4$, and hence v=0, and the minimum value of $S_j^o = -(2j+1)/4$. Operating on it successively with S_j^+ we see that for any even value of n there is a J=0 state with seniority v=0 in the j^n configuration. States with J=2,4,...2j-1 in the j^2 configuration have seniority v=2 and any j^n configuration with even value of $n \leq 2j-1$ has such states. The state of a single j nucleon has, by definition, v=1 and such a state with J=j, v=1 occurs in any j^n configuration with n odd.

The seniority scheme is very useful in constructing states and evaluating matrix elements of operators. To make full use of it we should consider the transformation of operators under SU(2) operations. In other words we should calculate the commutation relations between various operators and the quasi-spin components. We consider single nucleon operators which are irreducible tensors in ordinary space

$$T_\kappa^{(k)} = \sum_{mm'} (-1)^{j-m} \begin{pmatrix} j & k & j \\ -m & \kappa & m' \end{pmatrix} a_{jm}^+ a_{jm'}$$

for which $(j||T^{(k)}||j)=1$. Since such operators are products of a creation operator and annihilation operator they do not change the nucleon number and commute with S_j^o. This shows that all single nucleon operators are the $S_j^o=0$ components in quasi-spin space. To find the rank s in quasi-spin space of such operators we calculate

$$[S_j^+, T_\kappa^{(k)}] = -(2k+1)^{1/2} \sum_{mm'} (jmjm'|jjk\kappa) a_{jm}^+ a_{jm'}^+ \qquad (6)$$

Due to the symmetry properties of the Clebsch-Gordan coefficients the expression (6) vanishes for odd values of k. Components of such tensors are thus scalars (rank s=0) in quasi-spin space. They are diagonal in the seniority scheme. The commutation relations of the commutator in (6) with S_j^+ vanishes and hence, components of even rank tensor operators are the zero components of rank s=1 tensors. From these facts, relations of matrix elements of such operators for different values of n can be derived[7].

Let us now consider rotationally invariant Hamiltonians which are diagonal in the seniority scheme. Scalar products of single nucleon operators can be rewritten as a linear combination of two-body operators and single nucleon operators (the latter, being scalars in ordinary space, must be proportional to the number operator). Hence, any two body interaction has terms with ranks s=0, s=1 and s=2 in quasi-spin space. If it is diagonal in the seniority scheme the s=1 and s=2 parts must be equal to a linear combination of S_j^o and $(S_j^o)^2$. Such Hamiltonians have thus the most general form

$$H = H_{s=0} + \frac{1}{2} n\, \beta + \frac{1}{2} n\,(n-1)\alpha \tag{7}$$

Matrix elements of the scalar H_s=0 do not depend on S_j^o or n. Hence we obtain for even n

$$H_{s=0}|j^n v=0\ J=0> = H_{s=0}|n=0, v=0, J=0> = H_{s=0}|0> = 0$$

and the eigenvalues of H in such states are given by

$$\frac{1}{2} n\beta + \frac{1}{2} n\,(n-1)\alpha$$

For odd nuclei we obtain

$$H_{s=0}|j^n v=1\ J=j> = H_{s=0}|j^v v=1\ J=j> = (H - \frac{1}{2}\beta)|j^v v=1\ J=j> = -\frac{1}{2}\beta|j^v v=1\ J=j>$$

since the eigenvalue of the two body H is zero for the single nucleon state. Hence the eigenvalue of H in ground states of j^n configurations with n odd are given by

$$\frac{1}{2} n\,\beta + \frac{1}{2} n\,(n-1)\alpha - \frac{1}{2}\beta = \frac{1}{2}(n-1)\,\beta + \frac{1}{2} n\,(n-1)\alpha$$

Binding energies of odd semi-magic nuclei are obtained by adding n single nucleon energies C and the binding energies of the nucleus with no j nucleons. We thus obtain[7)]

$$\text{I} \qquad \text{B.E.}(j^n) = \text{B.E.}(n=0) + nC + \frac{1}{2} n\,(n-1)\alpha + \frac{n - \frac{1}{2}(1-(-1)^n)}{2}\beta \tag{8}$$

This simple result holds for <u>any</u> interaction which is diagonal in the seniority scheme. It has linear and quadratic terms as well as a pairing term. For the pairing operator (4), eq. (8) holds with $\alpha = -\frac{1}{2}$, $\beta = j+1$. For a δ-potential it holds with α=0. It was shown that if (8) holds for the lowest J=0 states of all j^n configurations then the Hamiltonian is diagonal in the seniority scheme[8)]. Any two body interaction with this property can be expanded in terms of scalar products of odd rank tensor operators and a k=0 tensor. Finally, the mass formulae (8) agrees well with experiment where the coefficient α is small and repulsve whereas the coefficient β of the pairing term is large and attractive[9)].

Another important property of Hamiltonians which are diagonal in the seniority scheme follows directly from (7). Differences of energy levels are determined only by $H_{s=0}$ since the other terms contribute the same amount to all levels. The eigenvalues of the quasi-spin scalar $H_{s=0}$ depend only on v and are independent of n. Hence

II Energy level spacings are independent of n.

Also this property agrees well with the experimental data.

Let us now look at measured spectra of actual nuclei where the seniority scheme can be utilized. This will also demonstrate how the shell model can be used in practice[9)]. We consider first ^{90}Zr with 50 neutrons in closed shells. The40 protons do not form a closed shell but their excitation energy is rather high and we are justified in considering just neutron configurations beyond 50 [10,11)]. The ground state of ^{91}Zr has $J=5/2^+$ due to a $2d_{5/2}$ single neutron. In ^{92}Zr there are 2^+ and 4^+ levels which may be due to the $(2d_{5/2})^2$ neutron configuration. Their positions above the ground state are determined by matrix elements of the <u>effective interaction</u> of the two valence neutrons. There is no reliable way to calculate those matrix elements from the interaction between free nucleons with techniques of many-body theory. Instead we determine them directly from the experimental levels of ^{92}Zr. How do we know that this procedure is consistent, that the configuration is the one assumed? There are several consistency checks that energies of other nuclei must satisfy. The simplest is to consider ^{94}Zr where the $(2d_{5/2})^4 \equiv (2d_{5/2})^{-2}$ is expected. Energy level spacings in a two hole configuration should be equal to those in the two particle configuration. Looking at the measured enregy levels of ^{94}Zr we see that this criterion is rather accurately satisfied.

In ^{95}Zr only the ground state beongs to the $(2d_{5/2})^5 \equiv (2d_{5/2})^{-1}$ configuration whereas in ^{96}Zr the $2d_{5/2}$ neutron orbit is completely filled. The nucleus of interest is ^{93}Zr where the $(2d_{5/2})^3$ configuration has states with J=3/2,5/2,9/2 allowed by the Pauli principle. To calculate their positions we notice that all $(2d_{5/2})^n$ levels are determined by three matrix elements $V_J = \langle (2d_{5/2})^2 JM|V|(2d_{5/2})^2 JM \rangle$ with J=0,2,4. <u>Any</u> two body interaction which reproduces the given V_o, V_2 and V_4 can be used to calculate all energy levels. We adopt the simple two-body interaction

$$V_{12} = a + 2b(j_1 \cdot j_2) + cq_{12} \tag{9}$$

where q_{12} is twice the pairing operator (4). The eigenvalues of the interaction (9) in a state with given v and J in the $(2d_{5/2})^n$ configuration are given by

$$\frac{n(n-1)}{2} a+b[J(J+1)-n\frac{35}{4}]+ c\frac{n-v}{2}(8-n-v) \qquad (10)$$

For n=2, putting in (10) J=0, v=0 we obtain V_o, for J=2, v=2 we obtain V_2 whereas putting J=4, v=2 yields V_4. Hence a,b and c are determined by V_o, V_2 and V_4. The parameter a contributing equally to all V_J, the energy differences in ^{92}Zr determine the following values of the parameters

$$b = 0.040 \text{ MeV} \qquad c = -0.116 \text{ MeV}$$

In the $(2d_{5/2})^3$ configuration, the J=5/2 state has seniority v=1 whereas the J=3/2 and J=9/2 states have v=3. Putting these values, as well as n=3, in (10) we obtain that the ground state is predicted to have J=5/2 and the J=3/2 state to be 0.26 MeV about it. This is in excellent agreement with experiment (0.267 MeV). The J=9/2 state is predicted at 1.10 MeV above the ground state and has not yet been discovered by experiment. Had the effective interaction been well approximated by the pairing interaction, the J=3/2 and J=9/2 should have been degenerate at 2/3 of the (degenerate) position of the J=2 and J=4 levels. This is far from the actual situation. The J=3/2 level with v=3 is rather close to the v=1, J=5/2 ground state. If we do not force naive ideas on the data, the position of the J=3/2 level (and J=j-1 levels in other cases) is simply explained by the effective interaction determined from experiment.

In addition to level spacings we can look at binding energies of neutron $(2d_{5/2})^n$ configurations. If we use the expression (8) we see that neutron <u>separation energies</u> lie on two straight and parallel lines, for even and odd values of n respectively. This prediction is in very good agreement with experiment. The odd-even variation is determined by the coefficient β of the pairing term. The value that fits the experimental binding energies is $\beta = -1.625$ MeV. It can be simply related to the level spacings by $\beta = \frac{7}{6}(V_o - \frac{5}{14}V_2 - \frac{9}{14}V_4)$. Using the ^{92}Zr levels we obtain for β the value -1.51 MeV, in fair agreement with the one dierived from binding energies. The odd even variation is thus related to a well defined experimental magnitude and not to some vague notion of an "energy gap".

The use of the seniority scheme greatly simplified the results obtained above. Still they could be derived directly for <u>any</u> two body interaction within $d^n_{5/2}$ configurations. There are no two states in such configurations with the same value of J. Thus, the seniority quantum number is unnecessary, or in other words, <u>any</u> two-body interaction is diagonal in the seniority scheme. In $(7/2)^n$

configurations (of identical nucleons) the situation is more complicated but it can be shown[7)] that also there any two body interaction is diagonal in the seniority scheme. To test the validity of seniority for the effective interaction, we must look at $j \geq 9/2$. Evidence for configurations with protons in the $1g_{9/2}$ orbit, in the $1h_{9/2}$ orbit and also in the $1h_{11/2}$ orbits was obtained by various experiments. The effective interactions derived from the experimental data have negligible matrix elements between states with different seniorities. Level spacings in even $(1g_{9/2})^n$ configurations in nuclei with 50 neutrons clearly display this property. Apart from the J=0 states which are strongly perturbed by $(2p_{1/2})^2$ pairs, spacings between v=2 levels with J=2,4,6,8 are independent of n to a good accuracy[11)].

In other semi-magic nuclei, the identical nucleons do not occupy a definite orbit and shell model configurations are strongly mixed. Still, the properties I and II obtained above for the seniority scheme are observed in actual even-even nuclei. Obviously a generalization of seniority is called for. The simplest generalization is to consider instead of the pair creation operator S_j^+, thesum over several orbits

$$S^+ = \Sigma S_j^+ \tag{11}$$

This sum is also a component of an angular momentum vector - the total quasi-spin of the system. The component (11) satisfies with ΣS_j^- and ΣS_j^o the same commutation relations as for a single j-orbit. Hence the seniority scheme can be defined by eigenstates of the pairing operator $(\Sigma S_j^+)(\Sigma S_j^-)$ and it has all properties discussed above. This quasi-spin scheme is however too restrictive, the coefficients of all S_j^+ in (11) being equal. In particular, level spacings of odd nuclei are also independent of n in sharp contradiction with experiment. A meaningful generalization of seniority would emerge if instead of (11) we consider the pair creation operator

$$S^+ = \Sigma \alpha_j S_j^+ \tag{12}$$

with unequal α_j coefficients. In that case, (12) is no longer an angular momentum component. The situation is more complicated mathematically but it is also of greater interest for nuclear physics.

In analogy with v=0 states in the seniority scheme we consider ground states of semi-magic nuclei to be well approximated by

$$(S^+)^N|0\rangle = (\Sigma \alpha_j S_j^+)^N|0\rangle \tag{13}$$

The wave function (13) looks like the part with 2N nucleons projected from the Bardeen-Cooper-Schrieffer (BCS) wave function of the ground state of a spherical

$$(S^+)^{N-1}D^+_M|0> \tag{19}$$

States like (19) were considered with variational parameters $\beta_{jj'}$, along with the α_j ("broken pair approximation")[12]. Also in this case we adopt a different approach. We consider the coefficients $_{jj'}$, to be constant independent of N and investigate the condition on shell model Hamiltonians for which (19) are eigenstates. We first obtain, for N=1,

$$HD^+_M|0> = V_2D^+_M|0> \tag{20}$$

and from N=2 we deduce the condition

$$[[H,S^+],D^+] = \quad S^+D^+ \tag{21}$$

For any case of interest it can be shown that the coefficient in (21) must be equal to in (16). Using (15) and (16) we now prove by induction the result

$$H(S^+)^{N-1}D^+|0> = E_{N-1}(S^+)^{N-1}D^+|0> + (S^+)^{N-1}HD^+|0> + (N-1)(S^+)^{N-2}[[H,S^+],D^+]|0>$$

If conditions (20) and (21) are satisfied we obtain for any N

II' $$H(S^+)^{N-1}D^+|0> = (E_N+V_2-V_o)(S^+)^{N-1}D^+|0> \tag{22}$$

with E_N given by (17). Thus, the energy separation between the ground state with generalized seniority v=0 and the first excited J=2 state is constant throughout the major shell. It is independent of N as is the case for seniority in a single j-orbit or the quasi-spin scheme. In the case of unequal α_j coefficients, however, this property is restricted to even nuclei. A beautiful example is furnished by Sn isotopes where neutron number changes from 52 to 80. The 0-2 separation is remarkably constant in sharp contrast to crossing of levels of odd Sn isotopes[11)]. In the case of unequal α_j coefficients the various orbits are not filled at the same rate even though α_j are constant.

The condition (16) as well as (21) put stringent restrictions on shell model Hamiltonians. Those conditions are equivalent to many non-linear equations which should be satisfied by the matrix elements of single nucleon terms and two-body interactions. There are non-trivial examples of such Hamiltonians but not any popular interaction can satisfy the conditions. For example, neither the pairing interaction nor the surface delta interaction satisfy those conditions with non-degenerate single nucleon energies. In fact, they do not yield a 0-2 separation independent of nucleon number. These conditions lead to a simple binding energy formula and constant 0-2 separation but the reverse is not strictly correct. For example, the form (17) of the nucleus (with $\alpha_j = v_j/u_j$). In fact, such wave functions were considered as variational functions for various interactions[12]. Our approach is very different. We take (13) with constant coefficients α_j, independent of N, to be the exact eigenstate of the one-body and two-body shell model Hamiltonian and see what the consequences are. Thus, we study the conditions on a shell model Hamiltonian H (normalized by $H|0\rangle = 0$) satisfying

$$H(S^+)^N|0\rangle = E_N(S^+)^N|0\rangle \tag{14}$$

Putting N=1 we obtain the condition

$$HS^+|0\rangle = V_o S^+|0\rangle \tag{15}$$

where $V_o = E_1$ is the eigenvalue for the state with one J=0 pair.

If we put N=2 in (14) we obtain

$$H(S^+)^2|0\rangle = [H,S^+]\ ^+|0\rangle + S^+HS^+|0\rangle = [[H,S^+],S^+]|0\rangle + 2V_o(S^+)^2|0\rangle$$

Hence a two pair state will be an eigenstate if

$$[[H,S^+],S^+] = \Delta(S^+)^2 \tag{16}$$

where Δ is a numerical coefficient. Once (15) and (16) are satisfied by H, the eigenstate equation (14) holds for any N. By induction we first derive

$$H(S^+)^N|0\rangle = N(S^+)^{N-1}[H,S^+]|0\rangle + \tfrac{1}{2}N(N-1)(S^+)^{N-2}[[H,S^+],S^+]|0\rangle$$

Using (15) and (16) we obtain for any N

I' $$H(S^+)^N|0\rangle = (NV_o + \tfrac{1}{2}N(N-1)\Delta)(S^+)^N|0\rangle \tag{17}$$

The eigengalues in (17) are the generalization of those in (8) for even nuclei. They give good description of binding energies. In particular, (17) demonstrates that when several orbits are being filled together according to the prescription (14), no subshell effects can be observed. This holds for any values of the α_j coefficients provided they are constant throughout the major shell. This feature, which has been known experimentally, is a remarkable consequence of the structure (14) of eigenstates.

To construct excited states with J=2 corresponding to seniority v=2 states we first define a J=2 pair creation operator

$$D_M^+ = \sum_{j \geq j'} \beta_{jj'} \sum_{mm'} (jm\ j'm'|jj'2M) a^+_{jm} a^+_{j'm'} \tag{18}$$

We now consider shell model Hamiltonians with exact eigenstates given by

number of fermion states]. Beyond N=Ω the boson states do not correspond to allowed fermion states.

The general idea is that we do not attempt to expand fermion states and operators in terms of boson states and operators. We establish a correspondence between fermion states (14) and (19) with boson states $(s^+)^N|0\rangle$ and $(s^+)^{N-1}d^+_M|0\rangle$. We then construct boson operators, like (23), which have the same matrix elements as fermion operators between corresponding fermion states.

Shell model states constructed with several D^+ operators should be more carefully handled. The state $(D^+\cdot D^+)|0\rangle$, for instance, is generally not orthogonal to the state $(S^+)^2|0\rangle$. To obtain an orthogonal scheme we subtract from any fermion state of the form $(S^+)^{n_s}(D^+)^{n_d}_{\gamma JM}|0\rangle$ any component with lower generalized seniority of the form $(S^+)^{n_s+1}B^+_{JM}|0\rangle$. The label γ indicates the various ways in which n_d operators D^+ can be coupled to yield a given J (and M) and B^+ creates a state with $2(n_d-1)$ nucleons. The resulting nucleon states may then correspond to boson states $(s^+)^{n_s}(d^+)^{n_d}_{\gamma JM}|0\rangle$ which belong to an orthognal set of states. Beyond the middle of the major shell, however, all nuclear states have the form $S^+B^+|0\rangle$. Hence, for $n_d>\Omega$, states of the form $(D^+)^{n_d}_{\gamma JM}|0\rangle$ will be completely removed by the seniority projection. As a result, using the symmetry between nucleons and holes, the boson states $(s^+)^{n_d}(d^+)^{n_d}_{\gamma JM}|0\rangle$ for $N=n_s+n_d$ correspond to fermion states constructed from J=0 and J=2 hole pairs.

The use of the boson model for identical valence nucleons does not simplify matters very much. In fact, the important results I' and II' were derived directly in the fermion space for shell model Hamiltonians. It is only in cases where there are both valence protons and valence neutrons outside closed shells that the boson model can demonstrate its great power.

The use of generalized seniority for semi-magic nuclei led to a very big simplification in the diagonalization of the Hamiltonian sub-matrix in the shell model space considered. The J=0 sub-matrix in the case of Sn isotopes can reach an order of more than 50,000 and J=2 matrix can be bigger than 250,000 × 250,000. The prescription (14) and (19) select one of all the myriads of states for which the Hamiltonian matrix is diagonal. Generalized seniority offers a very efficient truncation of the giant matrices of the full fledged shell model calculation. The need for truncation is even more severe in the case of nuclei with both valence protons and neutrons. The size of the matrices then becomes amazingly large, the order reaching 10^{15} and higher. In order to obtain the low-lying levels, which are the more collective ones, we must find a truncation scheme or a coupling scheme which will lead to a manageable problem. Such a coupling scheme should be constructed according to properties of the proton-neutron interaction (the T=0 part of the nuclear interaction). In most nuclei with collective spectra valence protons and valence neutrons occupy different major shells. The low-lying levels correspond then to states with definite isospin and thus there is no need nor advantage in using the isospin formalism.

In looking for an appropriate coupling scheme we can check whether it could be based on seniority. To check the applicability of seniority to the T=0 part of the nuclear interaction we can look at spectra of simple j^n configurations. Valence protons and neutrons in the $1g_{9/2}$ orbit furnish a nice example. If the interaction is diagonal in the seniority scheme spacings of levels with the same seniority quantum numbers should have the same spacings. We can compare the spacings of the J = 2,4,6,8 in ^{92}Mo and in ^{88}Y. The levels of ^{88}Y have very different spacings, the spectrum being almost inverted[11)]. This clearly demonstrates a drastic breakdown of the seniority scheme. This should be no surprise. Had seniority been a good quantum number also for T=0 interactions all level spacings would have been constant as in Sn isotopes. Once protons and neutrons occupy valence orbits there is a dramatic lowering of the J=2 level as seen in A=122 nuclei.

Seniority can be broken by a quadrupole-quadrupole interaction or other even rank tensor interactions. The importance of the quadrupole interaction was emphasized many years ago by Bohr and Mottelson. They argued that the "pairing interaction" leads to spherical nuclei whereas the quadrupole interaction leads to deformation and rotational spectra. They further argued that seniority prevails for a small number of valence nucleons whereas for a larger number the "long range" quadrupole interaction dominates. We agree that the nature of the spectrum is determined by the interplay of seniority conserving interactions and the quadrupole interaction. We can, however, add a precise definition of this interplay. As clearly seen from the data, interactions between identical nucleons (T=1 interactions) are diagonal in the seniority scheme. No matter how many identical valence nucleons are present, no deformation develops (like in Sn isotopes). The quadrupole interaction is effective only in the T=0 part and therefore can lead to deformation only when both valence protons and neutrons are active. The quadrupole term may be an important component in the interaction between identical nucleons but its seniority breaking effect is completely cancelled by other even rank interactions. In fact, any interaction diagonal in the seniority scheme can be expressed in terms of <u>odd</u> tensors and a scalar (k=0) only[7,8)]. Only when all states of a valence proton and a valence neutron are considered, is the expansion of the two body interaction in terms of scalar products of irreducible tensors unique. Only then can a quadrupole interaction have a decisive effect on the nature of the resulting spectrum.

eigenvalues of a shell model Hamiltonian for every N imposes many conditions on thematrix elements. A simple and natural solution of these conditions is given by (16). In fact, if the α_j coefficients are equal, the condition (16) is an exact consequence of the form (17) of eigenvalues. In general there may be other solutions (there are examples) but the conditions on matrix elements, while different, are as stringent.[8] There is no sense in checking the structure of groundstate wave functions to see whether they agree with the prescription (14). Instead any suggested effective interaction should be first tested whether it leads to a constant 0-2 separation with the given single nucleon energies. It seems that if we insist on a shell model description which fits the data of semi-magic nuclei in terms of valence identical nucleons, the adopted effective Hamiltonian should most probably satisfy conditions (16) and (21).

We now are in the position to make the first contact with the boson model. We can obtain the eigenvalues in (17) and in (22) from a rather simple boson Hamiltonian

$$V_o s^+ s + V_2 (d^+ \cdot \tilde{d}) + \tfrac{1}{2}\Delta (s^+)^2 s^+ + \Delta s^+ s (d^+ \cdot \tilde{d}) \tag{23}$$

In (23), s^+, d^+_μ are creation operators for $\ell=0$ and $\ell=2$ bosons, s and d_μ are the corresponding annihilation operators and $\tilde{d}_\mu = (-1)^\mu d_{-\mu}$. The eigenstates of (23) with eigenvalues (17) are the boson states $(s^+)^N|0>$ and those with eigenvalues (22) are $(s^+)^{N-1} d^+_\mu |0>$. It is remarkable that a (simple) boson Hamiltonian can reproduce eigenvalues of a rather complicated system of fermions. The reason is that the boson Hamiltonian (23) reproduces a tiny fraction of the eigenvalues of the shell model Hamiltonian.

It is important to state clearly the principles of introducing the boson model (23) for valence identical nucleons which may be very different from those of other approaches.

a) The expression (23) yields exactly certain eigenvalues of the fermion shell model Hamiltonian. It is not the first term in a "boson expansion" of that Hamiltonian.

b) The boson creation operators s^+ and d^+ do not create fermion pairs neither exactly nor approximately. They obey with s and d the usual Bose commutation relation whereas the commutation relations involving the fermion S^+ and D^+ are very complicated. The boson-boson terms in (23), as well as the single boson terms replace fermion-fermion interactions in the shell model Hamiltonian and not four fermion interactions.

c) In (23) only single boson and two boson terms appear but still it reproduces eigenvalues of fermion states (14) and (19) where the Pauli principle is strictly obeyed. This is true only as long as $N \leq \Omega$ $(2\Omega = \Sigma(2j+1)$ is the total

In view of the considerations made above we adopt a quadrupole-quadrupole interaction between protons and neutrons. A complete set of statesffor nuclei with valence protons and neutrons can be simply constructed. We first construct a complete set of states of the valence protons $|\alpha_\pi J_\pi M_\pi\rangle$ where α_π are the additional quantum numbers, or labels, which distinguish between the various orthogonal states with the same values of J_π and M_π. A similar complete set of states of the valence neutrons is given by $|\alpha_\nu J_\nu M_\nu\rangle$. There may be many thousands or even hundreds of thousands of states with a given value of J_π(or J_ν). We now couple all proton states with all neutron states by Clebsch-Gordan coefficients to obtain all states of the combined systems

$$|J_\pi J_\nu JM\rangle = \sum_{M_\pi M_\nu} (J_\pi M_\pi J_\nu M_\nu|J_\pi J_\nu JM)|\alpha_\pi J_\pi M_\pi\rangle|\alpha_\nu J_\nu M_\nu\rangle \quad (24)$$

The set of states (24) may be very large and may contain 10^{15} states or more for given total J and M. In this set matrix elements of the quadrupole interaction can be conveniently expressed by

$$\langle\alpha_\pi\alpha_\nu J_\pi J_\nu JM|Q_\pi^{(2)}\cdot Q_\nu^{(2)}|\alpha_\pi'\alpha_\nu' J_\pi' J_\nu' JM\rangle = (-1)^{J_\nu+J_\pi'+J} \begin{Bmatrix} J_\pi & J_\nu & J \\ J_\nu' & J_\pi' & 2 \end{Bmatrix} (\alpha_\pi J_\pi||Q_\pi^{(2)}||\alpha_\pi' J_\pi')(\alpha_\nu J_\nu||Q_\nu^{(2)}||\alpha_\nu' J_\nu') \quad (25)$$

The elements (25) belong to a huge matrix which must be drastically truncated if any reasonable attempt is made to diagonalize it.

An efficient truncation scheme for the low lying eigenstates should include those states (24) which have between them large matrix elements of the quadrupole interaction. The formula (25) indicates that those are the states constructed from proton states which have between them large matrix elements of the proton quadrupole operator and neutron states which have large matrix elements of the neutron quadrupole operator. In the absence of proton neutron interactions the lowest state would have been

$$(S_\pi^+)^{n_\pi}(S_\nu^+)^{n_\nu}|0\rangle$$

This state has non-vanishing matrix elements only with the state

$$Q_\pi^{(2)}(S_\pi^+)^{n_\pi}\cdot Q_\nu^{(2)}(S_\nu^+)^{n_\nu}|\rangle \quad (26)$$

If the quadrupole operators in (25) can be approximated by the quadrupole operators defined by[13,14)]

$$D_\pi^+ = \tfrac{1}{2}[Q_\pi, S_\pi^+] \qquad D_\nu^+ = \tfrac{1}{2}[Q_\nu, S_\nu^+] \quad (27)$$

Then the state (26) is equal to

$$4n_\pi n_\nu (D_\pi^+ \cdot D_\nu^+)(S_\pi^+)^{n_\pi -1}(S_\nu^+)^{n_\nu -1}|0\rangle \tag{28}$$

To see which states have non-vanishing matrix elements with the state (28) we consider separately proton and neutron states. We obtain in either case

$$(Q^{(2)} x D^+ (S^+)^{n-1})_M^{(J)}|0\rangle = 2(n-1)(S^+)^{n-2}(D^+ x D^+)_M^{(J)}|0\rangle + (S^+)^{n-1}(Q^{(2)} x D^+)_M^{(J)}|0\rangle \tag{29}$$

The first term on the r.h.s. of (29) is constructed with S^+ and D^+ operators. The second term may lead to states with a different structure. If the quadrupole operator satisfies also the relation

$$[Q_\mu^{(2)}, D_{\mu'}^+] = \sum_M [(2\mu 2\mu'|220M)C_o S^+ + (2\mu 2\mu'|222M)C_2 D_M^+] \tag{30}$$

then the operation of $Q_\pi^{(2)} \cdot Q_\nu^{(2)}$ will lead only to states constructed from S^+ and D^+ pairs of protons and neutrons (S-D space). In certain simple models (30) holds exactly[14,15]. In actual cases it could be a good approximation.

Replacing the huge matrices (or order 10^{15} or more) by matrices constructed in the S-D space gives a tremendous reduction in size. The matrices truncated in this way could be diagonalized without difficulty. Still, fermion states in the S-D space are difficult to handle and matrix elements difficult to calculate due to the Pauli principle. It is now that the boson model becomes very useful. Instead of using fermion states constructed with S_π^+, D_π^+, S_ν^+ and D_ν^+ operators we use corresponding boson states constructed with s_π^+, d_π^+, s_ν^+ and d_ν^+ operators. We saw how it can be done for the proton and neutron parts of the shell model Hamiltonian. We should now find a boson operator which will correspond to the proton neutron quadrupole interaction. Matrix elements of the boson quadrupole operator between boson states should be equal to those of the fermion quadrupole operator between corresponding fermion states. The Hamiltonian has the general form

$$H_\pi + H_\nu - Q_\pi^{(2)} \cdot Q_\nu^{(2)} \tag{31}$$

The proton and neutron parts, H_π and H_ν, can be taken from (23) or simpler still, taken as $\varepsilon(d_\pi^+ \cdot \tilde{d}_\pi)$ and $\varepsilon(d_\nu^+ \cdot \tilde{d}_\nu)$ where $\varepsilon = V_2 - V_o$. The simplest boson operator is given by

$$Q_\mu^{(2)} = \alpha_2 (d_\mu^+ s + s^+ \tilde{d}_\mu) + \beta_2 (d^+ x \tilde{d})_\mu^{(2)} = \chi [d_\mu^+ s + s^+ \tilde{d}_\mu + \chi (d^+ x \tilde{d})_\mu^{(2)}] \tag{32}$$

In some simple models the operator (32) with some constant coefficients α_2 and β_2 (or κ and χ) actually yields matrix elements equal to those in the fermion space. In other cases κ_π, χ_π must vary with n_π and κ_ν, χ_ν must vary with n_ν. It is impossible to determine this dependence without very detailed knowledge of the effective interactions in the shell model. Instead of making bad guesses we can determine them from experiment in a consistent fashion.

The shell model description of the boson model led us to a model with two kinds of bosons - proton s ,d and neutron s ,d bosons. It is called IBA-2 to distinguish it from the first verion IBA-1 with one kind of s,d bosons. It is more detailed than IBA-1 and has been successfully used to describe various collective spectra in many nuclei[2,16]. Its success is due to its incorporating the two important ingredients of the effective nuclear interaction. Those are seniority conserving interaction between identical nucleons and the strong quadrupole interaction between protons and neutrons. The space of IBA-1 cannot accommodate these two competing forces separately like the situation in the collective model. Still in certain cases, IBA-2 can reduce to IBA-1[4].

The interacting boson model IBA-2 has a shell model basis. It can be approximated by IBA-1 which is closely related to the collective model. This way a connection can be established between the two descriptions. There are many extensions of the boson model and it is encouraging that there is much space for improvements and further developments

References

1. A. Arima and F. Iachello, Phys. Rev. Lett. 35 (1975) 1069
2. Interacting Bosons in Nuclear Physics, F. Iachello ed., Plenum, New York (1979).
3. A. Arima and F. Iachello, Phys. Rev. Lett. 40 (1978) 385.
4. A. Arima, T. Otsuka, F. Iachello and I. Talmi, Phys. Lett. B66 (1977) 205; T. Otsuka, A. Arima, F. Iachello and I. Talmi, Phys. Lett. B76 (1978) 141.
5. G. Racah, Phys. Rev. 63 (1943) 367.
6. A. K. Kerman, Ann. Phys. (N.Y.) 12 (1961) 300.
7. Most spectroscopic derivations and results are found in A. de-Shalit and I. Talmi, Nuclear Shell Theory Academic Press, New York (1963).
8. I. Talmi, Nucl. Phys. 172 (1971) 1.
9. I. Talmi, Rev. Mod. Phys. 34 (1962) 704.
10. I. Talmi, Phys. Rev. 126 (1962) 2116.
11. I. Talmi in From Nuclei to Particles, A. Molinari Ed., North-Holland, Amsterdam (1982).
12. B. Lorazo, Nucl. Phys. A153 (1970) 255; Y.K. Gambhir, A. Rimini and T. Weber, Phys. Rev. 188 (1969) 1573.
13. S. Shlomo and I. Talmi, Nucl. Phys. A198 (1972) 81; I. Talmi, Riv. Nuovo Cimento 3 (1973) 85.
14. J.N. Ginocchio and I. Talmi, Nucl. Phys. A337 (1980) 431.
15. J.N. Ginnochi, Phys. Lett. B79 (1978) 173); B85 (1979) 9; Ann. Phys. (N.Y.) 126 (1980) 234.
16. Interacting Bose-Fermi Systems in Nuclei, F. Iachello Ed., Plenum, New York (1981).

AN INTRODUCTION TO THE INTERACTING BOSON-FERMION MODEL

F. Iachello

A.W. Whright Nuclear Structure Laboratory, Yale University, New Haven, Ct 06520

1. INTRODUCTION

Spectra of odd-even medium mass and heavy nuclei are rather complex since they arise from the interplay between collective and single particle degrees of freedom. Their properties can be discussed in terms of simple models only in a limited number of cases, as, for example, in spherical nuclei (where the shell model can be applied in a straight forward way), or in nuclei with a rigid axially symmetric deformation (where the deformed shell model, or Nilsson model [1], can be used). Neither of these models can, however, be applied to the large majority of nuclei, those forming the transitional classes. In the last few years, a model for odd-even nuclei has been introduced which is, on one side relatively simple, but which, on the other side, is able to describe the large variety of observed spectra. In this model, the collective degrees of freedom are described by bosons, while the single particle degrees of freedom are described by fermions, hence the name interacting boson-fermion model [2] given to it. In these lectures, I will describe the basic features of the model concentrating my attention to those cases that can be solved analytically, without resorting to numerical calculations. These analytic results are obtained by making use of group theory and in dealing with this subject I will draw some material from previous lectures given at the Dronten Summer School [3].

2. THE INTERACTING BOSON-FERMION MODEL

In its simplest form, the interacting boson-fermion model assumes that the low-lying spectra of odd-even nuclei are dominated by the excitations of the valence particles. Furthermore, it is assumed that, whenever possible, the valence particles join together into correlated pairs with angular momenta $J^P=0^+$ and 2^+. The pairs are treated as bosons, denoted respectively s and d, Fig.1. The subscript $\pi(\nu)$ distinguishes between protons and neutrons. In odd-even nuclei, at least one particle must be necessarily unpaired. Thus, a description of these nuclei requires the simultaneous introduction of pairs (bosons)

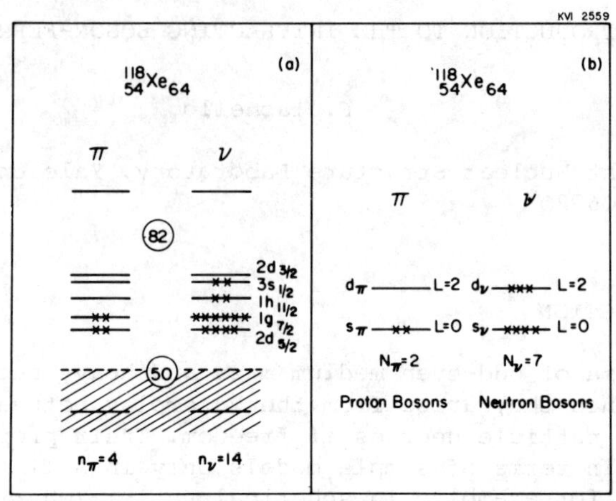

Fig. 1a. A schematic representation of the shell model problem for ${}^{118}_{54}Xe_{64}$ (n_π and n_ν are the numbers of protons and neutrons outside the major closed shell at 50).
b. The boson problem which replaces the shell model problem for ${}^{118}_{54}Xe_{64}$.

and single particles (fermions). Because of the particle-hole conjugation properties, bosons and fermions are counted from the nearest closed shells. Thus, in ${}^{115}_{55}Cs_{60}$, one has 5 valence protons, 55-50, and 10 valence neutrons, 60-50. This nucleus is treated as composed of N_π=2, N_ν=5 proton and neutron bosons and M_π=1, M_ν=0 proton and neutron fermions. On the other side, in ${}^{125}_{54}Xe_{71}$, one has 4 valence protons, 54-50, and 11 valence neutrons, 82-71. This nucleus is treated as composed of N_π=2, N_ν=5 bosons and M_π=0,M_ν=1 fermions. Since the neutrons are hole states, a bar is sometimes placed over their number, $N_\nu=\bar{5}$,$M_\nu=\bar{1}$. The version of the model in which a distinction is made between protons and neutrons is called interacting boson-fermion model-2. Sometimes, a simpler version is used, in which no distinction between proton and neutron degrees of freedom is made. In this version, called interacting boson-fermion model-1, the number of bosons and fermions is chosen to be $N=N_\pi+N_\nu$, $M=M_\pi+M_\nu$. In the remaining part of these lectures, I will discuss properties of the interacting boson-fermion model-1.

In order to calculate properties of nuclei, one needs to write down the appropriate operators. For energies, the appropriate operator is the Hamiltonian,H. For practical calculations, it has been found convenient to introduce a second quantized formalism. In this formalism, bosons are represented by creation ($s^\dagger, d^\dagger_\mu$ ($\mu = 0, \pm 1, \pm 2$)) and annihilation (s, d_μ) operators. The six operators $s^\dagger$, $d^\dagger_\mu$ will be altogether denoted by $b^\dagger_{\ell\mu}$, $\ell = 0, 2$. The commutation relations of these operators can be succinctly written as

$$[b_{\ell\mu}, b^\dagger_{\ell'\mu'}] = \delta_{\ell\ell'}\delta_{\mu\mu'} \tag{2.1}$$

It is well known that while the creation operators transform as spherical tensors under rotations, the annihilation operators do not. However, spherical tensors can be easily constructed by introducing the operators $\tilde{b}_{\ell,\mu} = (-)^{\ell-\mu} b_{\ell,-\mu}$. This gives $\tilde{d}_\mu = (-)^{-\mu} d_{-\mu}$ and $\tilde{s} = s$.

Similarly, one can introduce creation and annihilation operators for fermions. While in the case of the bosons the values of the angular momenta are always $J^P = 0^+$ and 2^+, in the case of the fermions they depend on the particular orbitals available to the single particle. I will therefore denote the creation operators in general as $a^\dagger_{j\mu}$ ($\mu = \pm j, \pm (j-1), \ldots, \pm \frac{1}{2}$) and the annihilation operators as $a_{j\mu}$. Here again tensor operators can be constructed from the annihilation operators by using $\tilde{a}_{j,\mu} = (-)^{j-\mu} a_{j,-\mu}$. In the situation of Fig.1, the values of j are j= 5/2, 7/2, 11/2, 3/2 and 1/2. The commutation relations of the fermion operators can be succinctly written as

$$\{a_{j\mu}, a^\dagger_{j'\mu'}\} = \delta_{jj'}\delta_{\mu\mu'} \tag{2.2}$$

With tensor operators one can construct tensor products. The tensor product of two operators $T^{(k_1)}_{\kappa_1}$, $T^{(k_2)}_{\kappa_2}$ is defined as

$$T^{(k_3)}_{\kappa_3} = \sum_{\kappa_1,\kappa_2} \langle k_1\,\kappa_1\,k_2\,\kappa_2 | k_3\,\kappa_3\rangle\, T^{(k_1)}_{\kappa_1}\, T^{(k_2)}_{\kappa_2} \quad , \tag{2.3}$$

and denoted by

$$T^{(k_3)} = [T^{(k_1)} \times T^{(k_2)}]^{(k_3)} \quad . \tag{2.4}$$

The scalar product of two tensor operators with integral k is defined as

$$(T^{(k)} \cdot U^{(k)}) = (-)^k \sqrt{2k+1}\,[T^{(k)} \times U^{(k)}]^{(0)}_0 = \sum_\kappa (-)^\kappa\, T^{(k)}_\kappa\, U^{(k)}_{-\kappa}. \tag{2.5}$$

In writing down explicitly the Hamiltonian and other operators it is convenient to introduce the bilinear products

$$B^{(\lambda)}_{\mu}(\ell,\ell') = (b^{\dagger}_{\ell} \times \tilde{b}_{\ell'})^{(\lambda)}_{\mu} \quad , \tag{2.6}$$

and

$$A^{(\lambda)}_{\mu}(j,j') = (a^{\dagger}_{j} \times \tilde{a}_{j'})^{(\lambda)}_{\mu} \quad . \tag{2.7}$$

If one assumes that the Hamiltonian contains at most one and two-body terms, one can then write its most general form as

$$H = H_B + H_F + V_{BF} \quad , \tag{2.8}$$

where

$$\begin{aligned} H_B &= E_O + \sum_{\ell} \varepsilon_{\ell}\, B^{(0)}_{O}(\ell,\ell) + \sum_{\lambda} \sum_{\ell_1,\ell_2,\ell_3,\ell_4} u^{(\lambda)}_{\ell_1\ell_2\ell_3\ell_4} \times \\ &\quad B^{(\lambda)}(\ell_1,\ell_2) \cdot B^{(\lambda)}(\ell_3,\ell_4), \\ H_F &= E'_O + \sum_{j} \eta_j\, A^{(0)}_{O}(j,j) + \sum_{\lambda} \sum_{j_1,j_2,j_3,j_4} v^{(\lambda)}_{j_1j_2j_3j_4} \times \\ &\quad A^{(\lambda)}(j_1,j_2) \cdot A^{(\lambda)}(j_3,j_4), \\ V_{BF} &= \sum_{\lambda} \sum_{\ell_1,\ell_2,j_1,j_2} w^{(\lambda)}_{\ell_1\ell_2j_1j_2}\, B^{(\lambda)}(\ell_1,\ell_2) \cdot A^{(\lambda)}(j_1,j_2). \end{aligned} \tag{2.9}$$

An even more explicit form can be obtained by writing down the B and A operators in terms of s,d bosons and a_j fermions. The boson part, H_B, has then the form

$$\begin{aligned} H_B &= E_O + \varepsilon_s\, s^{\dagger}s + \varepsilon_d \sum_{\mu} d^{\dagger}_{\mu} d_{\mu} + \sum_{L=0,2,4} \tfrac{1}{2}(2L+1)^{1/2} c_L [[d^{\dagger} \times d^{\dagger}]^{(L)} \times \\ &\quad \times [\tilde{d} \times \tilde{d}]^{(L)}]^{(0)} + \frac{1}{\sqrt{2}}\, \tilde{v}_2\, [[d^{\dagger} \times d^{\dagger}]^{(2)} \times [\tilde{d} \times \tilde{s}]^{(2)} + [d^{\dagger} \times s^{\dagger}]^{(2)} \times \\ &\quad \times [\tilde{d} \times \tilde{d}]^{(2)}]^{(0)} + \tfrac{1}{2}\, \tilde{v}_O [[d^{\dagger} \times d^{\dagger}]^{(0)} \times [\tilde{s} \times \tilde{s}]^{(0)} + [s^{\dagger} \times s^{\dagger}]^{(0)} \times \\ &\quad \times [\tilde{d} \times \tilde{d}]^{(0)}]^{(0)} + u_2 [[d^{\dagger} \times s^{\dagger}]^{(2)} \times [\tilde{d} \times \tilde{s}]^{(2)}]^{(0)} + \tfrac{1}{2}\, u_O [[s^{\dagger} \times s^{\dagger}]^{(0)} \times \\ &\quad \times [\tilde{s} \times \tilde{s}]^{(0)}]^{(0)} \quad . \end{aligned} \tag{2.10}$$

The explicit form of the fermion part, H_F, and of the boson-fermion interaction, V_{BF}, depends on the values of j. As an example, I shall consider the case in which j can take only one value , j=3/2. Then

$$H_F = E_0' + \eta \sum_\mu a_\mu^\dagger a_\mu + \sum_{L=1,3} \frac{1}{2}(2L+1)^{1/2} c_L' [[a^\dagger x a^\dagger]^{(L)} x$$

$$x\, [\tilde{a} x \tilde{a}]^{(L)}]^{(0)}, \qquad (2.11)$$

where $a_\mu^\dagger \equiv a_{3/2,\mu}^\dagger$, and

$$V_{BF} = w_0' [[s^\dagger x \tilde{s}]^{(0)} x [a^\dagger x \tilde{a}]^{(0)}]^{(0)} + w_2' [[d^\dagger x \tilde{s} + s^\dagger x \tilde{d}]^{(2)} x [a^\dagger x \tilde{a}]^{(2)}]^{(0)}$$

$$+ \sum_{L=0,1,2,3} w_L [[d^\dagger x \tilde{d}]^{(L)} x\, [a^\dagger x \tilde{a}]^{(L)}]^{(0)} . \qquad (2.12)$$

For given values of the coefficients ε_ℓ, $u^{(\lambda)}$; η_j, $v^{(\lambda)}$; $w^{(\lambda)}$ and of the boson and fermion numbers N and M (M=1 in odd-even nuclei) one must diagonalize the Hamiltonian, Eq.(2.8), numerically. When M=1, the two-body interaction terms $v^{(\lambda)}$ in Eq.(2.9) do not play any role and thus the only new parameters that one introduces when going from an even-even to an odd-even nucleus are the values of the single particle energies η_j. These can be taken from a shell model calculation. For example, the single proton energies in the shell 50-82 are shown in Fig.2. However, the number of new parameters contained in the boson-fermion interaction, V_{BF}, rapidly increases with the number of single particle levels considered, and phenomenological calculations become quickly impractical.

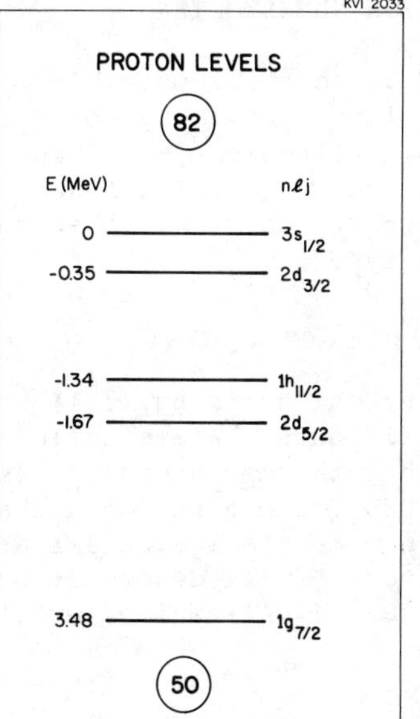

Two ways can be used to simplify the calculations.

(i) Semimicroscopic approach. Since the interacting boson-fermion model is a truncation of the shell model, one may attempt a microscopic derivation of the parameters appearing in the Hamilto

Fig.2. Proton levels in the shell 50-82. Energies are counted from the 3 $s_{1/2}$ level.

nian. This approach has suggested that three terms are particularly important in the boson-fermion interaction, called respectively monopole, quadrupole and exchange terms [2]

$$V'_{BF} = \sum_{j} A_j [[d^\dagger x\tilde{d}]^{(0)} x[a_j^\dagger \, x \, \tilde{a}_j]^{(0)}]^{(0)} +$$

$$+ \sum_{jj'j''} \Gamma_{jj'} [\{[s^\dagger x\tilde{d} + d^\dagger x\tilde{s}]^{(2)} + \chi[d^\dagger x\tilde{d}]\} x[a_j^\dagger x\tilde{a}_{j'}]^{(2)}]^{(0)}$$

$$+ \sum_{jj'j''} \Lambda_{jj'}^{j''} :[[a_j^\dagger x\tilde{d}]^{(j'')} x[d^\dagger x\tilde{a}_{j'}]^{(j'')}]^{(0)}: \quad . \qquad (2.13)$$

In this formula, the symbol (:) denotes normal ordering, that is all creation operators must be to the left of all annihilation operators when calculating matrix elements. The microscopic approach then suggests a simple dependence of the coefficients A_j, $\Gamma_{jj'}$, and $\Lambda_{jj'}^{j''}$, on the indices j,j',j" [4]. Examples of calculations of this type are given in Ref.[5].

(ii) Symmetry approach.

One may attempt to exploit the dynamic symmetries of the interacting boson-fermion Hamiltonian in order to construct analytic solutions. In these lectures, I will describe with one example how this approach is implemented.

3. GROUP STRUCTURE OF THE BOSON-FERMION HAMILTONIAN

I will keep the discussion here as short as possible referring the reader to the original articles. I begin by considering the group structure of the boson Hamiltonian, H_B. This can be studied by returning to the operators $B_\mu^{(\lambda)}(\ell,\ell')$ defined in Eq.(2.6). If one calls these operators X_a, one finds that they satisfy commutation relations of the type

$$[X_a, X_b] = \sum_c C_{ab}^c \, X_c \quad . \qquad (3.1)$$

Operators satisfying commutation relations as in Eq.(3.1) are said to form a Lie algebra. The coefficients C_{ab}^c are called Lie structure constants. One can prove that the operators $B_\mu^{(\lambda)}(\ell,\ell')$; ℓ,ℓ'=0,2 satisfy the Lie algebra of U(6), the group of unitary transformations in six dimensions. In order to indicate that this group is built with boson operators I will denote it $U^{(B)}(6)$. There are a total of 36=6^2 operators $B_\mu^{(\lambda)}(\ell,\ell')$ which, written down explicitly read

$$B_0^{(0)}(ss) = [s^\dagger x\tilde{s}]_0^{(0)} \qquad 1$$

$$B^{(0)}_{0}(dd) = [d^{\dagger} x \tilde{d}\,]^{(0)}_{0} \qquad 1$$

$$B^{(1)}_{\mu}(dd) = [d^{\dagger} x \tilde{d}\,]^{(1)}_{\mu} \qquad 3$$

$$B^{(2)}_{\mu}(dd) = [d^{\dagger} x \tilde{d}\,]^{(2)}_{\mu} \qquad 5$$

$$B^{(3)}_{\mu}(dd) = [d^{\dagger} x \tilde{d}\,]^{(3)}_{\mu} \qquad 7$$

$$B^{(4)}_{\mu}(dd) = [d^{\dagger} x \tilde{d}\,]^{(4)}_{\mu} \qquad 9$$

$$B^{(2)}_{\mu}(ds) = [d^{\dagger} x \tilde{s}\,]^{(2)}_{\mu} \qquad 5$$

$$B^{(2)}_{\mu}(sd) = [s^{\dagger} x \tilde{d}\,]^{(2)}_{\mu} \qquad \frac{5}{36=6^2}\,. \qquad (3.2)$$

I consider next the group structure of H_F. This is done by considering the operators $A^{(\lambda)}_{\mu}(j,j')$. These operators also satisfy a Lie algebra, which is that of U(m), where $m=\sum_j(2j+1)$.In order to indicate that the algebra is constructed with fermion operators, I will call it $U^{(F)}(m)$. As an example, consider the case j=3/2, m=4. In this case, there are a total of $16=4^2$ operators $A^{(\lambda)}_{\mu}(j,j')$, which, written down explicitly are

$$A^{(0)}_{0}(\tfrac{3}{2},\tfrac{3}{2}) = [a^{\dagger} x \tilde{a}]^{(0)}_{0} \qquad 1$$

$$A^{(1)}_{\mu}(\tfrac{3}{2},\tfrac{3}{2}) = [a^{\dagger} x \tilde{a}]^{(1)}_{\mu} \qquad 3$$

$$A^{(2)}_{\mu}(\tfrac{3}{2},\tfrac{3}{2}) = [a^{\dagger} x \tilde{a}]^{(2)}_{\mu} \qquad 5$$

$$A^{(3)}_{\mu}(\tfrac{3}{2},\tfrac{3}{2}) = [a^{\dagger} x \tilde{a}]^{(3)}_{\mu} \qquad \frac{7}{16=4^2}\,. \qquad (3.3)$$

Since the Hamiltonian in Eq.(2.9) is constructed with the operators B and A, one concludes that its group structure is that of the direct product $U^{(B)}(6)\otimes U^{(F)}(m)$. In particular, when j=3/2, the group structure of the interacting boson-fermion Hamiltonian is $U^{(B)}(6)\otimes U^{(F)}(4)$.

Once the full algebraic structure of the problem has been identified, the next step is to find all possible subalgebras of the full algebra. These are subsets of operators which are closed with respect to commutations. The subalgebras of the boson algebra $U^{(B)}(6)$ are discussed in Ref.[6]. Here they will be briefly reviewed for completeness. There are three possible chains of subalgebras.

Boson subalgebra I

Delete from the 36 operators in (3.2) the 11 operators $B^{(0)}_0(ss)$, $B^{(2)}_\mu(sd)$, $B^{(2)}_\mu(ds)$. The remaining 25 operators close under the algebra of U(5).

A) U(5)

$$
\begin{array}{lll}
B^{(0)}(dd) = [d^\dagger \times \tilde{d}]^{(0)}_0 & 1 & \\
B^{(1)}_\mu(dd) = [d^\dagger \times \tilde{d}]^{(1)}_\mu & 3 & \\
B^{(2)}_\mu(dd) = [d^\dagger \times \tilde{d}]^{(2)}_\mu & 5 & \\
B^{(3)}_\mu(dd) = [d^\dagger \times \tilde{d}]^{(3)}_\mu & 7 & \\
B^{(4)}_\mu(dd) = [d^\dagger \times \tilde{d}]^{(4)}_\mu & \underline{9} & (3.4) \\
 & 25=5^2 \quad . &
\end{array}
$$

Delete from the 25 operators in (3.4), the 15 operators $B^{(0)}_0(dd)$, $B^{(2)}_\mu(dd)$, $B^{(4)}_\mu(dd)$. The remaining 10 operators close under the algebra of O(5), the orthogonal group in five dimensions.

B) O(5)

$$
\begin{array}{lll}
B^{(1)}_\mu(dd) = [d^\dagger \times \tilde{d}]^{(1)}_\mu & 3 & \\
B^{(3)}_\mu(dd) = [d^\dagger \times \tilde{d}]^{(3)}_\mu & \underline{7} & (3.5) \\
 & 10=5\times4/2 \quad . &
\end{array}
$$

Delete from the 10 operators in (3.5), the 7 operators $B^{(3)}_\mu(dd)$. The remaining 3 operators close under the algebra of O(3), the ordinary rotation group

C) O(3)

$$B^{(1)}_{\mu}(dd) = [d^{\dagger} x \tilde{d}]^{(1)}_{\mu} \qquad \frac{3}{3=3x2/2} \quad . \qquad (3.6)$$

Finally, delete from the 3 operators in (3.6), the 2 operators $B^{(1)}_{\pm 1}(dd)$. The remaining operator is a generator of O(2), the group of rotations around the z-axis.

D) O(2)

$$B^{(1)}_{0}(dd) = [d^{\dagger} x \tilde{d}]^{(1)}_{0} \qquad 1 \quad . \qquad (3.7)$$

Thus, one possible chain of subalgebras is

$$U^{(B)}(6) \supset U^{(B)}(5) \supset O^{(B)}(5) \supset O^{(B)}(3) \supset O^{(B)}(2) . \quad (I) \quad (3.8)$$

Boson subalgebra II
Consider the operators

A) U(3)

$$B^{(0)}_{0}(ss) + \sqrt{5}\, B^{(0)}_{0}(dd) = [s^{\dagger} x \tilde{s}]^{(0)}_{0} + \sqrt{5}[d^{\dagger} x \tilde{d}]^{(0)}_{0} \qquad 1$$

$$B^{(1)}_{\mu}(dd) = [d^{\dagger} x \tilde{d}]^{(1)}_{\mu} \qquad 3$$

$$B^{(2)}_{\mu}(ds) + B^{(2)}_{\mu}(sd) - \frac{1}{2}\sqrt{7}\, B^{(2)}_{\mu}(dd) = [d^{\dagger} x \tilde{s} + s^{\dagger} x \tilde{d}]^{(2)}_{\mu} - \frac{1}{2}\sqrt{7}[d^{\dagger} x \tilde{d}]^{(2)}_{\mu} \qquad \frac{5}{9=3^2} \qquad (3.9)$$

These operators close under commutation and form the algebra of U(3).
Obvious subalgebras are now

B) O(3)

$$B^{(1)}_{\mu}(dd) = [d^{\dagger} x \tilde{d}]^{(1)}_{\mu} \qquad 3 \qquad (3.10)$$

and

C) O(2)

$$B^{(1)}_{0}(dd) = [d^{\dagger} x \tilde{d}]^{(1)}_{0} \qquad 1 \quad . \qquad (3.11)$$

Thus, a seconf possible chain of subalgebras is

$$U^{(B)}(6) \supset U^{(B)}(3) \supset O^{(B)}(3) \supset O^{(B)}(2) \quad . \qquad \text{(II)} \qquad (3.12)$$

Boson subalgebra III.
Consider the operators

A) O(6)

$$B^{(1)}_{\mu}(dd) = [d^{\dagger} x \tilde{d}]^{(1)}_{\mu} \qquad 3$$

$$B^{(3)}_{\mu}(dd) = [d^{\dagger} x \tilde{d}]^{(3)}_{\mu} \qquad 7$$

$$B^{(2)}_{\mu}(ds) + B^{(2)}_{\mu}(sd) = [d^{\dagger} x \tilde{s} + s^{\dagger} x \tilde{d}]^{(2)}_{\mu} \qquad \underline{5} \qquad (3.13)$$

$$15 = 6x5/2$$

These operators close under commutation, yielding the algebra of O(6). Obvious subalgebras are now

B) O(5)

$$B^{(1)}_{\mu}(dd) = [d^{\dagger} x \tilde{d}]^{(1)}_{\mu} \qquad 3$$

$$B^{(3)}_{\mu}(dd) = [d^{\dagger} x \tilde{d}]^{(3)}_{\mu} \qquad \underline{7} \qquad (3.14)$$

$$10 = 5x4/2 \quad ,$$

C) O(3)

$$B^{(1)}_{\mu}(dd) = [d^{\dagger} x \tilde{d}]^{(1)}_{\mu} \qquad 3 \qquad (3.15)$$

D) O(2)

$$B^{(1)}_{0}(dd) = [d^{\dagger} x \tilde{d}]^{(1)}_{0} \qquad 1 \quad . \qquad (3.16)$$

Thus, a third possible chain is

$$U^{(B)}(6) \supset O^{(B)}(6) \supset O^{(B)}(5) \supset O^{(B)}(3) \supset O^{(B)}(2) \quad , \text{(III)}. \qquad (3.17)$$

In a similar way, one can now investigate all possible subalgebras of the fermion algebra $U^{(F)}(m)$. In general, for m large, this is a complex problem. I will discuss here, as an example, the case j=3/2, m=4. This case has only one possible subalgebra.

Fermion subalgebra I
Consider the 15 operators

A) SU(4)

$$A^{(1)}_{\mu}(\tfrac{3}{2},\tfrac{3}{2}) = [a^{\dagger} x \tilde{a}]^{(1)}_{\mu} \qquad 3$$

$$A^{(2)}_{\mu}(\tfrac{3}{2},\tfrac{3}{2}) = [a^{\dagger} x \tilde{a}]^{(2)}_{\mu} \qquad 5$$

$$A^{(3)}_{\mu}(\tfrac{3}{2},\tfrac{3}{2}) = [a^{\dagger} x \tilde{a}]^{(3)}_{\mu} \qquad \frac{7}{15} \quad . \qquad (3.18)$$

These operators form the algebra of SU(4), the group of special unitary transformations in 4 dimensions.
Consider now the 10 operators

B) Sp(4)

$$A^{(1)}_{\mu}(\tfrac{3}{2},\tfrac{3}{2}) = [a^{\dagger} x \tilde{a}]^{(1)}_{\mu} \qquad 3$$

$$A^{(3)}_{\mu}(\tfrac{3}{2},\tfrac{3}{2}) = [a^{\dagger} x \tilde{a}]^{(3)}_{\mu} \qquad \frac{7}{10} \quad . \qquad (3.19)$$

These close under the algebra of Sp(4), the group of symplectic transformations in 4 dimensions. Obvious subalgebras are now

C) SU(2)

$$A^{(1)}_{\mu}(\tfrac{3}{2},\tfrac{3}{2}) = [a^{\dagger} x \tilde{a}]^{(1)}_{\mu} \qquad 3 \quad , \qquad (3.20)$$

D) SO(2)

$$A^{(1)}_{0}(\tfrac{3}{2},\tfrac{3}{2}) = [a^{\dagger} x \tilde{a}]^{(1)}_{0} \qquad 1 \quad . \qquad (3.21)$$

Thus, a fermionic chain is

$$U^{(F)}(4) \supset SU^{(F)}(4) \supset Sp^{(F)}(4) \supset SU^{(F)}(2) \supset SO^{(F)}(2) . \qquad (3.22)$$

4. CLASSIFICATION OF STATES

The group chains introduced in the previous section can be used to classify states and to provide a basis in which numerical calculations can be done. The classification of the boson states is discussed in Ref.[3] and will not be repeated here. I will only briefly review the quantum numbers labelling the states of the chain III.

Boson group chain III

The labels needed to classify the states of this chain are

$$
\begin{array}{ll}
U^{(B)}(6) & [N] \\
O^{(B)}(6) & \sigma \\
O^{(B)}(5) & \tau \\
O^{(B)}(3) & L \\
O^{(B)}(2) & M_L \quad .
\end{array} \tag{4.1}
$$

In general, there are six quantum numbers that label the representations of U(6),$[N_1,N_2,N_3,N_4,N_5,N_6]$. The fact that only one appears in Eq.(3.23) is a consequence of Bose statistics which implies that the states be totally symmetric and thus belonging to the Young tablean

$$[N] = \overbrace{\square\square \dots \square}^{\text{N-boxes}} \quad . \tag{4.2}$$

Similarly, there are three quantum numbers labelling the representations of O(6), $(\sigma_1,\sigma_2,\sigma_3)$, but for totally symmetric states only one is needed. The same is true for O(5), where two numbers are in general needed. The rules to calculate the values of the quantum numbers σ,τ,L,M_L contained in a given representation [N] are given in Refs.[3,6]. As explained there, a delicate problem arises in the reduction from O(5) to O(3), since sometimes several states with the same value of L are contained in a given representation of O(5). This problem is solved by introducing an additional quantum number, called ν_Δ. The complete classification is then

$$|N,\sigma,\tau,\nu_\Delta, L, M_L> \quad . \tag{4.3}$$

Fermion group chain

The labels needed to classify the representations of the groups in the chain, Eq.(3.22), are

$$
\begin{array}{ll}
U^{(F)}(4) \text{ and } SU^{(F)}(4) & \{M\} \\
Sp^{(F)}(4) & \Theta \\
SU^{(F)}(2) & J \\
SO^{(F)}(2) & M_J \quad .
\end{array} \tag{4.4}
$$

Here again, in general, there are four quantum numbers, $[M_1,M_2,M_3,M_4]$, that characterize the representations of U(4). However, we must consider representations which are totally antisymmetric, thus belonging to the Young tableau

$$\{M\} \equiv \left.\begin{array}{c}\square\\ \square\\ \cdot\\ \cdot\\ \cdot\\ \square\end{array}\right\} \text{M-boxes} \tag{4.5}$$

These are characterized by only one number. The same occurs for SU(4) and Sp(4). The rules that give the values of the quantum numbers contained in a given representation {M} are summarized in Table I.

Table I. Classification scheme for the fermion group chain, Eq.(3.22).

U(4)	Sp(4)	SU(2)
M	$(\Theta,0)$	J
0	(0,0)	0
1	(1,0)	3/2
2	(2,0)	1,3
$3\equiv\bar{1}$	(1,0)	3/2
$4\equiv 0$	(0,0)	0

5. DYNAMIC SYMMETRIES

Under particular conditions, the eigenvalue problem for the boson-fermion Hamiltonian H, Eq.(2.9), can be solved in closed form. These particular conditions occur whenever H can be written in terms of special operators, called Casimir operators, which are diagonal in a certain basis. For example, the

Hamiltonian

$$H = \alpha \vec{L}^2 \quad , \tag{5.1}$$

is diagonal in the basis $|L\rangle$, with eigenvalues

$$E(L) = \alpha L(L+1) \quad . \tag{5.2}$$

The situation in odd-even nuclei is by far more complex than in even-even nuclei since one is dealing simultaneously with bosons and fermions. However, it turns out that these cases can be treated using a technique which is an extension of a familiar concept known from simple angular momentum algebras. Consider in fact the Hamiltonian

$$H = \alpha_B \vec{L}^2 + \alpha_F \vec{S}^2 + \alpha_{BF} \vec{L}.\vec{S} \quad . \tag{5.3}$$

This Hamiltonian is diagonal in the basis $|L\ S\ J\rangle$ obtained by coupling the angular momenta

$$\vec{J} = \vec{L} + \vec{S} \quad , \tag{5.4}$$

with eigenvalues

$$E(L,S,J) = (\alpha_B - \frac{1}{2}\alpha_{BF})L(L+1) + (\alpha_F - \frac{1}{2}\alpha_{BF})S(S+1) + \frac{1}{2}\alpha_{BF}\ J(J+1) \quad . \tag{5.5}$$

The diagonalization is possible because H can be written in terms of the Casimir invariants $\vec{L}^2$, $\vec{S}^2$ and $\vec{J}^2$.

In group theoretical language, the addition of two angular momenta is the combination of operators having the same Lie algebra. This technique can thus be applied if boson and fermion group chains are identical (in group theoretical language their algebras must be isomorphic). It turns out that the example discussed in Sect.4 is indeed on of these cases. In fact, the algebras of $O^{(B)}(6)$ and $SU^{(F)}(4)$ are isomorphic and so are those of $O^{(B)}(5)$ and $Sp^{(F)}(4)$, of $O^{(B)}(3)$ and $SU^{(F)}(2)$, of $O^{(B)}(2)$ and $O^{(F)}(2)$. This isomorphism is written as

$$\begin{aligned} O^{(B)}(6) &\approx SU^{(F)}(4) \quad , \\ O^{(B)}(5) &\approx Sp^{(F)}(4) \quad , \\ O^{(B)}(3) &\approx SU^{(F)}(2) \quad , \\ O^{(B)}(2) &\approx O^{(F)}(2) \quad . \end{aligned} \tag{5.6}$$

A consequence of Eq.(5.6) is that one can combine boson and

fermion generators in a way similar to Eq.(5.4). This yields a group chain

$$U^{(B)}(6)\otimes U^{(F)}(4)\supset O^{(B)}(6)\otimes SU^{(F)}(4)\supset Spin(6)\supset Spin(5)\supset$$

$$Spin(3)\supset Spin(2) \quad , \tag{5.7}$$

where Spin(n) denotes a spinor group. The generators of the group Spin(6) are

$$G^{(1)}_{\mu} = B^{(1)}_{\mu}(dd) - \frac{1}{\sqrt{2}} A^{(1)}_{\mu}(\tfrac{3}{2},\tfrac{3}{2}) \quad , \qquad 3$$

$$G^{(2)}_{\mu} = B^{(2)}_{\mu}(ds) + B^{(2)}_{\mu}(sd) + A^{(2)}_{\mu}(\tfrac{3}{2},\tfrac{3}{2}) \quad , \qquad 5$$

$$G^{(3)}_{\mu} = B^{(3)}_{\mu}(dd) + \frac{1}{\sqrt{2}} A^{(3)}_{\mu}(\tfrac{3}{2},\tfrac{3}{2}) \quad . \qquad \frac{7}{15} \tag{5.8}$$

By deleting the 5 operators $G^{(2)}_{\mu}$ one obtains the generator of Spin(5); by deleting then the 7 operators $G^{(3)}_{\mu}$ one obtains the generators of Spin(3) and finally by deleting the 2 operators $G^{(1)}_{\pm 1}$ one obtains the generator of Spin(2).

Having found a common boson and fermion chain, Eq.(5.7), we need next to construct a basis. This is done by taking products of boson and fermion representations. In the example discussed above the basis is $|L\ S\ J>$. The values of J are given by the rule

$$|L-S| \le J \le |L+S| \quad . \tag{5.9}$$

In the present case, the rules for multiplying representations are more complex. First one needs to convert the representations of $SU^{(F)}(4)$ and $O^{(B)}(6)$ into representations of Spin(6). The representation of $SU^{(F)}(4)$ [1,0,0,0] yields, for example, the fundamental spinor representation $[\frac{1}{2},\frac{1}{2},\frac{1}{2}]$ of Spin(6), while the representation of $O^{(B)}(6)$ [N,0,0] yields the tensor representation [N,0,0] of Spin(6). Taking products of boson and fermion representations one can then construct a classification scheme for the combined chain, Eq.(5.7),

$U^{(B)}(6)$	[N]
$U^{(F)}(4)$	{M}
$O^{(B)}(6)$	Σ

Spin(6)	$(\sigma_1,\sigma_2,\sigma_3)$
Spin(5)	(τ_1,τ_2)
Spin(3)	J
Spin(2)	M_J

(5.10)

Group theoretical rules then give the values of the quantum numbers $\Sigma,(\sigma_1,\sigma_2,\sigma_3)$, (τ_1,τ_2), J,M_J contained in each representation $[N]\otimes\{M\}$ of $U^{(B)}(6)\otimes U^{(F)}(4)$. For M=1, the rules are particularly simple and are given in Ref.[7]. Due to the fact that the step from Spin(5) to Spin(3) is not fully reducible, one needs here an additional quantum number, ν_Δ, as in Eq.(4.3). The complete classification scheme is then

$$|N,M,\Sigma,(\sigma_1,\sigma_2,\sigma_3),\ (\tau_1,\tau_2),\ \nu_\Delta,J,\ M_J\ >. \tag{5.11}$$

Analytic solutions to the eigenvalue problem for the interacting boson-fermion Hamiltonian can now be found by writing H in terms of Casimir invariants of the group chain, Eq.(5.7). One begins with the groups $U^{(B)}(6)$ and $U^{(F)}(4)$. The linear and quadratic Casimir operators of these groups have, within the representations [N]and {M}, eigenvalues which depend only on the numbers N and M. Since, in calculating the spectrum of a nucleus, N and M are fixed, these operators contribute an overall constant. Thus, they can be neglected if one is interested only in excitation energies. It is therefore sufficient to consider the Hamiltonian

$$H=E_0+A\mathcal{C}_2(O^{(B)}(6))+A'\mathcal{C}_2(\mathrm{Spin}(6))+B\mathcal{C}_2(\mathrm{Spin}(5))+C\mathcal{C}_2(\mathrm{Spin}(3)). \tag{5.12}$$

In this expression, $\mathcal{C}_2$ denotes a quadratic Casimir operator.The Hamiltonian, Eq.(5.12), is diagonal in the basis (5.11) with eigenvalues

$$E(N,M,\Sigma,(\sigma_1,\sigma_2,\sigma_3),(\tau_1,\tau_2),\nu_\Delta,J,M_J)=E_0+A\Sigma(\Sigma+4)+$$
$$+\ A'[\sigma_1(\sigma_1+4)+\sigma_2(\sigma_2+2)+\sigma_3^2]+B[\tau_1(\tau_1+3)+\tau_2(\tau_2+1)]+C\ J(J+1). \tag{5.13}$$

In deriving Eq.(5.13) use has been made of the standard rules for eigenvalues of Casimir invariants. The structure of the spectrum corresponding to Eq.(5.13) is shown in Fig.3.

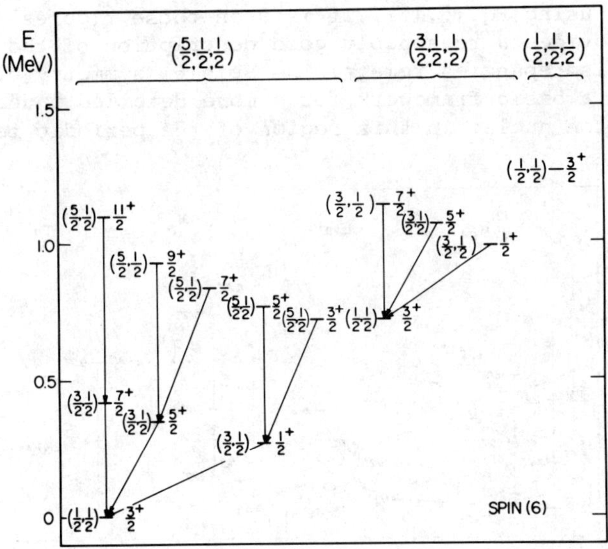

Fig.3. A typical spectrum of an odd-even nucleus (N=2,M=1) with Spin(6) symmetry. The energy levels are given by Eq. (5.13) with A=0, A'=90 keV, B=60 keV, C=10 keV. The numbers on top of the figure denote the Spin(6) quantum numbers $(\sigma_1,\sigma_2,\sigma_3)$. The numbers in parenthesis next to each level denote the Spin(5) quantum numbers (τ_1,τ_2). The ground state is taken as zero of the energy.

6. EXAMPLES OF SPECTRA WITH SPIN(6) SYMMETRY

The main advantage of the symmetry approach described above is that it provides simple expressions that can be easily compared with experiment. However, these simple expressions are obtained under certain conditions which must be, at least approsimately met in order that the theoretical expressions describe the experimental situation. The situation discussed here describes a particle with j=3/2 moving in a γ-unstable deformed nucleus. This situation occurs approximately in odd proton nuclei in the Os-Pt region. Here, in fact, even-even nuclei can be, to a good approximation, described by the O(6) symmetry [8], and, in addition, the odd particle occupies a single particle level with j=3/2, Fig.2. In Figs.4 and 5 the experimental energy levels of $^{191}_{77}Ir_{114}$ and $^{193}_{77}Ir_{116}$ are compared with those calculated using Eq.(5.13). It is seen these figures that the Spin(6) provides a reasonably good description of the spectra, although discrepancies remain. The Spin(6) symmetry can thus be used as a basic framework for a more detailed understanding of odd-proton nuclei in this region of the periodic table.

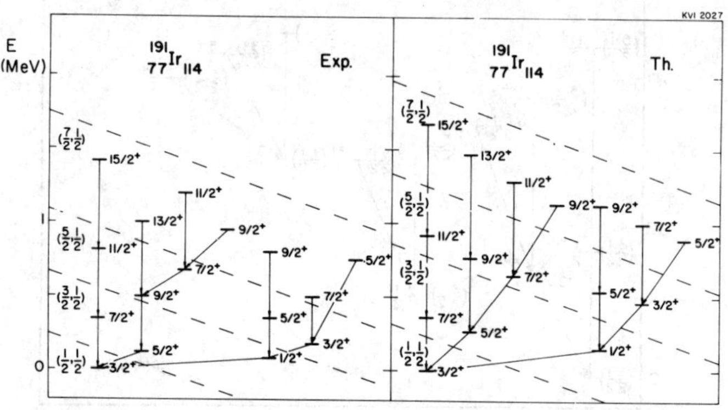

Fig.4. An example of a spectrum with Spin(6) symmetry: $^{191}_{77}Ir_{114}$ (N=8, M=1). The energy levels in the theoretical spectrum are calculated using Eq.(5.13), with B=40 keV and C=10 keV [7].

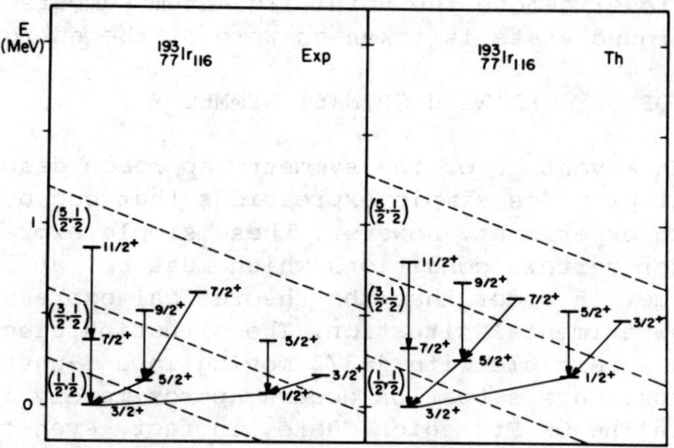

Fig.5. An example of a spectrum with Spin(6) symmetry: $^{193}_{77}Ir_{116}$ (N=7, M=1). The energy levels in the theoretical spectrum are calculated using Eq.(5.13), with B=40 keV and C=10 keV [7].

7. OTHER PROPERTIES. ELECTROMAGNETIC TRANSITION RATES

The group theoretic technique described above can be used to calculate other nuclear properties, such as electromagnetic transition rates. In order to do this, one must first write down the appropriate transition operators in terms of boson and fermion operators and then evaluate their elements between states, Eq. (5.11). In this article, I will not discuss how to evaluate matrix elements but only write down the appropriate form of the transition operators. If one assumes that these operators are at most one-body operators, one obtains, for a transition operator of multipolarity λ, the form

$$T^{(\lambda)} = T_B^{(\lambda)} + T_F^{(\lambda)} , \tag{7.1}$$

where

$$T_{B,\mu}^{(\lambda)} = \sum_{\ell\ell'} t_{\ell\ell'}^{(\lambda)} B_\mu^{(\lambda)}(\ell,\ell') ,$$

$$T_{F,\mu}^{(\lambda)} = \sum_{jj'} t_{jj'}^{(\lambda)} A_\mu^{(\lambda)}(j,j') . \tag{7.2}$$

A more explicit form can be obtained by expressing the B and A operators in terms of s,d bosons and a_j fermions. For example, for E2 transitions, the boson part of $T^{(\lambda)}$ has the form

$$T_{B,\mu}^{(E2)} = \alpha_2 (d^\dagger x\tilde{s} + s^\dagger x\tilde{d})_\mu^{(2)} + \beta_2 (d^\dagger x\tilde{d})_\mu^{(2)} . \tag{7.3}$$

The fermion part depends on the values of j. For j=3/2 it can be written as

$$T_{F,\mu}^{(E2)} = \gamma_2 (a^\dagger x\tilde{a})_\mu^{(2)} . \tag{7.4}$$

In the evaluation of the matrix elements of the operators $T^{(\lambda)}$ simplifications again occur whenever these operators can be written in terms of generators of a group. For example, if the E2 transition operator can be written in terms of the generator $G^{(2)}$ of Eq. (5.8)

$$T_\mu^{(E2)} = q_2 G_\mu^{(2)} , \tag{7.5}$$

the evaluation of its matrix elements between states Eq. (5.11) reduces to the computation of some group theoretic coefficients and it is relatively simple. The resulting B(E2) values are sum marized in Fig.6. The calculated values have been compared with

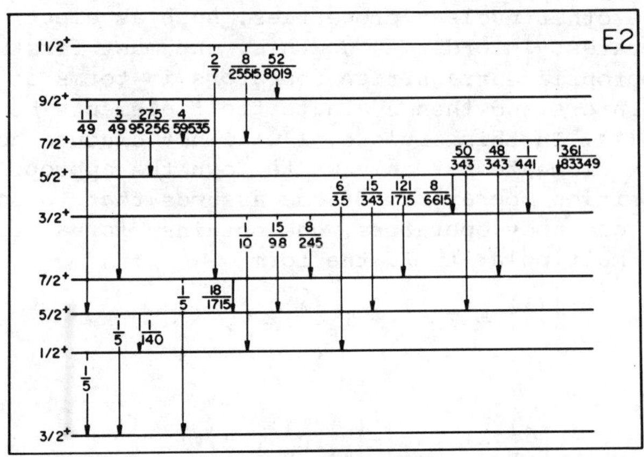

Fig.6. B(E2) values in units of $(q_2)^2$ in the Spin(6) limit. $\Delta\tau_1=0$ transitions should be multiplied by $(2N+5)^2$ and $\Delta\tau_1=1$ transitions by $(N-\tau_1+\frac{1}{2})$ $(N+\tau_1+\frac{9}{2})$ [7].

the experimental values in ${}^{191}_{77}Ir_{114}$ and ${}^{193}_{77}Ir_{116}$ in Ref.[7]. The agreement between the two sets is fair.

8. CONCLUSIONS

In these lectures, I have presented a brief introduction to the interacting boson-fermion model and used symmetry arguments to derive an analytic solution to it. These symmetry arguments are actually more general than described here and as a result it is possible to construct large classes of analytic solutions [9]. The analytic solutions are very useful, since they provide simple formulas that can be easily compared with experiments. Thus, they act as bench marks for a discussion of nuclear properties. In general, however, dynamic symmetries in nuclei are only approximate situations and for a more detailed description one must perform numerical calculations. A computer program has been written for this purpose [10]. Calculations using this program have already been done [5] and appear to describe, to a good approximation, the variety of observed properties of odd-even nuclei.

AKNOWLEDGEMENTS

I wish to thank R. Leonardi for his hospitality at the

University of Trento where this article was completed. This work was performed in part under the Department of Energy Contract No. DE-AC 02-76ER03074.

REFERENCES

[1] S.G. Nilsson, K.Dan.Vidensk. Selsk. Mat.Fys. Medd.29, No. 16 (1955).

[2] F. Iachello and O. Scholten, Phys. Rev. Lett. 43, 679 (1979); F. Iachello, Nucl. Phys. A347, 51 (1980).

[3] F. Iachello, in "Nuclear Structure", K. Abrahams, K.Allaart and A.E.L. Dieperink eds., Plenum Publ.Corp., New York (1981), p.53.

[4] O. Scholten and A.E.L. Dieperink, in "Interacting Bose-Fermi Systems in Nuclei", F. Iachello ed., Plenum Publ. Corp., New York (1981), p.343;
I. Talmi, in "Interacting Bose-Fermi Systems in Nuclei", F. Iachello ed., Plenum Publ.Corp., New York (1981),p.329.

[5] O. Scholten and N. Blasi, Nucl. Phys. A380, 509 (1982);
P.von Brentano, A.Gelberg and U. Kaup,in "Interacting Bose-Fermi Systems in Nuclei", F. Iachello ed., Plenum Publ.Corp., New York (1981), p.303;
R. Bijker and A.E.L. Dieperink, Nucl.Phys. A379, 221 (1982);
M.A. Cunningham, Nucl. Phys. A385, 204, 221 (1982).

[6] A. Arima and F. Iachello, Ann.Phys. (N.Y.) 99,253 (1976);
A. Arima and F. Iachello, Ann.Phys. (N.Y.) 111, 201 (1978);
A. Arima and F. Iachello, Ann.Phys. (N.Y.) 123, 468 (1979).

[7] F. Iachello and S. Kuyucak, Ann. Phys.(N.Y.) 136,19 (1981).

[8] J.A. Cizewski, R.F.Casten,G.J. Smith, M.L.Stelts, W.R.Kane, H.G. Börner and W.F. Davidson, Phys.Rev.Lett. 40,167(1978).

[9] A.B.Balantekin, I.Bars, R.Bijker and F. Iachello, Phys.Rev. C27, 1761 (1983);
D.D. Warner, R.F. Casten, M.L. Stelts, H.G. Börner and G. Barreau, Phys. Rev. C26, 1921 (1982);
Sun Hong Zhou, A. Frank and P.von Isacker, Phys.Lett.124B, 275 (1983);
Sun Hong Zhou, A. Frank and P.von Isacker, Phys.Rev.C27, 240 (1983);
R. Bijker and V.K.B.Kota,Ann.Phys.(N.Y.), in press;
R. Bijker and F. Iachello, in preparation.

[10] O. Scholten, Computer Program ODDA, University of Groningen, The Netherlands (1980).

TOWARD MICROSCOPIC UNDERSTANDING OF THE CONCEPT OF COLLECTIVE SUBSPACE. (I)

Toshio MARUMORI, Kenjiro TAKADA* and Fumihiko SAKATA**

Institute of Physics, University of Tsukuba, Ibaraki 305
*Department of Physics, Kyushu University, Fukuoka 812
**Institute for Nuclear Study, University of Tokyo, Tanashi, Tokyo 188

ABSTRACT

The series of lectures is divided into two parts. The purpose of this lecture, Part (I),*) is to review the elementary concepts necessary for the microscopic study of the low-lying collective excited states in spherical and transitional nuclei from a historical viewpoint, putting emphasis on explaining the microscopic investigation of the very concept of the collective subspace. A detailed discussion on the interacting boson model is also given in the light of the so-called modified Marumori boson mapping from the microscopic point of view.

---

*) The main part of (I) is based on Refs. 1) and 2).

§1. Independent-Particle and Collective Aspects of Nuclear Structure*)

Let us start with a short history on the collective motion. As we all know, the first step in the study of nuclear structure is characterized by the great success of the independent-particle model. A wide variety of experimental data demonstrated that, in the nucleus, the mean free path for collisions between constituent nucleons is large compared to the distance between them and is, under many circumstances, even larger than the dimensions of the nucleus.[7] Thus, under the name of the nuclear shell model, the independent-particle mode of motion had a firm basis by accounting for the main body of experimental evidence at that time.

The second step is characterized by a dramatic development of nuclear spectroscopy provided by the newly discovered Coulomb excitation reaction, and an extensive amount of data on low-energy excitation spectra displayed the occurrence of collective phenomena. The most striking evidence was the discovery of characteristic patterns of rotation in a class of nuclei with the large stable deformations in shape. Shortly afterward, it was also found that a significant class of low-energy excitation spectra exhibited characteristic patterns of quadrupole (shape) vibrations about a spherical equilibrium.

---

*) The title of this section is taken from that of the original paper of the collective model of Bohr and Mottelson,[3] "Collective and Individual-Particle Aspects of Nuclear Structure". (It is important to emphasize that it is quite wrong to call the Bohr-Mottelson model the "liquid drop" model.) As for the explanation of the basic concepts which have played important roles in arriving at the present understanding of the nuclear structure, there are very excellent articles by A. Bohr and B. R. Mottelson.[4)∼6)] The main part of this section and a part of the next section are based on these articles as well as Bohr--Mottelson's books.[7]

Looking back into the history of development of nuclear structure theory over the past quarter century, it may be possible to say that "a central theme has been the struggle to find the proper place for the complementary concepts referring to the independent motion of the individual nucleons and the collective behaviour of the nucleus as a whole."[5]

A starting point in exploring the properties of nuclear system as a result of the "nuclear dynamics" of interplay between the two modes of motion was the collective model of Bohr and Mottelson.[3] The basic idea underlying this model is the following: In the shell model, the average potential in which particles move independently is assumed to be of spherical shape and to be impossible to deform its shape. If we allow the average potential to be time-dependent and deformable, then the average potential is capable by itself to move self-consistently with action of the particles. This motion is just the collective mode of motion, which may be phenomenologically simulated as oscillations of the liquid drop introduced in the earliest discussion of nuclei by N. Bohr.

Thus the Hamiltonian of a nuclear system in the collective model is given by

$$H_{BM} = H_{coll} + H_p + H_{int}^{(BM)} \tag{1.1}$$

where H_{coll} represents the surface vibrational motion of an imcompressible liquid drop simulating the collective motion of the average potential and H_p is the individual-particle Hamiltonian consisting of the spherical average potential and certain residual interactions. The term $H_{int}^{(BM)}$ denotes the coupling between the two modes of motion, through which the action of the particles is delivered to the average potential and the effect of the variations of the average potential is transmitted to the motion of particles. When the average potential has an equilibrium shape with a large deformation, a part of $H_{int}^{(BM)}$ is renormalized into H_p so as to produce the deformed average potential for the particle motion, and the system exhibits the collective rotational spectra. In the collective model of Bohr and Mottelson, therefore, the nucleus with stable deformation is assumed to have the average nuclear potential of deformed equilibrium shape, i.e., the Nilsson potential.[8)]

A very sensitive experimental test for the assumption of one-particle orbits in the deformed average potential was provided by the recognition of the spectra of band heads of various rotational bands in odd-A nuclei. It was found that the one particle orbits in the Nilsson potential were nicely able to explain the experimental data. Here it must be emphasized that, in nuclei with large equilibrium deformations, the concept of the closed-shell core which plays a crucial role in spherical nuclei loses its own clear-cut physical meaning. This is because the crossing of levels belonging to the different major shells is essential for the onset of the large equilibrium deformations as first pointed out by Mottelson and Nilsson.[9)] Any models and theories that attempt to fit data for nuclei in the deformed region would have to be consistent with this fact.[10)]

A striking success of the phenomenological collective model of Bohr and Mottelson in interpreting the low-energy nuclear excitation spectra stimulated a new subject in the study of nuclear structure. The subject is to analyse the collective motion in terms of the particle degrees of freedom on the basis of the nuclear many-body problem, and it has been called the <u>microscopic description of collective motion</u>.

A starting point for this subject was provided by the "cranking model" of Inglis[11)] for the rotational motion. Through the analyses of the moment of inertia which is one of the crucial parameters in the phenomenological collective model, it was recognized that, in the particle motion inside the nucleus, there were two types of basic correlations arising from the residual interactions: The first type is the <u>pairing correlation</u> producing the tendency to bind the particles into pairs with angular momentum J=0 so as to stabilize the spherical equilibrium shape. The other type is the quadrupole correlation producing the tendency toward the deformation with dominant quadrupole component. Thus, it was clarified that the main features of the low-energy excitation spectra could be qualitatively understood from the viewpoint of a competition between these two types of correlation effects in the particle motion.

The first task of the nuclear many-body problem is therefore to derive the phenomenological Hamiltonian (1.1) of the Bohr-Mottelson collective model, starting with the many-body nuclear Hamiltonian with the so-called pairing plus quadrupole force.

§2. Elementary Modes of Collective Excitations and "Phase Transition"

Soon after the dynamics of low-energy excitations was qualitatively understood, there was a period of rapidly expanding development of nuclear many-body problem. It was recognized[12)∼14)] that the new idea of Bardeen, Cooper and Schrieffer and the concept of quasiparticles of Bogoliubov and Valatin for superconductors provided the necessary approximation tools for describing the pairing correlations to produce the J=0 coupled pairs of particles. It was also recognized that the quanta of the collective quadrupole vibrations were built out of coherently correlated two-quasiparticle excitations;[*),15)∼17)]

$$\hat{D}^{\dagger}_{\mu} = \sum_{a,b}\psi(ab;J=2)[a^{\dagger}_{a}a^{\dagger}_{b}]_{J=2,\mu} \qquad (\mu = -2,-1,0,1,2)$$

$$[a^{\dagger}_{a}a^{\dagger}_{b}]_{J,\mu} = \sum_{m_{\alpha}m_{\beta}}\langle j_{a}m_{\alpha}j_{b}m_{\beta}|J\mu\rangle a^{\dagger}_{\alpha}a^{\dagger}_{\beta} \qquad (2.1a)$$

(the Tamm-Dancoff approximation (TDA)),

---

*) The single-particle states in the spherical shell model are characterized by a set of quantum numbers $\alpha = \{n,l,j,m$ and charge $q\}$. In association with the Greek letter α, we use the Roman letter a to denote the same set except the magnetic quantum number m.

or

$$\hat{X}^{\dagger}_{\rho,J=2,\mu} = \sum_{ab}\{\psi_{\rho}(ab)[a^{\dagger}_{a}a^{\dagger}_{b}]_{J=2,\mu} - \phi_{\rho}(ab)(-)^{2-\mu}[a_{b}a_{a}]_{J=2,-\mu}\},$$

$$[\hat{H}, \hat{X}^{\dagger}_{\rho,J\mu}] = \omega_{\rho,J}\hat{X}^{\dagger}_{\rho,J\mu} + :\hat{Z}: \ , \quad (\omega_{\rho,J} > 0) \ , \tag{2.1b}$$

(the random phase approximation (RPA)),

where $:\hat{Z}:$ consists of the terms with $a^{\dagger}a$ and those with the normal-ordered 4-quasiparticle operators.

The correlated ground-state $|\Psi_0(\text{RPA})\rangle$ in the quasiparticle RPA, which may be defined by

$$\hat{X}_{\rho,J\mu}|\Psi_0(\text{RPA})\rangle = 0 \ , \tag{2.2}$$

is more accurate than the BCS ground state $|\phi_{BCS}\rangle$ in the quasiparticle TDA, since it includes the ground-state correlation in a form with 0,4,8, ··· quasiparticle states:

$$|\Psi_0(\text{RPA})\rangle = \chi_0|\phi_{BCS}\rangle + \Sigma\chi_4(\alpha\beta\gamma\delta)a^{\dagger}_{\alpha}a^{\dagger}_{\beta}a^{\dagger}_{\gamma}a^{\dagger}_{\delta}|\phi_{BCS}\rangle + \cdots \ . \tag{2.3}$$

An important concept derived from the quasiparticle RPA is the stability of the BCS ground state $|\phi_{BCS}\rangle$. It was shown that the appearance of a complex root of the eigenvalue $\omega_{\rho,J}$ in the quasiparticle RPA implies the instability of $|\phi_{BCS}\rangle$: When the pairing correlation which favours the spherical shape is dominant over the quadrupole correlation, the BCS ground state $|\phi_{BCS}\rangle$ is stable. As the quadrupole correlation increases, i.e., as we go into the region of the transitional nuclei, the lowest excitation energy ω_{ρ_0} of the elementary collective mode $\hat{X}^{\dagger}_{\rho_0,J=2}$ tends to zero, and the ground-state correlations, i.e., χ_4, χ_8,··· terms in Eq. (2.3) become large. When ω_{ρ_0} becomes complex beyond zero, i.e., χ_4, χ_8,··· in the true ground state become larger than χ_0, the BCS ground state $|\phi_{BCS}\rangle$ as the approximate ground state is unstable toward quadrupole deformation.

This implies that, because of the strong quadrupole correlation, <u>the true ground state is too far from the approximate ground state</u> $|\phi_{BCS}\rangle$ <u>to be obtained by means of the perturbation method</u>. In this case we have to choose another stable approximate ground state $|\phi_{\beta_0}\rangle$ with an equilibrium deformation β_0, which specifies new single-particle orbits, in terms of the generalized Hartree-Fock-Bogoliubov (HFB) method by taking account of the dominant quadrupole correlation from the beginning.

In analogy to solid state physics, we call such a change of the approximate ground state specifying the single-particle orbits a <u>phase transition of the system to a symmetry-violating state</u> i.e., <u>a deformed state</u>. After the phase transition, of course, the elementary modes of collective motion, which essentially depend on what kind of the HFB ground state is used, manifest themselves in different forms. One of the most typical is the collective rotational mode of motion which restores the broken symmetry. The above-mentioned mechanism relating to the symmetry-violating phase transition is just the physical mechanism which has been emphasized in the collective model of Bohr and Mottelson in a phenomenological way.

Here it must be stressed that there is an essential difference between the phase transitions in the nuclear physics and those in the solid-state physics: In a finite system such as the nucleus, of course, we can not expect phase transitions with sharp singularities as in macroscopic systems. As has been mentioned, the most important in nuclear structure is the concept of a <u>mean field</u>, which convincingly has a firm basis in connection with the individual-particle mode of motion. It must be stressed, however, that the dynamics of the occurence of the <u>stable deformation of the mean field</u> has no correspondence in such a macroscopic system as the liquid drop: This results from the specific geometry of the <u>quantized orbits</u> of the individual particles, called the effects of the shell structure. Thus, the nucleus provides us an important possibility of investigating the phase transition in terms of the individual quantum states.

At present such an investigation of phase transitions is one of the challenging open problems. Characteristic difficulties come from the following facts:

(i) The single-particle motion and so the quasiparticle quantum numbers have to alter their features self-consistently in accordance with the evolution of the collective coordinates associated with the deformation of the mean field.

(ii) The nuclear collective motion is of large amplitude and highly non-linear, so that there will be large fluctuations in shape. The stable equilibrium deformation of the mean field can manifest itself, only if its deviation from spherical symmetry is of an amount exceeding the zero-point shape fluctuations. (Then it becomes possible to separate the rotational motion from the others.)

In the nucleus, thus the phase transition is not sharp but always smeared out. In fact, we find a gradual transition in medium-heavy nuclei from the conservation to the strong breaking of spherical symmetry.[18] This clearly implies an essential difference between the phase transitions in the nuclear physics and in the solid-state physics.

§3. Boson Mapping and Collective Subspace

In the viewpoint of the gradual phase transition from the spherical to the deformed shape, we expect that, for nuclei in the transitional region, we have to go beyond the simple mean field approach by taking account of complicated large quantum fluctuations properly. The structure of the fluctuations reflects characteristics of the dynamical mechanism of the phase transition.

The in-beam spectroscopy invented by Morinaga and Gugelot[19] made it possible to disclose the features of the complicated fluctuations in the transitional nuclei, and Sakai's proposal of the quasi-band structure[18] stimulated the microscopic investigation of the very concept of the collective subspace, which was introduced phenomenologically in the Bohr-Mottelson model as a simple boson (i.e., phonon) space.

The first task in the microscopic investigation of the collective subspace was to develop a method which precisely maps fermion pairs onto bosons. The basic idea of the boson mapping is quite simple.*) Now, let $\{|p>\}$ denote an orthonormal basis for the fermion-state space of the even-fermion system and $\{|p)_B\}$ a <u>corresponding</u> orthonormal basis in an <u>ideal boson space</u>. The states $\{|p)_B\}$ span a linear subspace in the ideal boson space, called the <u>physical subspace</u>. The rest of the boson space is called the unphysical subspace, because the states $\{|u)_B\}$ in the subspace have nothing to do with the fermion system under consideration. Thus, the mapping from a fermion state $|p>$ to the corresponding physical boson state $|p)_B$ has an exact one-to-one correspondence (Fig.1[20]);

$$|p> \longleftrightarrow |p)_B \tag{3.1}$$

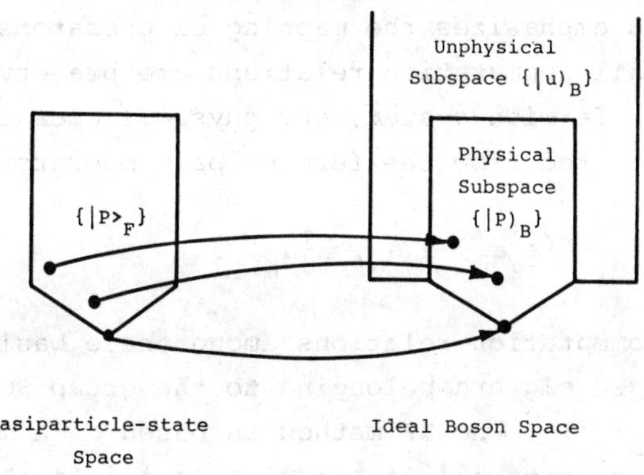

Fig.1

Mapping of the quasiparticle-state space onto the ideal boson space. This figure is taken from Ref.20)

---

*) As for the methods of the boson mapping, there are excellent review articles by Marshalek[20] and Ring and Schuk.[21]

The boson mapping requires that any fermion operator $\hat{T}$ be mapped onto an operator T in the ideal boson space

$$\hat{T} \rightarrow \mathbf{T} \tag{3.2}$$

so as to preserve the relation

$$\langle p|\hat{T}|p'\rangle = (p|\mathbf{T}|p')_B \ . \tag{3.3}$$

In addition, it is also required that

$$(u|\mathbf{T}|p)_B = (p|\mathbf{T}|u)_B = 0 \ , \tag{3.4}$$

which implies that $\mathbf{T}$ leaves the physical boson subspace invariant.

There are two basic general methods to introduce this mapping explicitly:

i) Belyaev-Zelevinsky method (BZ method)[22]: This method emphasizes the mapping of operators in such a way that all commutation relations are preserved. In the even fermion system, any physical operators can be constructed from the fermion pair operators

$$\{a_\alpha^\dagger a_\beta^\dagger,\ a_\beta a_\alpha,\ a_\alpha^\dagger a_\beta\} \ . \tag{3.5}$$

The commutation relations among these basis operators form a closed algebra belonging to the group SO(2Ω) with $2\Omega = \sum_a (2j_a + 1)$. The BZ method is based on a mapping of the basis operators $\{a_\alpha^\dagger a_\beta^\dagger,\ a_\beta a_\alpha,\ a_\alpha^\dagger a_\beta\}$ into the ideal boson space so as to preserve the SO(2Ω) algebra. The boson mapping of the BZ method is explicitly performed by requiring that the mapped operators are expressed as Taylor-series expansion

$$\mathbf{T}^{[BZ]} = \mathbf{T}^{(0)} + \mathbf{T}^{(1)} + \cdots + \mathbf{T}^{(n)} + \cdots \ , \tag{3.6}$$

where $\mathbf{T}^{(n)}$ is the n-th term in the Taylor series. Thus, any commutation relation $[\hat{A}, \hat{B}] = \hat{C}$ must be fulfilled in each order of expansion in the boson mapping:

$$[\mathbf{A}^{(1)}, \mathbf{B}^{(1)}] = \mathbf{C}^{(0)} ,$$

$$[\mathbf{A}^{(1)}, \mathbf{B}^{(2)}] + [\mathbf{A}^{(2)}, \mathbf{B}^{(1)}] = \mathbf{C}^{(1)} ,$$

$$\cdots\cdots . \tag{3.7}$$

ii) <u>Marumori-Yamamura-Tokunaga method</u> (MYT method)[23]: This method emphasizes the mapping of state vectors, and introduce a "unitary" mapping operator

$$\hat{\mathbf{U}}_M = |0\rangle\cdot\{\Sigma_p |p)_B \cdot \langle p|\}\cdot{}_B(0| \tag{3.8}$$

which satisfies

$$\hat{\mathbf{U}}_M^\dagger \hat{\mathbf{U}}_M = 1_F \cdot \mathbf{\Gamma}_0 , \quad \hat{\mathbf{U}}_M \hat{\mathbf{U}}_M^\dagger = \mathbf{P}\cdot\hat{\Gamma}_0$$
$$\mathbf{\Gamma}_0 \equiv |0)_B \cdot {}_B(0| , \quad \hat{\Gamma}_0 \equiv |0\rangle\cdot\langle 0| , \quad \mathbf{P} = \sum_p |p)_B \cdot {}_B(p| , \tag{3.9}$$

where 1_F is the identity in the fermion space and $\mathbf{P}$ is the projection operator to the physical boson subspace. By definition, the MYT mapping operator $\hat{\mathbf{U}}_M$ satisfies

$$\hat{\mathbf{U}}_M |p\rangle \otimes |0)_B = |0\rangle \otimes |p)_B , \quad \hat{\mathbf{U}}_M^\dagger |p)_B \otimes |0\rangle = |0)_B \otimes |p\rangle ,$$
$$\hat{\mathbf{U}}_M^\dagger |u)_B \otimes |0\rangle = 0 , \tag{3.10}$$

and we have

$$\hat{\mathbf{U}}_M \hat{T} \hat{\mathbf{U}}_M^\dagger = \hat{\Gamma}_0 \cdot \mathbf{T}^{[MYT]} = \hat{\Gamma}_0 \cdot \Sigma_{p,p'} \langle p|\hat{T}|p'\rangle \cdot |p)_B \cdot {}_B(p'| , \tag{3.11}$$

which satisfies the conditions (3.3) and (3.4). Since any boson state can be represented by a repeated application of boson creation operators on the boson vacuum $|0)_B$, the completeness relation in the boson space, $|0)_B \cdot {}_B(0|+\Sigma_{p\neq 0}|p)_B \cdot {}_B(p| +\Sigma_u|u)_B \cdot {}_B(u| = \mathbf{1}$, makes it possible to express $|0)_B \cdot {}_B(0|$ in terms of the boson operators through an iteration. Thus, the operator $|p)_B \cdot {}_B(p'|$ in $\mathbf{T}^{[MYT]}$ can be written in terms of an infinite series of boson expansion.

It has been verified[24),25)] that there is a relation between the two methods:

$$\mathbf{T}^{[MYT]} = \mathbf{p}\mathbf{T}^{[BZ]}\mathbf{p} \tag{3.12}$$

It turned out that the boson mapping of the pure fermion-pair basis operators $\{a^\dagger_\alpha a^\dagger_\beta,\ a_\beta a_\alpha,\ a^\dagger_\alpha a_\beta\}$ was not very useful in practical applications because the corresponding boson expansion did not converge rapidly. It is therefore crucial to choose some <u>correlated fermion operators</u> which are mapped in the first approximation onto bosons in such a proper way that the rapid convergence is guaranteed. The phenomenological collective model of Bohr and Mottelson strongly suggests that only <u>some special subspace</u> of fermion-state space can be mapped onto the

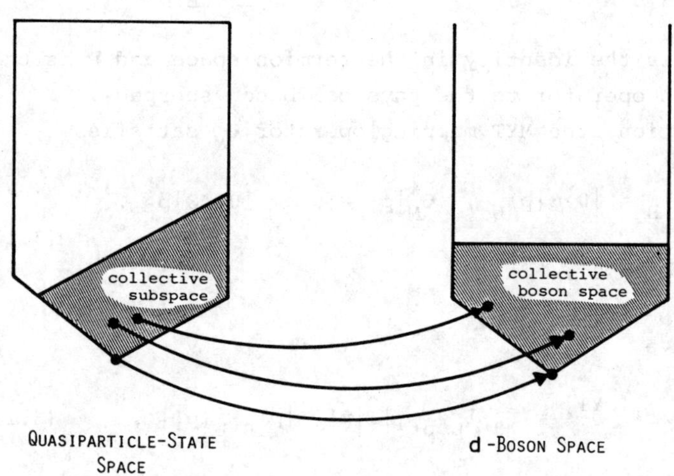

Fig. 2
The boson mapping of the collective subspace.

collective boson space. We have already seen that the operators characterizing such a <u>collective subspace</u> in the fermion-state space are the collective 2-quasiparticle 2(QP) mode $\hat{D}_{\mu}^{\dagger}$ (in the TD method) in Eq.(2.1a) or the collective eigenmode $\hat{X}_{\rho_0,2\mu}^{\dagger}$ (in the RPA) in Eq.(2.1b). We may thus expect that, only within the collective (fermion) subspace, the boson mapping with rapid convergence is possible. (Fig.2)

The boson mapping of the collective subspace along the line of the BZ method has been made by B. Sorensen[26)] and by Kishimoto and Tamura.[27)] The problem in this case is that, within the collective subspace, the closed commutator algebra of the pair operators (belonging to the SO(2Ω)), on which the BZ method is essentially based, is not satisfied.[*)] On the contrary, the MYT method has no such problems because it maps the state vectors, and the boson mapping of the collective subspace along the line of the MYT method has been done by Kleber,[28)] Li-Dreizler-Klein,[29)] Lie-Holzwarth,[30)] and recently by Iwasaki-Sakata-Takada[31)]. This is often called the <u>modified Marumori boson-mapping (expansion) method</u>. Thus, it turned out that, as we expected, the boson mapping of the collective subspace by the modified Marumori method led us to a rapid convergence of the basis operators characterizing the collective subspace.

In the following, we illustrate some examples of the modified Marumori method.

[Example 1] <u>Boson Mapping of J=0-Coupled Nucleon (Cooper) Pairs in a Single-j-Shell</u>

In this simple case, the collective subspace of fermion-state space is spanned by the orthonormal set of states $\{|n\rangle; n = 0,1,2,\cdots\Omega;\ \Omega \equiv j+1/2\}$ with

$$|n\rangle = C_F(n) \frac{1}{\sqrt{n!}} (\hat{\mathcal{S}}_+)^n |0\rangle , \tag{3.13}$$

*) Concerning this problem and the relation with the modified Marumori method, see the very recent investigation by T. Tamura.[32,33)]

where $C_F(n)$ is the normalization constant given by $C_F(n) = \{\langle 0|(\hat{\mathcal{S}}_-)^n(\hat{\mathcal{S}}_+)^n|0\rangle/n!\}^{-1/2}$ and the operator $\hat{\mathcal{S}}_+$ is the J=0-coupled nucleon pair

$$\hat{\mathcal{S}}_+ = \sqrt{\frac{\Omega}{2}}\Sigma_{m,m'}\langle jmjm'|00\rangle c^\dagger_{jm}c^\dagger_{jm'} \equiv \sqrt{\frac{\Omega}{2}}[c^\dagger_j c^\dagger_j]_0 \ , \ c_{jm}|0\rangle = 0 \ , \quad (3.14)$$

obeying the commutation relations of the group SU(2)

$$[\hat{\mathcal{S}}_+, \hat{\mathcal{S}}_-] = 2\hat{\mathcal{S}}_0 \ , \quad [\hat{\mathcal{S}}_0, \hat{\mathcal{S}}_\pm] = \pm\hat{\mathcal{S}}_\pm \ ,$$
$$\hat{\mathcal{S}}_- = (\hat{\mathcal{S}}_+)^\dagger \ , \quad \hat{\mathcal{S}}_0 = \frac{1}{2}(\Sigma_m c^\dagger_{jm}c_{jm} - \Omega) \ . \quad (3.15)$$

The BZ method for this case simply corresponds to the perturbative expansion (based on the Taylor series in the small parameter of Ω^{-1}) of the Holstein-Primakoff representation of the SU(2) algebra

$$\hat{\mathcal{S}}_+ \to \mathbf{s}_+^{[BZ]} = s^\dagger\sqrt{\Omega-\mathbf{n}_s} \ , \quad \hat{\mathcal{S}}_- \to \mathbf{s}_-^{[BZ]} = \sqrt{\Omega-\mathbf{n}_s}\,s \ ,$$
$$\hat{\mathcal{S}}_0 \to \mathbf{s}_0^{[BZ]} = s^\dagger s - \Omega/2 \ , \quad (3.16)$$

where $(s^\dagger, s)$ are the s-boson operators satisfying

$$[s, s^\dagger] = 1 \ , \quad \mathbf{n}_s \equiv s^\dagger s \ . \quad (3.17)$$

Making correspondence to the orthonormal set $\{|n\rangle\}$ in Eq.(3.13), in the modified Marumori method, we explicitly introduce an ideal s-boson space spanned by the orthonormal set $\{|n)_B; n=0,1,2,\cdots,\infty\}$ with

$$|n)_B = \frac{1}{\sqrt{n!}}(s^\dagger)^n|0)_B \ . \quad (3.18)$$

The MYT-type mapping operator in this case is then given by

$$\hat{\mathbf{U}}_M = |0\rangle\{\sum_{n=0}^{\Omega}|n)_B\langle n|\}\cdot{}_B(0| \ , \quad (3.19)$$

and we obtain the boson mapping

$$\hat{\mathcal{S}}_+ \to \hat{\mathbf{U}}_M \hat{\mathcal{S}}_+ \hat{\mathbf{U}}_M^\dagger = \hat{\Gamma}_0 \cdot \sum_{n=1}^{\Omega} \langle n|\hat{\mathcal{S}}_+|n-1\rangle \cdot |n)_B \cdot {}_B(n-1| \ . \tag{3.20}$$

Since the straightforward calculation shows

$$\langle n|\hat{\mathcal{S}}_+|n-1\rangle = \frac{1}{\sqrt{n!}} C_F(n) \langle 0|(\hat{\mathcal{S}}_-)^n(\hat{\mathcal{S}}_+)^n|0\rangle \frac{1}{\sqrt{(n-1)!}} C_F(n-1)$$

$$= \sqrt{n}\{C_F(n-1)/C_F(n)\} = \sqrt{n}\sqrt{\Omega-n+1} \equiv \sqrt{n} f(n) \ , \tag{3.21}$$

Eq. (3.20) is written with the use of Eq. (3.18) and of the formula $|0)_B \cdot {}_B(0| = :\exp(-s^\dagger s): \equiv \Sigma_{\gamma=0}^{\infty}(-1)^r (s^\dagger)^r (s)^r/r!$ as a <u>normal ordered</u> infinite series of boson expansion, whose rapid convergence was emphasized by Kleber.[28] With the use of Eq. (3.21), Eq. (3.20) can also be written as

$$\hat{\mathbf{U}}_M \hat{\mathcal{S}}_+ \hat{\mathbf{U}}_M^\dagger = \hat{\Gamma}_0 \Sigma_{n=1}^{\Omega} f(n)\sqrt{n}|n)_B \cdot {}_B(n-1|$$

$$= \hat{\Gamma}_0 \Sigma_{n=1} |n)_B \cdot {}_B(n| f(\mathbf{n}_s) s^\dagger = \mathbf{P} \cdot s^\dagger f(\mathbf{n}_s+1)$$

$$= \hat{\Gamma}_0 \cdot \mathbf{P} \cdot s^\dagger \sqrt{\Omega - \mathbf{n}_s} \equiv \hat{\Gamma}_0 \cdot \mathbf{S}_+^{[MYT]} \ , \tag{3.22}$$

where $\mathbf{P}$ is the projection operator to the physical boson subspace, $\mathbf{P} = \Sigma_{n=0}^{\Omega} |n)_B \cdot {}_B(n|$. This expression is simply the Holstein-Primakoff representation $\mathbf{S}_+^{[BZ]}$ with the precise restriction to the physical boson subspace $\{|n)_B; n=0,1,2,\cdots,\Omega\}$, i.e., $\mathbf{S}_+^{[MYT]} = \mathbf{P} \cdot \mathbf{S}_+^{[BZ]}$.

[Example 2] <u>Boson Mapping of J=2-Coupled Quasiparticle Pairs</u>[1),35)]

In order to define the MYT-type boson mapping in this case, we have to have the orthonormal basis of the collective subspace. This is done in the following way: (i) We introduce an orthonormal set of multi-d-boson states, $\{|n\beta JM)_B;\ n=0,1,2,\cdots,\infty\}$, to specify an ideal d-boson space with

$$|n\beta JM)_B \equiv |(d^\dagger)^n\beta JM)_B \ , \qquad d_\mu|n=0)_B = 0 \ , \tag{3.23}$$

where $\{d_\mu,\ d_\mu^\dagger;\ \mu=0,\pm1,\pm2\}$ are d-boson operators obeying

$$[d_\mu,\ d_{\mu'}^\dagger] = \delta_{\mu\mu'} \ , \qquad [d_\mu^\dagger,\ d_{\mu'}^\dagger] = 0 \ . \tag{3.24}$$

(ii) Then we define a set of states $\{|(\hat{D}^\dagger)^n\beta JM\rangle:\ n=0,1,2,\cdots\}$ in the fermion space by replacing the d-boson operators $d_\mu^\dagger$ in the state $|(d^\dagger)^n\beta JM)_B$ by the $\hat{D}_\mu^\dagger$ operators defined in Eq.(2.1a). In contrast with orthonormal boson states $|(d^\dagger)^n\beta JM)_B$, the states $|(\hat{D}^\dagger)^n\beta JM\rangle$ are neither orthogonal nor normalized. Furthermore they are <u>overcomplete</u> in general. (iii) Next we consider the eigenvalue equation for the norm matrix $M^{(n,J)}(\beta,\beta')$;

$$\Sigma_{\beta'} M^{(n,J)}(\beta,\beta')u^{(n,J)}(\beta',\gamma) = M(nJ\gamma)u^{(n,J)}(\beta,\gamma) \ , \tag{3.25}$$

$$M^{(n,J)}(\beta,\beta') \equiv \langle(\hat{D}^\dagger)^n\beta JM|(\hat{D}^\dagger)^n\beta' JM\rangle \ , \tag{3.26}$$

where the eigenvectors satisfy the orthonormality relation

$$\Sigma_\beta u^{(n,J)}(\beta,\gamma)u^{(n,J)}(\beta,\gamma') = \delta_{\gamma\gamma'} \ . \tag{3.27}$$

(iv) Denoting the eigenvectors with the zero eigenvalue $M(nJ\gamma) = 0$ by $\gamma = \gamma_0$, we can define the orthonormal set $\{|n\gamma JM\rangle;n=0,1,2,\cdots\}$ by

$$|n\gamma JM\rangle = M(nJ\gamma)^{-1/2}\sum_\beta u^{(n,J)}(\beta,\gamma)|(\hat{D}^\dagger)^n\beta JM\rangle \ , \text{ for } \gamma \neq \gamma_0 \ . \tag{3.28}$$

The corresponding boson basis $\{|n\gamma JM)_B, n=0,1,2,\cdots\}$ is given by

$$|n\gamma JM)_B = \Sigma_\beta u^{(n,J)}(\beta,\gamma)|(d^\dagger)^n\beta JM)_B \ . \tag{3.29}$$

With the use of the orthonormal basis vectors (3.28) and (3.29), we can define the MYT-type boson mapping by

$$\hat{\mathbf{U}}_M = |0\rangle \cdot \{ \sum_{\gamma \neq \gamma_0} \sum_{nJM} |n\gamma JM)_B \langle n\gamma JM| \} \cdot {}_B(0| \ . \tag{3.30}$$

Using this mapping, we obtain the boson mapping result[1)]

$$\hat{D}^{\dagger}_{\mu} \rightarrow \hat{\mathbf{U}}_M \hat{D}^{\dagger}_{\mu} \hat{\mathbf{U}}^{\dagger}_M = \hat{\Gamma}_0 \cdot \{ \mathbf{P}\mathbf{M}^{1/2} d^{\dagger}_{\mu} \mathbf{M}^{-1/2} \} \equiv \hat{\Gamma}_0 \cdot \mathbf{D}^{\dagger}_{\mu} \ , \tag{3.31}$$

$$\mathbf{P} = \sum_{\gamma \neq \gamma_0} \sum_{nJM} |n\gamma JM)_B \cdot {}_B(n\gamma JM| \ , \tag{3.32}$$

where the operator $\mathbf{M}$ in the boson space is defined by

$$(n\gamma JM|\mathbf{M}|n'\gamma' J'M')_B \equiv M(nJ\gamma)\delta_{nn'}\delta_{\gamma\gamma'}\delta_{JJ'}\delta_{MM'} \ . \tag{3.33}$$

In the modified Marumori method, therefore, the problem is reduced to calculate the norm matrix (3.26).

Holzwarth, Janssen and Jolos[34)] derived a recursion formula for the norm matrix with a kinematical assumption

$$[\hat{D}_{\mu}, [\hat{D}_{\nu}, \hat{D}^{\dagger}_{\mu'}]] \cong -\Sigma_{\nu'} \Gamma^{\mu\nu}_{\mu'\nu'} \hat{D}_{\nu'} \ , \tag{3.34}$$

which means that the double commutator does not scatter out of the collective subspace. Starting from the two-phonon norm $M^{(2J)}(\beta,\beta')$, which involves the c-number quantity $\Gamma^{\mu\nu}_{\mu'\nu'}$, we are therefore able to express all higher norm matrix elements $M^{(nJ)}(\beta,\beta')$ by $\Gamma^{\mu\nu}_{\mu'\nu'}$. Iwasaki, Sakata and Takada[31)] found a method to obtain an exact recursion formula for the norm matrix $M^{(nJ)}(\beta,\beta')$. It turned out that the norm matrix elements droped rapidly for large values of n due to the effect of the Pauli principle, so that the mapped boson Hamiltonian converged rapidly. In a similar way to Iwasaki, Sakata and Takada's, Tamura[32)] formulated a method to obtain $\mathbf{D}^{\dagger}_{\mu}$ in Eq. (3.31) in a boson-expansion form to any desired orders.

From the discussions so far, we arrive at the important conclusion that, if once the collective subspace in the fermion-state space is given from the outset, an exact boson image of it is easily obtained by the modified Marumori method. Furthermore, the algebra we need in performing it is straightforward.

§4. SU(6)-Approximation and the Interacting Boson Model

[A] SU(6)-Approximation

The operator $\mathbf{D}_\mu^\dagger$ in Eq.(3.31) is generally given in a form of an infinite series of d-boson operators. As is shown later, however, the operator $\mathbf{D}_\mu^\dagger$ is approximately expressed in a compact form when the SU(6)-approximation[35] is adopted.[1),36]

Based on this SU(6)-approximation, Janssen, Jolos and Dönau[35] have for the first time introduced the SU(6) description of nuclear quadrupole degrees of freedom in terms of interacting bosons. This idea in terms of the symmetry representation of the group SU(6) has been extensively studied by Arima and Iachello[37] in their phenomenological interacting boson model (IBM).

The SU(6)-approximation is based on the following two assumptions: (i) All the double commutators among $\hat{D}_\mu^\dagger$ and $\hat{D}_\mu$ do not scatter out of the collective subspace (i.e., Eq.(3.34)), and (ii) the J-dependence of C_J which appears in the double commutator is neglected in such a way that

$$[[\hat{D}_\mu, \hat{D}_{\mu'}^\dagger], \hat{D}_{\mu''}^\dagger] \cong -\bar{C}(\delta_{\mu\mu'}\hat{D}_{\mu''}^\dagger+\delta_{\mu\mu''}\hat{D}_{\mu'}^\dagger) , \qquad (4.1)$$

where $\bar{C}$ is an average value of the quantity

$$\bar{C} = 1 - \langle(\hat{D}^\dagger)^2 JM|(\hat{D}^\dagger)^2 JM\rangle , \quad J = 0,2,4 , \qquad (4.2)$$

$|(\hat{D}^\dagger)^2 JM\rangle$ being a fermion state obtained by replacing the d-boson operator $d_\mu^\dagger$ in the normalized d-boson state $|(d^\dagger)^2,JM)_B$ by the collective 2QP-mode $\hat{D}_\mu^\dagger$ in Eq.(2.1a). With the SU(6)-approximation, Janssen, Jolos and Dönau[35] showed that the collective 2QP-mode operators $\hat{D}_\mu^\dagger$ and $\hat{D}_\mu$ and their commutators

$[\hat{D}_\mu, \hat{D}^\dagger_{\mu'}]$ formed 35 linearly-independent basis operators of the Lie algebra of the group SU(6)

$$\tilde{D}^\dagger_\mu \equiv (1/\bar{C})\hat{D}^\dagger_\mu , \quad \tilde{F}_{\mu\mu'} \equiv [\tilde{D}_\mu, \tilde{D}^\dagger_{\mu'}] ,$$

$$[\tilde{D}^\dagger_\mu, \tilde{F}_{\mu'\mu''}] = \delta_{\mu'\mu''}\tilde{D}^\dagger_\mu + \delta_{\mu\mu'}\tilde{D}^\dagger_{\mu''}$$

$$[\tilde{D}_\mu, \tilde{F}_{\mu'\mu''}] = -\delta_{\mu'\mu''}\tilde{D}_\mu - \delta_{\mu\mu'}\tilde{D}_{\mu''} ,$$

$$[\tilde{F}_{\mu\mu'}, \tilde{F}_{\mu''\mu'''}] = -\delta_{\mu\mu'''}\tilde{F}_{\mu''\mu'} + \delta_{\mu'\mu''}\tilde{F}_{\mu\mu'''} \tag{4.3}$$

They also found a boson representation which exactly satisfies the Lie algebra of the group SU(6), Eq.(4.3). The representation is given by

$$\tilde{D}^\dagger_\mu \rightarrow d^\dagger_\mu\sqrt{\mathcal{N}-n_d} , \quad \tilde{D}_\mu \rightarrow \sqrt{\mathcal{N}-n_d}\cdot d_\mu$$
$$\tilde{F}_{\mu\mu'} \rightarrow -d^\dagger_\mu d_{\mu'} + (\mathcal{N}-n_d)\delta_{\mu\mu'} , \quad n_d \equiv \Sigma_\mu d^\dagger_\mu d_\mu . \tag{4.4}$$

For arbitrary value of the constant $\mathcal{N}$, as far as it is larger than the maximum number of d-bosons for the physical boson states, the representation (4.4) exactly reproduces the algebra (4.3). Arima and Iachello[37] in their first paper of the IBM have introduced a very elegant way for treating the square root factor in (4.4) by making a transformation

$$d^\dagger_\mu\sqrt{\mathcal{N}-n_d} \rightarrow d^\dagger_\mu s , \quad \sqrt{\mathcal{N}-n_d}\cdot d_\mu \rightarrow s^\dagger d_\mu , \tag{4.5a}$$

$$\mathcal{N}-n_d = s^\dagger s \equiv n_s , \tag{4.5b}$$

where $s^\dagger$ is an s-boson with angular momentum J=0.

The relation between the Modified Marumori method and this algebraic method of the SU(6)-approximation can be understood in the following way[1),36)]. When the SU(6)-approximation (4.3) is applied to Iwasaki-Sakata-Takada's recurrence formula for the norm matrix, we have a simple relation[31)]

$$M(nJ\gamma) = \prod_{m=1}^{n} \{1-(m-1)\bar{C}\} , \tag{4.6}$$

which depends only on n. In this case, Eq.(3.31) is drastically simplified to be

$$\hat{D}^{\dagger}_{\mu} \to \hat{\mathbf{U}}_M \hat{D}^{\dagger} \hat{\mathbf{U}}^{\dagger}_M = \hat{\Gamma}_0 \cdot \{\mathbf{P} \cdot d^{\dagger}_{\mu} \sqrt{1-\bar{C} \cdot \mathbf{n}_d}\} . \tag{4.7}$$

Comparing Eq.(4.4) with Eq.(4.7), we can easily see that the Janssen-Jolos-Dönau algebraic boson representation is completely equivalent to the modified Marumori boson-mapping result except for the projection operator $\mathbf{P}$ onto the physical collective boson subspace. Thus, we can <u>determine</u> the arbitrary constant $\mathcal{N}$ in Eq.(4.4) so as to satisfy

$$\mathcal{N} = 1/\bar{C} . \tag{4.8}$$

The mapped collective Hamiltonian under the SU(6)-approximation is then written as

$$\hat{H} \to \hat{\mathbf{U}}_M \hat{H} \hat{\mathbf{U}}^{\dagger}_M = \hat{\Gamma}_0 (\mathbf{P} \cdot \mathbf{H})$$

$$\begin{aligned} \mathbf{H} = h_0 &+ \varepsilon_s s^{\dagger} s + \varepsilon_d [d^{\dagger} d]_0 + \Sigma_{L=0,2,4} C_L [[d^{\dagger} d^{\dagger}]_L [dd]_L]_0 \\ &+ \{V_2 [[d^{\dagger} d^{\dagger}]_2 [ds]_2]_0 + V_0 [d^{\dagger} d^{\dagger}]_0 ss \\ &+ u_2 [[d^{\dagger} s^{\dagger}]_2 [ds]_2]_0 + \text{h.c.}\} \end{aligned} \tag{4.9}$$

with

$$\mathcal{N} = \mathbf{n}_d + \mathbf{n}_s , \tag{4.10}$$

where we have used Eq.(4.5) and the symbol $[\quad]_{JM}$ denotes the angular momentum coupling. This form of the collective Hamiltonian is simply the same as one which the <u>phenomenological</u> IBM[37] (renamed the IBM-1 later) starts from. It must be emphasized, however, that the s- boson here is a correspondent to the square root factor $\sqrt{1 - \bar{C}\mathbf{n}_d}$ in Eq.(4.7) (which originates from the Pauli-principle between the collective eigen-mode operators $D^{\dagger}_{\mu}$) and does not mean any introduction of a new degree of freedom under the condition (4.5b). Therefore, it has nothing to do with the s-boson in the IBM,[37] which is identified with the boson image of the J=0- coupled nucleon pair. Correspondingly, the constant $\mathcal{N}$ in Eq. (4.10) has nothing to do with the nucleon number identified by the IBM. As has been shown in Eq.(4.8), it depends strongly on the structure of the basis operators $\hat{D}^{\dagger}_{\mu}$.

[B] The IBM in the Light of the Modified Marumori Method

In the previous discussion [A], we have employed the quasi-particle representation based on the BCS ground state. In order to relate the IBM with the modified Marumori boson mapping, it is necessary to discuss the boson mapping of the nucleon-state subspace, which is built up with correlated nucleon pairs coupled to either J = 0 and J = 2. This is because the phenomenological IBM Hamiltonian (e.g. the IBM-2 Hamiltonian) is given by

$$\mathbf{H}_{IBM-2} = \varepsilon(\mathbf{n}_{d\pi}+\mathbf{n}_{d\nu}) + \kappa\mathbf{Q}_{\pi}\cdot\mathbf{Q}_{\nu} + \mathbf{M}_{\pi\nu} + \mathbf{V}_{\pi\pi} + \mathbf{V}_{\nu\nu}{}^{*)} \tag{4.11a}$$

$$(\mathbf{Q}_{\rho})_{2\mu} = [d^{\dagger}_{\rho}s_{\rho}]_{2\mu} + [s^{\dagger}_{\rho}d_{\rho}]_{2\mu} + \chi_{\rho}[d^{\dagger}_{\rho}d_{\rho}]_{2\mu} \tag{4.11b}$$

$$\mathbf{V}_{\rho\rho} = \sum_{L=0,2,4} (\sqrt{2L+1}/2)\cdot C^{(\rho)}_{L}[[d^{\dagger}_{\rho}d^{\dagger}_{\rho}]_{L}[d_{\rho}d_{\rho}]_{L}]_{00}; \quad (\rho=\pi,\nu) \;, \tag{4.11c}$$

and the IBM identifies the s and d bosons with the correlated nucleon pairs coupled to J = 0 and J = 2, respectively. Thus, the IBM asserts that

(i) The sum of the number operators of s and d bosons must be equal to the number of fermion pairs in the open

---

*) We do not write down the so-called Majorana term $\mathbf{M}_{\pi\nu}$ explicitly, because it is lengthy and its explicit form is not important in the following discussions.

shell; $n_s + n_d = N$.

(ii) The Hamiltonian terminates in a quartic form.

(iii) The Hamiltonian and some other physical operators must contain only terms that conserve the boson number N.

In this model, therefore, the collective subspace may be spanned by a set of basis states;

$$|N;\beta\rangle = (1/\sqrt{N!})\{\hat{P}_0 (\hat{\mathcal{D}}^\dagger)^q\}(\hat{\mathcal{S}}^\dagger)^{N-q}|0\rangle \ , \qquad q = 0,1,2,\cdots,N \tag{4.12a}$$

$$\hat{\mathcal{D}}^\dagger_\mu = \sum_{ab} \psi(ab;J=2)[c_a^\dagger c_b^\dagger]_{J=2} ,$$

$$\hat{\mathcal{S}}^\dagger = \sum_a \psi(aa;J=0)[c_a^\dagger c_a^\dagger]_{J=0} \ . \tag{4.12b}$$

where $|0\rangle$ is the vacuum of nucleons, $c_\alpha|0\rangle=0$, and β is a set of additional quantum numbers to specify the state. The operator $\hat{P}_0$ is the projection operator to remove all J=0-coupled nucleon pairs arising from the recoupling of $(\hat{\mathcal{D}}^\dagger)^q$. Evidently, the collective subspace defined by the set of states (4.12a) is the same as that adopted by Otsuka, Arima Iachello and Talmi[38] in their work to give a microscopic interpretation of the IBM.

The states $|N;\beta\rangle$ are neither orthogonal nor normalized, and may generally be overcomplete. The norm matrix is defined by

$$\langle N;\beta|N;\beta'\rangle = M^{(N)}(\beta,\beta') \ . \tag{4.13}$$

In the same way as Eq. (3.28), we then obtain an orthonormal set by

$$|N;\gamma\rangle = M(N,\gamma)^{-1/2} \sum_\beta u^{(N)}(\beta,\gamma)|N;\beta\rangle \quad \text{for} \quad \gamma \neq \gamma_0 \ . \tag{4.14}$$

where $u^{(N)}(\beta,\gamma)$'s are the components of an eigenvector of the norm matrix (4.13) belonging to the non-zero eigenvalues $M(N,\gamma)$, and satisfy the orthonormality relation (3.27). (γ_0 denotes the eigenvectors with the zero eigenvalue $M(N,\gamma)=0$.) The corresponding orthonormal set of boson states is then written as

$$|N;\gamma)_B = \sum_{\beta} u^{(N)}(\beta,\gamma)|N;\beta)_B \,, \qquad \gamma \neq \gamma_0$$
$$|N;\beta)_B = \frac{1}{\sqrt{N!}}(d^\dagger)^q(s^\dagger)^{N-q}|0)_B \qquad (q = 0,1,2,\cdots N) \,. \tag{4.15}$$

Thus, the MYT-type mapping operator in this case is defined by

$$\hat{U}_M = |0>\{\sum_{\gamma\neq\gamma_0} |N;\gamma)_B \cdot <N;\gamma|\} \cdot {}_B(0| \,. \tag{4.16}$$

The modified Marumori mapping (4.16), however, yields a quite different boson Hamiltonian from that of the IBM.[39),1),10)] This will be discussed in the next section.

§5. Intrinsic and Pairing Subspaces in the Fermion-State Space

In order to see the physical situation in the mapping with Eq.(4.16) in a transparent way, it is decisive to define intrinsic and pairing subspaces in the nucleon-state space.[40),41),42)] The intrinsic subspace consists of the states which never involve the J=0-coupled nucleon pairs: Thus, any state in this intrinsic subspace has to satisfy the condition

$$\hat{\mathcal{S}}_-(a)|\psi_{intr}> = 0 \,, \quad \text{for all } a \,,$$
$$\hat{\mathcal{S}}_-(a) = \sqrt{\Omega_a/2}\cdot[c_a c_a]_{J=0} \,, \quad \Omega_a \equiv j_a + \frac{1}{2} \,. \tag{5.1}$$

The orthonormal basis vectors in the intrinsic subspace are therefore given by

$$\{|n;\Gamma>> = K_n \cdot \hat{P}_0[c^\dagger_{\gamma_1} c^\dagger_{\gamma_2} \cdots c^\dagger_{\gamma_n}]|0>; \; n=0,1,2,\cdots\} \tag{5.2}$$

where K_n is the normalization constant and $\hat{P}_0$ is the projection operator by which all the J=0-coupled nucleon pairs are completely removed out of the operator $c^\dagger_{\gamma_1} c^\dagger_{\gamma_2} \cdots c^\dagger_{\gamma_n}$. (The eigenvectors of the projection operator $\hat{P}_0$ are therefore related to the coefficients of the fractional parentage(cfp) with seniority classification.) By definition, in the intrinsic subspace the number of the nucleons n exactly corresponds to the seniority number v, i.e., n=v.

Contrary to the intrinsic subspace, the pairing subspace[40),41),42)] consists of the states which involve only the J=0-coupled nucleon pairs, so that any state in this subspace has the seniority v=0. The basis vectors of the pairing subspace are expressed as

$$\{\prod_a [\hat{\mathcal{S}}_+(a)]^{q_a}|0\rangle;\ \Sigma q_a = 1,2,\cdots\}\ ,$$
$$\hat{\mathcal{S}}_+(a) = \sqrt{\Omega_a/2}\,[c_a^\dagger c_a^\dagger]_{J=0} = \{\hat{\mathcal{S}}_-(a)\}^\dagger\ . \tag{5.3}$$

The concepts of the intrinsic and pairing subspace make the physical situations in the boson mapping (4.16) transparent. To see this, we perform the boson mapping through the following two-step procedure.(Fig.3)

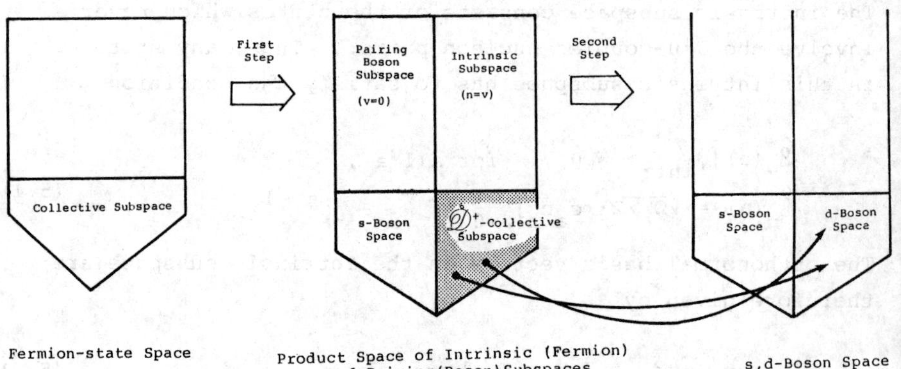

FIG. 3. STRUCTURE OF THE BOSON MAPPING OF THE COLLECTIVE SUBSPACE
$(\hat{P}_0(\hat{\mathcal{D}}^+)^q(\hat{\mathcal{S}}^+)^{n-q}|0\rangle) \rightarrow ((d^+)(s^+)^{n-q}|0)_B)$

As a first step, we employ the transformation[42),43),44)] which enables us to replace all the pairing degrees of freedom $\hat{\mathcal{S}}_+(a)$ by bosons $s_a^\dagger$ through

$$\hat{\mathcal{S}}_+(a) \rightarrow s_a^\dagger\sqrt{\Omega_a - \hat{v}_a - s_a^\dagger s_a} \tag{5.4a}$$

$$[s_a, s_b^\dagger] = \delta_{ab} ,$$

where $\hat{v}_a$ is the seniority number operator of the orbit a defined by ideal quasiparticles[44)] $(\mathring{c}_\alpha^\dagger, \mathring{c}_\alpha)$ in the intrinsic subspace

$$\hat{v}_a = \sum_{m_\alpha} \mathring{c}_\alpha^\dagger \mathring{c}_\alpha , \qquad \mathring{c}_\alpha^\dagger \equiv \hat{\Gamma}_{intr} c_\alpha^\dagger \hat{\Gamma}_{intr} , \tag{5.4b}$$

$\hat{\Gamma}_{intr}$ being the projection operator into the intrinsic subspace,

$$\hat{\Gamma}_{intr} = \sum_{n,\Gamma} |n;\Gamma\rangle\rangle\langle\langle n;\Gamma| . \tag{5.5}$$

Since the intrinsic and the pairing subspaces are orthogonal to each other by definition, the nucleon-state space after the transformation is replaced by a product space of a boson space and a fermion space: The boson space is spanned by a orthonormal set

$$\{\prod_a \frac{1}{\sqrt{q_a!}}(s_a^\dagger)^{q_a}|0)_B ; \sum_a q_a = 0,1,2,\cdots,\} \tag{5.6}$$

and exactly corresponds to the pairing subspace, and the fermion space corresponds to the intrinsic subspace which is described in terms of only the ideal quasiparticles $(\mathring{c}_\alpha^\dagger, \mathring{c}_\alpha)$. Within the pairing boson space spanned by Eq. (5.6), we can easily define the s-boson subspace employed in Eq. (4.15),

$$\{(s^\dagger)^p|0)_B ; p = 0,1,2,\cdots,N\} ,$$
$$s^\dagger = \sum_a \psi(aa;J=0)\cdot\sqrt{2/\Omega_a}\cdot s_a^\dagger . \tag{5.7}$$

where the correlation amplitude $\psi(aa;J=0)$ are the same as one given in Eq. (4.12b).

As the second step, we make the d-boson mapping,

$$\{[\hat{P}_0(\hat{\mathfrak{D}}^\dagger)^q]|0\rangle\cdot(s^\dagger)^{N-q}|0)_B;\ q=0,1,2,\cdots,N\}$$

$$\longrightarrow \{(d^\dagger)^q(s^\dagger)^{N-q}|0)_B;\ q=0,1,2,\cdots,N\}\ . \qquad (5.8)$$

This mapping can be easily made in a way formally parallel to that given in [Example 2] in §3. Only one essential difference in this case is that, instead of $|(\hat{D}^\dagger)^n\beta JM\rangle$ in the definition of the norm matrix (3.26), we have to use $|[\hat{P}_0(\hat{\mathfrak{D}}^\dagger)^n]\beta JM\rangle$, which is written as $|(\mathring{\mathfrak{D}}^\dagger)^n\beta JM\rangle$ in terms of the ideal quasiparticles $(\mathring{c}_\alpha^\dagger,\ \mathring{c}_\alpha)$ with

$$\mathring{\mathfrak{D}}_\mu^\dagger \equiv \sum_{ab}\psi(ab;J=2)\,[\mathring{c}_a^\dagger\mathring{c}_b^\dagger]_{J=2,\mu}\ . \qquad (5.9)$$

We thus obtain the d-boson mapping corresponding to Eq. (3.31),

$$\mathring{\mathfrak{D}}_\mu^\dagger \rightarrow \mathring{\mathbf{D}}_\mu^\dagger = \mathring{\mathbf{M}}^{1/2}d_\mu^\dagger\mathring{\mathbf{M}}^{-1/2}\ , \qquad (5.10)$$

where the operator $\mathring{\mathbf{M}}$ in the d-boson space is defined by

$$(n\gamma JM|\mathring{\mathbf{M}}|n'\gamma'JM')_B = \mathring{M}(nJ\gamma)\,\delta_{nn'}\delta_{\gamma\gamma'}\delta_{JJ'}\delta_{MM'}\ , \qquad (5.11)$$

$\mathring{M}(nJ\gamma)$ being the non-zero eigenvalue of the norm matrix $\mathring{M}^{(nJ)}(\beta,\beta') = \langle(\mathring{\mathfrak{D}}^\dagger)^n\beta JM|(\mathring{\mathfrak{D}}^\dagger)^n\beta' JM\rangle$.

It is now transparent that the modified Marumori mapping with (4.16) leads us to a quite a different boson Hamiltonian from that of the IBM. Contrary to the IBM, the resultant boson Hamiltonian never terminates in a quartic form, because the operator $\mathring{\mathbf{D}}_\mu^\dagger$ in Eq. (5.10) is generally expressed in a form of an infinite series of d-boson operators. Of oucrse, we can also use the SU(6) approximation discussed §4, in order to express the operator $\mathring{\mathbf{D}}_\mu^\dagger$ in a compact form approximately. Corresponding to Eq. (4.7), then the SU(6) approximation leads to

$$\mathring{D}^{\dagger}_{\mu} \simeq d^{\dagger}_{\mu}\sqrt{1 - \bar{\mathring{C}}\cdot n_d} , \qquad (5.12)$$

where $\bar{\mathring{C}}$ is an average value of $\mathring{C}_J = 1-\mathring{M}$ $(n=2,J,\gamma)$ with $J=0,2,4$. In this case, however, the resultant Hamiltonian must have operators due to the square root factors in place of the c-number parameters of the original IBM. When we use the Arima-Iachelle transformation (4.5) instead of such replacements, we are forced to introduce another type of s-boson in addition to the s-boson defined by Eq. (5.7).

§6. The BCS Approximation and the Quasiparticle Representation

In the first-step transformation discussed in §5, which transforms the nucleon-state space into the product space of the pairing (boson) space and the intrinsic (fermion) space, we have defined the ideal quasiparticle operator $(\mathring{c}^{\dagger}_{\alpha}, \mathring{c}_{\alpha})$ by Eq. (5.4b). Suzuki and Matsuyanagi[44] have found that, after the first step transformation, the nucleon operators $(c^{\dagger}_{\alpha}, c_{\alpha})$ can be expressed in terms of the ideal quasiparticle operators and the s_a-boson operators as

$$c^{\dagger}_{\alpha} \rightarrow \mathring{c}^{\dagger}_{\alpha}\hat{u}_a + \hat{v}_a\mathring{c}_{\tilde{\alpha}} \qquad (6.1a)$$

$$\hat{u}_a = \sqrt{1 - \frac{s^{\dagger}_a s_a}{\Omega_a - \hat{\nu}_a}} , \quad \hat{v}_a = \frac{s_a}{\sqrt{\Omega_a - \hat{\nu}_a}} \qquad (6.2a)$$

$$\hat{u}^{\dagger}_a\hat{u}_a + \hat{v}^{\dagger}_a\hat{v}_a = 1 ,$$

where $\hat{\nu}_a = \Sigma_{m_\alpha} \mathring{c}^{\dagger}_{\alpha}\mathring{c}_{\alpha}$. This expression is very much like the well-known Bogoliubov transformation

$$c^{\dagger}_{\alpha} = u_a a^{\dagger}_{\alpha} + v_a a_{\tilde{\alpha}} . \qquad (6.3)$$

Indeed, as we all know, the Bogoliubov quasiparticles $(a^{\dagger}_{\alpha}, a_{\alpha})$ have been introduced as substantiation of the seniority concept, in such a way that the quasiparticle numbers n_a

correspond to the seniority numbers v_a. Thus, the Bogoliubov quasiparticles $(a^\dagger_\alpha, a_\alpha)$ can be regarded as an approximate solution of the ideal quasiparticle operators $(\overset{\circ}{c}{}^\dagger_\alpha, \overset{\circ}{c}_\alpha)$, which is obtained by replacing $\hat{u}_a$ and $\hat{v}_a$ with the c-number u_a and v_a under the assumption[39)] that the s_a-bosons can be decomposed into static and fluctuation parts

$$s^\dagger_a = \nu_a + \tilde{s}^\dagger_a \tag{6.4}$$

where ν_a are real parameters specifying a coherent state of the s_a-bosons. The coherent state may be determined as an approximate solution of the ground state in the pairing boson space

$$|\Phi_0)_B = \frac{1}{\sqrt{N!}} (s^\dagger)^N |0)_B \,, \tag{6.5}$$

where $s^\dagger$ is defined by Eq. (5.7).

The BCS approximation is just one of the simplest and convenient way to determine the approximate ground state with <u>the full account of the Pauli-principle</u>. With the correspondence Eq. (6.5), the ground state in the pairing subspace is given by

$$|\Phi_0> \propto (\hat{\mathcal{S}}^\dagger)^N |0> \tag{6.6}$$

where $\hat{\mathcal{S}}^\dagger$ is given by Eq. (4.12b) with the <u>lowest energy distribution</u> $\psi(aa;J=0)$ of the J=0-coupled nucleons pairs $\hat{\mathcal{S}}_+(a)$ of the orbit a. Because the calculation of the variational parameter $\psi(aa;J=0)$ is practically difficult, in the BCS approximation, the state $|\Phi_0>$ is approximated to be of the form

$$|\Phi_0> \Rightarrow |\phi_{BCS}> = \prod_{\alpha>0} (u_a + v_a c^\dagger_\alpha c^\dagger_{\tilde{\alpha}})|0>$$

$$= (\prod_{\alpha>0} u_a) \exp\{\sum_a \frac{v_a}{u_a} \hat{\mathcal{S}}_+(a)\}|0> \,. \tag{6.7}$$

The parameters u_a and v_a are determined by the variational principle with the constraint condition that the expectation value of the particle number $\hat{N} = \sum_\alpha c^\dagger_\alpha c_\alpha$ is the given value 2N, i.e., $<\phi_{BCS}|\hat{N}|\phi_{BCS}> = \sum_a v_a^2 = 2N$. Comparing Eq. (6.6) with Eq. (6.7), we can easily see that the operator $\hat{\mathcal{S}}^\dagger$ in (6.6) is now approximated by

$$\hat{\mathcal{S}}^\dagger \Rightarrow \sum_a \frac{v_a}{u_a} \hat{\mathcal{S}}_+(a) \ . \tag{6.8}$$

The use of $|\phi_{BCS}>$ as a condensate of J=0-coupled pairs $\hat{\mathcal{S}}^\dagger$ for $|\Phi_0>$ has a disadvantage of violation of the particle-number conservation. However, it has a great advantage to implement the Pauli-principle exactly. It is also knownthat its violation of number conservation produces no major errors, as long as the exact energy is a smooth function of the particle number. Indeed, Vincent[45)] has recently demonstrated that the BCS approximation is much more reliable than the so-called number-operator approximation proposed by Otsuka and Arima,[46)] where the average number of particles in each orbit can easily violates the Pauli-principle.

In the quasiparticle representation based on $|\phi_{BCS}>$, the state with seniority v=2 is simply approximated by

$$\hat{\mathcal{D}}_\mu^\dagger (\hat{\mathcal{S}}^\dagger)^{N-1}|0> \Rightarrow \hat{D}_\mu^\dagger |\phi_{BCS}> \ ,$$
$$\hat{D}_\mu^\dagger = \sum_{ab} \psi(ab;J=2)[a_a^\dagger a_b^\dagger] \ , \tag{6.9}$$

This is because we have

$$a_\alpha^\dagger a_\beta^\dagger |\phi_{BCS}> = c_\alpha^\dagger c_\beta^\dagger \prod_{\gamma\neq\alpha,\beta} (u_c + v_c c_\gamma^\dagger c_{\tilde{\gamma}}^\dagger)|0> \ , \ (\beta\neq-\alpha) . \tag{6.10}$$

In the same way, the state $\{\hat{P}_0(\hat{\mathcal{D}}^\dagger)^q\}(\hat{\mathcal{S}}^\dagger)^{N-q}|0>$ with seniority v= 2q in Eq. (4.12a) is approximated as a 2q-quasiparticle state

$$\{\hat{P}_0(\hat{\mathcal{D}}^\dagger)^q\}\cdot(\hat{\mathcal{S}}^\dagger)^{N-q}|0> \Rightarrow \{\hat{P}_0(\hat{D}^\dagger)^q\}|\phi_{BCS}> \ . \tag{6.11}$$

Thus, in the quasiparticle representation, the intrinsic subspace is spanned by a set of orthonormal basis vectors

$$\{|n;\Gamma_{BCS}\rangle\rangle = K_n \cdot \hat{P}_0 [a^\dagger_{\gamma_1} a^\dagger_{\gamma_2} \cdots a^\dagger_{\gamma_n}] |\phi_{BCS}\rangle; n=0,2,4,\cdots\}. \quad (6.12)$$

Contrary to the ideal quasiparticles ($\overset{\circ}{c}{}^\dagger_\alpha$, $\overset{\circ}{c}{}^\dagger_\alpha$), which define only the intrinsic subspace by definition, the use of the Bogoliubov quasiparticles ($a^\dagger_\alpha$, a_α) makes it possible to represent the pairing subspace*) in terms of the basis vectors

$$\begin{aligned} &\{\prod_a [\hat{S}_+(a)]^{q_a} |\phi_{BCS}\rangle; \ \sum_a q_a = 1,2,\cdots\} , \\ &\hat{S}_+(a) = \sqrt{\Omega_a/2} \cdot [a^\dagger_a a^\dagger_a]_{J=0} = \{\hat{S}_-(a)\}^\dagger . \end{aligned} \quad (6.13)$$

In the pairing boson space spanned by Eq. (5.6), the choice of the basis bectors (6.13) corresponds to that of the basis vectors

$$\{\prod_a \frac{1}{\sqrt{q_a!}} (\tilde{s}^\dagger_a)^{q_a} |\phi)_B; \ \sum_a q_a = 0,1,2,\cdots\} \quad (6.14)$$

where the operator $\tilde{s}^\dagger_a$ is the fluctuation part of $s^\dagger_a$ operator defiend in Eq. (6.4), and $|\phi)_B$ is the coherent boson state satisfying $s_a|\phi)_B = \nu_a|\phi)_B$ i.e., $\tilde{s}_a|\phi)_B = 0$. (Needless to say, the coherent boson state $|\phi)_B$ corresponds to the BCS ground state $|\phi_{BCS}\rangle$.) Thus, the pairing subspace in terms of the quasiparticles <u>physically</u> describes various excited states with the <u>same seniority number v=0</u> as that of the ground state given by Eq. (6.5). One of the important excitation modes within the pairing subspace is the <u>pairing vibrational mode</u> of which importance has, for the first time, pointed out by Bés and Broglia.[47] It is well-known to play a decisive role in the two-nucleon transfer reactions.

Here it must be emphasized that, in the IBM, such a type of excitations in the pairing subspace are completely disregarded: In the IBM, once the J=0-coupled nucleon pair $\hat{\mathcal{S}}^\dagger$ is broken, the

---

*) Needless to say, all spurious states resulting from the violation of the nucleon-number conservation belong to this pairing subspace.

two-nucleons in the pair always have to create $\hat{\mathfrak{D}}_\mu^\dagger$, while the pairing excitation means that the J=0-coupled nucleon pair $\hat{\mathcal{S}}^\dagger$ is broken and the two-nucleons in the pair create another excited pair $\tilde{\mathcal{S}}^\dagger$ with J=0.

§7. Limit of Applicability of the SU(6)-Approximation

Through the discussions so far, it has been clarified that the SU(6)-approximation in the intrinsic subspace seems to be necessary for obtaining a (hermitian) boson Hamiltonian which terminates in a quartic form such as that of the IBM by means of the modified Marumori method. Thus, it becomes quite important to investigate to what extent the SU(6)-approximation in the intrinsic subspace can be reliable.

This can be performed in the following way:

i) We first construct elementary modes of excitation with seniority n[40,41,42] within the collective subspace $\{[\hat{P}_0(\hat{D}^\dagger)^q]|\phi_{BCS}\rangle$ in Eq. (6.11) by means of the Tamm-Dancoff (TD) method or the New-Tamm-Dancoff (NTD) method with the ground-state correlation. The same elementary modes are also constructed under the SU(6)-approximation.

ii) We then compare numerical results of the excitation energies as well as the B(E2) values obtained by the elementary modes with those obtained by the same elementary modes under the SU(6)-approximation.

Thus, the following answer on the limit of applicability of the SU(6)-approximation has been obtained:[48,49,50]

i) Within the framework of the TD method, in so far as the low-lying states in the ground-state (yrast) band are concerned, the SU(6)-approximation is good and can well reproduce the characteristic feature of the phenomenological "phonon" picture. For higher excited states in this band, however, the SU(6) approximation becomes worse.

ii) Within the framework of the NTD method, the more important the ground-state correlation becomes, the worse the SU(6) approximation is.

iii) The collective subspace $\{[\hat{P}_0(\hat{D}^\dagger)^q]|\phi_{BCS}\rangle$ and therefore the SU(6)-approximation cannot realize the quasi-side-band structure[18] characterized by the experimental facts.

This fact implies that there exists a significant deviation from the SU(6) symmetry in the microscopic structure of the lower boson-seniority states in the quasi-side bands.

In order to see the limit of applicability of the SU(6) approximation, as is seen from Eq. (4.2), it is more transparent to judge whether the relations

$$\bar{\overset{\circ}{c}} \approx \overset{\circ}{c}_0 \approx \overset{\circ}{c}_2 \approx \overset{\circ}{c}_4 \ , \qquad \overset{\circ}{c}_J \equiv 1 - \langle \hat{P}_0 \ (\hat{D}^\dagger)^2; JM | \hat{P}_0 \ (\hat{D}^\dagger)^2; JM \rangle \tag{7.1}$$

are satisfied or not. Suzuki, Fuyuki and Matsuyanagi[51] showed that the property $\overset{\circ}{c}_2 \approx \overset{\circ}{c}_4 \ll \overset{\circ}{c}_0$ always holds for spherical and transitional nuclei although the magnitudes of $\overset{\circ}{c}_J$ are strongly dependent on the internal structure of the collective eigenmode operator $\hat{D}^\dagger_\mu$. This implies that the condition (7.1) for the SU(6) approximation <u>is not generally satisfied</u>, with the exception of the low-lying states in the ground-state band, consisting of the highest boson-seniority states in which the quantity $\overset{\circ}{c}_0$ has no contribution by definition.

§8. Dynamical Interplay between Pairing and Quadrupole Correlations

We are now in a position to discuss the significant deviation from the SU(6) symmetry in the microscopic structure of the lower boson-seniority states in the quasi-side bands. Let us first remember that the K-quantum number is the concept defined in the <u>axially symmetrical</u> deformed nuclei. Therefore, Sakai's proposal[18] of the concept of "quasi-K-band" implies that a kind of <u>symmetry-keeping ingredient</u> plays an important role in the process of the phase transition to the deformed state.

As has been discussed at the end of §1, the dynamics of the spherical-deformed phase transition in medium-heavy nuclei is considered to be governed by an interweaving between the two different types of correlations, i.e., the pairing correlation to keep the spherical symmetry and the quadrupole correlation to produce the deformation. Thus, the significant deviation from the SU(6) symmetry implies that, even if the collective subspace defined in Eq. (4.12a),

$$\{[\hat{P}_0(\hat{\mathcal{D}}^\dagger)^q](\hat{\mathcal{S}}^\dagger)^{N-q}|0\rangle;\ q=0,1,2\cdots N\}\ , \tag{8.1}$$

is properly chosen, <u>there still remains an inter-space coupling</u> between the intrinsic subspace and the pairing subspace (involving the pairing correlations), which is <u>never</u> described within the collective subspace (8.1).

As going from the spherical to the transitional nuclei, the inter-space coupling becomes more and more <u>substantial</u> in causing the large and complicated quantum fluctuations in both subspaces, and finally we are led to reconstruct a new collective subspace for the deformed nuclei so as to provide the collective rotational motion.

In order to see the interspace coupling in a transparent way, let us make the transformation (6.1) with (6.2) for the two-body interaction

$$\hat{H}_{int} = \sum_{\alpha\beta\gamma\delta} v_{\alpha\beta\gamma\delta} c_\alpha^\dagger c_\beta^\dagger c_\delta c_\gamma\ . \tag{8.2}$$

We then have

$$\hat{H}_{int} \rightarrow \hat{\mathbf{H}}_{int} = \hat{\mathbf{H}}_X + \hat{\mathbf{H}}_V + \hat{\mathbf{H}}_Y\ , \tag{8.3}$$

where $\hat{\mathbf{H}}_X$, $\hat{\mathbf{H}}_V$ and $\hat{\mathbf{H}}_Y$ are <u>schematically</u> expressed as

$$\begin{aligned}
\hat{\mathbf{H}}_X &= \sum_{\alpha\beta\gamma\delta} \mathbf{V}_X(\alpha\beta\gamma\delta;\hat{\mathbf{u}},\hat{\mathbf{v}})\, \mathring{c}_\alpha^\dagger \mathring{c}_\beta^\dagger \mathring{c}_\delta \mathring{c}_\gamma\ ,\\
\hat{\mathbf{H}}_V &= \sum_{\alpha\beta\gamma\delta} \mathbf{V}_V(\alpha\beta\gamma\delta;\hat{\mathbf{u}},\hat{\mathbf{v}})\, \mathring{c}_\alpha^\dagger \mathring{c}_\beta^\dagger \mathring{c}_{\tilde{\delta}}^\dagger \mathring{c}_{\tilde{\gamma}}^\dagger + \text{h.c.}\\
\hat{\mathbf{H}}_Y &= \sum_{\alpha\beta\gamma\delta} \mathbf{V}_Y(\alpha\beta\gamma\delta;\hat{\mathbf{u}},\hat{\mathbf{v}})\, \mathring{c}_\alpha^\dagger \mathring{c}_\beta^\dagger \mathring{c}_{\tilde{\delta}}^\dagger \mathring{c}_\gamma + \text{h.c.}
\end{aligned} \tag{8.4}$$

Inserting Eq. (6.4) into Eq. (8.3), we can decompose $\hat{\mathbf{H}}_{\text{int}}$ into the following static and fluctuation parts

$$\hat{\mathbf{H}}_{\text{int}} = \hat{H}^{(0)}_{\text{int}} + \tilde{\mathbf{H}}_{\text{int}} \tag{8.5}$$

where $\hat{H}^{(0)}_{\text{int}}$ denotes an expression of the interaction (8.2) in terms of the Bogoliubov quasiparticle $(a^{\dagger}_{\alpha}a_{\alpha})$ within the intrinsic subspace (6.12). The other part $\tilde{\mathbf{H}}_{\text{int}}$ always contains $\tilde{s}_a$-bosons specifying the basis vectors (6.14), and just represents the interspace coupling. As has been emphasized at the end of §6, the interspace coupling $\tilde{\mathbf{H}}_{\text{int}}$ can never be described within the collective subspace (8.1) underlying the IBM.

Sakata, Iwasaki, Marumori and Takada[52] have shown that this interspace coupling $\tilde{\mathbf{H}}_{\text{int}}$ plays a decisive role in preparing the "anomalous" low-lying 0^+_2 states which have been found in even-even nuclei with N or Z $\cong$ 40, through interactions between the pairing vibrational modes[47] and the collective 2-"phonon" modes with $J^{\pi} = 0^+$. Their calculation successfully reproduces the main features of the experimental systematics of the excitation energies as well as $B(E2;\ 0^+_2 \to 2^+_1)$. (Fig. 4). Such

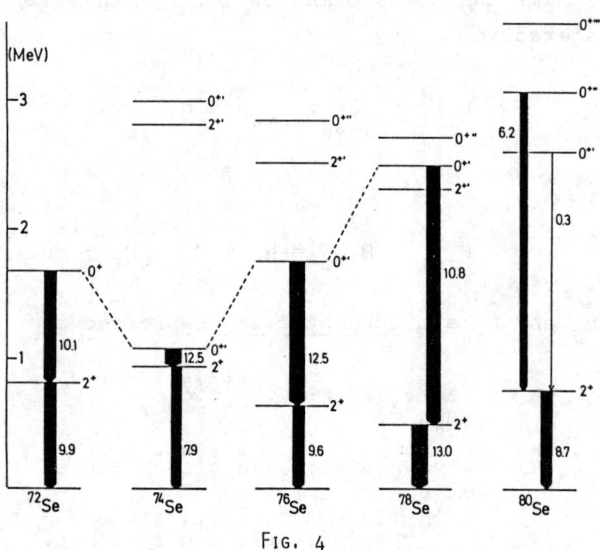

FIG. 4

CALCULATED ENERGY EIGENVALUES AND THE B(E2) OF EVEN-N Se-ISOTOPES. THE B(E2) VALUES ARE GIVEN IN UNITS OF $10^{-50}e^2cm^4$. THE FIGURE IS TAKEN FROM REF.52)

characteristic behaviour of the 0_2^+ state is not specific for this region but is generally found over wide range of isotopes and isotones. Particularly the regions of N∿60 (e.g., the neutron-rich Mo isotopes) and N∿66 (the Cd, Sn and Te isotopes) are of especially appreciable. It must be noticed that these regions lie just in the transitional region or in the vicinity of rotational one. In the first region N∿40, some of the collective excited states have been interpreted with the phenomenological co-existence model[53)] because the nuclei in this region seem to show vibrational character as well as rotational one. The second region N∿60 is at the beginning of neutron-rich new deformed region. The third region lies in the entrance of neutron-deficient quasi-rotational region.[18)] This indicates that the characteristic behaviour of the 0_2^+ states seems to relate to the phase transition from spherical to deformed nuclei.

Takada, Tazaki, Kaneko and Sakata[54)] have given a consistent interpretation of the characteristic behaviour of the 0_2^+ state in these regions through the interactions between the pairing vibrational mode and the $J^\pi = 0^+$ 2-"phonon" mode based on the interspace coupling $\tilde{H}_{int}$. With the very similar standpoint, Weeks, Tamura, Udagawa and Hahne[55)] have discussed the 0_2^+ states in the Ge isotopes. These microscopic studies conclude that these 0_2^+ states are strongly mixed states of the pairing- vibrational states and the $J^\pi = 0^+$ 2-"phonon" states, and the electric transition probabilities concerning the 0_2^+ states are largely influenced by this mixing property. Sakata and Holzwarth[56)] have also discussed a possible mechanism that the pairing vibrational mode mediates a new coupling between the collective 2^+ mode and the non-collective states in the intrinsic subspace through the interspace coupling $\tilde{H}_{int}$.

We may thus suppose that the process of reconstruction of the new collective subspace in the phase transition from the spherical to the deformed may be caused by interactions between the collective quadrupole mode and the non-collective intrinsic excitations with the intermediation of the pairing excitations through the interspace coupling $\tilde{H}_{int}$. Investigation of such a dynamical mechanism is one of the important open problems in the nuclear phase transition.

Summarizing this lecture of the first part (I), we arrive at the following conclusion. Once the nuclear system is given in the fermion-state space and the microscopic structure of s and d bosons is specified, the parameters of the IBM can be always calculated with desired accuracy in terms of the modified Marumori boson mapping from the microscopic point of view. Every formula to calculate them has already been prepared. In order to avoid unnecessary confusion misled and obtain further understanding for the nuclear collective dynamics, therefore, it is strongly desired that any works concerning the IBM have to be done at least at the level of the microscopic theory.

References

1) T. Marumori, K. Takada and F. Sakata, Prog. Theor. Phys. Suppl. No.71 (1981), 1.
2) T. Marumori, K. Takada and F. Sakata, Proceedings of the Symposium on Nuclear Collectivity (Dōgashima, Japan 1981), 137.
3) A. Bohr, Mat. Fys. Medd. Dan. Vid. Selsk. 26 (1952), No.14.
A. Bohr and B. R. Mottelson, Mat. Fys. Medd. Dan. Vid. Selsk. 27 (1953), No.16.
4) A. Bohr, Rev. Mod. Phys. 48 (1976), 365.
5) B. R. Mottelson, Rev. Mod. Phys. 48 (1976), 375.
6) A. Bohr and B. R. Mottelson, Ann. Rev. Nucl. Sci. 23 (1973), 363.
A. Bohr, Science 172 (1971), 17.
7) A. Bohr and B. R. Mottelson, Nuclear Structure (W. A. Benjamin Inc.) Vol.I (1969) and Vol.II (1975).
8) S. G. Nilsson, Mat. Fys. Medd. Dan. Vid. Selsk. 29 (1955), No.16.
9) B. R. Mottelson and S. G. Nilsson, Phys. Rev. 99 (1955), 1615.
10) T. Tamura, "Boson Expansion Theories in the Particle-Particle and Particle-Hole Modes" (1984), to be published.
11) D. R. Inglis, Phys. Rev. 96 (1954), 1054.
12) A. Bohr, B. R. Mottelson and D. Pines, Phys. Rev. 110 (1958), 936.
B. R. Mottelson, in The Many-Body Problem (John Wiley & Sons, Inc., New York, 1959).
13) V. G. Soloviev, Nucl. Phys. 9 (1958), 665.
14) S. T. Belyaev, Mat. Fys. Medd. Dan. Vid. Selsk. 31 (1959), No.11; in The Many-Body Problem (John Wiley & Sons, Inc., New York, 1959).
15) M. Kobayashi and T. Marumori, Prog. Theor. Phys. 23 (1960), 387.
T. Marumori, Prog. Theor. Phys. 24 (1960), 331.
16) R. Arview and M. Veneroni, Compt. rend. 250 (1960), 992.
17) M. Baranger, Phys. Rev. 120 (1960), 957.

18) M. Sakai, Nucl. Phys. A104 (1967), 301; INS-Rep.-493 (1984).

19) H. Morinaga and P. C. Gugelot, Nucl. Phys. 46 (1963), 210.

20) E. R. Marshalek, Nucl. Phys. A347 (1980), 253.

21) P. Ring and P. Schuck, The Nuclear Many-Body Problem (Springer-Verlag, 1980) Chap.9.

22) S. T. Belyaev and V. G. Zelevinsky, Nucl. Phys. 39 (1962), 582.

23) T. Marumori, M. Yamamura and A. Tokunaga, Prog. Theor. Phys. 31 (1964), 1009.
T. Marumori, M. Yamamura, A. Tokunaga and K. Takada, Prog. Theor. Phys. 32 (1964), 726.

24) E. R. Marshalek and J. Weneser, Phys. Rev. C2 (1970), 1682.
E. R. Marshalek, Nucl. Phys. A161 (1971), 401.

25) D. Janssen, F. Dönau, S. Frauendorf and R. V. Jolos, Nucl. Phys. A172 (1971), 145.

26) B. Sorensen, Nucl. Phys. A97 (1967), 65; A142 (1970), 392, 411.

27) T. Kishimoto and T. Tamura, Nucl. Phys. A192 (1972), 246; A270 (1976), 317.
T. Tamura, K. J. Weeks and T. Kishimoto, Phys. Rev. C20 (1979), 307.

28) M. Kleber, Phys. Letters 30B (1969), 588.

29) S. Y. Li, R. M. Dreizler and A. Klein, Phys. Rev. C4 (1971), 1571.

30) S. G. Lie and G. Holtzwarth, Phys. Rev. C12 (1975), 1035.

31) S. Iwasaki, F. Sakata and K. Takada, Prog. Theor. Phys. 57 (1977), 1289.

32) T. Tamura, Proceedings of the 1982 INS International Symposium on Dynamics of Nuclear Collective Motion (at the foot of Mt. Fuji, Japan 1982), 400.
T. Tamura and T. Kishimoto, Prog. Theor. Phys. Suppl. 74 & 75 (1983), 282.

33) T. Tamura, Phys. Rev. C28 (1983), 2154.

34) G. Holzwarth, D. Janssen and R. V. Jolos, Nucl. Phys. A261 (1976), 1.

35) D. Janssen, R. V. Jolos and F. Dönau, Nucl. Phys. A224 (1974), 93.

36) K. Takada, K. Kaneko, F. Sakata and S. Tazaki, Prog. Theor. Phys. Suppl. 71 (1981), 71.

37) A. Arima and F. Iachello, Phys. Rev. Letters 35 (1975), 1069; Phys. Letters 57B (1975), 39; Ann. of Phys. 99 (1976), 253; 111 (1978), 201; Phys. Rev. Letters 40 (1978), 385.

38) T. Otsuka, A. Arima, F. Iachello and I. Talmi, Phys. Letters 76B (1978), 139.
T. Otsuka, A. Arima, F. Iachello, Nucl. Phys. A309 (1978), 1.

39) T. Suzuki, M. Fuyuki and K. Matsuyanagi, Prog. Theor. Phys. 61 (1979), 1082.
K. Matsuyanagi, Proceedings of the 1981 Trieste Workshop on Nucl. Physics (North Holland, 1982, Edit. by C. H. Dasso), 29.

40) A. Kuriyama, T. Marumori and K. Matsuyanagi, Prog. Theor. Phys. 45 (1971), 784.

41) N. Kanesaki, T. Marumori, F. Sakata and K. Takada, Prog. Theor. Phys. 49 (1973), 181; 50 (1973), 867.

42) A. Kuriyama, T. Marumori and K. Matsuyanagi, F. Sakata and T. Suzuki, Prog. Theor. Phys. Suppl. No.58 (1975), Chaps. 1 and 7.

43) F. Sakata, T. Marumori and K. Takada, Prog. Theor. Phys. Suppl. 71 (1981), Chap. 2.

44) T. Suzuki and K. Matsuyanagi, Prog. Theor. Phys. 56 (1976), 1156.
T. Suzuki, Prog. Theor. Phys. 60 (1978), 1366.

45) C. M. Vincent, Phys. Rev. C27 (1983), 426.

46) T. Otsuka and A. Arima, Phys. Letters 77B (1978), 1.

47) D. R. Bés and R. A. Broglia, Nucl. Phys. 80 (1966), 289.
As a review, see R. A. Broglia, O. Hansen and C. Riedel, Adv. in Nucl. Phys. 6 (1973), 287.

48) S. Iwasaki, T. Marumori, F. Sakata and K. Takada, Prog. Theor. Phys. 56 (1976), 846.

49) K. Takada, K. Kaneko, F. Sakata and S. Tazaki, Prog. Theor. Phys. Suppl. No.71 (1981), 71.

50) K. Takada, submitted to Nucl. Phys.

51) T. Suzuki, M. Fuyuki and K. Matsuyanagi, Prog. Theor. Phys. 61 (1979), 1682.

52) F. Sakata, S. Iwasaki, T. Marumori and K. Takada, Z. Phys. A286 (1978), 195; Prog. Theor. Phys. 56 (1976), 1140; Proceedings of the International Conference on Nuclear Structure, Tokyo, 1977 (Suppl. J. Phys. Soc. Japan 44 (1981)), 520.

53) J. H. Hamilton, A. V. Ramayya, W. T. Pinkston, R. M. Ronningen, G. Garcia-Bermudez, H. K. Carter, R. L. Robinson, H. J. Kim and R. O. Sayer, Phys. Rev. Lett. 32 (1974), 239.
R. B. Piercy, A. V. Ramayya, R. M. Ronningen, J. H. Hamilton, V. Maruhn-Rezwani, R. L. Robinson and H. J. Kim, Phys. Rev. C19 (1979), 1344.

54) K. Takada and S. Tazaki, Prog. Theor. Phys. 61 (1979), 1666.
K. Takada, S. Tazaki and K. Kaneko, Prog. Theor. Phys. 62 (1979), 440.
S. Tazaki, K. Takada, K. Kaneko and F. Sakata, Prog. Theor. Phys. Suppl. No.71 (1981), 123.

55) K. J. Weeks, T. Tamura, T. Udagawa and F. J. W. Hahne, Phys. Rev. C24 (1981), 703.

56) F. Sakata and G. Holzwarth, Prog. Theor. Phys. 61 (1979), 1649.

TOWARD MICROSCOPIC UNDERSTANDING OF THE CONCEPT OF COLLECTIVE SUBSPACE. (II)

Toshio MARUMORI, Fumihiko SAKATA*, Yukio HASHIMOTO
Tsutomu UNE and Masanori OGURA

Institute of Physics, University of Tsukuba, Ibaraki 305
*Institute for Nuclear Study, University of Tokyo, Tanashi, Tokyo 188

ABSTRACT

In this lecture, Part (II), a survey is given of basic ideas of a new microscopic theory, which goes beyond the random phase approximation and is capable by itself of choosing a "maximally-decoupled" collective subspace appropriate for the large amplitude collective motion of "soft" spherical nuclei. Emphasis is put on clarifying the very concept of maximally-decoupled collective subspace with the use of the framework of the time-dependent Hartree-Fock (TDHF) theory. The maximal-decoupling condition on large-amplitude collective motion is formulated in a general form called the invariance principle of the time-dependent Schrödinger equation. In sharp contrast with the conventional a priori assumption that the microscopic structure of the collective bosons is supposed to be unchanged, it is shown that dynamical change of the internal structure of the bosons due to the maximal-decoupling condition becomes more and more serious for higher collective excitations.

---

*) The main part of (II) is based on Ref.1).

§1. Introduction

In Part (I), we have arrived at the conclusion that, once the collective subspace in the fermion-state space is specified with reasonable assumptions on the microscopic structure of the collective bosons, a precise collective boson Hamiltonian can be always obtained with desired accuracy in terms of the modified Marumori boson-mapping from the microscopic point of view.

Then, the problem naturally arises of how to properly treat the inter-space coupling between the collective subspace and its orthogonal complement: In §8 of Part (I), we have already demonstrated that the inter-space coupling between the collective subspace (I,8.1), i.e.,

$$\{[\hat{P}_0(\hat{\mathfrak{D}}^\dagger)^q](\hat{\mathfrak{S}}^\dagger)^{N-q}|0>;q=0,1,2,\cdots N\} , \tag{1.1}$$

and its orthogonal complement may play the decisive role in dynamical change of the structure of the collective subspace from the spherical one to the deformed one.

The first important task toward solving this problem is to find a method of how to determine an "optimum" collective subspace rather than assuming it a priori. The optimum collective subspace has to satisfy a maximal - decoupling condition that the Hamiltonian provides no serious coupling between the collective subspace and its orthogonal complement (called an "intrinsic" subspace hereafter). Namely, the collective subspace has to be an approximate invariant subspace of the Hamiltonian. How can we determine the optimum collective subspace self-consistently in terms of the particle excitations?

A clue for such an investigation may be provided by the time-dependent Hartree-Fock (TDHF) theory. As you all know, the TDHF theory with the basic equation

$$\delta<\phi(t)|\{i\frac{\partial}{\partial t} - \hat{H})|\phi(t)>\} = 0^{*)} \quad \text{and} \quad \text{h.c.} \tag{1.2}$$

is an approximation theory to a many-body system obeying the Schrödinger equation

$$(i\frac{\partial}{\partial t} - \hat{H})|\Psi(t)\rangle = 0 \tag{1.3}$$

under the Hartree-Fock approximation

$$|\Psi(t)\rangle \to |\phi(t)\rangle \ , \tag{1.4}$$

where $|\phi(t)\rangle$ is a single Slater determinant. Now, it is well known that, from the TDHF equation (1.2), we can derive the RPA-eigenvalue equation under the small-amplitude approximation, and the eigenvalue equation gives us the normal collective vibrational modes which satisfy the maximal-decoupling condition. It is also known that the TDHF equation (1.2) leads us to the "classical" collective Hamiltonian of the large-amplitude collective motion such as the rotational motion of strongly deformed nuclei which satisfies the maximal-decoupling condition. Therefore, if once we can recognize a "basic principle" underlying these derivations of the collective modes of motion, we may have a key to construct a quantum theory of the maximally-decoupled large-amplitude collective motion, by directly applying the basic principle to the many-body Schrödinger equation (1.3).

§2. Invariance Principle of the Time-Dependent Schrödinger Equation

[2A] TDHF Derivation of the RPA Collective Modes[3)]

Our first task is thus to find the basic principle underlying the TDHF derivations of the maximally-decoupled collective modes of motion. For this purpose, let us start with

---

*) Throughout this lecture, the convention of using $\hbar = 1$ is adopted.

remembering the TDHF derivation of the RPA collective modes. According to the Thouless theorem,[2)] the TDHF wave function for Eq.(1.2) is generally given in the form

$$|\phi(t)\rangle = \exp\{i\hat{G}_0(t)\}|\phi_0(t)\rangle \ , \qquad (2.1a)$$

where $|\phi_0(t)\rangle = |\phi_0\rangle e^{-i\varepsilon_0 t}$ is the Hartree-Fock ground state satisfying

$$\delta\langle\phi_0(t)|(i\frac{\partial}{\partial t} - \hat{H})|\phi_0(t)\rangle = 0, \text{ i.e., } \delta\langle\phi_0|\hat{H}|\phi_0\rangle = 0 \ , \qquad (2.1b)$$

and $\hat{G}_0(t)$ is a one-body hermitian operator.

The essential point of the RPA derivation is to assume $\hat{G}_0(t)$ to be of the form

$$\hat{G}_0(t) = i\{\eta^*(t)\hat{X} + \hat{X}^\dagger\eta(t)\} \ ,$$

$$\hat{X}^\dagger = \sum_{\mu i}\{\psi(\mu i)c_\mu^\dagger c_i - \phi(\mu i)c_i^\dagger c_\mu\} \ ,^{*)} \qquad (2.2a)$$

$$\eta(t) = \eta_0 \cdot e^{-i\omega t} \qquad (2.2b)$$

The assumption that the collective mode of motion under consideration is of small amplitude is taken into account in the form of $\eta(t)$ in Eq.(2.2b) with samll quantity η_0. Then, using the formulae

$$e^{-i\hat{G}_0}\hat{H}e^{i\hat{G}_0} = \hat{H}+[\hat{H},i\hat{G}_0]+\frac{1}{2!}[[\hat{H},i\hat{G}_0],i\hat{G}_0]+ \cdots \ , \qquad (2.3)$$

---

*) Throughout the lecture, we use the convention of denoting occupied single-particle states of $|\phi_0\rangle$ by indices i,j, ···, and unoccupied single-particle states by μ,ν, ···.

$$e^{-i\hat{G}_0}\frac{\partial}{\partial t}e^{i\hat{G}_0}=i\frac{\partial\hat{G}_0}{\partial t}+\frac{1}{2!}[i\frac{\partial\hat{G}_0}{\partial t},i\hat{G}_0]+\frac{1}{3!}[[i\frac{\partial\hat{G}_0}{\partial t},i\hat{G}_0],i\hat{G}_0]+\cdots ,$$

the TDHF equation (1.2) is reduced, within the first order of η_0, to

$$\delta\langle\phi_0|[\hat{H},i\hat{G}_0]+\frac{\partial}{\partial t}\hat{G}_0|\phi_0\rangle = 0 ,$$

i.e.,

$$\delta\langle\phi_0|[\hat{H},\hat{X}^\dagger]-\omega\hat{X}^\dagger|\phi_0\rangle = 0 \quad \text{and} \quad \text{h.c.} , \tag{2.4}$$

which simply leads us to the well-known RPA eigenvalue equation to determine the amplitudes $\psi(\mu i)$ and $\phi(\mu i)$ and the eigenvalue ω.

In this TDHF treatment, the collective Hamiltonian is usually obtained as follows: Within the second order of (η^*,η), we have the collective vibrational energy

$$\langle\phi(t)|\hat{H}|\phi(t)\rangle-\langle\phi_0|\hat{H}|\phi_0\rangle = \omega\eta^*\eta \equiv \mathcal{H}_c^{(0)}(\eta^*,\eta) . \tag{2.5}$$

This is regarded as the c-number version of the collective "phonon" (boson) Hamiltonian,

$$\mathcal{H}_c^{(0)}(\eta^*,\eta) \leftrightarrow \mathbf{H}_B^{(0)}(b^\dagger,b) = \omega b^\dagger b$$

$$[b,b^\dagger] = 1 , \tag{2.6}$$

which implies that the time-dependent parameters $(\eta^*(t),\eta(t))$ correspond to the c-number version of the boson operators in the Heisenberg representation;

$$\eta^*(t) \leftrightarrow b^\dagger(t) = \exp(i\mathbf{H}_B^{(0)}t)b^\dagger\exp(-i\mathbf{H}_B^{(0)}t) ,$$

$$\eta(t) \leftrightarrow b(t) = \exp(iH_B^{(0)} t) b \exp(-iH_B^{(0)} t) . \qquad (2.7)$$

The physical correspondents in the fermion-state space of the boson oprators are easily obtained from the result in the first order of (η^*, η);

$$<\phi(t)|\hat{X}^\dagger|\phi(t)> = \eta^* , \quad \text{and} \quad \text{h.c.} , \qquad (2.8)$$

which implies that

$$\hat{X}^\dagger \leftrightarrow b^\dagger , \qquad \hat{X} \leftrightarrow b . \qquad (2.9)$$

[2.B] the TDHF and the Rotational Motion[3)]

Nextly, let us remind you of the TDHF derivation of the rotational motion of strongly deformed nuclei. Experimentaly, the rotational motion is specified in the following way. Let $E_{rot}(I)$ and $\mathcal{H}_{rot}(I)$ be the observed energy of the state with spin I in a rotational band and the expectation value of the rotational Hamiltonian, respectively. We then set up two fundamental suppositions:

(i) The rotational motion is of the maximal decoupling, i.e.,

$$E_{rot}(I) = \mathcal{H}_{rot}(I). \qquad (2.10)$$

(ii) The rotational motion obeys the canonical equation of motion,

$$\dot{\theta} \equiv \omega = \frac{\partial}{\partial I}\mathcal{H}_{rot}(I) , \quad \dot{I} = -\frac{\partial}{\partial \theta}\mathcal{H}_{rot}(I) = 0 , \qquad (2.11)$$

where I is the component of the angular momentum to the axis of rotation and ω is the rotational frequency about the axis of rotation. Thus, the rotational frequency is determined experimentally by

$$\omega(I) = \frac{\partial}{\partial I}\mathcal{H}_{rot}(I) \approx \frac{1}{2}\{E_{rot}(I+1) - E_{rot}(I-1)\}. \quad (2.12)$$

In investigating the microscopic dynamics underlying the collective modes of motion, an essential element is to find "particle-collective coupling", which is just defined as <u>what organizes the collective modes out of the particle excitations</u>. This implies that, even in a completely decoupled collective mode, we need the information on the "particle-collective coupling" organizing the collective mode out of the particle excitations. How can we find such a coupling?

A clue for this problem is provided by the cranking model of Inglis.[4)] In the cranking model, one considers the particle motion in a mean field with a deformed equilibrium which is uniformly rotating with frequency ω. In this (time-dependent) rotating coordinate frame, the time-displacement operator is written as

$$\hat{H}' = \hat{H} - \omega\hat{J} \quad (2.13)$$

where $\hat{J}$ is the component of the total angular-momentum operator of particles to the rotational axis, and the "particle-collective coupling" which organizes the collective rotation manifests itself as the Coriolis coupling $-\omega\hat{J}$. This coupling in the rotating frame gives rise to an increase in the energy of the particle motion, which is identified with the collective rotational energy,

$$\mathcal{H}_{rot}(\omega) \equiv \langle\phi(\omega)|\hat{H}|\phi(\omega)\rangle_\beta - \langle\phi_0|\hat{H}|\phi_0\rangle_\beta . \quad (2.14)$$

Here $|\phi_0\rangle_\beta$ is the Hartree-Fock ground state with the deformation β and $|\phi(\omega)\rangle_\beta$ is the Hartree-Fock state in the rotating frame satisfying

$$\delta\langle\phi(\omega)|\hat{H} - \omega\hat{J}|\phi(\omega)\rangle_\beta = 0 . \quad (2.15)$$

[2.C] Generalized Moving Frame Associated with the Time-Dependent Self-Consistent Mean Field[3),5)]

The important fact which we have learned from the cranking model is that the "particle-collective coupling" organizing the "maximally-decoupled" collective motion manifests itself only in the moving frame associating with the time-dependent self-consistent mean field. Now, let us formulate this idea in order to describe the generalized large-amplitude collective motion about the spherical Hartree-Fock ground state $|\phi_0\rangle$. In order to find the "particle-collective coupling" in this case, we introduce a generalized moving frame by a time-dependent unitary transformation with a set of complex collective parameters $\{\eta_r^*(t),\eta_r(t);\ r=1,2,\cdots,k\}$,

$$|\phi(\eta^*(t),\eta(t))\rangle = \hat{U}^{-1}(\eta^*(t),\eta(t))|\phi_0\rangle\ ,$$

$$\hat{U}^{-1}(\eta^*,\eta) = \exp\{i\hat{G}(\eta^*,\eta)\}\ , \qquad (2.16)$$

where $|\phi(\eta^*,\eta)\rangle$ is a Hartree-Fock state in the moving frame and $|\phi_0\rangle$ is the Hartree-Fock ground state satisfying

$$\delta\langle\phi_0|\hat{H}|\phi_0\rangle = 0\ , \qquad \langle\phi_0|\hat{H}|\phi_0\rangle \equiv \varepsilon_0\ . \qquad (2.17)$$

The set of parameters $\{\eta_r^*(t),\eta_r(t);\ r=1,2,\cdots,k\}$ specifies the time-dependent variations of the mean field, associated with the collective motion described by a set of collective coordinates $q_r(t)$ and their canonically conjugate momenta $p_r(t)$,

$$q_r = \frac{1}{\sqrt{2}}(\eta_r^* + \eta_r),\quad p_r = \frac{i}{\sqrt{2}}(\eta_r^* - \eta_r),\ (r=1,2,\cdots,k). \qquad (2.18)$$

Namely, the parameters $\{\eta_r^*,\eta_r;\ r=1,2,\cdots,k\}$ can be regarded as c-number correspondents of the collective bosons $\{b_r^\dagger(t),b_r(t);\ r=1,2,\cdots,k\}$ in the Heisenberg picture

$$\eta_r^*(t) \leftrightarrow b_r^\dagger(t) \equiv e^{i\mathbb{H}_B t} b_r^\dagger e^{-i\mathbb{H}_B t} ,$$

$$[b_r, b_s^\dagger] = \delta_{rs} \qquad (2.19)$$

where the collective boson Hamiltonian $\mathbb{H}_B(b^\dagger, b)$ has to be determined self-consistently later.

Since the time-dependence of the parameters $\{\eta_r^*(t), \eta_r(t)\}$ completely specifies the time-dependence of the state $|\phi(\eta^*,\eta)\rangle$, we have

$$i\frac{\partial}{\partial t}|\phi(\eta^*,\eta)\rangle = \sum_r \{i\dot{\eta}_r \hat{O}_r^\dagger(\eta^*,\eta) - i\dot{\eta}_r^* \hat{O}_r(\eta^*,\eta)\}|\phi(\eta^*,\eta)\rangle, \qquad (2.20)$$

where the operator $\hat{O}_r^\dagger(\eta^*,\eta)$ is local infinitesimal generator with respect to η_r, and is defined by

$$\hat{O}_r^\dagger(\eta^*,\eta) \equiv \{\frac{\partial}{\partial \eta_r}\hat{U}^{-1}(\eta^*,\eta)\}\hat{U}(\eta^*,\eta) . \qquad (2.21)$$

[2.D] Invariance Principle of the Time-Dependent Schrödinger Equation

The next task is to determine the moving frame as well as the structure of the operator $\hat{O}_r^\dagger(\eta^*,\eta)$. For this purpose, we demand a principle, which we call hereafter the invariance principle of the time-dependent Schrödinger equation.*),6),7) This principle can be considered as a generalization of the important fact that the well-isolated collective modes such as the translational and rotational modes of motion are generally accompanied by the invariance properties of the Schrödinger equation. The principle is simply stated as follows: The time-dependence of the parameters $\{\eta_r^*(t), \eta_r(t)\}$ as well as the

*) A detailed and self-contained explanation on the invariance principle is geven in Ref.3).

structure of the local infinitesimal generators $\{\hat{O}_r^\dagger(\eta^*,\eta), \hat{O}_r(\eta^*,\eta)\}$ (characterizing the unitary transformation $\hat{U}(\eta^*,\eta)$ in Eq. (2.16)) must be determined so as to keep the time-dependent Schrödinger equation invariant. In the TDHF variational form, the principle is expressed as

$$\delta_0 \langle\phi(\eta^*,\eta)|\{(i\frac{\partial}{\partial t} - \hat{H})|\phi(\eta^*,\eta)\rangle\} = 0 \quad \text{and} \quad \text{h.c.} \tag{2.22}$$

with the boundary condition at $\eta_r = \eta_r^* = 0$,

$$\delta\langle\phi_0|\hat{H}|\phi_0\rangle = 0 \ , \quad \delta\langle\phi_0|\phi_0\rangle = 0$$

$$\hat{U}^{-1}(\eta^* = 0, \eta = 0) = 1 \ , \tag{2.23}$$

where the variation $|\delta_0\phi(\eta^*,\eta)\rangle$ is defined by

$$|\delta_0\phi(\eta^*,\eta)\rangle \equiv \hat{U}^{-1}(\eta^*,\eta)|\delta\phi_0\rangle \ . \tag{2.24}$$

The principle is also written as

$$\delta\langle\phi_0(t)|\{(i\frac{\partial}{\partial t} - \hat{H})|\phi_0(t)\rangle\} = 0, \quad |\phi_0(t)\rangle \equiv |\phi_0\rangle e^{-i\varepsilon_0 t} \ ,$$

$$\delta\langle\phi_0(t)|\{\hat{U}(\eta^*,\eta)(i\frac{\partial}{\partial t} - \hat{H})\hat{U}^{-1}(\eta^*,\eta)|\phi_0(t)\rangle\} = 0 \ , \tag{2.25}$$

which implies that <u>the Hartree-Fock ground state</u> $|\phi_0\rangle$ <u>must be a variational solution of both the Hamiltonian</u> $\hat{H}$ <u>and the operator</u> $\hat{U}(\eta^*,\eta)(i\partial/\partial t - \hat{H})\hat{U}^{-1}(\eta^*,\eta)$. As will be discussed later, it turns out that the invariance principle of the time-dependent Schrödinger equation is simply a formulation of the maximal-decoupling condition on the collective motion under consideration.

Now, with the use of Eq. (2.20), Eq. (2.22) can be written as

$$\delta_0 \langle\phi(\eta^*,\eta)|\hat{H}-\sum_r\{i\dot{\eta}_r\hat{O}_r^\dagger(\eta^*,\eta)-i\dot{\eta}_r^*\hat{O}_r(\eta^*,\eta)\}|\phi(\eta^*,\eta)\rangle = 0. \quad (2.26)$$

This equation just corresponds to the generalization of the cranking model into the generalized moving frame. The operator

$$\hat{H}' \equiv \hat{H} - \sum_r\{i\dot{\eta}_r\hat{O}_r^\dagger(\eta^*,\eta)-i\dot{\eta}_r^*\hat{O}_r(\eta^*,\eta)\} \quad (2.27)$$

corresponds to the time-displacement operator describing the time-evolution of the system in the moving frame. In this moving frame, therefore, the "particle-collective coupling" which is highly non-linear in general manifests itself as the coupling $\hat{H}'-\hat{H} = -\sum_r\{i\dot{\eta}_r\hat{O}_r^\dagger(\eta^*,\eta)-i\dot{\eta}_r^*\hat{O}_r(\eta^*,\eta)\}$. This coupling gives rise to an increase in the energy of the particle motion, which is identified with the collective vibrational energy of the large-amplitude collective motion under consideration;

$$\mathcal{H}_c(\eta^*,\eta) \equiv \langle\phi(\eta^*,\eta)|\hat{H}|\phi(\eta^*,\eta)\rangle - \langle\phi_0|\hat{H}|\phi_0\rangle . \quad (2.28)$$

This corresponds to the c-number version of the collective boson Hamiltonian $\mathbb{H}_B(b^\dagger,b)$ defined in Eq. (2.19).

$$\mathcal{H}_c(\eta^*,\eta) \leftrightarrow \mathbb{H}_B(b^\dagger,b). \quad (2.29)$$

[2.E] Canonical-Variables Condition[3),5)]

The discussion so far is not affected by any variable transformation $\eta'_r = f_r(\eta^*,\eta)$. Using this freedom at this stage, we require the following condition on the collective variables $\{\eta_r^*,\eta_r;\ r=1,2,\cdots,k\}$: Corresponding to the Heisenberg equation of motion of the boson operators $i\dot{b}_r(t) = [b_r(t),\mathbb{H}_B(b^\dagger,b)]$, the time dependence of the variables $\{\eta_r^*(t),\eta_r(t)\}$ has to be determined in terms of the canonical equation of motion

$$i\dot{\eta}_r = \frac{\partial}{\partial \eta_r^*}\mathcal{H}_c(\eta^*,\eta) \ , \quad -i\dot{\eta}_r^* = \frac{\partial}{\partial \eta}\mathcal{H}_c(\eta^*,\eta) \tag{2.30a}$$

which is written as

$$\dot{q}_r = \frac{\partial}{\partial p_r}\mathcal{H}_c(p,q) \ , \quad -\dot{p}_r = \frac{\partial}{\partial q_r}\mathcal{H}_c(p,q) \ . \tag{2.30b}$$

It can be verified[3),5)] that this condition is fulfilled when we choose the variables $\{\eta_r^*, \eta_r\}$ in such a way that the local infinitesimal generators satisfy the "weak" boson-like commutation relations

$$<\phi(\eta^*,\eta)|[\hat{O}_r(\eta^*,\eta),\hat{O}_s^\dagger(\eta^*,\eta)]|\phi(\eta^*,\eta)> = \delta_{rs} \ ,$$

$$<\phi(\eta^*,\eta)|[\hat{O}^\dagger(\eta^*,\eta),\hat{O}_s^\dagger(\eta^*,\eta)]|\phi(\eta^*,\eta)> = 0 \ . \tag{2.31}$$

It can also be shown[3),5)] that one of the simplest choices of the variables $\{\eta_r^*, \eta_r\}$, which satisfy the weak boson-like commutation relation (2.31), is to determine the variables through relations

$$<\phi(\eta^*,\eta)|\hat{O}_r^\dagger(\eta^*,\eta)|\phi(\eta^*,\eta)> = \frac{1}{2}\,\eta_r^* \quad \text{and} \quad \text{h.c.}, \tag{2.32}$$

which we call hereafter the <u>canonical-variables condition</u>. It is also proved[5)] that such a choice of variables is generally possible.

§3. A Set of Basic Equations of the Theory and its Solution for a Simple Model

[3.A] A Self-Consistent Solution of the Basic Equations[3),5)]

As has been shown in the above discussion, a set of basic equations in our theory, which has to be solved self-consistently, consists of (i) <u>the invariance principle</u> (2.26), i.e,

$$\delta\langle\phi_0|\hat{U}(\eta^*,\eta)\{\hat{H} - \sum_r \frac{\partial\mathcal{H}_c}{\partial\eta^*_r}\hat{O}^\dagger_r(\eta^*,\eta) - \sum_r \frac{\partial\mathcal{H}_c}{\partial\eta_r}\hat{O}_r(\eta^*,\eta)\}\hat{U}^{-1}(\eta^*,\eta)|\phi\rangle = 0 \qquad (3.1)$$

with the aid of Eq. (2.30a) and (ii) <u>the canonical-variables condition</u> (2.32). It is now easily seen that the problem to solve the set of basic equations self-consistently can be reduced to finding the hermitian operator $\hat{G}(\eta^*,\eta)$ in Eq. (2.16) so as to satisfy the set of basic equations. In this case, it is necessary to set up a specific "boundary" condition appropriate for the collective motion under consideration. This is an important physical input for our theory. For example, we can set up the following boundary condition on the collective motion:

$$\mathcal{H}_c(\eta^*,\eta) \to \mathcal{H}_{RPA}(\eta^*,\eta) \equiv \omega_{\lambda_0}\eta^*\eta, \quad \text{for} \quad \eta \to \text{small} \qquad (3.2)$$

where ω_{λ_0} is the lowest eigenvalue of the RPA mode and, for convenience' sake , we have restricted ourselves to the simplest case of r = 1 with a single pair of variables (η^*,η). The condition (3.2) implies that our large-amplitude collective motion under consideration is connected with the RPA-"phonon" mode in the "small-amplitude" (i.e., small-η) limit.

With this boundary condition, we can uniquely determine the hermitian operator $\hat{G}(\eta^*,\eta)$ in the following way: Since $\hat{G}(\eta^*,\eta)$ is a one-body operator by definition, we can express it in the form

$$\hat{G}(\eta^*,\eta) = \sum_\lambda\{g_\lambda(\eta^*,\eta)\hat{X}_\lambda + g^*_\lambda(\eta^*,\eta)\hat{X}^\dagger_\lambda\} , \qquad (3.3)$$

where $\{\hat{X}^\dagger_\lambda,\hat{X}_\lambda\}$ is a complete set of creation and annihilation operators of the RPA modes

$$\hat{X}_\lambda^\dagger = \sum_{\mu i} \{\psi_\lambda(\mu i) c_\mu^\dagger c_i + \phi_\lambda(\mu i) c_i^\dagger c_\mu\} , \tag{3.4}$$

consisting of the particle-hole pair operators $\{c_\mu^\dagger c_i, c_i^\dagger c_\mu\}$. Then, the coefficients $g_\lambda(\eta^*,\eta)$ in Eq.(3.3) are expanded as a power series of (η^*,η),

$$g_\lambda(\eta^*,\eta) = \sum_{n\geq 1} g_\lambda(n) , \quad g_\lambda(n) \equiv \sum_{\substack{r,s \\ (r+s=n)}} g_{rs}^{(\lambda)} (\eta^*)^r (\eta)^s . \tag{3.5}$$

Inserting the expression (3.3) with (3.5) into the set of basic equations and evaluating the coefficients of each power of (η^*, η) in these equations step by step, we can uniquely determine the unknown coefficients $g_{rs}^{(\lambda)}$ in Eq.(3.5) as well as the collective Hamiltonian $\mathcal{H}_c(\eta^*, \eta)$ up to the desired order.

For example, with the boundary condition

$$g_\lambda(1) = i\eta^* \cdot \delta_{\lambda\lambda_0} \tag{3.6}$$

which corresponds to Eq. (3.2), we obtain the collective Hamiltonian of the form

$$\mathcal{H}_{coll}(\eta^*,\eta) = \omega_{\lambda_0} \cdot \eta^*\eta + \sum_{n\geq 3} h(n) , \quad h(n) \equiv \sum_{\substack{rs \\ (r+s=n)}} h_{rs} \cdot (\eta^*)^r (\eta)^s . \tag{3.7}$$

where

$$h(3) = \frac{1}{3!} \langle\phi_0 | [[[\hat{H}, i\hat{G}(1)], i\hat{G}(1)], i\hat{G}(1)] | \phi_0\rangle ,$$

$$\begin{aligned} h(4) = {} & \omega_0 \{ i g_{\lambda_0}^*(3) \eta^* - i g_{\lambda_0}(3) \eta \} + \sum_{\lambda\neq\lambda_0} \omega_\lambda g_\lambda(2) g_\lambda^*(2) \\ & + \frac{1}{4!} \langle\phi_0 | [[[[\hat{H}, i\hat{G}(1)], i\hat{G}(1)], i\hat{G}(1)], i\hat{G}(1)] | \phi_0\rangle \\ & + \frac{1}{3!} \langle\phi_0 | \{ [[[\hat{H}, i\hat{G}(2)], i\hat{G}(1)], i\hat{G}(1)] + [[[\hat{H}, i\hat{G}(1)], i\hat{G}(2)], i\hat{G}(1)] \\ & + [[[\hat{H}, i\hat{G}(1)], i\hat{G}(1)], i\hat{G}(2)] \} | \phi_0\rangle , \end{aligned} \tag{3.8}$$

with

$$\hat{G}(n) \equiv \sum_{\lambda}\{g_{\lambda}(n)\hat{X}_{\lambda}+g^{*}_{\lambda}(n)\hat{X}^{\dagger}_{\lambda}\} . \qquad (3.9)$$

From Eq. (3.8) we can easily find an essential difference of our theory from the conventional boson-mapping theories: In sharp contrast with the conventional boson-mapping theories where the microscopic structure of the boson is assumed to be unchanged from the outset, our collective Hamiltonian automatically includes the effects due to the non-collective modes with $\lambda \neq \lambda_0$, which become more and more important for the higher order terms of the collective Hamiltonian, displaying dynamical change of the internal structure of the boson-correspondents (η^*,η) due to the maximal-decoupling condition.

[3.B] Application to A Simple Three-Level Model[3),8)]

In the TDHF theory, we introduce the TDHF wave function

$$|\phi(t)\rangle = \exp\{i\hat{F}(t)\}|\phi_0\rangle e^{i\varepsilon_0 t} ,$$

$$\hat{F}(t) = \sum_{\mu i}\{f_{\mu i}(t)c^{\dagger}_{\mu}c_{i} + f^{*}_{\mu i}(t)c^{\dagger}_{i}c_{\mu}\} \qquad (3.10)$$

with M×N pairs of complex parameters $\{f^{*}_{\mu i}(t),f_{\mu i}(t)\}$,where M and N mean a number of particle states $\{\mu\}$ and of the hole states $\{i\}$, respectively. Thus, the TDHF theory gives trajectories in a 2MN dimensional phase space which we call hereafter a TDHF manifold. In order to get a TDHF trajectory corresponding to the large-amplitude collective motion in the TDHF manifold, one has to face a difficult problem of solving the equation of motion which consists of a huge 2MN-dimensional coupled differential equations. Without having to solve the coupled differential equation directly, our theory intends to dynamically extract a so-called hypersurface (i.e., a 2k-dimensional collective submanifold) characterized by the collective variables $\{\eta_r^*, \eta_r;\ r=1,2,\cdots,k\}$ out of the TDHF manifold, by demanding the invariance principle of the time-dependent Schrödinger equation. Does this collective hypersurface really satisfy the maximal-decoupling condition in such a way that the collective trajectory under consideration can be reproduced on the collective hypersurface as precisely as possible?

In order to see this problem, it is interesting to compare our theory with the exact TDHF calculation by using a simple model. For this aim, let us employ the following simple model Hamiltonian

$$\hat{H} = E_0\hat{K}_{00}+E_1\hat{K}_{11}+E_2\hat{K}_{22}+ \frac{V_1}{2}\{\hat{K}_{10}\hat{K}_{10}+\text{h.c.}\}$$

$$+ \frac{V_2}{2}\{\hat{K}_{20}\hat{K}_{20}+\text{h.c.}\} + V_3\{(\hat{K}_{10}+\hat{K}_{20})(\hat{K}_{12}+\hat{K}_{21})+\text{h.c.}\}. \quad (3.11)$$

There are three levels with energies $E_0<E_1<E_2$ and each level has N-hold degeneracy. The fermion pair operators $\hat{K}_{\alpha\beta}$ are defined as

$$\hat{K}_{\alpha\beta} = \sum_{m=1}^{N} c_{\alpha m}^{\dagger} c_{\beta m}\ , \quad (\alpha,\beta = 0,1,2)$$

$$[\hat{K}_{\alpha\beta},\hat{K}_{\gamma\delta}] = \delta_{\beta\gamma}\hat{K}_{\alpha\delta} - \delta_{\alpha\delta}\hat{K}_{\gamma\beta}\ . \quad (3.12)$$

We consider a system with N particles and the lowest energy state $|\phi_0\rangle$ without the interaction is given by

$$|\phi_0\rangle = \prod_{m=1}^{N} c_{0m}^{\dagger}|0\rangle\ , \quad (3.13)$$

where $|0\rangle$ denotes the vacuum of the fermion operator $c_{\alpha m}$. (See Fig. 1). The Hamiltonian in the original three level model[9] corresponds to the case with $V_3 = 0$ in Eq. (3.11). However,

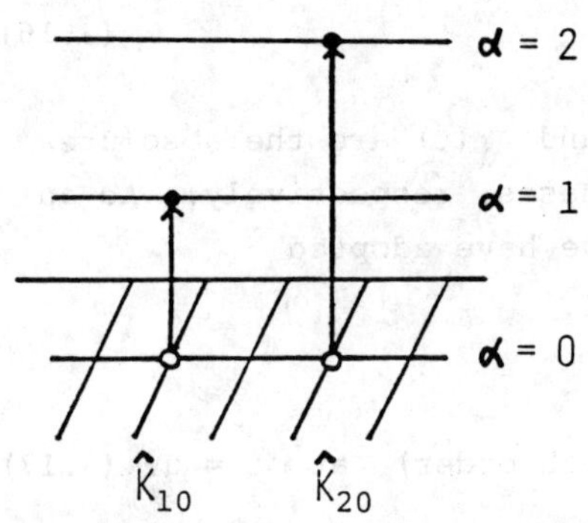

Fig. 1. Three level model.

such a model Hamiltonian is too simple to check up our theory. We therefore added to it the interaction with $V_3 \neq 0$, which is notorious for its role in causing the complicated dynamical anharmonicity on the collective vibrations.

In our three-level model, we have two independent RPA modes $\{\hat{X}^{\dagger}_{\lambda}, \hat{X}_{\lambda}; \lambda=1,2\}$ and the full TDHF theory requires two pairs of time-dependent complex parameter $\{f_{\lambda}(t), f^{*}_{\lambda}(t); \lambda=1,2\}$ through

$$|\phi(t)\rangle = \exp\{i\hat{F}(t)\}|\phi_0\rangle$$
$$\hat{F}(t) = \sum_{\lambda=1}^{2} \{f_{\lambda}(t)\hat{X}_{\lambda} + f^{*}_{\lambda}(t)\hat{X}^{\dagger}_{\lambda}\} . \tag{3.14}$$

Contrary to the full TDHF theory, our theory in this case introduces only one pair of the complex parameters $(\eta^{*}(t), \eta(t))$ through

$$|\phi(\eta^{*},\eta) = \exp\{i\hat{G}(\eta^{*},\eta)\}|\phi_0\rangle$$
$$\hat{G}(\eta^{*},\eta) = \sum_{\lambda=1}^{2} \{g_{\lambda}(\eta^{*},\eta)\hat{X}_{\lambda} + g^{*}_{\lambda}(\eta^{*},\eta)\hat{X}^{\dagger}_{\lambda}\} , \tag{3.15}$$

where $g_{\lambda}(\eta^{*},\eta)$ is expanded as a power series of (η^{*},η) according to Eq. (3.5). With the boundary condition that the collective motion under consideration is connected with the RPA-phonon mode in the small amplitude (i.e., small-η) limit, we have solved the time-dependece of $f_{\lambda}(t)$ and $g_{\lambda}(\eta^{*},\eta)$;

$$f_\lambda(t) \equiv R_\lambda(t)\ \exp\{i\Phi_\lambda(t)\},\quad g_\lambda(\eta^*(t),\eta(t)) \equiv \tilde{R}_\lambda(t)\cdot\exp\{i\tilde{\Phi}_\lambda(t)\}, \tag{3.16}$$

where $R_\lambda(t)$ and $\tilde{R}_\lambda(t)$ and $\Phi_\lambda(t)$ and $\tilde{\Phi}_\lambda(t)$ are the absolute values of f_λ and g_λ and their phases, respectively. As an initial condition for our theory we have adopted

$$\eta^*(t)\eta(t) = 1 \quad \text{at} \quad t = 0,$$

$$f_\lambda(t) = g_\lambda(\eta^*(t),\eta(t)) \quad \text{(in the 4th order)} \quad \text{at} \quad t = 0. \tag{3.17}$$

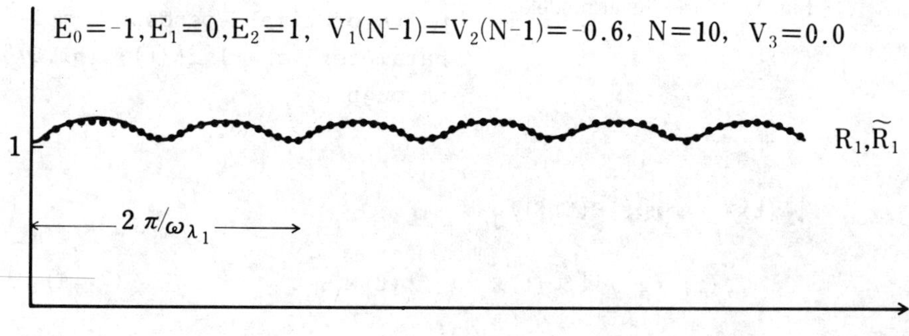

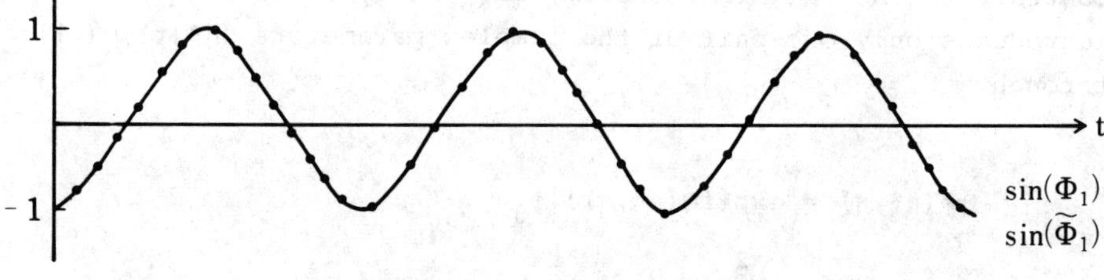

•••••• OUR METHOD(4th order)

—— FULL TDHF

Fig. 2 Numerical results for the case V_3= 0

In Fig. 2, the numerical results for the case with $V_3 = 0$ are shown. In this case, within the 4th order, the RPA-phonon mode λ_1 does not couple with the other non-collective RPA mode λ_2, so that we always have $g_{\lambda_2}(\eta^*(t),\eta(t)) = 0$ and only $g_{\lambda_1}(\eta^*(t),\eta(t))$ is given in Fig. 2. Here, the dotted lines shows the 4th order (η^*,η)-expansion solution and the solid line shows the exact TDHF solution. The time scale is $2\pi/\omega_{\lambda_1}$, where ω_{λ_1} is the frequency of the RPA- phonon mode. Fig. 2 demonstrates that, in the case with $V_3 = 0$, we cannot distinguish our 4th order (η^*,η)-expansion solution from the exact TDHF solution.

In Fig. 3, the numerical resutls for the case with $V_3 \neq 0$ are shown. We can see that our theory well reproduces a trajectory obtained by the exact TDHF theory. In Fig. 4, the numerical results for the case with a much larger value of V_3 are

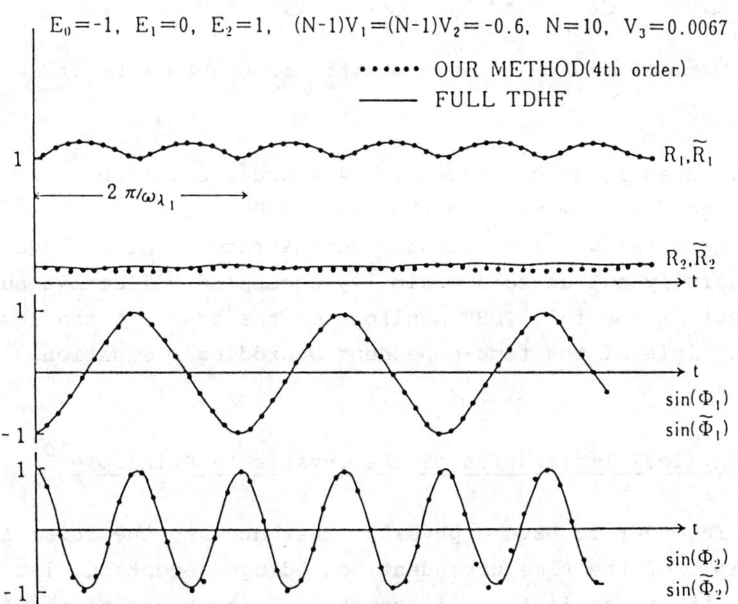

Fig.3 Numerical results for the case with a small value of V_3

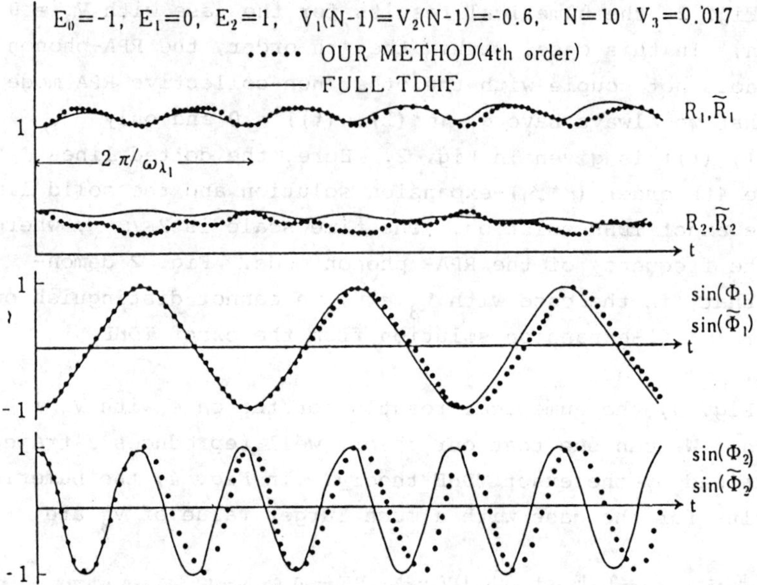

Fig.4 Numerical results for the case with a much larger value of V_3

shown. Even in such a case, the 4th order solution gives a fairly good agreement with the exact TDHF solution.

These calculations clearly demonstrates that we have successfully extracted a maximally-decoupled collective submanifold out of the full TDHF manifold on the basis of the invariance principle of the time-dependent Schrödinger equation.

§4. Physical Implication of the Invariance Principle[10)]

In order to have a physical insight into the invariance principle of the time-dependent Schrödinger equation, let us start with investigation of geometrical structure of the TDHF theory in the TDHF manifold associated with the 2MN parameters

$\{f_{\mu i}(t),\ f^*_{\mu i}(t)\}$ defined by Eq. (3.10). Now, it is always possible[10)] to find that, by using a suitable variable transformation

$$\{f^*_{\mu i}(t), f_{\mu i}(t)\} \rightarrow \{C^*_{\mu i}(t), C_{\mu i}(t)\} \ , \tag{4.1}$$

the TDHF equation (1.2) can be simply reduced to the following set of equations of motion;

$$i\dot{C}_{\mu i} = \frac{\partial H}{\partial C^*_{\mu i}} \ , \qquad i\dot{C}^*_{\mu i} = -\frac{\partial H}{\partial C_{\mu i}} \ , \tag{4.2a}$$

i.e.,

$$\dot{q}_{\mu i} = \frac{\partial H}{\partial p_{\mu i}} \ , \qquad \dot{p}_{\mu i} = -\frac{\partial H}{\partial q_{\mu i}} \ ,$$

$$q_{\mu i} = \frac{1}{\sqrt{2}}(C^*_{\mu i}+C_{\mu i}) \ , \quad p_{\mu i} = \frac{i}{\sqrt{2}}(C^*_{\mu i}-C_{\mu i}) \ , \tag{4.2b}$$

where

$$H \equiv \langle\phi(t)|\hat{H}|\phi(t)\rangle - \langle\phi_0|\hat{H}|\phi_0\rangle \ . \tag{4.3}$$

The existence of such a set of <u>canonical variables</u> $\{C^*_{\mu i}(t), C_{\mu i}(t)\}$ demonstrates that the TDHF trajectories, which represent the time-evolution of the system specified by the time-dependent single Slater determinants (3.10), are simply described by the <u>canonical equations of motion (4.2a) in classical mechanics</u> in the TDHF manifold as a symplectic manifold M^{2MN}. It is also shown[10)] that the canonical variables $\{C^*_{\mu i}, C_{\mu i}\}$ are nothing but the c-number version of the boson operators $\{B^\dagger_{\mu i}, B_{\mu i}\}$;

$$B_{\mu i} \leftrightarrow iC_{\mu i} \ , \qquad B^\dagger_{\mu i} \leftrightarrow -iC^*_{\mu i} \ ,$$

$$[B_{\mu i}, B^\dagger_{\nu j}] = \delta_{\mu\nu}\delta_{ij}, \quad [B_{\mu i}, B_{\nu j}] = 0 \ , \tag{4.4}$$

which are used in the boson mapping [11),12)] of the fermion-pair operators $\{c^\dagger_\mu c_\nu,\ c_k c^\dagger_\ell,\ c^\dagger_\mu c_k,\ c^\dagger_k c_\mu\}$ through

$$
\begin{aligned}
c^\dagger_\mu c_\nu &\to \hat{\mathbf{U}}_M c^\dagger_\mu c_\nu \hat{\mathbf{U}}^\dagger_M = \hat{\Gamma}_0 \cdot \mathbf{P} \cdot \sum_j B^\dagger_{\mu j} B_{\nu j} \\
c_k c^\dagger_\ell &\to \hat{\mathbf{U}}_M c_k c^\dagger_\ell \hat{\mathbf{U}}^\dagger_M = \hat{\Gamma}_0 \cdot \mathbf{P} \cdot \sum_\mu B^\dagger_{\mu k} B_{\mu \ell} \\
c^\dagger_\mu c_k &\to \hat{\mathbf{U}}_M c^\dagger_\mu c_k \hat{\mathbf{U}}^\dagger_M = \hat{\Gamma}_0 \cdot \mathbf{P} \cdot (B^\dagger \sqrt{\mathbb{1}_M - A})_{k\mu} , \\
c^\dagger_k c_\mu &\to \hat{\mathbf{U}}_M c^\dagger_k c_\mu \hat{\mathbf{U}}^\dagger_M = \hat{\Gamma}_0 \cdot \mathbf{P} \cdot (\sqrt{\mathbb{1}_M - A}\ B)_{\mu k} , \\
A_{\nu\mu} &\equiv \sum_j B^\dagger_{\mu j} B_{\nu j} .
\end{aligned}
\tag{4.5}
$$

The mapping operator $\hat{\mathbf{U}}_M$ and the projection operators $\hat{\Gamma}_0$ and $\mathbf{P}$ in Eq. (4.5) are given by Eq. (I.3.8) and (I.3.9) in Part (I), and $\mathbb{1}_M$ is an M by M unit matrix in the boson space.

The use of the canonical variables $\{C^*_{\mu i}(t), C_{\mu i}(t)\}$ enables us to investigate geometrical structure of the invariance principle of the time-dependent Schrödinger equation (formulated within the framework of the TDHF theory), by using the language of the classical mechanics.

If there exists an (approximate) invariant subspace of the Hamiltonian, i.e., the third integral of the motion, a certain kind of trajectories in the TDHF manifold M^{2MN} may be described by only a few "collective" canonical variables $\{\eta^*_r(t), \eta_r(t); r=1,2,\cdots,k<<MN\}$. In order to avoid unnecessary complexity of the presentation, we restrict ourselves to the simplest case with r = 1, i.e., a single pair of collective variables (η^*,η). We then consider a general variable transformation from the canonical variables $\{C^*_{\mu i}, C_{\mu i}\}$ to a set of variables $\{\eta^*,\eta,\xi^*_\alpha,\xi_\alpha;\ \alpha= 1,2,\cdots,MN-1\}$ including the collective variables (η^*,η):

$$
C_{\mu i} = C_{\mu i}(\eta^*,\eta;\xi^*_\alpha,\xi_\alpha), \quad C^*_{\mu i} = C^*_{\mu i}(\eta^*,\eta;\xi^*_\alpha,\xi_\alpha). \tag{4.6}
$$

Now, let us introduce the Taylor expansion of $\{C^*_{\mu i}, C_{\mu i}\}$ with respect to the (2MN-2) variables $\{\xi^*_\alpha, \xi_\alpha\}$, which are called <u>non-collective variables</u>, around a point $(\eta^*, \eta; \xi^*_\alpha = 0,\ \xi_\alpha = 0)$;

$$C_{\mu i} = C^{(0)}_{\mu i} + C^{(1)}_{\mu i} + \cdots + C^{(n)}_{\mu i} + \cdots ,$$

$$C^{(0)}_{\mu i} = [C_{\mu i}] , \quad C^{(1)}_{\mu i} = \sum_\alpha \{ [\frac{\partial C_{\mu i}}{\partial \xi_\alpha}] \xi_\alpha + [\frac{\partial C_{\mu i}}{\partial \xi^*_\alpha}] \xi^*_\alpha \}$$

$$C^{(2)}_{\mu i} = \frac{1}{2!} \sum_{\alpha,\beta} \{ [\frac{\partial^2 C_{\mu i}}{\partial \xi_\alpha \partial \xi_\beta}] \xi_\alpha \xi_\beta + 2 [\frac{\partial^2 C_{\mu i}}{\partial \xi_\alpha \partial \xi^*_\beta}] \xi_\alpha \xi^*_\beta + [\frac{\partial^2 C_{\mu i}}{\partial \xi^*_\alpha \partial \xi^*_\beta}] \xi^*_\alpha \xi^*_\beta \} ,$$

etc., (4.7)

where the symbol [F] for any function $F(\eta^*, \eta, \xi^*_\alpha, \xi_\beta)$ denotes the value at the expansion point $(\eta^*, \eta; \xi^*_\alpha = 0,\ \xi_\alpha = 0)$ and is a function of the collective variables (η^*, η) alone. Then, we can express the invariance principle of the time-dependent Schrödinger equation (formulated in the TDHF framework) in the form

$$i \frac{\partial [C_{\mu i}]}{\partial t} = [\frac{\partial H}{\partial C^*_{\mu i}}] , \quad -i \frac{\partial [C^*_{\mu i}]}{\partial t} = [\frac{\partial H}{\partial C_{\mu i}}] , \tag{4.8}$$

whose formal structure is the same as that of the original equations of motion (4.2a).

The canonical-variables condition corresponding to Eq. (2.32), which requires that the pair of the collective variables (η^*, η) has to be <u>canonical</u> so as to satisfy

$$[\frac{\partial H}{\partial \eta}] = -i\dot{\eta}^* , \quad [\frac{\partial H}{\partial \eta^*}] = i\dot{\eta} , \tag{4.9}$$

is shown[10] from Eq. (4.8) to be

$$\mathrm{Tr}\{ [\mathbb{C}^\dagger] \frac{\partial [\mathbb{C}]}{\partial \eta} - \frac{\partial [\mathbb{C}^\dagger]}{\partial \eta} [\mathbb{C}] \} = \eta^*$$

$$\mathrm{Tr}\{ [\mathbb{C}^\dagger] \frac{\partial [\mathbb{C}]}{\partial \eta^*} - \frac{\partial [\mathbb{C}^\dagger]}{\partial \eta^*} [\mathbb{C}] \} = -\eta , \tag{4.10}$$

where $\mathbb{C}$ and $\mathbb{C}^\dagger$ denote M by N matrices defined by $(\mathbb{C})_{\mu i} = C_{\mu i}$ and $(\mathbb{C}^\dagger)_{i\mu} = C^*_{\mu i}$, respectively, and the notation $\mathrm{Tr}\{[\mathbb{C}^\dagger]\cdot\partial[\mathbb{C}]/\partial\eta\}$ means

$$\mathrm{Tr}\{[\mathbb{C}^\dagger]\frac{\partial[\mathbb{C}]}{\partial\eta}\} \equiv \sum_{\mu i}\{[C^*_{\mu i}]\frac{\partial[C_{\mu i}]}{\partial\eta}\} . \tag{4.11}$$

Provided that $[C_{\mu i}]$ and $[C^*_{\mu i}]$ are analytic function of η^* and η, the condition (4.10) is easily rewritten as

$$\mathrm{Tr}\{\frac{\partial[\mathbb{C}^\dagger]}{\partial\eta^*}\frac{\partial[\mathbb{C}]}{\partial\eta} - \frac{\partial[\mathbb{C}^\dagger]}{\partial\eta}\frac{\partial[\mathbb{C}]}{\partial\eta^*}\} = 1 \tag{4.12}$$

which corresponds to Eq. (2.31).

The following is now clear. Both Eq. (4.8) and Eq. (4.10) play a role to specify the functional forms of $[C_{\mu i}]$ and $[C^*_{\mu i}]$, which depend on the collective variables (η^*,η) alone and define a collective submanifold (hypersurface) Σ imbedded in the 2MN-dimensional TDHF manifold M^{2MN}. The use of a pair of the

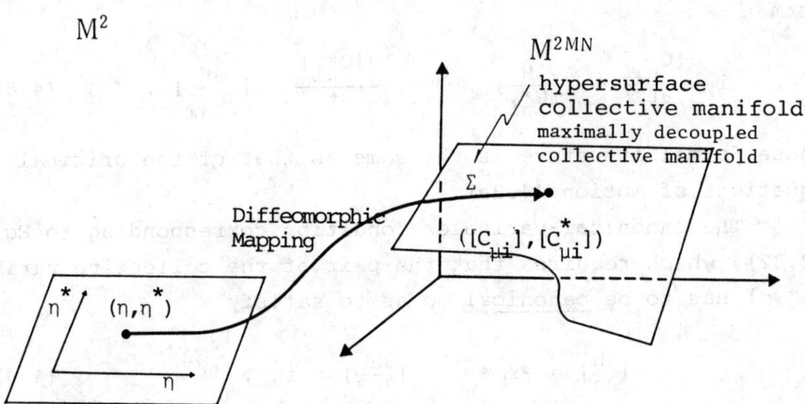

Fig. 5 Diffeomorphic Mapping $M^2 \to \Sigma$

canonical collective variables is simply to introduce a two-dimensional "collective" symplectic manifold M^2 which is a diffeomorphic manifold of the collective submanifold Σ. (See Fig. 5.) The time-evolution of the collective motion on the "collective" symplectic manifold M^2, which is described by Eq. (4.9), is mapped into the hypersurface Σ by the functions $[C_{\mu i}]$ and $[C^*_{\mu i}]$.

In order to demonstrate, from the above geometrical point of view, that the invariance principle (4.8) satisfies the maximal-decoupling condition, it is decisive to define "local" canonical non-collective variables which are dependent on each point of the hypersurface Σ and are "locally" canonical in the neighborhood of the respective point. The "local" canonical non-collective variables $\{\xi^*_\alpha, \xi_\alpha; \alpha=1,2,\cdots,MN-1\}$ are defined through the following "local" canonical-variables condition

$$\mathrm{Tr}\{[\frac{\partial\mathbb{C}^\dagger}{\partial\xi_\alpha}]\frac{\partial[\mathbb{C}]}{\partial\eta} - \frac{\partial[\mathbb{C}^\dagger]}{\partial\eta}[\frac{\partial\mathbb{C}}{\partial\xi_\alpha}]\} = 0, \quad \mathrm{Tr}\{[\frac{\partial\mathbb{C}^\dagger}{\partial\xi^*_\alpha}]\frac{\partial[\mathbb{C}]}{\partial\eta} - \frac{\partial[\mathbb{C}^\dagger]}{\partial\eta}[\frac{\partial\mathbb{C}}{\partial\xi^*_\alpha}]\} = 0,$$

$$\mathrm{Tr}\{[\frac{\partial\mathbb{C}^\dagger}{\partial\xi_\alpha}]\frac{\partial[\mathbb{C}]}{\partial\eta^*} - \frac{\partial[\mathbb{C}^\dagger]}{\partial\eta^*}[\frac{\partial\mathbb{C}}{\partial\xi_\alpha}]\} = 0, \quad \mathrm{Tr}\{[\frac{\partial\mathbb{C}^\dagger}{\partial\xi^*_\alpha}]\frac{\partial[\mathbb{C}]}{\partial\eta^*} - \frac{\partial[\mathbb{C}^\dagger]}{\partial\eta^*}[\frac{\partial\mathbb{C}}{\partial\xi^*_\alpha}]\} = 0,$$

$$\mathrm{Tr}\{[\frac{\partial\mathbb{C}^\dagger}{\partial\xi^*_\beta}][\frac{\partial\mathbb{C}}{\partial\xi_\alpha}] - [\frac{\partial\mathbb{C}^\dagger}{\partial\xi_\alpha}][\frac{\partial\mathbb{C}}{\partial\xi^*_\beta}]\} = \delta_{\alpha\beta}, \quad \mathrm{Tr}\{[\frac{\partial\mathbb{C}^\dagger}{\partial\xi_\alpha}][\frac{\partial\mathbb{C}}{\partial\xi_\beta}] - [\frac{\partial\mathbb{C}^\dagger}{\partial\xi_\beta}][\frac{\partial\mathbb{C}}{\partial\xi_\alpha}]\} = 0. \tag{4.13}$$

It is easily shown[10)] that, with the aid of Eq. (4.13), the invariance principle (4.8) leads us to the maximal-decoupling condition

$$[\frac{\partial H}{\partial\xi_\alpha}] = 0, \quad [\frac{\partial H}{\partial\xi^*_\alpha}] = 0. \quad (\alpha = 1,2,\cdots,MN-1). \tag{4.14}$$

It is now clear that the invariance principle of the time-dependent Schrödinger equation (formulated within the TDHF framework) is just equivalent to the requirement that a certain kind of trajectories described by the canonical equations of motion (4.2a) in the TDHF manifold is approximately reproduced on the hypersurface Σ[13].

In order to make clear a physical meaning of the condition (4.13) to define the local canonical variables, let us employ the notations

$$\psi_{coll}(\mu i) \equiv \frac{\partial[C_{\mu i}]}{\partial\eta}, \qquad \phi_{coll}(\mu i) \equiv \frac{\partial[C^*_{\mu i}]}{\partial\eta},$$

$$\psi_\alpha(\mu i) \equiv [\frac{\partial C_{\mu i}}{\partial\xi_\alpha}], \qquad \phi_\alpha(\mu i) \equiv [\frac{\partial C^*_{\mu i}}{\partial\xi_\alpha}]. \tag{4.15}$$

Then, the condition together with Eq. (4.12) can be expressed as

$$\sum_{\mu i}\{\psi_{coll}(\mu i)\psi^*_{coll}(\mu i) - \phi_{coll}(\mu i)\phi^*_{coll}(\mu i)\} = 1,$$

$$\sum_{\mu i}\{\psi_{coll}(\mu i)\psi^*_\alpha(\mu i) - \phi_{coll}(\mu i)\phi^*_\alpha(\mu i)\}$$

$$= \sum_{\mu i}\{\psi_{coll}(\mu i)\phi_\alpha(\mu i) - \phi_{coll}(\mu i)\psi_\alpha(\mu i)\} = 0,$$

$$\sum_{\mu i}\{\psi_\alpha(\mu i)\psi^*_\beta(\mu i) - \phi_\alpha(\mu i)\phi^*_\beta(\mu i)\} = \delta_{\alpha\beta},$$

$$\sum_{\mu i}\{\psi_\alpha(\mu i)\phi_\beta(\mu i) - \phi_\alpha(\mu i)\psi_\beta(\mu i)\} = 0. \tag{4.16}$$

Equation (4.16) simply has the same structure as the orthonormality relation appearing in the RPA;

$$\hat{X}^\dagger_{coll} = \sum_{\mu i}\{\psi_{coll}(\mu i)\alpha^\dagger_\mu\beta^\dagger_i + \phi_{coll}(\mu i)\beta_i\alpha_\mu\} \equiv :\hat{O}^\dagger(\eta^*,\eta): \tag{4.17a}$$

$$\hat{X}^\dagger_\alpha = \sum_{\mu i}\{\psi_\alpha(\mu i)\alpha^\dagger_\mu\beta^\dagger_i + \phi_\alpha(\mu i)\beta_i\alpha_\mu\}. \tag{4.17b}$$

Here

$$\alpha_\mu^\dagger \equiv \exp\{i[\hat{F}]\}c_\mu^\dagger\exp\{-i[\hat{F}]\}$$

$$\beta_i^\dagger \equiv \exp\{i[\hat{F}]\}c_i\exp\{-i[\hat{F}]\} \ , \tag{4.18}$$

and the operator $[\hat{F}]$ is a function of (η^*,η) alone, which is obtained from the operator $\hat{F}$ in Eq. (3.10) according to the Taylor expansion (4.7).

§5. Toward a Quantum Theory of "Maximally-Decoupled" Collcetive Motion

Through the discussion so far, it has been recognized that the basic principle underlying the TDHF derivation of the maximally-decoupled collective motion is just the invariance principle of the time-dependent Schrödinger equation. The next task is to directly apply the basic invariance principle to the many-body Schrödinger equation (1.3) in order to obtain a full quantum theory of "maximally-decoupled" collective motion.

To formulate this idea in a transparent way, it is convenient to employ the so-called "unitary-transformation method with auxiliary collective variables (, i.e., auxiliary bosons)",[14,15,16)*)] which was proposed in 1950's for the aim to give a microscopic foundation of the collective model of Bohr and Mottelson. At that time, the applications of this method were only feasible provided that one had some idea about the functional forms of the collective variables in terms of the particle variables. This is because we could not recognize, at that time, the basic principle to specify the "optimum"

---

*) The first transformation discussed in §5 of Part (I), which leads us to Eq.(I.5.4a), can be just performed by this method. See Refs. 42), 43) and 44) in Part (I).

collective variables, i.e., the invariance principle of the time-dependent Schrödinger equation. If once, the unitary-transformation method with auxiliary collective variables is associated with the invariance principle, therefore, we can formulate the full quantum theory of maximally-decoupled collective motion.

This program has been done in Refs. 3), 17) and 18) and the following facts have been demonstrated:

i) The unitary transformation operator with auxiliary bosons is nothing but an extention of the mapping operator $\hat{U}_M$ of the modified Marumori boson-mapping method, into the whole fermion-boson product space.

ii) The requirement of the invariance principle of the time-dependent Schrödinger equation is compeletely equivalent to demanding the existence of the invariant collective subspace of the Hamiltonian.

iii) The unitary-transformation method with auxiliary bosons associated with the invariance principle leads us to the collective boson Hamiltonian in the invariant collective subspace.

iv) A set of the basic equations in this theory can be reduced to the set of the basic equations (3.1) and (2.31) obtained within the TDHF framework under an approximation called a "classical approximation".

References

1) T. Marumori, F. Sakata, T. Une and Y. Hashimoto; invited talk presented at The Asia Pasific Physics Conference, Singapore 12-18 June 1983.

2) D. J. Thouless, Nucl. Phys. 21 (1960), 225.

3) T. Marumori, F. Sakata, T. Maskawa, T. Une and Y. Hashimoto, "Nuclear Collective Dynamics", Lectures of 1982 International Summer School of Nuclear Physics, Poiana Brasov, Romania 1982, (World Scientific Publishing Co. Pte. Ltd. 1983), p1.

4) D. R. Inglis, Phys. Rev. 96 (1954), 1054.

5) T. Marumori, T. Maskawa, F. Sakata and A. Kuriyama, Prog. Theor. Phys. 64 (1980), 1294.

6) T. Marumori, Prog. Theor. Phys. 57 (1977), 112.

7) T. Marumori, A. Hayahsi, T. Tomoda, A. Kuriyama and T. Maskawa, Prog. Theor. Phys. 63 (1980), 1576.

8) F. Sakata, Y. Hashimoto, T. Marumori and T. Une, Prog. Theor. Phys. 70 (1983), 163.

9) S. Y. Li, A. Klein and R. M. Dreizler, Journal Math. Phys. 11 (1970), 975.

10) F. Sakata, T. Marumori, Y. Hashimoto and T. Une, Prog. Theor. Phys. 70 (1983), 424.

11a) S. T. Beliaev and V. G. Zelevinsky, Nucl. Phys. 39 (1962), 582.

11b) T. Marumori, M. Yamamura and T. Tokunaga, Prog. Theor. Phys. 31 (1964), 1009.

12) E. R. Marshalek and J. Weneser, Phys. Rev. C2 (1970), 1682.
E. R. Marshalek, Nucl. Phys. A161 (1971), 401.
D. Janssen, F. Dönau, S. Frauendorf and R. V. Jolos, Nucl. Phys. A172 (1971), 145.

13) J. da Providencia and J. N. Urbano, Proceedings of the 1982 INS International Symposium on "Dynamics of Nuclear Collective Motion" (INS, 1982, July 6-10), 361:

"Nuclear Collective Dynamics", Lectures of 1982 Brasov International Summer School of Nuclear Physics, Poiana Brasov, Romania 1982, (World Scientific Publishing Co. Pte. Ltd.), p.

14) T. Marumori, J. Yukawa and R. Tanaka, Prog. Theor. Phys. 13 (1955), 442.
T. Marumori, Prog. Theor. Phys. 14 (1955), 608.
T. Marumori and E. Yamada, Prog Theor. Phys 14 (1955), 1557.
S. Hayakawa and T. Marumori, Prog. Theor. Phys. 18 (1957), 396.

15) S. Tomonaga, Prog. Theor. Phys. 13 (1955), 467.
T. Miyazima and T. Tamura, Prog. Theor. Phys. 15 (1956), 255.

16) H. Lipkin, A. de Shalit and I. Talmi, Phys. Rev. 103 (1956), 1773.
F. Villars, Ann. Rev. Nucl. Sci. 7 (1957), 185.

17) T. Marumori, F. Sakata, T. Une, Y. Hashimoto and T. Maskawa, Prog. Theor. Phys. 66 (1981), 1651.

18) T. Marumori, F. Sakata, T. Une and Y. Hashimoto, "Microscopic Theories of Nuclear Collective Motions", Proceedings of the 5th Kyoto Summer Institute, Kyoto, 1982, Prog. Theor. Phys. Supplement 74 & 75 (1983), 221.

FLUID DYNAMICAL METHODS FOR THE DESCRIPTION OF NUCLEAR GIANT MULTIPOLE RESONANCES

Gottfried Holzwarth
Universität Siegen, FB 7
59 Siegen, FRG

INTRODUCTION

In the following lectures we shall be concerned with the "Giant Multipole Resonances" (GMR) observed in almost all atomic nuclei which may be interpreted as collective oscillations of the nucleon fluid. Many different types of GMR have been theoretically suggested, distinguished by their isospin T (protons and neutrons moving in phase (T=O), or in opposite phase (T=1)), their parity Π ("Electric modes" with $\Pi = (-)^L$, "Magnetic modes" with $\Pi = (-)^{1+L}$, where L characterizes the multipolarity of the GMR), "surface modes" characterized by a flow-velocity field $\vec{v}$ satisfying $\vec{\nabla}\cdot\vec{v} \equiv 0$, or "compressional" modes which involve density changes in the nuclear interior. Quite a number of these have been experimentally identified and successfully interpreted in the microscopic formalism of the "Random Phase Approximation" (RPA). The outstanding feature of GMR is that they exhaust a large fraction of corresponding sum rules, which means that many nucleons move in a coherent way. This leads to the expectation that it should be possible to describe their basic properties in terms of simple field variables, like the density $\rho(\vec{x},t)$, or the velocity $\vec{v}(\vec{x},t)$.

Especially after the successful application of energy-density functionals

$$E[\rho] = \int(\tau[\rho] + V[\rho])\, d^3x \tag{1}$$

for deriving static nuclear properties[2] through the variational equation

$$\frac{\delta}{\delta\rho} E[\rho] - \lambda_F = 0 \tag{2}$$

one tried to use the same functional (1) also in dynamical calculations. For the potential part $V[\rho]$ in (1) the Skyrme parametrization[3] has proven very useful, we shall, however, for simplicity discuss here a simplified version

$$V[\rho] = \sum_\sigma a_\sigma \rho^\sigma + a_s(\vec{\nabla}\rho)^2 \tag{3}$$

which consists of a volume part $\Sigma a_\sigma \rho^\sigma$ and a surface part $a_s(\vec{\nabla}\rho)^2$ with constants a_σ and a_s. The local density $\rho(\vec{x},t)$ is the sum of proton and neutron single-particle densities as we shall consider here mainly T=O states of spin-saturated systems.

In order to obtain also the kinetic energy density τ as a functional of the local density ρ it is necessary to make assumptions about the distribution of particle momenta $\vec{p}$ in phase space. Assuming isotropic momentum distribution and the occupation of available states up to $|\vec{p}| = p_F$, i.e.

$$f_o(\vec{x},\vec{p}) = f_o(\vec{x},p) \propto \Theta(p_F - p) \tag{4}$$

for the groundstate distribution function f_o, leads to the kinetic energy functional approximated by[4]

$$\tau[\rho] = \frac{\hbar^2}{2m}\left(\frac{3}{5}\left(\frac{3}{2}\pi^2\right)^{2/3}\rho^{5/3} + \frac{1}{4}\eta\,\frac{(\nabla\rho)^2}{\rho} + \zeta\Delta\rho\right). \tag{5}$$

The volume term $\rho^{5/3}$ is the usual Thomas Fermi expression. Together with the surface terms with parameters η and ζ we shall refer to (5) as the "Extended Thomas Fermi" (ETF) approximation.

Once we accept the equation of state (1) to hold also for

time-dependent densities, conservation of mass and momentum flow leads to the coupled set of fluid-dynamical equations

$$\frac{\partial}{\partial t}\rho + \vec{\nabla}\cdot\vec{j} = 0 \tag{6}$$

$$\frac{\partial}{\partial t}\vec{j} = -\frac{1}{m}\rho\vec{\nabla}\frac{\delta}{\delta\rho}E[\rho] . \tag{7}$$

In the Euler equation (7) we have already linearized the left-hand side as we shall consider only small oscillations. The current $\vec{j}$ is connected with the velocity field $\vec{v}(\vec{x},t)$ as usual

$$\vec{j} = \rho\vec{v} . \tag{8}$$

We shall refer to the set (6), (7) as "hydrodynamics" or "first sound" propagation. It may now be easily seen that this approach must fail to reproduce essential features of surface modes (satisfying $\vec{\nabla}\cdot\vec{v} \equiv 0$) encountered in droplets of Fermi fluids:

Frequencies ω for a velocity field $\vec{v}(\vec{x},t) = \dot{\alpha}(t)\,\vec{u}(\vec{x})$ are given by[5,6]

$$\omega^2 = C/B \tag{9}$$

with the inertia parameter

$$B = \frac{1}{2} m \int \rho_0\, u^2\, d^3x \tag{10}$$

and the force constant

$$C = \frac{1}{2}\int\left(\frac{\delta^2}{\delta\rho^2}E[\rho]\right)_{\rho=\rho_0}(\delta\rho)^2\, d^3x = \frac{1}{2}\frac{\partial^2}{\partial\alpha^2}E[\rho]. \tag{11}$$

where

$$\rho = \rho_0 + \alpha\,\delta\rho\ ,\quad \delta\rho = -\vec{\nabla}\cdot(\rho_0\vec{u}) .$$

Evidently, volume parts in $E[\rho]$ cannot contribute to C for a velocity field $\vec{u}$ satisfying $\vec{\nabla}\cdot\vec{u} = 0$, because for the contributions of the volume part E_{vol} of E to C we have

$$C = \frac{1}{2} \int \left(\frac{\partial^2}{\partial\rho_o^2} E_{vol}[\rho_o]\right) (\vec{u}\cdot\vec{\nabla}\rho_o)^2 d^3x = \frac{1}{2}\int (\vec{u}\cdot\vec{\nabla}\rho_o)(\vec{u}\cdot\vec{\nabla}\frac{\partial}{\partial\rho_o} E_{vol}[\rho_o])d^3x.$$

According to (2) $\partial/\partial\rho_o\ E_{vol}\ [\rho_o]$ may be completely expressed through surface terms and the constant λ_F therefore C can be expressed through surface terms alone. As an example we consider the Giant Quadrupole Resonance (GQR) with

$$\vec{u} = \vec{\nabla}Q(\vec{x}), \quad Q(\vec{x}) = r^2 Y_{20} \tag{12}$$

which leads to[6]

$$\omega^2 = \frac{4\ E_{sur}[\rho_o]}{m\ A\ \langle r^2\rangle} . \tag{13}$$

The surface part of the groundstate energy $E_{sur}[\rho_o]$ in the numerator and the mean square radius $\langle r^2\rangle$ in the denominator are both proportional to $A^{2/3}$, therefore we obtain $\omega \propto A^{-1/2}$. This result is in serious conflict with the experimentally observed GQR which by its formfactors is clearly identified as a surface mode but occurs near $\hbar\omega = 60\ A^{-1/3}$ MeV[1].

Quantum mechanical sum rules[7] may be used to estimate frequencies of modes which exhaust large portions of the sums:

$$\hbar^2\omega^2 \approx m_3/m_1 \tag{14}$$

with

$$m_1 = \sum_n \hbar\omega_n |\langle n|Q|O\rangle|^2 = \frac{1}{2} \langle O|\,[Q,[H,Q]]\,|O\rangle$$

$$m_3 = \sum_n \hbar^3\omega_n^3 |\langle n|Q|O\rangle|^2 = -\frac{1}{2} \langle O|\,[[Q,H],[H,[Q,H]]]\,|O\rangle .$$

Using (12) it turns out that the numerator of (14) contains also the total ground-state kinetic energy $\langle T\rangle$ together with the surface part of the potential

$$m_3/m_1 = \hbar^2 \frac{4(\langle T\rangle + \langle V\rangle_{sur})}{m\ A\ \langle r^2\rangle} . \tag{15}$$

This expression gives the right $A^{-1/3}$ dependence for the frequency because the volume part of <T> dominates in the numerator. Comparison of (13) and (15) gives clear evidence that the kinetic energy is not treated correctly in the energy-density formalism if we use the static equation of state (1) with (5) in studying the dynamical behaviour of the system.

THE SCALING APPROXIMATION

In the following we shall discuss a simple extension of the set (6), (7) of the form[8,9]

$$\rho \frac{\partial}{\partial t} \vec{s} + \vec{j} = 0 \tag{16}$$

$$\frac{\partial}{\partial t} \vec{j} = \frac{1}{m} \frac{\delta}{\delta \vec{s}} E[\vec{s}] \ . \tag{17}$$

We shall call this the "Scaling Approximation" (SCA) because the modified Euler equation (17) results from the quantum-mechanical time-dependent variational principle

$$\delta \int \langle\psi| i\partial_t - H|\psi\rangle \, dt = 0$$

if the deviations of the time-even part $\hat{\rho}_+$ of the single-particle density matrix $\hat{\rho}$ from its ground-state values $\hat{\rho}_0$ are restricted to the form

$$\hat{\rho}_+(\vec{x},\vec{x}',t) = \exp(\tfrac{1}{2}(\vec{s}(\vec{x},t)\cdot\vec{\nabla} + \vec{\nabla}\cdot\vec{s}(\vec{x},t))) \times$$
$$\times \exp(\tfrac{1}{2}(\vec{s}(\vec{x}',t)\cdot\vec{\nabla}' + \vec{\nabla}'\cdot\vec{s}(\vec{x}',t))) \ \hat{\rho}_0(\vec{x},\vec{x}') . \tag{18}$$

This means that the time-dependence is introduced into $\hat{\rho}_+$ by scaling the arguments $\vec{x}$ and $\vec{x}'$ of $\hat{\rho}_0$ with an arbitrary (variational) time-dependent scaling field $\vec{s}(\vec{x},t)$. If we let $\vec{x}=\vec{x}'$ in (18) we obtain for the local density

$$\rho(\vec{x},t) = \exp(\vec{\nabla}\cdot\vec{s}(\vec{x},t))\,\rho_0(\vec{x}) \ . \tag{19}$$

(In (18) and (19) the gradients are understood to act on all functions standing to the right of the $\vec{\nabla}$). Equation (16) defines the velocity in terms of the scaling field

$$\vec{v} = -\frac{\partial}{\partial t}\vec{s}. \tag{20}$$

From (19) follows $\dot{\rho} = \vec{\nabla}\cdot(\rho\dot{\vec{s}})$ therefore the continuity equation (6) is satisfied. It should, however, be noted that (6) equates only the longitudinal parts of velocity and scaling field, while (16) requires also the transverse parts to be equal.

Through (18) the kinetic energy part in E becomes a functional of $\vec{s}$ which to any desired order is obtained from

$$\tau[\vec{s}] = \frac{\hbar^2}{2m}(-\Delta_y\hat{\rho}_+(\vec{x},\vec{x}',t))_{\vec{y}=0} \quad , \quad (\vec{y} = \vec{x} - \vec{x}') \ . \tag{21}$$

As we shall be concerned here with the harmonic approximation we give only the second order term[8] $E^{(2)}[\vec{s}]$:

$$E^{(2)}[\vec{s}] = \frac{1}{2}\int\tau^{(o)}_{ik}(s_{ij}s_{jk}+s_{ij}s_{kj}-s_{\ell}s_{ik\ell})d^3x + \frac{\hbar^2}{8m}\int s_{ik\ell}s_{\ell ki}\rho_o d^3x$$
$$+ \int V^{(2)}[\vec{s}]d^3x \tag{22}$$

where

$$s_{ijk..} = ..\frac{\partial}{\partial x_k}\frac{\partial}{\partial x_j}s_i \ ,$$

$$\tau^{(o)}_{ik} = -\frac{\hbar^2}{m}\left(\frac{\partial}{\partial y_i}\frac{\partial}{\partial y_k}\hat{\rho}_o(\vec{x},\vec{x}')\right)_{\vec{y}=o} \tag{23}$$

and $V^{(2)}$ is the second order term in an expansion of (19) and (3) in terms of $\vec{s}$. It is this functional (22) which now enters into the equation of motion (17).

It is easily seen that the functional (22) resolves the problem encountered in the first sound dynamics. In order to show this it is sufficient to consider only the volume terms in (22) and to omit also the quantum correction term containing the square of second order derivatives $s_{ik\ell}s_{\ell ki}$. The volume terms in the static quantity $\tau^{(o)}_{ik}$ must be diagonal ($\tau^{(o)}_{ik} = \delta_{ik}\frac{2}{3}\tau^{(o)}_{vol}$) therefore we have from (22)

$$E_{vol}[\vec{s}] = \int\{\frac{1}{3}\tau^{(o)}_{vol}(s_{ij}s_{ji} + s_{ij}s_{ij} + s_{ii}s_{jj}) + \frac{1}{2}(\frac{\partial^2}{\partial\rho_o^2}V_{vol}(\rho_o))\rho_o^2\, s_{ii}s_{jj}\}\, d^3x.$$

For $\vec{s} = \alpha(t)\vec{\nabla}r^2_{20}$ we obtain

$$E_{vol} = \alpha^2 \frac{5}{4\pi} \cdot 4\int\tau^{(o)}_{vol}\, d^3x, \quad B = \frac{1}{2}\, m\, \frac{5}{4\pi}\, 2 \int \rho_o r^2\, d^3x.$$

Therefore we get for the frequency (with $C = \frac{1}{2}(\partial^2/\partial\alpha^2 E)$)

$$\omega^2 = \frac{C}{B} = 4\int\tau^{(o)}_{vol}d^3x/m\, A\, \langle r^2\rangle$$

in agreement with the sum-rule value (15). The physical reason for this result is that through (18) we have allowed for dynamical distortions of the local momentum distribution which can be easily demonstrated by evaluating the Wigner transform of the scaled density matrix (18)[8].

We have obtained the above results without really solving the dynamical problem (16), (17) but just by making use of the energy functionals (1) or (22) together with an imposed flow field (12). This is not unphysical but rather corresponds to the physical situation where a certain flow pattern is forced on the system through a given excitation mechanism. However, for studying the detailed response of the system to a given external field we must solve the set (16), (17). After variation of (22) with respect to $\vec{s}$ and making use of the equilibrium condition

$$\partial_k\tau^{(o)}_{ik} + \partial_i \sum_\sigma a_\sigma\, (\sigma-1)\, \rho_o^\sigma - 2a_s\, \rho_o\, \partial_i\, \Delta\rho_o = 0 \qquad (24)$$

the basic fluid-dynamical equation of motion in SCA reads[8]

$$-m\omega^2\rho_o s_i = \partial_j(\tau^{(o)}_{ik}(s_{jk}+s_{kj})) + s_{jki}\tau^{(o)}_{jk} - \frac{\hbar^2}{4m}\partial_k\partial_\ell(s_{\ell ki}\rho_o) + \rho_o\partial_i(\rho_o s_{jj}\frac{\partial^2}{\partial\rho_o^2}\sum_\sigma a_\sigma\rho_o^\sigma) + 2a_s\rho_o\partial_i(\Delta\partial_j(s_j\rho_o) - s_j\Delta\partial_j\rho_o). \qquad (25)$$

The quantum correction and the surface term of the potential make it a fourth-order differential equation for the scaling field $\vec{s}$. It should be stressed that in contrast to the density-functional method (6), (7) (where we are forced to use an approximation like ETP for τ (5)) the static input functions ρ_o and $\tau_{ik}^{(o)}$ in (25) can be calculated in any desired approximation. A very good choice would be to take ρ_o and $\tau_{ik}^{(o)}$ from a Hartree-Fock calculation with the self-consistent field $w[\rho]$ given by the functional derivative

$$w[\rho] = \frac{\delta}{\delta\rho} \int V[\rho]\, d^3x = \sum_\sigma \sigma\, a_\sigma \rho^{\sigma-1} - 2\, a_s \Delta\rho. \qquad (26)$$

Then ρ_o and $\tau_{ik}^{(o)}$ would reflect the detailed single-particle structure of the groundstate which, as we shall see, is especially important for the asymptotic ($r\to\infty$) behaviour of the solutions of (25). Much more crude and much simpler would be to take $\tau_{ik}^{(o)}$ in the ETF approximation and to obtain ρ_o from the variational problem (2). One might even obtain analytical solutions of (25) by using a simple square density $\rho_o(r) = \rho_{oo}\Theta(R-r)$ and take the TF value for $\tau_{ik}^{(o)}$

$$(\tau_{ik}^{(o)})_{TF} = \delta_{ik}\, \frac{1}{5}\, \rho_{oo}\, \frac{p_F^2}{m}\,. \qquad (27)$$

Only in this last case it will be necessary to explicitly supply boundary conditions for $\vec{s}$ at the nuclear surface r=R. In all other cases the normalization conditions

$$\int \rho_o\, \vec{s}^{(n)} \cdot \vec{s}^{(m)}\, d^3x = \delta_{nm} \qquad (28)$$

for solutions with discrete eigenfrequencies $\omega_n^2 \neq \omega_m^2$ or

$$\int \rho_o\, \vec{s}^{(\omega)} \cdot \vec{s}^{(\omega')}\, d^3x = \delta(\omega-\omega') \qquad (29)$$

for solutions with continuous eigenvalues prove to be sufficient to completely determine the solutions of (25)[10]. The orthogonality of solutions with different eigenfrequencies is due to the fact that (25) may be written in the form

$$m\omega^2 \rho_o \vec{s}(\vec{x}) = \vec{\mathcal{L}}\, [\vec{s}(\vec{x})]$$

where the differential (vector) operator

$$\vec{\mathcal{L}}[\vec{s}] \equiv \frac{\delta}{\delta \vec{s}} E[\vec{s}]$$

is hermitian for any two vector fields $\vec{\varphi}$ and $\vec{\psi}$ for which the integrals exist:

$$\int \vec{\varphi} \cdot \vec{\mathcal{L}}[\vec{\psi}]\, d^3x = \int \vec{\psi} \cdot \vec{\mathcal{L}}[\vec{\varphi}]\, d^3x \quad .$$

After these general remarks about the SCA we shall discuss at first a few analytical properties of (25) for the case of a simple square density.

SQUARE DENSITY MODEL[11 12]

We omit surface terms from (3) and introduce the abbreviation

$$v(\rho_o) = \sum_\sigma a_\sigma \rho_o^\sigma \qquad (\rho_o \equiv \rho_{oo} = \text{const.})$$

and the Landauparameter

$$F_o = \frac{3m}{p_F^2} \rho_o \frac{\partial^2}{\partial \rho_o^2} v(\rho_o) \quad . \tag{30}$$

First we notice that (25) implies a quite different sound speed for compressional modes as compared to the first sound propagation. From (6), (7) we obtain for the density change $\delta\rho = \rho - \rho_o$

$$- m\omega^2 \delta\rho = \rho_o \left(\frac{\partial}{\partial \rho_o^2} E(\rho_o)\right) \Delta\delta\rho = \frac{p_F^2}{3m} (1+F_o)\, \Delta\delta\rho \;, \tag{31}$$

i.e., the first sound speed is $c_L^2 = v_F^2(1+F_o)/3$. For the square density (25) simplifies (with (27)) to

$$- m\omega^2 \vec{s} = \frac{p_F^2}{m}\left(\frac{2}{5} + \frac{F_o}{3}\right) \vec{\nabla}(\vec{\nabla}\cdot s) + \frac{p_F^2}{5m} \Delta\vec{s} \;. \tag{32}$$

Therefore the longitudinal part of $\vec{s}$ is determined by

$$- m\omega^2 \vec{\nabla}\cdot\vec{s} = \frac{p_F^2}{m}\left(\frac{3}{5} + \frac{F_o}{3}\right) \Delta \; (\vec{\nabla}\cdot\vec{s}) \tag{33}$$

i.e. the longitudinal scaling sound speed is $c_L^2 = v_F^2(3/5+F_o/3)$. We see that the contribution from the kinetic energy to the longitudinal sound speed is almost twice as large as in the first sound result. Interestingly (33) allows for transverse modes with a transverse sound speed $c_T^2=v_F^2/5$ while in first sound propagation such modes are excluded.

Another decisive difference concerns the boundary conditions which solutions of (31) or (32) should satisfy at the nuclear surface r=R. The physical requirement is that no forces should be exerted on the freely moving surface. In the density-functional method the pressure is, of course, proportional to the density change $\delta\rho$, i.e. we must require for solutions of (31) (which are of the form $\delta\rho_L = j_L(kr)Y_{LM}$)

$$j_L(kR) = 0. \tag{34}$$

The pressure tensor underlying the dynamics (32), however, is

$$P_{ik} = \rho_o \frac{p_F^2}{m^2} \{ (\frac{F_o}{3} + \frac{1}{5}) \delta_{ik} s_{jj} + \frac{1}{5} (s_{ik} + s_{ki}) \} \tag{35}$$

which can be seen by writing (32) in the Euler form

$$\frac{\partial}{\partial t} j_i = - \partial_k P_{ik} , \quad \text{with } P_{ik} = P_{ki} \quad . \tag{36}$$

On the free surface we therefore must require

$$x_k P_{ik}|_R = 0. \tag{37}$$

This condition is in general much more complicated than (34) because it mixes the transverse and longitudinal components in the solutions of (32) which are of the form (for Electric modes)

$$\vec{s}_L = \alpha_L \vec{\nabla} (j_L(k_{||} r) Y_{LM}) + \beta_L \vec{\nabla} \times (j_L(k_\perp r)\vec{Y}_{LLM}) . \tag{38}$$

As we have seen, in first sound dynamics the restoring forces for surface modes originate in the surface energy alone, therefore the square-density model contains no first sound surface modes. The scaling mechanism, however, does provide for the volume terms which comprise the essential part of the restoring force for surface modes. Therefore one might wonder to which extent the solutions (38) may represent surface modes: In fact, the lowest energy eigenvalue obtained from (37), (38) for L=2 is $\hbar\omega \approx 56\ A^{-1/3}$ MeV in rather close agreement with our previous result. On the other hand, it is clear that (38) cannot really represent pure surface modes unless $k_{||}R \to 0$, which is not possible because $k_{||}$ is tied to the frequency through the fixed sound speed. The actual values of $k_{||}R$ turn out to be of the order of 1 (or 2 for $k_{\perp}R$) so that in the nuclear interior the Bessel-functions in (38) are still well approximated by powers r^L. In this limit there are no transverse components in (38) (because of the identity $\vec{\nabla} \times r^L \vec{Y}_{LLM} \propto \vec{\nabla}\, r^L Y_{LM}$), but the deviations of $j_L(k_{\perp}r)$ from $(k_{\perp}r)^L$ near the nuclear surface represent genuine transverse components. It is striking to notice that for smooth self-consistent densities $\rho_o(r)$ the combination of longitudinal and transverse components acts to reestablish the Tassie transition density

$$\delta\rho_L = \vec{\nabla} \cdot (\rho_o(r) \vec{\nabla} r^L Y_{Lo}) \qquad (39)$$

throughout the nuclear surface with very good accuracy. On the other hand one may notice that (33) does allow for pure surface modes if we restrict $\vec{s}$ from the outset to be irrotational

$$\vec{s} = \vec{\nabla}\,\phi \qquad (40)$$

and solve (33) for the displacement potential ϕ:

$$\phi_L = (j_L(kr) + \alpha_L r^L)\, Y_{LM}\ . \qquad (41)$$

In this case the transverse part of the boundary condition (37) is replaced by another condition[13] due to additional surface terms which arise from the partial integration of the gradient in (40), such that both constants k and α_L in (41) are uniquely determined.

Eq. (32) contains also purely transverse "Magnetic" modes

$$\vec{s}_L = j_L\ (k_{\perp}r)\ \vec{Y}_{LLM}\ . \qquad (42)$$

Writing the excitation operator $B_L^+ = \vec{s}_L \cdot \vec{\nabla}$ with (42) in the form

$$B_L^+ = j_L(k_\perp r)\vec{Y}_{LL0} \cdot \vec{\nabla} = \frac{1}{r} j_L(k_\perp r) F_L(\Omega)\hat{L}_z \ , \quad (\hat{L}_z = i\vec{x} \times \vec{\nabla})$$

gives a very intuitive picture of these Magnetic modes: They are rotations around the z-axis by coordinate-dependent angles[8]. A most interesting type is the Magnetic Quadrupole mode (L=2) for which one has $F_2(\Omega) = \cos\Theta$, therefore (for $k_\perp R \approx 1$)

$$B_2^+ \approx z\hat{L}_z = [\vec{x} \otimes \vec{L}]_{20}. \tag{43}$$

Evidently, the 2^- mode represents a twisting motion of the nucleus where the top is rotated in opposite phase to the bottom. In the square density model the twist frequency occurs near $\hbar\omega \approx 50 A^{-1/3}$ MeV (depending on ρ_0). For general smooth surface densities $\rho_0(r)$ it is, however, quite sensitive to the surface profile and lies near $40\ A^{-1/3}$ for realistic nuclei. This is in good agreement with experimental results for the center of the 2^- strength distribution. Microscopically (43) shows that B_2^+ involves 1 $\hbar\omega_0$ (one shell distance) transitions, therefore we expect this mode to be strongly mixed with the 1 $\hbar\omega_0$ spin-flip mode $[\vec{x} \otimes \vec{\sigma}]_2$ in a realistic nucleus. This makes a clear experimental identification of the twist components difficult which, however, would be highly desirable because the simple density-functional method does not contain these modes.

Before we present some detailed results of the scaling approach in more realistic cases we shall discuss its connection with the RPA in a classical analogue. This will show how the scaling concept (18) emerges as a step in a systematic expansion and allows us to go beyond it in order to check its validity.

CONNECTION WITH RPA IN THE CLASSICAL LIMIT.

In the microscopic RPA method the normal coordinates $\hat{P}, \hat{Q}$ for collective modes are obtained by solving the variational problem

$$\delta \langle 0 | [\hat{P}, \hat{H}] - \hbar^2\omega^2 \frac{M}{i\hbar} \hat{Q} | 0 \rangle = 0, \tag{44a}$$

$$\delta\langle O|[\hat{Q},\hat{H}] - \frac{i\hbar}{M}\hat{P}|O\rangle = 0. \tag{44b}$$

Variation with respect to the ground-state determinant $|O\rangle$ leads to the RPA equations for the particle-hole and hole-particle matrix elements of $\hat{P}$ and $\hat{Q}$. (M is the collective inertia parameter). In the classical limit we replace the commutators by Poisson brackets $[\ ,\ ] \to i\hbar\{\ ,\ \}$, operators $\hat{A}$ by their Wigner transforms $A(\vec{x},\vec{p})$, and ground-state expectation values by averaging in phase space over the single-particle distribution function $f(\vec{x},\vec{p})$:

$$\delta\int (\{P(\vec{x},\vec{p}),\, h(\vec{x},\vec{p})\} + M\omega^2\, Q(\vec{x},\vec{p}))\, f(\vec{x},\vec{p})\, d^3x\, d^3p = 0 \tag{45a}$$

$$\delta\int (\{Q(\vec{x},\vec{p}),\, h(\vec{x},\vec{p})\} - \frac{1}{M}\, P(\vec{x},\vec{p}))\, f(\vec{x},\vec{p})\, d^3x\, d^3p = 0\ . \tag{45b}$$

We define local tensor fields $\chi(\vec{x})$, $s_\alpha(\vec{x})$, $\phi_{\alpha\beta}(\vec{x})$, $\Theta_{\alpha\beta\gamma}(\vec{x})$, etc. by expanding the time-even Q and time-odd P in powers of the momentum variable $\vec{p}$:

$$\sqrt{\frac{M\omega}{2\hbar}}\, Q(\vec{x},\vec{p}) = \chi(\vec{x}) + \frac{1}{2}\, p_\alpha p_\beta\, \phi_{\alpha\beta}(\vec{x}) + \ldots \tag{46}$$

$$-i\sqrt{\frac{1}{2M\hbar\omega}}\, P(\vec{x},\vec{p}) = s_\alpha(\vec{x})\, p_\alpha + \frac{1}{6}\, \Theta_{\alpha\beta\gamma}(\vec{x})\, p_\alpha p_\beta p_\gamma + \ldots\ . \tag{47}$$

and use for h the self-consistent single-particle Hamiltonian

$$h(\vec{x},\vec{p}) = \frac{p^2}{2m} + \frac{\partial}{\partial\rho}\, v(\rho(\vec{x}))\ . \tag{48}$$

The coupled equations for P and Q are then obtained by considering variations $\delta f = f - f_o$, where δf is restricted to the Fermi surface, i.e.

$$\delta f(\vec{x},\vec{p}) = \delta(p-p_F)\sum_{\kappa\mu} \nu_{\kappa\mu}(\vec{x})\, Y_{\kappa\mu}(\hat{p}) \tag{49}$$

with arbitrary functions $\nu_{\kappa\mu}(\vec{x})$.

For an infinite system this would correspond to solving the Landau-Vlassov equation[14]. The fact that in a finite system boundary conditions have to be specified for each of the tensor fields χ, s_α, $\phi_{\alpha\beta}$, ... makes it necessary to truncate the expansions (46), (47). To the order indicated in (46), (47) one obtains the following equations of motion:

$$i\omega\bar{\chi} = \frac{1}{m}\frac{p_F^2}{3}(1+F_o)\,\partial_\alpha\bar{s}_\alpha \tag{50a}$$

$$i\omega\bar{s}_\alpha = \frac{1}{m}(\partial_\alpha\bar{\chi} + \frac{p_F^2}{5}\partial_\beta\bar{\phi}_{\alpha\beta}) \tag{50b}$$

$$i\omega\bar{\phi}_{\alpha\beta} = \frac{1}{m}(-\frac{2}{3}\delta_{\alpha\beta}\partial_\gamma\bar{s}_\gamma+\partial_\alpha\bar{s}_\beta+\partial_\beta\bar{s}_\alpha+\frac{p_F^2}{7}\partial_\gamma\bar{\Theta}_{\gamma\alpha\beta}) \tag{50c}$$

$$i\omega\bar{\Theta}_{\alpha\beta\gamma} = \frac{1}{m}(-\frac{2}{5}\delta_{\alpha\beta}\partial_\mu\bar{\phi}_{\gamma\mu}+\partial_\alpha\bar{\phi}_{\beta\gamma})_{\text{symm.}} \tag{50d}$$

with

$$\bar{\chi} \equiv \chi + \frac{p_F^2}{6}\phi_{\alpha\alpha}\,, \qquad \bar{s}_\mu \equiv s_\mu + \frac{p_F^2}{10}\Theta_{\mu\alpha\alpha}\,,$$

$$\bar{\phi}_{\mu\nu} \equiv \phi_{\mu\nu} - \frac{1}{3}\delta_{\mu\nu}\phi_{\alpha\alpha}\,, \qquad \bar{\Theta}_{\mu\nu\rho} \equiv \Theta_{\mu\nu\rho} - \frac{1}{5}(\delta_{\mu\nu}\Theta_{\rho\alpha\alpha})_{\text{symm.}}\,.$$

In the square-density model the free variation of the nuclear surface at r=R in the coordinate space integration in (45) gives the scalar, vector, and tensor boundary conditions to be satisfied by the solutions of (50):

$$(\bar{\chi} - \frac{p_F^2}{15}\phi_{\alpha\alpha})_{r=R} = 0, \tag{51a}$$

$$(x_\alpha\phi_{\alpha\beta})_{r=R} = 0, \tag{51b}$$

$$(x_\alpha(\Theta_{\alpha\beta\gamma} - \frac{1}{3}\delta_{\beta\gamma}\Theta_{\alpha\nu\nu}))_{r=R} = 0\,. \tag{51c}$$

It is easy to see that (50) contains the first sound propagation as the lowest approximation: Keeping only χ and $\vec{s}$ in (50a,b) yields

$$-m\omega^2\chi = \frac{p_F^2}{3m}(1 + F_o)\,\Delta\chi$$

in agreement with (31). The scalar boundary condition (51a) reduces to $\chi|_R = 0$ in agreement with (34).

We now show that inclusion of the tensor $\phi_{\mu\nu}$ in (50a,b,c) is equivalent to the scaling approximation. Inserting (50a) and (50c) into (50b) yields

$$-m\,\omega^2 s_\mu = \frac{p_F^2}{m}(\frac{2}{5} + \frac{F_o}{3})\,\partial_\mu\vec{\nabla}\cdot\vec{s} + \frac{p_F^2}{5m}\Delta s_\mu$$

in agreement with (32). Inserting (50c) into the vector boundary condition (51b) yields

$$x_\nu \phi_{\nu\mu}\big|_R = x_\mu\left(\frac{1}{3}\ \phi_{\alpha\alpha} - \frac{2}{3im\omega}\ (\vec{\nabla}\cdot\vec{s})\right) + \frac{1}{im\omega}\ x_\nu(\partial_\nu s_\mu + \partial_\mu s_\nu)\big|_R \ .$$

Eliminating the trace $\phi_{\alpha\alpha}$ through the scalar boundary condition (51a) and eq. (50a) and comparing with the pressure tensor (35) we have

$$x_\nu \phi_{\nu\mu}\big|_R = \frac{5\ m}{i\omega\rho_o p_F^2}\ x_\nu P_{\nu\mu}\big|_R \quad .$$

Thus we have obtained equation of motion and boundary condition of the SCA.

Including finally $\Theta_{\alpha\beta\gamma}$ in (50 a-d) allows to investigate the validity of the SCA by going one step further. The details of such an investigation are somewhat involved[15] and we shall present only some general results here.

One obtains two values for the longitudinal sound speed

$$c_L^2 = v_F^2\ \frac{1}{2}\left(\frac{6}{7} + \frac{F_o}{3} \pm \left(\left(\frac{F_o}{3}\right)^2 + \frac{1}{35}\left(8F_o + \frac{96}{7}\right)\right)^{1/2}\right). \qquad (52)$$

This allows to describe in addition to the giant electric modes also low-lying collective modes. This is an especially interesting possibility for octupole (E3) modes where the low-lying component is well known. It is also helpful in the case of quadrupole modes where the existence of a low-lying collective E2 state depends on the specific nucleus. However, for a very large nucleus (which the square density model corresponds to) there will always be a low-lying state and therefore inclusion of $\Theta_{\alpha\beta\gamma}$ allows to separate it from the giant state.

The transverse sound speed is also quite different from its value in the SCA: One obtains

$$c_T^2 = v_F^2 \cdot \frac{3}{7} \ . \qquad (53)$$

Together with these changes the modified boundary conditions (51) act, however, in such a way as to reestablish the results of the SCA for frequencies, transition densities and flow patterns to a remarkable degree for strongly collective giant states in all cases where low-lying collective states of the same multipolarity do not exist (i.e. for Magnetic and Isovector Electric modes).

It is an instructive exercise to put the set of equations (50a-d) in fluiddynamical form. For that purpose we write the change δf_{B^+} in the distribution function f_o caused by the excitation B^+

$$\delta f_{B^+} \equiv \{f_o, B^+\} \equiv \{f_o, \sqrt{\frac{M\omega}{2\hbar}}\, Q - i\sqrt{\frac{1}{2\hbar M\omega}}\, P\} \tag{54}$$

and evaluate the density change $\delta\rho$

$$\delta\rho \equiv \int \delta f_{B^+}\, d^3p = \partial_\alpha(\rho_\alpha \bar{s}_\alpha), \tag{55}$$

the current j_α

$$j_\alpha \equiv \int \delta f_{B^+} \frac{p_\alpha}{m}\, d^3p = \frac{1}{m}\rho_o(\partial_\alpha\bar{\chi} + \frac{p_F^2}{5}\partial_\beta \bar{\Phi}_{\beta\alpha}), \tag{56}$$

and the kinetic pressure tensor $P^{kin}_{\alpha\beta}$

$$P^{kin}_{\alpha\beta} \equiv \int \delta f_{B^+} \frac{p_\alpha p_\beta}{m^2}\, d^3p = \frac{\rho_o p_F^2}{5m^2}(\delta_{\alpha\beta}\,\partial_\gamma\bar{s}_\gamma + \partial_\alpha\bar{s}_\beta + \partial_\beta\bar{s}_\alpha + \frac{p_F^2}{7}\partial_\gamma\bar{\Theta}_{\gamma\alpha\beta}). \tag{57}$$

Evidently, by comparing (56) with (50b), the velocity field v_α is given by the time derivative of the scaling field (modified through $\Theta_{\mu\alpha\alpha}$)

$$j_\alpha = \rho_o\, v_\alpha, \quad v_\alpha = i\omega\bar{s}_\alpha \quad . \tag{58}$$

This is the relation (16) postulated in the SCA (i.e. in the approximation scheme with $\Theta_{\alpha\beta\gamma} = 0$). The pressure tensor $p_{\alpha\beta}$ in

$$i\omega j_\alpha = \partial_\beta p_{\alpha\beta}$$

is, by comparison of (57) with (50,b,c) given by the sum of the kinetic part (57) and the part due to the selfconsistent field

$$p_{\alpha\beta} = \rho_o \frac{p_F^2}{m^2} \frac{F_o}{3} \delta_{\alpha\beta} \partial_\gamma \bar{s}_\gamma + p_{\alpha\beta}^{kin}$$
$$= \frac{i\omega}{m} \rho_o (\delta_{\alpha\beta} \bar{\chi} + \frac{p_F^2}{5} \bar{\Theta}_{\alpha\beta}) \quad . \tag{59}$$

It displays clearly the diagonal "liquid drop" part $\bar{\chi}$ and the traceless tensor part $\bar{\Theta}_{\alpha\beta}$ which comprises the single-particle effects.

UNBOUND RESONANCES IN THE SCALING APPROACH

For existing nuclei the GMR lie generally above the neutron-emission threshold and should therefore be considered as bound states embedded in the particle continuum. It is a very appealing feature of the SCA equations (25) that they naturally yield the GMR as resonances embedded in a continuum of solutions if one uses as an input into (25) smooth selfconsistent ground-state densities. At this point, however, it again turns out that one has to be extremely cautious with the use of density functionals:

As we have said the input for (25) (namely $\rho_o(r)$ and $\tau_{ik}^{(o)}(r)$) may be taken from any feasible model (HF, ETF, ..). Any selfconsistent theory will lead to an exponentially decreasing ground-state density

$$\rho_o(r \to \infty) \to e^{-\mu r}/r^2 \quad . \tag{60}$$

In HF-theory μ reflects the binding energy λ_F of the last bound particle, i.e.

$$\mu = \frac{2}{\hbar}(2m|\lambda_F|)^{1/2} \quad ,$$

while in ETF (with the kinetic energy density approximated by (5)) the self-consistent solution of (2) decreases like (60) with

$$\mu = \frac{2}{\hbar}(2m|\lambda_F|/\eta)^{1/2}$$

as if the last particle was bound with λ_F/η instead of λ_F. For static properties this difference is quite unimportant because it appears only for r much larger than the nuclear radius. For the dynamical calculations considered here the difference in the asymptotic behaviour of ρ_o is, however, crucial because it determines the asymptotic form of the equation of motion (25) which reads in the asymptotic limit[11]

$$- m\omega^2 \vec{s} = - \frac{\hbar^2}{4m}(\partial_r^2 - \mu\partial_r)^2 \vec{s}$$

With $\exp(\gamma r)$ for the asymptotic radial part of $\vec{s}$ we have

$$(\gamma^2 - \mu\gamma) = \pm 2m\omega/\hbar$$

or $\hbar\omega < |\lambda_F|$ for HF , (61a)

$\hbar\omega < |\lambda_F|/\eta$ for ETF (61b)

for bound discrete solutions. For ω values above these critical limits the basic equation (25) has a continuous spectrum, corresponding to single-particle escape. Evidently, the threshold for the onset of the continuum is correctly reproduced with a HF-density $\rho_o(r)$, while in the ETF case it is shifted to much higher values (commonly used values for η are 1/9 or 4/9) corresponding to the seemingly stronger bound last particle. It turns out that this difference is not very important for calculating the position of a specific resonance, nor for the form of the corresponding transition densities or currents. However, obviously, if we calculate the width (= escape width) of a specific resonance it will make a decisive difference how far above threshold the peak of the resonance occurs. If we denote the thresholds (61) as $h\omega_{crit}$ one finds approximately

$$\Gamma^{\uparrow} \propto (\omega - \omega_{crit})^{5/3} .$$

For instance, in ETF with the value $\eta = 4/9$ it will frequently happen that a GMR appears as a discrete bound state, while with the HF-input one has to calculate a strength function

$$S_Q(\omega) = (\frac{\hbar}{2M\omega})(\int \delta\rho_\omega(\vec{x})Q(\vec{x})\ d^3x)^2 \tag{62}$$

which characterizes the dynamical response of the system to an external field $Q(\vec{x})$, as a function of the continuous variable ω. The strength function (62) will then display a resonance structure around the energy value obtained in ETF as discrete eigenvaules. This structure may be directly compared with corresponding RPA results calculated with an identical potential-energy function. A few examples of such comparisons are shown in the figures[16]. We conclude that it is crucial for the calculation of strength functions to use an HF-input for the scaling functional. But then it is very gratifying to see how well the simple SCA reproduces essential features of transition densities, currents and strength distributions which otherwise may only be obtained through microscopic continuum RPA calculations which also yield a lot of intermediate (and often rather arbitrary and accidental) single-particle structure in the functions $S(\omega)$.

We may therefore conclude that for low-multipolarity Giant Resonances in finite Fermi systems the successful use of functionals of the local density for the potential part of the total energy may be complemented by the scaling functional (22) for the kinetic part to obtain a fluid-dynamical set of equations which determine the main features of physically relevant field variables and strength functions.

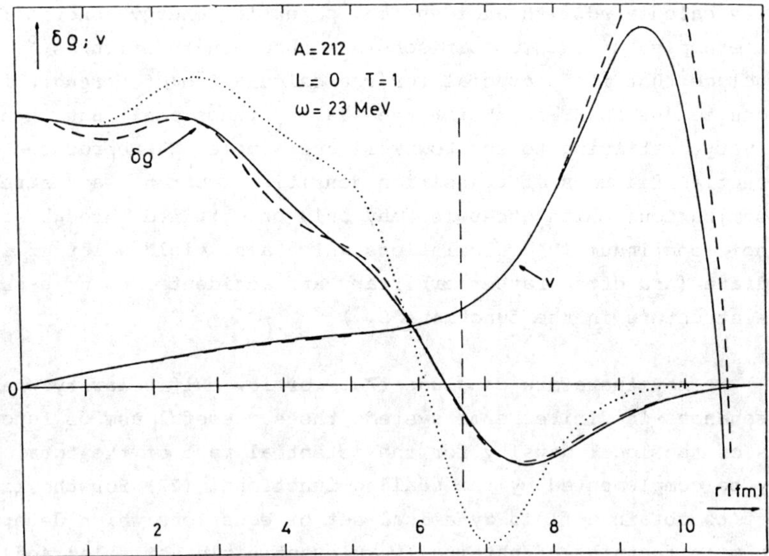

Fig. 1. Comparison of the transition density $\delta\rho(r)$ and radial flow velocity $v(r)$ for the isovector monopole resonance in an N=Z nucleus with A=212 (without Coulomb forces). The full curves are results of a microscopic RPA calculation, the dashed curves are SCA results. The dotted curve is the Tassie transition density (i.e. it corresponds to $v(r) \equiv r$). The vertical dashed line indicates the half-density radius. The energy for this comparison is chosen near the peak of the resonance at $\hbar\omega$ = 23 MeV (cf. Fig. 3).

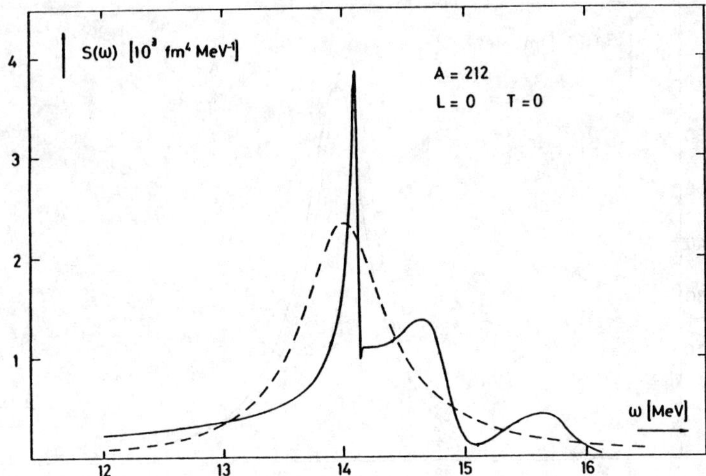

Fig. 2. Strengthfunction S(ω) for isoscalar monopole excitation (full line: RPA, dashed line: SCA)

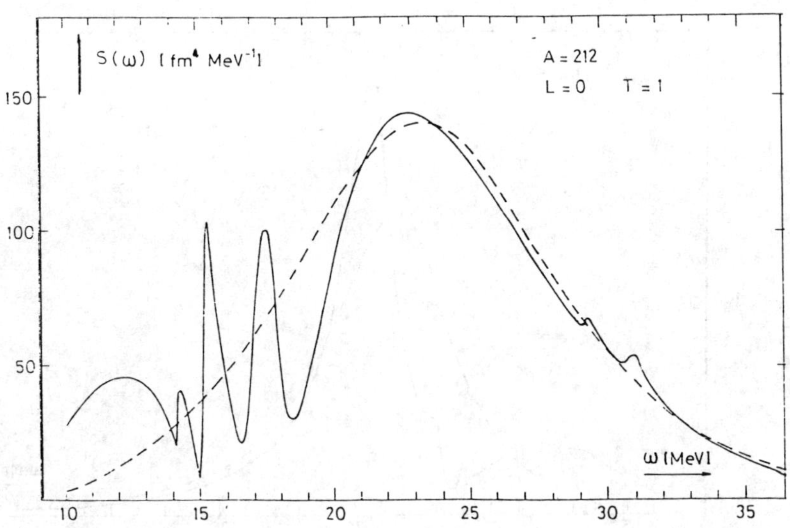

Fig. 3. Strengthfunction S(ω) for isovector monopole excitation (full line: RPA, dashed line: SCA)

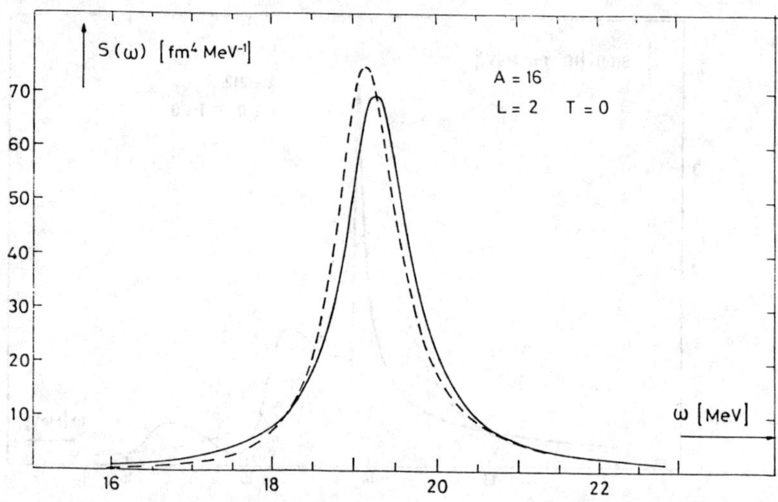

Fig. 4. Strengthfunction S(ω) for isoscalar quadrupole excitation (full line: RPA, dashed line: Irrotational SCA)

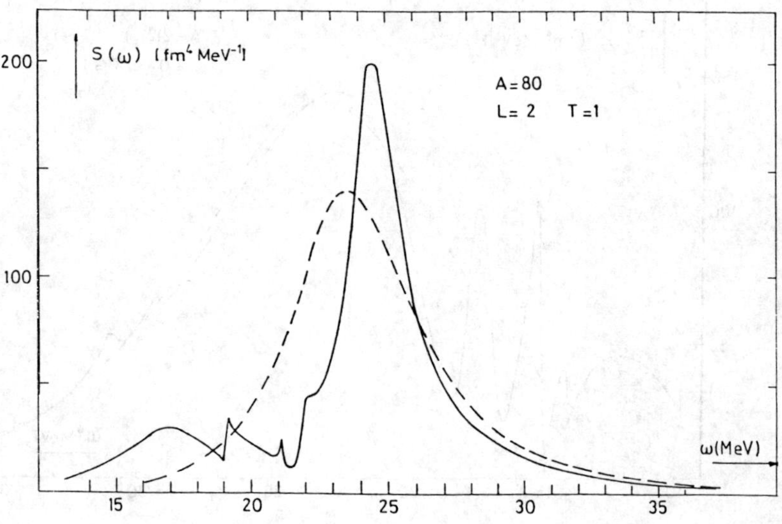

Fig. 5. Strengthfunction S(ω) for isovector quadrupole excitation (full line: RPA, dashed line: Irrotational SCA).

References

1. F. E. Bertrand (Editor), "Giant Multipole Resonances", Nuclear Science Research Conference Series, Vol. 1 Harwood Academic Publishers 1980.

J. Speth and A. van der Woude, Rep. Prog. Phys. 44:719 (1981)

2. O. Bohigas, X. Campi, H. Krivine and J. Treiner, Phys. Letters 64 B:381(1976).

G. Eckart and G. Holzwarth, Z. Phys. A 281:385(1977).

3. M. Breiner, H. Flocard, N. Van Giai and P. Quentin, Nucl. Phys. A 238:29(1975).

H. Krivine, J. Treiner and O. Bohigas, Nucl. Phys. A 336: 155(1980).

4. D. A. Kirshnits, "Field theoretical methods in many-body systems", Pergamon, London 1967.

M. Brack, B. K. Jennings and Y. H. Chu, Phys. Letters 65 B: 1(1976).

H. Krivine and J. Treiner, Phys. Letters 88 B:212(1979).

5. A. Bohr and B. R. Mottelson, "Nuclear Structure", Vol. 2, Ch. 6 A, Benjamin, Reading 1975.

6. H. Sagawa and G. Holzwarth, Prog. Theor. Phys. 59:1213 (1978).

7. J. Martorell, O. Bohigas, S. Fallieros and A. M. Lane, Phys. Letters 60 B:313(1976).

O. Bohigas, A. M. Lane and J. Martorell, Phys. Rep. 51: 267(1979).

8. G. Holzwarth and G. Eckart, Nucl. Phys. A 325:1(1979).

9. S. Stringari, Nucl. Phys. A 279:454(1977).

10. G. Eckart and G. Holzwarth, Phys. Letters 118 B:9(1982).

11. G. Eckart, G. Holzwarth and J. P. da Providencia, Nucl. Phys. A 364:1(1981)

12. F. E. Serr, Phys. Letters 97 B:180(1980).

13. K. Andō and S. Nishizaki, Prog. Theor. Phys. 68:1196(1982).

14. T. Yukawa and G. Holzwarth, Nucl. Phys. A 364:29 (1981).

15. H. Thorn, Diploma thesis, Siegen University 1983.

16. K. Andō and H. Eckart, Nucl. Phys. A(1984), in press

Liquid-Particle Model For Nuclear Dynamics

V. Strutinsky, A. Magner, M. Brack[+]
Institute for Nuclear Research, Kiev, USSR;
Regensburg University, FRG[+]

Abstract: (A) Equations of macroscopic nuclear dynamics are derived which take into account the interaction with the quasi-particle component (the LiPa-model). (B) A Sharp Effective Nuclear Surface (SENS) is introduced according to the max. density-gradient positions and boundary conditions are obtained which relate the SENS-dynamics to that of the volume distributions. (C) In the semi-classical limit, the LiPa-equation for the gas component inside the nuclear volume turns into the Landau's 0th-sound equation leading to a simple description of nuclear isoscalar giant resonances as due to the 0th-sound modes coupled to SENS-distortions.

(A)

Macroscopic quantities are introduced through statistically averaged quantal Wigner distribution function. Their dynamics is coupled to the quantal single-particle motion which can be described in a manner close to some traditional models. We start from the Wigner-transform

$$f(rpt) = \int d^3 s \exp(-\frac{i}{\hbar} ps)\, n(r+s/2, r-s/2, t)$$

for the solution $n(r_1 r_2 t)$ to the single-particle problem

$$i\hbar d' n(r_1 r_2)/dt = [H, n] , \qquad V = tr(vn)$$

(d'/dx is partial derivative) and define formally the macro-distribution as a statistical average, e.g. in the single-particle phase space,

$$\bar{f}(rpt) = \langle f(rpt) \rangle_{(\Delta r \Delta p) > (2\pi\hbar)^3}$$

The following macro-quantities are determined then:

density $\bar{n}(rt) = (2\pi\hbar)^{-3} \int d^3 p \bar{f}(rpt)$,
current $\bar{j}(rt) = (2\pi\hbar)^{-3} \int d^3 p M^{-1} p \bar{f}(rpt)$,
average velocity $u(rt) = \bar{j}/\bar{n}$,
pressure tensor $\bar{P}_{kl}(rut) = (2\pi\hbar)^{-3} M^{-1} \int d^3 p (p-Mu)_k (p-Mu)_l \bar{f}(rpt)$.

The quantity $\bar{f}$ is assumed to represent the locally equilibrated component and is, therefore, taken in the form

$$\bar{f}(rpt) = F(r,(p-Mu)^2,t).$$

It leads to diagonal $\bar{P}_{kl}(rt) = (2/3)\mathcal{E}_{kin}\,\delta_{kl}$. It is further assumed that the macro-component has the properties of a liquid and its dynamics can be described in terms of the zeroth and first moments of $\bar{f}$.

The well-known equation for the exact f(rpt),

$$d'f/dt + M^{-1}p\cdot grad(f) - (2/\hbar)\sin((\hbar/2)grad_r Vgrad_p f)Vf = 0$$

leads to the following equation for the first p-moment,

$$d'\bar{j}_k/dt + \sum_l d'(\bar{n}u_k u_l)/dr_l + M^{-1}\sum_l d'\bar{P}_{kl}/dr_l + (\bar{n}/M)d'V/dr_k =$$
$$= \frac{1}{i\hbar}(2\pi\hbar)^{-3}\int d^3p(p/M)\int d^3s\exp(ips/\hbar)(-i\hbar d'n_1/dt + [\bar{H}_t,n_1] + [V_1,n_1]). \tag{1}$$

Here, $n(r_1r_2t) = \bar{n}(r_1r_2t) + n_1(r_1r_2t)$, where $\bar{n}(r_1r_2t)$ is defined as the Wigner inverse of $\bar{f}$ and

$$V_1 = tr(vn_1).$$

The quantity

$$\bar{H}_t = T + \bar{V},\ \text{where}\ \bar{V} = tr(v\bar{n}),$$

may be considered as some not-selfconsistent, time-dependent single-particle Hamiltonian, such as, e.g. the deformed Woods-Saxon potential and it is reasonable to define n_1 as a solution to the single-particle problem

$$ihd'n_1/dt = [\bar{H}_t,n_1],\quad tr(n_1) = 0. \tag{2}$$

Neglecting the 2nd-order term in the r.h.s. of (1), one obtains, then, corrected fluid-dynamics equations for the macro-quantities,

$$d'\bar{n}(rt)/dt = -div(\bar{j}) = -div(u\bar{n}),$$
$$d'u(rt)/dt + (ugrad)u = -M^{-1}grad(\delta\mathcal{E}(\bar{n})/\delta\bar{n} + V_1). \tag{3}$$

where $\mathcal{E}(\bar{n})$ is the phenomenological energy-density functional and the coupling term V_1 is of the 1st order in n_1. The 2nd-order term in the r.h.s. of (1) must be included in (2) to determine the quasi-particle correlations such as in the RPA-phonon states or the zeroth sound.

It is shown that:

1) The sum of the macro-energy

$$\bar{E}(t) = \int d^3r(M\bar{n}u^2/2 + \mathcal{E}(\bar{n})),$$

and the microscopic quantal gas energy correction

$$E_1(t) = tr(\bar{H}_t n_1); \quad dE_1/dt = d^3rn_1(d'\bar{V}/dt) = d^3rV_1(d'\bar{n}/dt)$$

is conserved, although each of the components is not.

2) The simplest, quasi-static solution to $n_1 \approx n_1(r_1r_2;\bar{n}_t)$ (n_1 is real) adds the shell energy correction to the liquid drop deformation energy through V_1.

3) Adiabatic (complex) $n_1(r_1r_2t;\bar{n}_t)$ such as in the cranking model changes the macroscopic inertia, as well. The LiPa-equations couple the macro-dynamics, which may include the surface as well as volume density distortions, to quasi-particle modes of excitations.

(B)

Solving the dynamics problem for the bulk density $\bar{n}(rt)$ is helped by introducing Sharp Effective Nuclear Surface (SENS) as a dynamic variable. The SENS is defined by positions of max$|$grad $\bar{n}(rt)|$ and varies with time. The problem is split into two parts, one for the nuclear interior and the other for the edge region. In the interior, simplified equations can be used analogous to the infinite matter theories. Near the edge the equations are resolved by using a special SENS-based coordinate system. However, instead of solving the detailed equations in this region, where the density gradient is large and microscopic descriptions become much speculative, one may use proper boundary conditions for the interior problem set at the dynamic SENS. The boundary conditions involve certain phenomenological parameters, such as the surface tension constant or other. In this way, equations are obtained which describe the nuclear shape dynamics in large-scale processes, such as fission or fusion, and its coupling to the intrinsic modes.

The boundary conditions set at the SENS are trivial for the isoscalar density modes:

1. Normal-to-SENS component of the mean velocity should equal the velocity of the normal displacement of the SENS itself.
2. Normal-to-SENS component of the stress tensor caused by density distortion should compensate the excess pressure due to the surface tension σ at the bent surface equal to

$$\sigma(1/R_1 + 1/R_2),$$

where R_1 and R_2 are the two curvature radii.

Eigen-frequencies of collective modes can then be determined, particularly those that involve the surface dynamics.

The simplified form of the dynamic-surface approach valid for small distortions of the spherical distribution has been used in the Bohr-Mottelson's vol. 2 to describe the interplay between the surface distortion and the volume compressions, or the sound. It is essential that the SENS is always directly coupled to whatever takes place inside the nucleus. In particular, the surface oscillations frequencies are identical to those of the volume density modes and vice versa. Note also that the surface region contributes only a half to the liquid drop surface energy, the other coming from the volume compression due to the surface tension.

(C)

Although being an integral-differential, the 0th-sound equation in the interior region is homogeneous and linear. The solution can thus be found as a superposition of Landau's plane-wave solutions for the infinite matter. For the 2^L-pole mode of the density oscillations the solution is of the form

$$f_L^{(o)}(rpt) = \alpha^{(o)} \exp(-iwt)\, \delta(e-e_F) \int dO_k Y_{LO}(O_k) \exp(ikr\cos(kr))$$

$$(\cos(pk)/(s-\cos(pk)))(1+(F_1 s/F_o(1+F_1/3))\cos(pk))$$

Here, $w = skv_F$, $s = u^{(o)}/v_F$ is the 0th-sound velocity in units of v_F and s is determined by the algebraic equation,

$$Z(s) = ((s/2)\ln((s+1)/s-1))-1)^{-1} = F_o + (F_1 s^2/(1+F_1/3)), \tag{4}$$

valid for the quasi-particle interaction amplitude

$$F(pp') = F_o + F_1 \cos(pp').$$

The quantity $f_L^{(o)}(rpt)$ corresponds to $n_1(r_1 r_2 t)$ of (A). The normal-to-SENS components of the velocity and pressure are found from $f_L^{(o)}$ without difficulty and used in the dynamic-SENS boundary conditions stated in (B). As derived from the LiPa-model, the stress tensor appearing in the boundary condition should also include a component due to the quasi-particle interaction. The characteristic eigen-frequency equation for combined, Oth-sound + surface modes is

$$j_L'(x) = xK\cdot((1-s^2+(F_o+Z)/3)j_L(x)+(1-3s^2+Z)j_L''(x)), \qquad (5)$$

$$K = 3A^{1/3}(e_F/b_{surf})/((L-1)(L+2))$$

Here, $x = kR_o$, j_L is the spherical Bessel function, e_F is the Fermi energy, $b_{surf} = 4\pi r_o^2 \sigma \approx 17$ MeV. This equation turns into the Bohr-Mottelson's for the 1st sound + surface modes at $F_o = F_1 = 0$, $s = 3^{-1/2}$. In the LiPa-approach such are the independent modes. Eq. (5) has no solution at $x \ll 1$ which would correspond to nearly pure shape distortion. The lowest values of x obtained from (5) are in the range 1.3 - 2.5, the higher ones approach roots of the Bessel functions.

The resonance energies $E_L^{(o)} = q_L \hbar v_F/R_o = D/A^{-1/3}$. The values of D obtained in these calculations are rather insensitive to parameters of (5). Particularly insignificant is the A-dependence, so the energies follow the $A^{-1/3}$-law. For the quadrupole and octupole density vibrations the calculated energies agree well with the experiment. (D = 62-78, 110-130 and 150-180 MeV for L = 2,3 and 4 with $\mathcal{E}_F = 40$ MeV, $r_o = 1.2$ Fermi). The calculated monopole energies exceed significantly the experimental values which suggests, possibly, that in this instance the observed resonances are the 1st-sound modes.

The multipole moments derived for the lowest resonances are contributed essentially by the distortion of SENS, i.e. they correspond to surface-located transition densities, in agreement with some microscopic calculations. Although the surface dynamics is of little significance for the frequencies of higher resonances, the surface contribution nearly cancels their multipole strengths, as in the case of the 1st-sound + surface modes.

COLLECTIVE EXCITATIONS IN CLOSED-SHELL NUCLEI

NGUYEN VAN GIAI

Division de Physique Théorique*, Institut de Physique Nucléaire
91406 Orsay Cedex, France

Abstract : We discuss some recent applications of the RPA response function method to the study of collective excitations. The first example concerns transition densities and transition currents of low lying states and giant resonances. Definite differences appear, which should show up in transverse form factors as measured in inelastic electron scattering. The second example deals with high lying structures observed in heavy ion reactions. Within a semi-classical description of the collision, it is possible to understand these structures in terms of multiphonon excitations where the RPA phonons serve as building blocks.

---

* Laboratoire Associé au C.N.R.S.

I. Introduction

The existence of collective excitations in nuclei has been recognized long ago. The first observation dates back to the early (γ,n) experiments [1] on ^{12}C and ^{63}Cu in which the isovector dipole excitation appeared as a giant resonance in the measured cross-sections. This collective mode was immediately interpreted as an oscillation of the protons as a whole against the neutrons [2]. It was later shown that the collective character of the giant dipole resonance could be understood microscopically in the framework of the Random Phase Approximation (RPA) [3]. By now, a large variety of collective modes of different spin and isospin characters (isoscalar or isovector, spin-flip or non spin-flip) have been measured. In parallel, the RPA has proved to be a good theoretical tool for the microscopic study of these modes. The adequacy of RPA means that the excitations under consideration are basically collective vibrations of small amplitude. Of course, the actual states are more complicated than that, and the RPA solution is just a doorway state which will be coupled to the more complex configurations. This coupling which leads to the fragmentation and spreading of the observed strengths will be discussed by other speakers [4]. Here, our scope is limited to the RPA framework.

In these lectures, we put the emphasis on the consistency when performing RPA studies. By this we mean that the single particle spectrum and the residual particle-hole (p-h) interaction should not be chosen as independent inputs. This is clear from the fact that the RPA originates from the linearization of the time-dependent Hartree-Fock (TDHF) equations, and therefore a definite relationship exists between the interaction giving rise to the static HF field and the p-h interaction.

Adopting this point of view has the disadvantage that one cannot handle easily realistic N-N interactions because of the difficulty of performing static HF calculations with a fully realistic interaction or a G-matrix derived from it. There are cases, however, where it may be sufficient to work with a parametrized effective interaction which reproduces the relevant properties of a G-matrix. Such cases are, for instance, the non spin-flip modes which depend mostly on some bulk properties like the nuclear incompressibility (monopole) or the nuclear surface (multipole surface vibrations). Then, the self-consistent RPA has the advantage of making a contact with the sum-rule approaches which also make use of the same effective interactions [5] , and with the fluid-dynamical models where one starts from an energy density functional analogous to the HF one [6] .

In the following, we wish to show some recent applications of the self-consistent RPA method. One interesting problem is the study of current distributions of collective excitations. The RPA predicts various types of current distributions which can be experimentally checked if one is able to measure transverse form factors at the relevant momentum transfer values in (e,e') scattering. Another problem that we shall discuss here is the possibility of exciting multiphonon states when bombarding a target nucleus with heavy ions. Under specific kinematical conditions, the excitation of the phonon corresponding to the giant quadrupole resonance (GQR) is large and dominates over other phonons, and this leads to a sizable probability of exciting states made of several GQR phonons. Before taking up these two problems, we shall briefly recall in the next section the general method we use to calculate the RPA nuclear response function.

II. Outline of the RPA response function method

Our main requirement is to work with an effective interaction which gives a satisfactory description of basic ground state properties such as total binding energies and nuclear densities in the HF approximation. Very successful interactions in this respect are the Skyrme-type forces [7] which are widely used in HF calculations. These forces have a simple analytical form, being of zero-range but containing a momentum dependence to mock up finite range effects, and also a density dependence to insure correct saturation properties. They can be considered as simple parametrizations of a G-matrix [8] . The main advantage of such forces is that RPA calculations can be carried out directly in coordinate space [9] without any expansion on a discrete single particle basis as it is generally the case when one is working with a finite range force. The first consequence is that one can handle the complete p-h space and avoid problems related to space truncations. The second consequence is that the continuous part of the single particle spectrum is fully treated and therefore the RPA states above nucleon emission threshold appear as resonances with their natural escape widths [10] .

Having solved the static HF problem and obtained the HF hamiltonian H_0, one can construct the non-interacting p-h Green function $G^{(0)}$. It is sufficient to restrict the p-h pair to be at the same point, $\vec{r}$ or $\vec{r}\,'$, as long as one is concerned with transitions induced by local one-body operators. Then, the coordinate representation of $G^{(0)}$ is :

$$G^{(0)}(\vec{r},\vec{r}\,';\omega) = \sum_i \varphi_i^*(\vec{r}) \langle \vec{r} \,| \frac{1}{H_0 - \varepsilon_i - \omega - i\eta} + \frac{1}{H_0 - \varepsilon_i + \omega - i\eta} | \vec{r}\,' \rangle \varphi_i(\vec{r}\,')$$

where ω is the excitation energy, φ_i and ε_i are the HF wave function and energy of an occupied state i and the sum runs over all occupied states. We omit to write explicitly spin and isospin variables to keep simple notations. If one expands the operators $[H_o - \varepsilon_i \mp \omega - i\eta]^{-1}$ on a complete (discrete and continuous) set of eigenstates, one can see that the hole-hole terms cancel out and $G^{(o)}$ takes the familiar form of a sum of p-h terms. Furthermore, the HF potential U in H_o is local for Skyrme-type forces, the non-locality of the HF field being all contained in an $\vec{r}$-dependent effective mass $m^*(\vec{r})$. Therefore, the quantities $\langle \vec{r} | [H_o - \varepsilon_i \mp \omega - i\eta]^{-1} | \vec{r}' \rangle$ can be constructed from linearly independent eigenfunctions of H_o satisfying appropriate boundary conditions at the origin and at infinity [10] .

The RPA equations can be derived by linearizing the TDHF equation. This is explicitly done for instance in ref. [9] . One then sees that the p-h residual interaction (including exchange terms) is $V_{ph}\, \delta(\vec{r}_1 - \vec{r}_2)$, where :

$$V_{ph} = \frac{\partial U}{\partial \rho} \quad , \tag{2}$$

i.e. V_{ph} is the functional derivative of the HF potential with respect to the density matrix. If the original effective force has no density dependence, the residual interaction is equivalent to the original force. Any density dependence will introduce rearrangement contributions to the residual interaction.

The RPA Green function $G(\vec{r}, \vec{r}\,' ; \omega)$ satisfies the following integral equation :

$$G(\vec{r},\vec{r}';\omega) = G^{(0)}(\vec{r},\vec{r}';\omega) + \int G^{(0)}(\vec{r},\vec{r}'';\omega)\, V_{ph}(\vec{r}'')\, G(\vec{r}'',\vec{r}';\omega)\, d^3r'' . \quad (3)$$

A more transparent expression of the Green function G is obtained by writing its spectral representation. Denoting by $|\Phi_0\rangle$, $|\Phi_N\rangle$ the RPA ground and excited states, one has :

$$G(\vec{r},\vec{r}';\omega) = \sum_N \langle\Phi_0|\psi^\dagger(\vec{r})\psi(\vec{r})|\Phi_N\rangle\langle\Phi_N|\psi^\dagger(\vec{r}')\psi(\vec{r}')|\Phi_0\rangle$$

$$\times \left[\frac{1}{\omega_N - \omega - i\eta} + \frac{1}{\omega_N + \omega - i\eta} \right] , \quad (4)$$

where $\psi^\dagger(\vec{r})$ and $\psi(\vec{r})$ are the field operators which respectively create and annihilate a nucleon at point $\vec{r}$, ω_N is the excitation energy of the state $|\Phi_N\rangle$, and the sum runs over all positive energy RPA states (this sum therefore becomes an integral when the states $|\Phi_N\rangle$ form a continuum). The quantities $\langle\Phi_0|\psi^\dagger(\vec{r})\psi(\vec{r})|\Phi_N\rangle \equiv \rho_N(\vec{r})$ are recognized to be the transition densities. For a discrete state $|\Phi_N\rangle$, the transition density is simply related to the residue of G at the pole $\omega = \omega_N$.

A very convenient quantity to introduce is the nuclear response function. Suppose that we perturb the nucleus by turning on an external field $A(\vec{r})e^{-i\omega t} + c.c.$ which we choose weak enough so that the nucleus responds linearly. The ground state expectation value of any one-body operator $Q(\vec{r})$ will be affected, and this change is characterized by the response function [11] :

$$R(\omega) = \int d^3r\, d^3r'\, A(\vec{r})\, G(\vec{r},\vec{r}';\omega)\, Q(\vec{r}') \qquad [\omega > 0]. \tag{5}$$

For instance, if we choose $A = Q$, we can see from eqs. (4) and (5) that the residue of $R(\omega)$ at a discrete pole $\omega = \omega_N$ is just the transition strength $|\langle \Phi_0 | Q | \Phi_N \rangle|^2$.

One is generally interested in the distribution of transition strength as a function of excitation energy ω . This strength distribution $S(\omega)$ is immediately given in terms of $R(\omega)$ by :

$$S(\omega) \equiv \sum_N |\langle \Phi_0 | Q | \Phi_N \rangle|^2\, \delta(\omega - \omega_N) = \frac{1}{\pi}\, \mathrm{Im}\, R(\omega) \; . \tag{6}$$

The response function method has been used to calculate various multipole strength distributions in spherical nuclei, and we shall not dwell on this here (see, for instance, ref. [11]).

III. Transition densities and current distributions

III.1. Classical and sum-rule predictions.

Before looking at the transition densities and transition currents calculated in RPA, let us briefly recall the predictions of the classical hydrodynamical model and how they can be interpreted when one makes use of the sum-rule method.

The classical hydrodynamical model considers the motion of nucleons in the nucleus as that of a fluid (or of two fluids if one deals with the isovector case). One generally makes the following assumptions :

i) the motion corresponds to an incompressible flow, i.e. $\frac{d\rho}{dt} = 0$. By using the continuity equation :

$$\frac{\partial \rho}{\partial t} + \mathrm{div}\,\vec{j} = 0 \quad , \tag{7}$$

where $\vec{j} = \rho \vec{v}$ is the current while $\vec{v}$ is the velocity field, one sees that assumption i) is equivalent to say that the field $\vec{v}$ is divergenceless : $\mathrm{div}\,\vec{v} = 0$.

ii) the flow is also irrotational, i.e.

$$\mathrm{rot}\,\vec{v} = 0 \ .$$

These two assumptions then lead to the transition densities derived by Tassie [12] . For a multipole $\lambda \neq 0$, one has :

$$\rho_{tr}^{(\lambda)}(\vec{r}) \propto r^{\lambda-1} \frac{d\rho_0(r)}{dr} Y_{\lambda 0}(\hat{r}) \quad , \tag{8}$$

where ρ_0 is the ground state density.

It is quite interesting to observe that these classical results can be obtained if one starts from a microscopic hamiltonian H and makes some specific assumptions. To see that, it is convenient to use the sum rule method to derive expressions for the transition and current densities [13] . Let us first define the density operator $\hat{\rho}$, convection current operator $\hat{j}$, and a general one-body operator $\hat{F}$ independent of momentum by :

$$\hat{\rho}(\vec{r}) \equiv \sum_{i=1}^{A} \delta(\vec{r}-\vec{r}_i)$$

$$\hat{\jmath}(\vec{r}) \equiv \frac{1}{2m} \sum_i \left\{ \delta(\vec{r}-\vec{r}_i)\vec{p}_i - \overleftarrow{p}_i\,\delta(\vec{r}-\vec{r}_i) \right\} \qquad (9)$$

$$\hat{F} \equiv \sum_i f(\vec{r}_i) = \int \hat{\rho}(\vec{r})\, f(\vec{r})\, d^3r \quad .$$

We can now write down the three following sum rules. The non energy-weighted sum rule (NEWSR) is readily obtained by using the completeness property of the states $|N\rangle$:

$$\sum_N \langle 0|\hat{\jmath}(\vec{r})|N\rangle\langle N|\hat{F}|0\rangle = \frac{1}{2}\langle 0|[\hat{\jmath}(\vec{r}),\hat{F}]|0\rangle$$

$$= -\frac{i}{2m}\rho_0(\vec{r})\,\nabla f(\vec{r}) \quad , \qquad (10)$$

where $\rho_0(\vec{r}) \equiv \langle 0|\hat{\rho}(\vec{r})|0\rangle$.

The continuity equation

$$-i[H,\hat{\rho}(\vec{r})] = \nabla.\hat{\jmath}(\vec{r}) \quad , \qquad (11)$$

which is the analog of eq. (7), leads to the two energy-weighted sum rules (EWSR) :

$$\sum_N \omega_N \langle 0|\hat{\rho}(\vec{r})|N\rangle\langle N|\hat{F}|0\rangle = -\frac{1}{2m}\nabla.\left\{\rho_0(\vec{r})\nabla f(\vec{r})\right\} \quad , \qquad (12a)$$

$$\sum_N \omega_N \langle 0|\hat{F}|N\rangle\langle N|\hat{F}|0\rangle = \frac{1}{2m}\int \rho_0(\vec{r})\,|\nabla f(\vec{r})|^2\, d^3r \quad . \qquad (12b)$$

It is now straightforward to obtain the transition densities $\rho_N = \langle 0|\hat{\rho}|N\rangle$ and transition currents $\vec{J}_N = \langle 0|\hat{J}|N\rangle$ within the following hypothesis :

a) Choose the operator $\hat{F}$ such that :

$$f(\vec{r}) = r^\lambda Y_{\lambda 0}(\hat{r}) \qquad [\lambda \neq 0] \quad . \tag{13}$$

b) Assume that one state $|N\rangle$ exhausts the strength of this operator $\hat{F}$.

Then, eq. (12b) gives us the total energy-weighted strength S_λ :

$$S_\lambda = \omega_N |\langle 0|\hat{F}|N\rangle|^2 = \frac{\lambda(2\lambda+1)}{8\pi m} A \langle r^{2\lambda-2}\rangle \quad . \tag{14}$$

The transition density ρ_N is obtained from (12a) and (14) :

$$\rho_N(\vec{r}) = -\frac{\lambda}{2m\sqrt{\omega_N S_\lambda}} r^{\lambda-1} \frac{d\rho_0(r)}{dr} Y_{\lambda 0}(\hat{r}) \quad , \tag{15}$$

while eq. (10) gives us the transition current :

$$\vec{J}_N(\vec{r}) = -\frac{i}{2m}\left(\frac{\omega_N}{S_\lambda}\right)^{1/2} \rho_0(\vec{r})\ \nabla\left(r^\lambda Y_{\lambda 0}(\hat{r})\right) \quad . \tag{16}$$

Therefore, the two assumptions a) and b) have led us to the Tassie transition densities. Furthermore, one can see that the velocity field is divergenceless and irrotational. Indeed, by making the usual identification of the coefficient of ρ_0 in $\vec{J}_N(\vec{r})$ with the velocity $\vec{v}$, we obtain from eq. (16) that :

$$\text{rot} \langle 0 | \vec{v}(\vec{r}) | N \rangle = 0 \quad ,$$

$$\text{div} \langle 0 | \vec{v}(\vec{r}) | N \rangle = 0 \,.$$

III.2. Transition densities in RPA.

RPA transition densities can be calculated from the Green function $G(\vec{r},\vec{r}';\omega)$. We just need to evaluate the response function (5) where Q is chosen of the form (13) whereas A is the density operator $\hat{\rho}$ of eq. (9). We then obtain :

$$R(\vec{r},\omega) = \int d^3r' \; G(\vec{r},\vec{r}';\omega)\, Q(\vec{r}')$$

$$= \sum_N \rho_N(\vec{r}) \langle \Phi_N | Q | \Phi_0 \rangle \left[\frac{1}{\omega - \omega_N - i\eta} + \frac{1}{\omega + \omega_N - i\eta} \right] , \qquad (17)$$

where we have used eq. (4). If $|\Phi_N\rangle$ is a discrete state, its transition density is given by the residue of $R(\vec{r},\omega)$ at the pole $\omega = \omega_N$. Above particle threshold, transition densities of resonances are obtained from the imaginary part of $R(\vec{r},\omega)$. The transition densities are normalized according to :

$$\langle \Phi_N | Q | \Phi_0 \rangle = \int \rho_N(\vec{r})\, Q(\vec{r})\, d^3r . \qquad (18)$$

As an example, let us look at the case of isoscalar quadrupole states in ^{208}Pb. Two collective excitations are known experimentally : a low lying discrete 2^+ state at E_X = 4.1 MeV with B(E2) = (3.18 $\pm$ 0.15) x 10^3 e^2fm^4 [14] , and the GQR around 10.9 MeV which contains about 2/3 of the EWSR [15] . In ref. [16] , an RPA calculation with the Skyrme interaction SGII [17] was performed, giving a discrete state at E_x = 5.1 MeV with B(E2) = 2.82 x 10^3 e^2fm^4 (i.e. 14 % of EWSR) and a giant resonance at E_x = 11.4 MeV with a narrow escape width of 300 keV and a strength B(E2) = 5.16 x 10^3 e^2fm^4 (i.e. 70 % of EWSR).

In Fig. 1 are shown the radial parts of the transition charge densities for the two calculated excitations [16] and for the discrete 2^+ state experimentally determined by (e,e') scattering [14] . There is no experimental transition density for the GQR. One can see that the GQR transition density has a Tassie-type surface peak with additional small bumps around 2 and 4 fm. Since this excitation exhausts a large fraction of EWSR, we are in a situation where the assumptions leading to the Tassie model are well fulfilled. On the other hand, the low lying state takes a modest fraction of EWSR although it is quite collective. Consequently, its calculated transition density deviates somewhat from the Tassie prediction. This is in fair agreement with experiment.

III.3. Transition current densities.

The transition current densities can also be calculated from the Green function $G(\vec{r},\vec{r}';\omega)$. They are obtained by suitably choosing the operator A in eq. (5). The convection charge current operator $\hat{J}^{cc}$ is analogous to the operator $\hat{J}$ of eq. (9) with a factor $e(\frac{1}{2}-t_z(i))$ multiplying the bracket. We also need the magnetization current operator :

$$\hat{J}^{MC}(\vec{r}) = \frac{e}{2m} \sum_i g_i \, \delta(\vec{r}-\vec{r}_i) \, \vec{\nabla} \times \vec{S}_i \quad , \tag{19}$$

where $\vec{s}_i$ is the spin of nucleon i, g_p = 5.58 and g_n = - 3.82. The total charge current is the sum : $\hat{J}^T = \hat{J}^{cc} + \hat{J}^{MC}$. Details are given in the Appendix for calculating the transition current densities $\langle 0|\hat{J}(\vec{r})|N\rangle$.

It is customary to expand $\langle 0|\hat{J}(\vec{r})|N\rangle$ on the basis of vector spherical harmonics :

$$\langle 0|\hat{J}(\vec{r})|N\rangle = -i \sum_{\ell} J_{\lambda\ell}(r)\, \mathbf{Y}^{\mu *}_{\lambda\ell 1}(\hat{r}) \quad , \tag{20}$$

where (λ, μ) is the angular momentum of state $|N\rangle$. For an unnatural parity state (magnetic transition) only $\ell = \lambda$ contributes to (20), whereas for a natural parity state (electric transition) the contributions come from $\ell = \lambda \pm 1$. Here, we shall restrict ourselves to the latter case.

In principle, the continuity equation (11) must be obeyed in self-consistent RPA [18] . We can check it numerically by taking its matrix elements between states $|0\rangle$ and $|N\rangle$:

$$\omega_N \rho_N(r) = \left(\frac{\lambda}{2\lambda+1}\right)^{1/2} \left[\frac{\lambda-1}{r} - \frac{d}{dr}\right] J_{\lambda\,\lambda-1}(r)$$
$$+ \left(\frac{\lambda+1}{2\lambda+1}\right)^{1/2} \left[\frac{\lambda+2}{r} + \frac{d}{dr}\right] J_{\lambda\,\lambda+1}(r) \quad . \tag{21}$$

For a velocity-dependent interaction like the Skyrme force, additional terms should appear on the r.h.s. of eq. (11) or (21) in the isovector case [19] , but they vanish in the isoscalar case. The lower half of Fig. 1 shows the r.h.s. of eq. (21) calculated for the discrete isoscalar 2^+ state and the GQR in ^{208}Pb [16] . Only $\mathcal{J}^{cc}$ contributes to the continuity equation because the magnetization current is divergenceless. One can see that the continuity equation is obeyed to a good accuracy. The deviations to it can be attributed to the effect of the spin-orbit interaction.

In Fig. 2 are shown the convection current patterns and the corresponding components $\mathcal{J}^{cc}_{\lambda\lambda\pm1}$ for the isoscalar, $\lambda = 2$ states in ^{208}Pb [16] . An irrotational, incompressible flow would have an identically zero $\mathcal{J}^{cc}_{\lambda\lambda+1}$ component. This is not the case for any of the states in Fig. 2. However, for the GQR the $\mathcal{J}^{cc}_{\lambda\lambda-1}$ component is largely dominant and consequently its flow pattern is close to that of the Bohr-Tassie model. On the other hand, the current components $\mathcal{J}^{cc}_{\lambda\lambda\pm1}$ of the low lying 2^+ state are of comparable magnitude but with opposite signs at the nuclear surface, and this gives rise to vortices in the current pattern in that region. This confirms the conclusion of the previous subsection, namely that the assumptions of the hydrodynamical model are justified for the state exhausting a large fraction of EWSR (the GQR in the present case), but the low lying collective state is of a quite different nature. Actually, the RPA model predicts a wide variety of flow patterns for vibrational states, ranging from almost purely irrotational to quite complex behaviours [20, 21] . Current distributions for rotational states have also been studied in the framework of the cranking model [22] .

III.4. Transverse form factors.

The total transition current $\langle 0 | \hat{\mathcal{J}}^{T}(\vec{r}) | N \rangle$ is a quantity of interest for inelastic electron scattering. It determines the transverse form factor $F_T(q)$ which can be measured in backward-angle (e,e') experiments [23] . In plane wave Born approximation, the transverse form factor for a natural parity transition is :

$$F_T(q) = \frac{1}{q} \int [\nabla \times j_\lambda(qr) Y^{\mu}_{\lambda\lambda 1}(\hat{r})] . \langle 0 | \hat{J}^T(\vec{r}) | N \rangle \, d^3r , \qquad (22)$$

where q is the momentum transfer. This can be expressed in terms of the current components $J^T_{\lambda\lambda\pm1}$ as :

$$F_T(q) = \int \Big\{ \left(\frac{\lambda+1}{2\lambda+1}\right)^{1/2} j_{\lambda-1}(qr) \, J^T_{\lambda\lambda-1}(r) - \left(\frac{\lambda}{2\lambda+1}\right)^{1/2} j_{\lambda+1}(qr) \, J^T_{\lambda\lambda+1}(r) \Big\} r^2 dr \quad . \qquad (23)$$

In the case of 2^+ states in ^{208}Pb, one finds that the contribution of the magnetization current to the total current is relatively small compared to that of the convection current [16] . One therefore expects that the low lying 2^+ state and the GQR will have different transverse form factors because of their different convection current patterns. This is illustrated in Fig. 3 where we show $|F_T(q)|^2$ for the two collective 2^+ states in ^{208}Pb. Experimental values for the low lying state [24] are also shown. There are no measured values for the GQR. Two main differences in the calculated form factors can be noted : i) the first minimum in the form factor of the GQR is at q = 0.8 fm^{-1} whereas the corresponding minimum for the low lying state is at q = 1.05 fm^{-1} ; ii) the ratio of the first to the second maximum is about 40 for the GQR and only 13 for the low lying state.

This example shows that the RPA model leads to definite predictions for the current distributions of collective states. Some of these predictions agree with the fluid dynamical results, others don't. Measurements of transverse form factors by (e,e') scattering, although quite delicate, are very useful by allowing further confrontation with theory.

IV. Multiphonon excitations

In this section we discuss the possibility of reaching excited states having a multiphonon structure in a target nucleus via heavy ion reactions. We briefly summarize in the first subsection the experimental evidence which indicates that high lying excitations have been observed in heavy ion collisions. In the next subsections, we then present a theoretical approach for analyzing the observed structures in the cross-sections.

IV.1. Experimental situation.

The first experimental indications of regularly spaced structures in the excitation functions come from the inclusive measurements of N. Frascaria et al. [25] done on symmetrical systems (^{40}Ca + ^{40}Ca, ^{63}Cu + ^{63}Cu). Since then, many experiments have been carried out for symmetrical as well as non-symmetrical systems. They include inclusive reactions and also coincidence measurements of light emitted particles. In Fig. 4 are shown typical inelastic spectra recently measured in the ^{36}Ar + ^{208}Pb reaction at an incident energy of 11 MeV/nucleon, and in the ^{20}Ne + ^{208}Pb reaction at 30 MeV/nucleon [26] . The target nucleus ^{208}Pb will serve as a test case for our discussion in the following subsections.

The general conclusions that one can draw from the experimental studies of ref. [26] and of previous coincidence experiments performed with various systems and at different energies are :

i) The excitation functions exhibit regularly spaced structures up to rather high excitation energies. The positions of the structures do not depend on the nature of the heavy ion projectile, its incident energy or the scattering angle (but the intensities of the structures do depend on these variables). This is a clear indication that the structures correspond to excitations in the target nucleus.

ii) The widths of these structures grow moderately like $\sqrt{E_x}$, where E_x is the excitation energy. For the ^{208}Pb target, for instance, the widths vary from 4 MeV for $E_x \lesssim 20$ MeV to 15 MeV for $E_x \sim 120$ MeV.

iii) The structures are enhanced, especially those at lower energies, when the scattering angle is around the value of the grazing angle. This strongly suggests that their excitation is achieved by a direct process. One must also note that the same structures are observed in the few-nucleon transfer channels.

iv) The capture-evaporation process, where the projectile would capture a few nucleons and then emit them statistically, can contribute to some background but would not lead to structures having the properties i), ii) and iii). Also, explanations in terms of knock-out processes are ruled out by the results of coincidence experiments.

These characteristic features lead one to think that the passing-by projectile might be exciting the target in the form of single phonon states, or possibly multiphonon states. In the following, we first derive the external field responsible for the transitions, and then proceed to calculate the excitation probabilities of multiphonon states, using the RPA phonons as building blocks.

IV.2. First order calculation of the one-phonon excitation amplitude.

The picture we adopt is quite analogous to that used in Coulomb excitation. The collision is described semi-classically, as in ref. [27] . The projectile follows a classical trajectory characterized by the position of its center $\vec{R}(t)$ (the origin of coordinates is chosen at the center of the target nucleus). We assume that this trajectory is little affected by the target excitation, and we are mainly interested in the trajectories close to the grazing one. The projectile carries along a perturbing, time-dependent external field $V(\vec{r}-\vec{R}(t))$ where $\vec{r}$ is the coordinate of a target nucleon. The first order transition amplitude to a one-phonon state $|N\rangle$ is, in time-dependent perturbation theory :

$$A_N = \hbar^{-1} \int_{-\infty}^{\infty} \langle N| V(\vec{r}-\vec{R}(t))|0\rangle \, e^{i\omega_N t} \, dt \, , \qquad (24)$$

and we can see that its calculation will involve, apart from a time integration, the nuclear response function introduced in sect. II.

The external field V is assumed real and of a fixed Woods-Saxon shape, with parameters corresponding to the projectile. Its general multipole expansion is :

$$V(\vec{r}-\vec{R}) = 4\pi \sum_{\lambda\mu} (2\lambda+1)^{-1} v_\lambda(r,R) \, Y_{\lambda\mu}(\hat{r}) \, Y^*_{\lambda\mu}(\hat{R}) \, . \qquad (25)$$

We could use this expansion to calculate directly eq. (24), but a great simplification occurs if one restricts the problem to peripheral collisions, because in this case the target nucleons feel mainly the tail of the potential V. Then, we can replace in (24) and (25) the Woods-Saxon potential by a Yukawa potential having the same asymptotic behaviour. This leads to the following approximate factorization of the multipoles $\mathcal{V}_\lambda(r,R)$:

$$\mathcal{V}_\lambda(r,R) \simeq \mathcal{V}_\lambda(r,d)\,\frac{d}{R}\exp\left(\frac{d-R}{a}\right) , \qquad (26)$$

where d is the distance of closest approach corresponding to the trajectory $\vec{R}(t)$, and a is the diffuseness parameter of the Woods-Saxon potential. This approximate form (26) has been checked to be quite accurate numerically for the calculation of (24) [26], and it was used in refs. [26, 28]. The evaluation of the amplitude A_N is now of the same level of simplicity as in the case of Coulomb excitation where such a factorization occurs exactly. One obtains :

$$A_N = \hbar^{-1}\langle N|\mathcal{V}_\lambda(r,d)\,Y_{\lambda\mu}(\hat{r})|0\rangle\, T^*_{\lambda\mu}(\omega_N) , \qquad (27)$$

where

$$T_{\lambda\mu}(\omega_N) = \frac{4\pi}{2\lambda+1}\int_{-\infty}^{\infty}\left\{\frac{d}{R}\,e^{\frac{d-R}{a}}\,Y_{\lambda\mu}(\hat{R})\right\}e^{-i\omega_N t}\,dt . \qquad (28)$$

Our first remark is that the transition amplitudes A_N depend on the nuclear response to the multipole operators $Q_{\lambda\mu} = \mathcal{V}_\lambda(r,d)\,Y_{\lambda\mu}(\hat{r})$.

In Fig. 5 are shown, for $\lambda = 2$, the shapes of $v_\lambda(r, R)$ for various choices of R. They are compared to the operators r^λ and $j_\lambda(qr)$ which are often used in the study of nuclear strength distributions. The strength distribution of a given multipole operator depends very much on its radial shape. For an operator like r^λ which is sensitive mostly to the outer tail of nuclear densities, the strength is concentrated to energies below 3-4 $\hbar\omega$ for all λ [29]. If the radial operator is peaked at the nuclear surface, like in the case of $j_\lambda(qr)$ with $q \sim$ 1-2 fm^{-1}, the low lying strength becomes less important but high lying components can become dominant [30, 31]. In the present case, we are in an intermediate situation and we expect that the low lying components will still be strong like in the r^λ case, but also that the tails of the strength distributions will decrease more slowly with increasing excitation energy than for r^λ. This is actually confirmed by the calculations [26]. Some of the calculated strength distributions are shown in Figs. 6-8.

Our second remark is that the excitation amplitude A_N is much influenced by the factor $T_{\lambda\mu}(\omega_N)$. The significance of this factor is the following : it describes classically the matching between the two quantities transferred to the target, namely the energy ω_N and the angular momentum (λ, μ). This can be seen by the following argument : in the integrand of the r.h.s. of eq. (28), only the values of $R(t)$ close to the distance of closest approach d contribute. We can therefore replace $\vec{R}(t)$ by the tangent $\vec{v}t + \vec{d}$ to the classical trajectory at the point $\vec{d}$ ($\vec{v}$ and $\vec{d}$ are perpendicular), and consider $\frac{vt}{d}$ as a small quantity. The bracket in eq. (28) can then

be approximated by $\exp(-\frac{v^2t^2}{2ad})\, Y_{\lambda\mu}(\frac{\pi}{2},0)\exp(\frac{i\mu v t}{d})$.
This gives an analytical expression for $T_{\lambda\mu}(\omega_N)$:

$$T_{\lambda\mu}(\omega_N) \simeq \frac{4\pi}{2\lambda+1}\frac{\sqrt{2\pi a d}}{v}\, Y_{\lambda\mu}(\frac{\pi}{2},0)\exp\left[-\frac{(\mu-\bar{\mu})^2}{2\Delta^2}\right], \quad (29)$$

where $\Delta = \sqrt{\frac{d}{a}}$, and $\bar{\mu} = \frac{\omega_N d}{v}$. A direct evaluation of (28) using the exact classical trajectory $\vec{R}(t)$ shows that the expression (29) is accurate up to 2 % [26] . Eq. (29) clearly expresses that the z-component of the transferred angular momentum must be close to the classical angular momentum transfer $\bar{\mu}$.

In principle, the Coulomb excitation amplitude should be added to the nuclear amplitude (27). Its calculation presents no special difficulty. However, it turns out that Coulomb excitation is quite negligible if one is interested in trajectories around the grazing one, as it is the case here.

We apply the above formalism to the ^{36}Ar + ^{208}Pb system at E_{inc} = 11 MeV/nucleon. The grazing angle is θ_{gr} = 28°, which corresponds to $d \simeq 12$ fm in our classical description of the ^{36}Ar motion. In Fig. 9 is shown as a function of excitation energy E_x the quantity $P(E_x)$ which is the squared amplitudes $|A_N|^2$ summed over all isoscalar excitations up to λ = 6 in ^{208}Pb. Fig. 9 corresponds to d = 11.35 fm . The quantity $P(E_x)$ represents the first order probability density of exciting a one-phonon state. Contrarily to the strength distributions (see Figs. 6-8), $P(E_x)$ drops very rapidly beyond 20 MeV. This reflects the influence of the energy-angular momentum matching factor $T_{\lambda\mu}(E_x)$. Indeed, for the present value of E_{inc}, one finds that $\bar{\mu}$ would be 25 if E_x is about 50 MeV. The main point to notice is the large value of $P(E_x)$ around 12 MeV, whose origin comes from the very strong contribution of the GQR, and also to a lesser extent from $\lambda = 4$ and $\lambda = 6$ excitations. This feature points to the possibility of exciting multiphonon states built on 12 MeV phonons, as we shall see now.

IV.3. Poisson distribution of multiphonon states.

Because of the very large values of $P(E_x)$, it is clear that a first order perturbation treatment is insufficient to correctly describe the transition probabilities. An elegant method has been used in refs. [26, 32] to calculate these probabilities to any order, in the quasi-boson approximation. The hamiltonian of the target is the sum of a time-independent term representing non-interacting bosons B_N^+, and a time-dependent term due to the external field $V(t)$. The phonons B_N^+ are just the RPA excitations discussed earlier. One has :

$$H = \sum_N \omega_N B_N^\dagger B_N + \sum_N \left\{ \langle 0|V(t)|N\rangle B_N^\dagger + h.c. \right\} . \tag{30}$$

The solution $|\Phi(t)\rangle$ of this hamiltonian such that $|\Phi(t)\rangle \xrightarrow[t\to-\infty]{} |0\rangle$, where $|0\rangle$ is the RPA ground state, is a quasi-classical coherent state :

$$|\Phi(t)\rangle = \left\{ \prod_N \hat{D}(A_N(t)) \right\} |0\rangle \qquad , \tag{31}$$

where the operator $\hat{D}$ is :

$$\hat{D}(A_N) = e^{(A_N B_N^+ - A_N^* B_N)} = \exp(-\frac{|A_N|^2}{2}) e^{A_N B_N^+} e^{-A_N^* B_N} . \qquad (32)$$

In eqs. (31-32), $A_N(t)$ is the Fourier transform of the first order amplitude $A_N(\omega_N)$ of eq. (24). Then, the probability $P_n(N)$ for finding a n-phonon state of type N is given by a Poisson distribution law :

$$P_n(N) = \frac{(|A_N(\omega_N)|^2)^n}{n!} e^{-|A_N(\omega_N)|^2} . \qquad (33)$$

This result can be extended to the case where the phonons N form a continuous spectrum, by performing convolution integrals on the energy variable. The probability density $P_n(E_x)$ for a transition to a n-phonon state at energy E_x is expressed in terms of the quantity P(E) introduced in the previous subsection as :

$$P_n(E_x) = \frac{\exp[-\int P(E)\,dE]}{n!} \int dE_1 \int dE_2 \ldots \int dE_{n-1}$$

$$\times P(E_1)\, P(E_2 - E_1) \ldots P(E - E_{n-1}) . \qquad (34)$$

The probability densities $P_n(E_x)$ have been calculated [26] for the $^{36}Ar + {}^{208}Pb$ system, for various values of the distance of closest approach d corresponding to scattering angles θ around the grazing angle θ_{gr} = 28° (d = 12 fm). The results are shown in Fig. 10, where the partial components $P_n(E_x)$ for n up to 9 are displayed, together with their sum. For very peripheral collisions ($d \gtrsim 13$ fm), i.e. at forward scattering angles, only the one- and two-phonon components are strong. As θ approaches θ_{gr} , multiphonon components become important and give rise to characteristic regularly spaced structures. At larger angles ($d \simeq 10$ fm) the spectrum is dominated by multiphonon components. Here, one is dealing with the deep inelastic regime, and the multiphonon mechanism seems to account for the large energy transfer to the target.

The evolution of the calculated spectra as a function of θ (or d) agrees well with experiment [26] . In the present calculation, the RPA phonons have escape widths but no spreading widths have been included. The latter would broaden the structures, but they would still persist. Also, the $\sqrt{E_x}$ law for the observed widths supports the assumption of non-interacting phonons. In conclusion, the picture of multiphonon excitations seems adequate and promising for the understanding of the high energy structures observed in heavy ion reactions. This picture might even apply to the deep inelastic regime for explaining part of the energy dissipation.

Finally it should be worthwhile to study the connection between the present approach and TDHF calculations(33) which also lead to the appearance of high energy structures .

Appendix

Calculation of transition current components.

If one performs the RPA diagonalization in a discrete configuration space $\{mi\}$, the transition densities and transition currents are obtained by taking matrix elements of the operators (9) and (19) between states $|0\rangle$ and $|N\rangle$, and the results are expressed in terms of the RPA amplitudes $X_{mi}^{(N)}$, $Y_{mi}^{(N)}$. In our case, it is possible to calculate directly the transition densities and currents from the Green function $G(\vec{r},\vec{r}\,';\omega)$.

To solve the integral equation (3) with a Skyrme-type interaction, one has to generalize the quantities $G(\vec{r}_1,\vec{r}_2;\omega)$ and $G^{(0)}(\vec{r}_1,\vec{r}_2;\omega)$ by defining :

$$G_{\alpha\beta}(\vec{r}_1,\vec{r}_2;\omega) \equiv K_\alpha(1)\, G(\vec{r}_1,\vec{r}_2;\omega)\, K_\beta^\dagger(2) \quad , \qquad \text{(A1)}$$

and a similar definition for $G_{\alpha\beta}^{(0)}$. In (A1), $K_\alpha(1)$ and $K_\beta^\dagger(2)$ are tensor operators acting respectively to the right on space and spin coordinates 1 and to the left on coordinates 2. For a given total angular momentum (λ,μ) , the following set of operators K_α is required in order to treat the velocity dependence of the interaction [9,34] :

α	K_α
1	$\mathbb{1}$
2	$\nabla_p^2 + \nabla_h^2$
3	$\sqrt{2}\left(\frac{2\lambda+1}{4\pi}\right)^{1/2} \mathcal{D}_{\mu 1}^{\lambda *} (\nabla_p + \nabla_h)_1$
4	$\sqrt{2}\left(\frac{2\lambda+1}{4\pi}\right)^{1/2} \mathcal{D}_{\mu 1}^{\lambda *} (\nabla_p - \nabla_h)_1$
5	$\left(\frac{2\lambda+1}{4\pi}\right)^{1/2} \mathcal{D}_{\mu 0}^{\lambda *} (\nabla_p + \nabla_h)_0$
6	$\left(\frac{2\lambda+1}{4\pi}\right)^{1/2} \mathcal{D}_{\mu 0}^{\lambda *} (\nabla_p - \nabla_h)_0$

In the above definitions, $\mathcal{D}_{\mu\nu}^{\lambda}$ is a rotation matrix, and the subscripts p and h indicate that the operators act on the particle or the hole wave function, respectively. The spin-dependence of the interaction just adds more operators to the set $\{K_\alpha\}$.

Let us consider for example the convection current operator $\hat{\jmath}(\vec{r})$ of eq. (9). We expand it on the basis of vector spherical harmonics :

$$\hat{\jmath} = \sum_{\lambda \ell \mu} \hat{\jmath}_{\lambda\ell}^{\mu} \, \mathcal{Y}_{\lambda\ell 1}^{\mu}(\hat{r}) \quad . \tag{A2}$$

The components $\hat{\jmath}_{\lambda\ell}^{\mu}$ are :

$$\hat{\jmath}_{\lambda\ell}^{\mu} = \int \hat{\jmath} \cdot \mathcal{Y}_{\lambda\ell_1}^{\mu\dagger}(\hat{r})\, d\hat{r}$$

$$= \frac{1}{2mi} \sum_{i=1}^{A} \frac{\delta(r-r_i)}{r r_i} \left[Y_{\ell}^{*}(\hat{r}_i) \otimes \left(\vec{\nabla}(i) - \overleftarrow{\nabla}(i) \right) \right]_{\lambda\mu} . \quad \text{(A3)}$$

It is equivalent to write $\nabla_p(i)$ and $\nabla_h(i)$ instead of $\vec{\nabla}(i)$ and $\overleftarrow{\nabla}(i)$, in eq. (A3). One can use the helicity representation to rewrite $\hat{\jmath}_{\lambda\ell}^{\mu}$ as :

$$\hat{\jmath}_{\lambda\ell}^{\mu} = \frac{1}{2mi} \sum_{i=1}^{A} \frac{\delta(r-r_i)}{r r_i} \sum_{\nu} \left(\frac{2\ell+1}{4\pi} \right)^{1/2} \langle \ell\, 1\, 0\, \nu \,|\, \lambda \nu \rangle\, \mathcal{D}_{\mu\nu}^{\lambda *} \left(\nabla_p(i) - \nabla_h(i) \right)_{\nu} . \quad \text{(A4)}$$

This gives explicitly :

$$\hat{\jmath}_{\lambda\ell}^{\mu} = \frac{1}{2mi} \sum_{i=1}^{A} \frac{\delta(r-r_i)}{r r_i}$$

$$\begin{cases} -\left(\frac{\lambda+1}{4\pi}\right)^{1/2} \mathcal{D}_{\mu 0}^{\lambda *} \left(\nabla_p(i) - \nabla_h(i)\right)_0 + \sqrt{2} \left(\frac{\lambda}{4\pi}\right)^{1/2} \mathcal{D}_{\mu 1}^{\lambda *} \left(\nabla_p(i) - \nabla_h(i)\right)_1 & \text{if } \ell = \lambda+1 \\ -\sqrt{2} \left(\frac{2\lambda+1}{4\pi}\right)^{1/2} \mathcal{D}_{\mu 1}^{\lambda *} \left(\nabla_p(i) - \nabla_h(i)\right)_1 & \text{if } \ell = \lambda \\ \left(\frac{\lambda}{4\pi}\right)^{1/2} \mathcal{D}_{\mu 0}^{\lambda *} \left(\nabla_p(i) - \nabla_h(i)\right)_0 + \sqrt{2} \left(\frac{\lambda+1}{4\pi}\right)^{1/2} \mathcal{D}_{\mu 1}^{\lambda *} \left(\nabla_p(i) - \nabla_h(i)\right)_1 & \text{if } \ell = \lambda-1 \end{cases} \quad \text{(A5)}$$

For an unnatural parity transition, only $\ell = \lambda$ contributes whereas for a natural parity transition, only $\ell = \lambda \pm 1$ must be considered. The components $\hat{\mathcal{J}}^{\mu}_{\lambda\ell}$ have the same structure as the operators K_4 and K_6. In the natural parity case, the components $\mathcal{J}_{\lambda\,\lambda\pm1}(r)$ of eq. (20) for the convection current are therefore given by :

$$\mathcal{J}_{\lambda\lambda-1}(r) = \frac{1}{\pi}\,\mathrm{Im}\left\{\frac{1}{2mi}\int\left[\left(\frac{\lambda+1}{2\lambda+1}\right)^{1/2} G_{41}(\vec{r},\vec{r}\,';\omega) + \left(\frac{\lambda}{2\lambda+1}\right)^{1/2} G_{61}(\vec{r},\vec{r}\,';\omega)\right]\right.$$
$$\left.\times\; Q(\vec{r}\,')\, d\hat{r}\, d^3r'\right\} ,$$

$$\mathcal{J}_{\lambda\lambda+1}(r) = \frac{1}{\pi}\,\mathrm{Im}\left\{\frac{1}{2mi}\int\left[\left(\frac{\lambda}{2\lambda+1}\right)^{1/2} G_{41}(\vec{r},\vec{r}\,';\omega) - \left(\frac{\lambda+1}{2\lambda+1}\right)^{1/2} G_{61}(\vec{r},\vec{r}\,';\omega)\right]\right.$$
$$\left.\times\; Q(\vec{r}\,')\, d\hat{r}\, d^3r'\right\} . \qquad \text{(A6)}$$

References

1. G.C. Baldwin and G.S. Klaiber, Phys. Rev. 73 (1948) 1156.
2. M. Goldhaber and E. Teller, Phys. Rev. 74 (1948) 1046 ;
   A. Steinwedel and J. Jensen, Z. Naturforsch. 5 (1950) 413.
3. G.E. Brown and M. Bolsterli, Phys. Rev. Lett. 3 (1959) 472.
4. P.F. Bortignon, lectures at this meeting ;
   S. Adachi, talk at this meeting.
5. O. Bohigas, A.M. Lane and J. Martorell, Phys. Reports 51 (1979) 267 ;
   S. Stringari, lectures at this meeting.
6. G. Holzwarth, lectures at this meeting.
7. M. Beiner et al., Nucl. Phys. A238 (1975) 29.
8. J.W. Negele and D. Vautherin, Phys. Rev. C5 (1972) 472.
9. G.F. Bertsch and S.F. Tsai, Phys. Reports 18C (1975) 126.
10. S. Shlomo and G.F. Bertsch, Nucl. Phys. A243 (1975) 507 ;
    K.F. Liu and Nguyen Van Giai, Phys. Lett. 65B (1976) 23.
11. Nguyen Van Giai, in Nuclear Collective Dynamics, Eds. D. Bucurescu, V. Ceausescu and N.V. Zamfir (World Scientific, 1983) p. 356.
12. L.J. Tassie, Australian J. of Phys. 9 (1956) 407.
13. T. Suzuki, lecture notes, Saclay (1983).
14. J. Heisenberg et al., Phys. Rev. C25 (1982) 2292.
15. C. Djalali et al., Nucl. Phys. A380 (1982) 42.
16. H. Sagawa and Nguyen Van Giai, Phys. Lett. 127B (1983) 393.
17. Nguyen Van Giai and H. Sagawa, Phys. Lett. 106B (1981) 379.
18. G.F. Bertsch, Suppl. Prog. Theor. Phys. 74-75 (1983) 115.
19. Y.M. Engel et al., Nucl. Phys. A249 (1975) 215.

20. T. Suzuki and D.J. Rowe, Nucl. Phys. A286 (1977) 307.

21. F.E. Serr et al., Nucl. Phys. A404 (1983) 359 ;
T.S. Dumitrescu and Toru Suzuki, Copenhagen preprint (1983).

22. A. Schuh et al., Suppl. Prog. Theor. Phys. 74-75 (1983) 402.

23. J. Heisenberg, Adv. Nucl. Phys. 12 (1981) 61.

24. R.S. Hicks et al., Phys. Rev. C26 (1982) 920.

25. N. Frascaria et al., Phys. Rev. Lett. 39 (1977) 918 ; Z. für Physik A294 (1980) 167.

26. P. Chomaz, Thèse de 3e cycle, Université Paris-Sud (1984) ;
P. Chomaz et al., to be published.

27. R.A. Broglia, C.H. Dasso and A. Winther, Varenna Course LXXVII, North-Holland, Amsterdam (1981) 327.

28. P. Chomaz and D. Vautherin, to appear in Phys. Lett. B.

29. Nguyen Van Giai, Phys. Lett. 105B (1981) 11.

30. J. Dechargé et al., Phys. Rev. Lett. 49 (1982) 982.

31. Nguyen Van Giai, Suppl. Prog. Theor. Phys. 74-75 (1983) 330.

32. H. Tricoire, C. Marty and D. Vautherin, Phys. Lett. 100B (1981) 109.

33. H.Flocard and M.Weiss, Phys, Lett. 105 B (1981) 16 .

34. S.F.Tsai, Phys. Rev. C17 (1978) 1862.

Figure captions

Fig. 1 : Upper half : transition charge densities of isoscalar 2^+ states in ^{208}Pb. The solid and dashed curves correspond to the calculated states at 11.4 MeV and 5.1 MeV, respectively. The dotted curve is the experimental transition density of the observed state at 4.1 MeV. Lower half : divergence of transition currents for the same states.

Fig. 2 : Convection current patterns in the (x, z) plane, and the corresponding current components $J^{cc}_{\lambda\lambda\pm1}$. The length of the arrow is proportional to the current strength.

Fig. 3 : Transverse form factors of the GQR at E_x = 11.4 MeV (solid curve) and of the 2^+ state at E_x = 5.1 MeV (dashed curve). Experimental values are from ref. [24] .

Fig. 4 : Experimental inelastic spectra at various angles, measured in different reactions on the same ^{208}Pb target.

Fig. 5 : (a) Radial part of the excitation operator $\mathcal{V}_L(r, R)$ for L = 2, calculated for a ^{40}Ca + ^{40}Ca system ; (b) Comparison between $\mathcal{V}_L(r, R)$ and other operators often used, for the same system.

Fig. 6 : L = 2 strength distributions in ^{208}Pb calculated with the operators $v_L(r, R=11.35)\, Y_{L0}(\hat{r})$ and $r^L\, Y_{L0}(\hat{r})$. Solid and dot-and-dashed curves correspond respectively to RPA and HF results.

Fig. 7 : Same as Fig. 6 for L = 4.

Fig. 8 : Same as Fig. 6 for L = 6.

Fig. 9 : The first order probability density $P(E_x)$ of exciting a one-phonon state.

Fig. 10 : Probability densities $P_n(E_x)$ of exciting n-phonon states, for various values of d. The heavy lines show the summed probabilities.

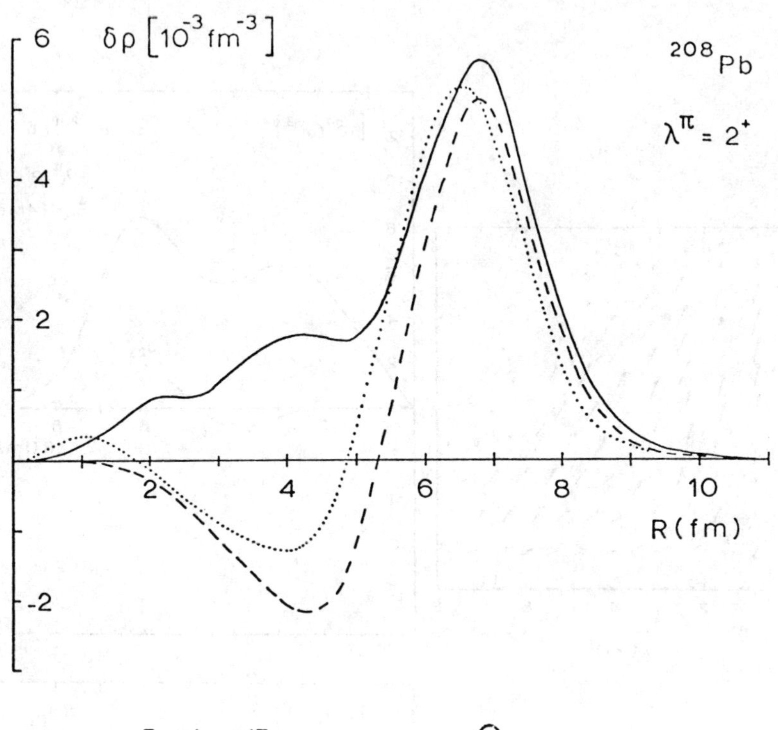
6
δρ [10^{-3} fm^{-3}]
^{208}Pb
$\lambda^{\pi} = 2^{+}$
4
2
-2
2
4
6
8
10
R(fm)

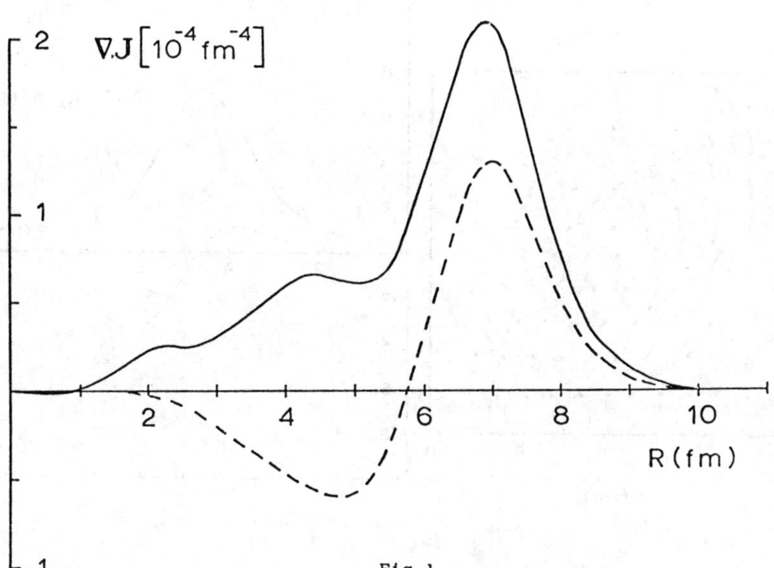
2
∇.J [10^{-4} fm^{-4}]
1
-1
2
4
6
8
10
R(fm)

Fig.1

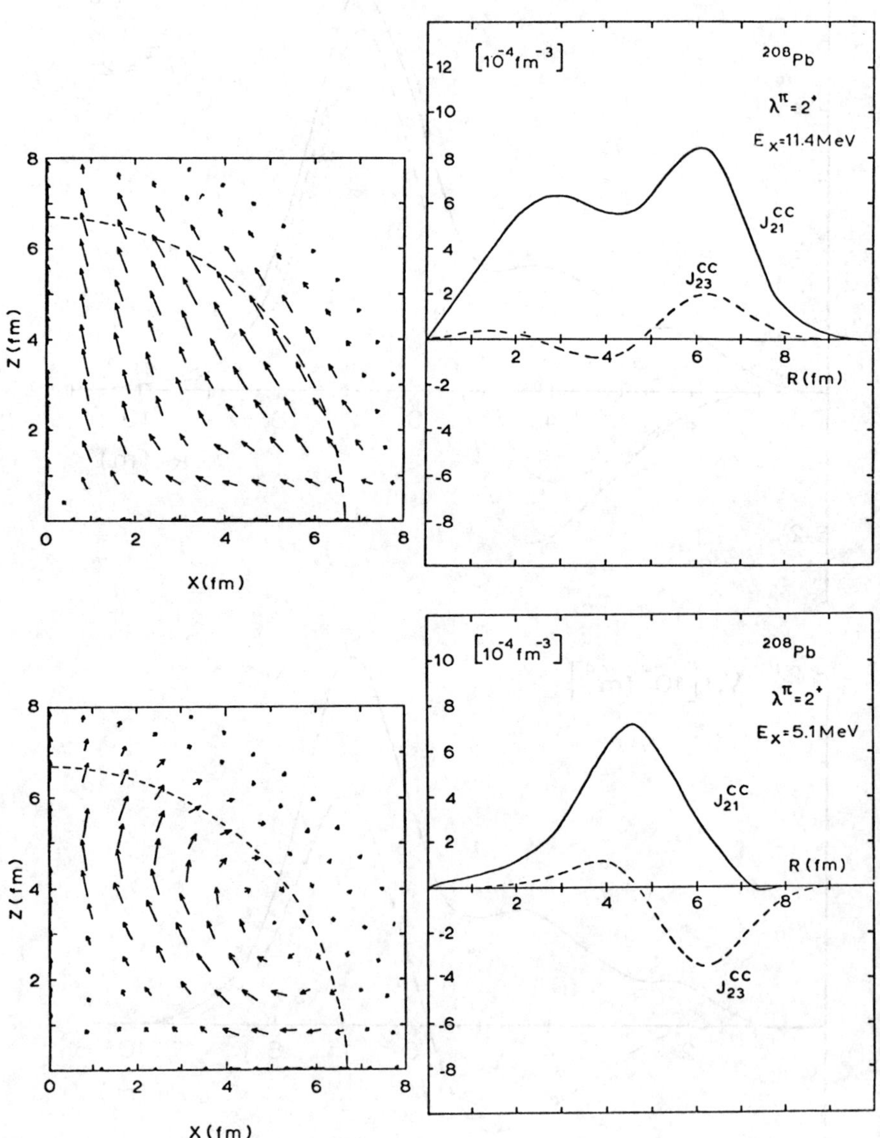

$[10^{-4}\,fm^{-3}]$
^{208}Pb
$\lambda^{\pi}=2^{+}$
$E_x=11.4\,MeV$
J_{21}^{CC}
J_{23}^{CC}
R(fm)
Z(fm)
X(fm)
$[10^{-4}\,fm^{-3}]$
^{208}Pb
$\lambda^{\pi}=2^{+}$
$E_x=5.1\,MeV$
J_{21}^{CC}
J_{23}^{CC}
R(fm)
Z(fm)
X(fm)

Fig.2

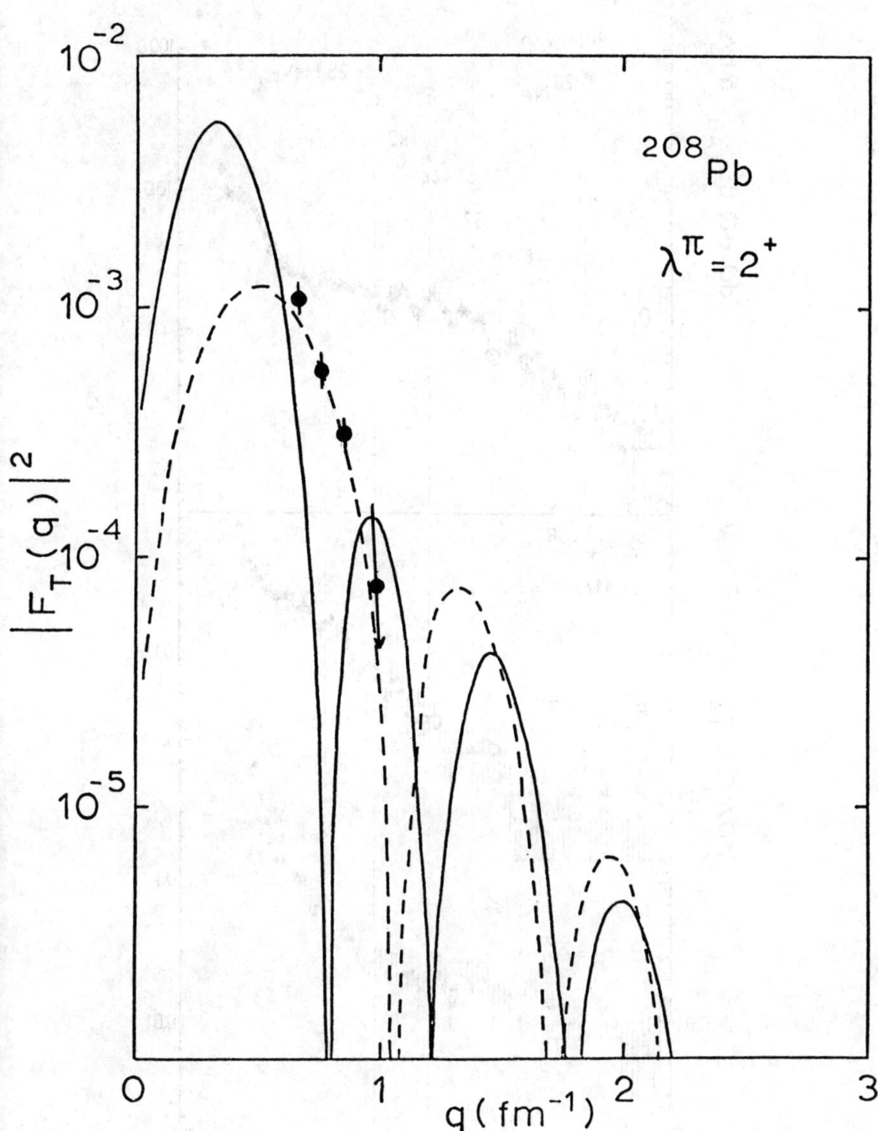

Fig. 3

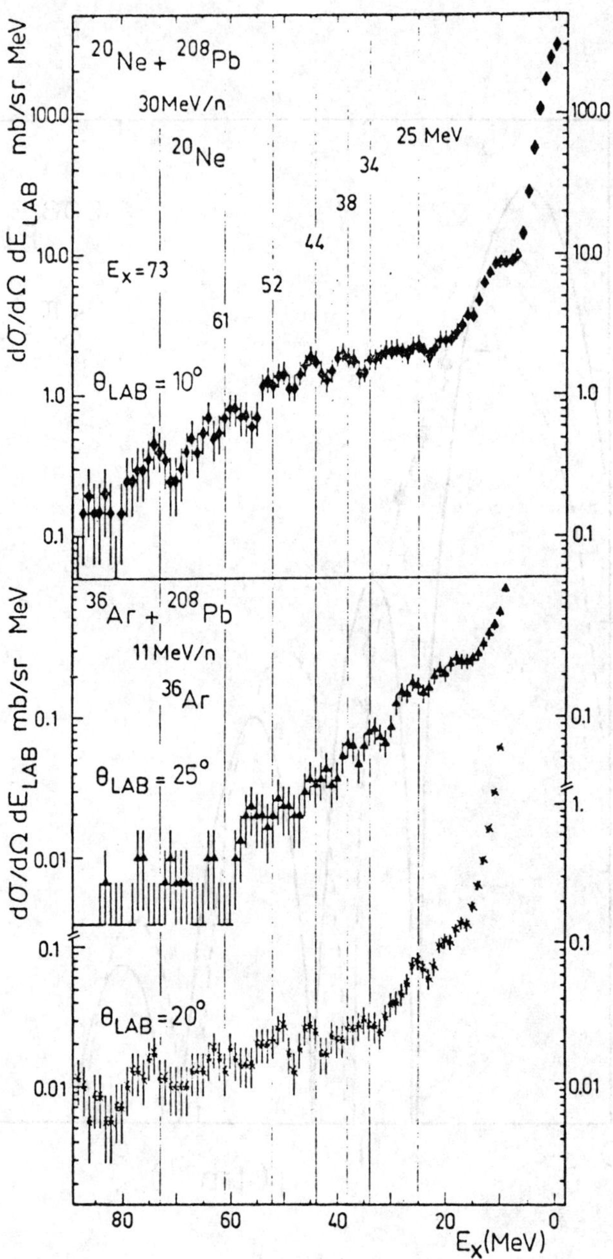
^{20}Ne + ^{208}Pb
30MeV/n
^{20}Ne
25 MeV
34
38
44
52
61
E_x=73
θ_{LAB} = 10°
dσ/dΩ dE$_{LAB}$ mb/sr MeV
100.0
10.0
1.0
0.1
^{36}Ar + ^{208}Pb
11MeV/n
^{36}Ar
θ_{LAB} = 25°
θ_{LAB} = 20°
dσ/dΩ dE$_{LAB}$ mb/sr MeV
0.1
0.01
1.
0.1
0.01
80
60
40
20
0
E_x(MeV)

Fig. 4

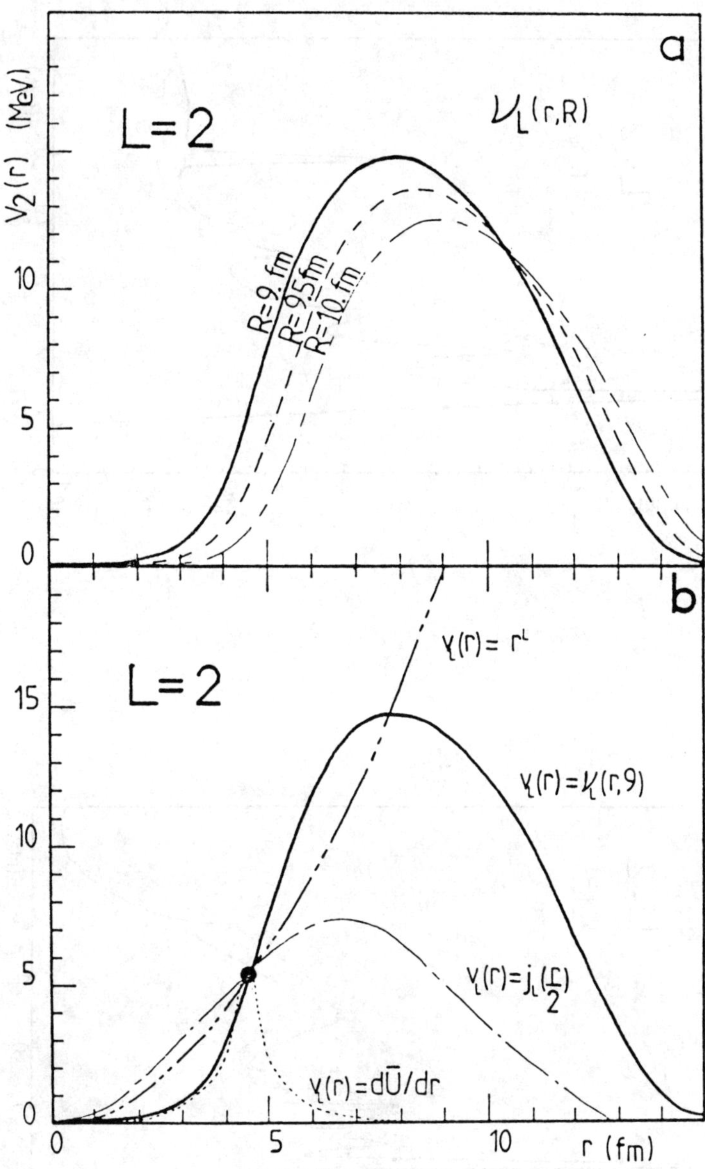
a
L= 2
$\mathcal{V}_L(r,R)$
$V_2(r)$ (MeV)
R=9. fm
R=9.5fm
R=10. fm
b
L= 2
$v_l(r) = r^L$
$v_l(r) = \mathcal{V}_l(r,9)$
$v_l(r) = j_L(\frac{r}{2})$
$v_l(r) = d\bar{U}/dr$
r (fm)

Fig. 5

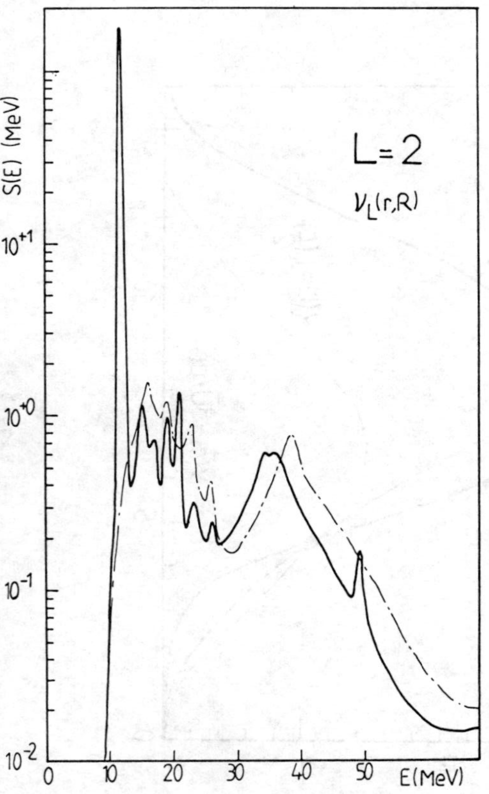

L=2
$\nu_L(r,R)$
S(E) (MeV)
10+1
10+0
10-1
10-2
0
10
20
30
40
50
E(MeV)

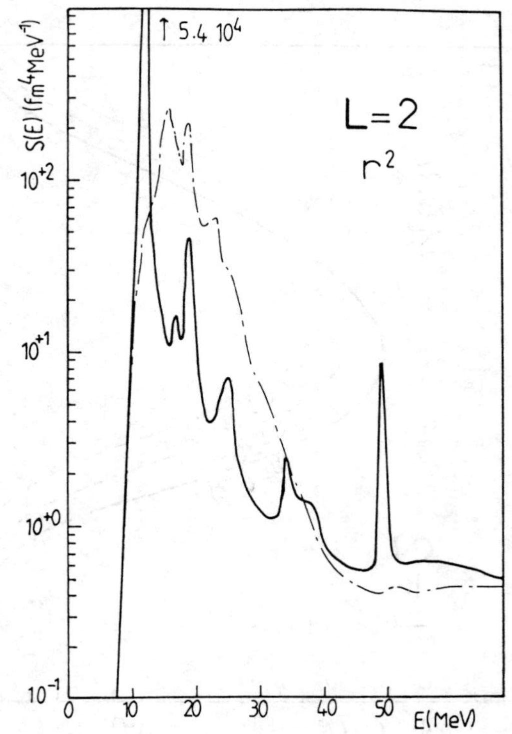

L=2
r2
↑ 5.4 10⁴
S(E) (fm4MeV-1)
10+2
10+1
10+0
10-1
0
10
20
30
40
50
E(MeV)

Fig. 6

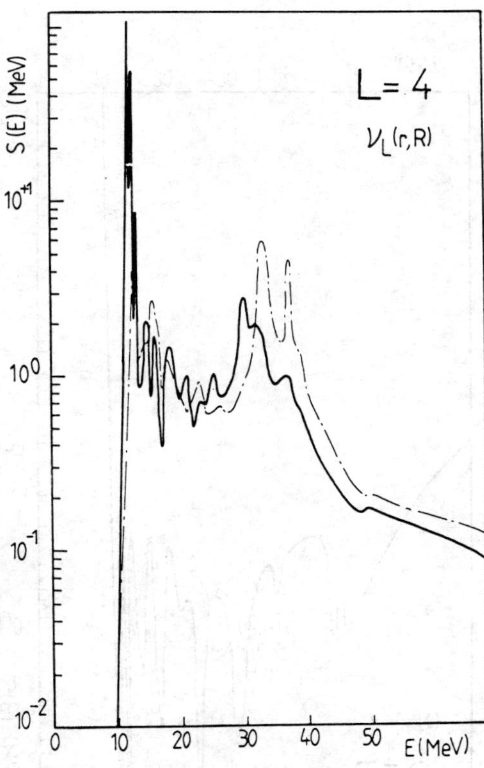
S(E) (MeV)
L = 4
$\mathcal{V}_L(r,R)$
10^{+1}
10^{0}
10^{-1}
10^{-2}
0
10
20
30
40
50
E(MeV)

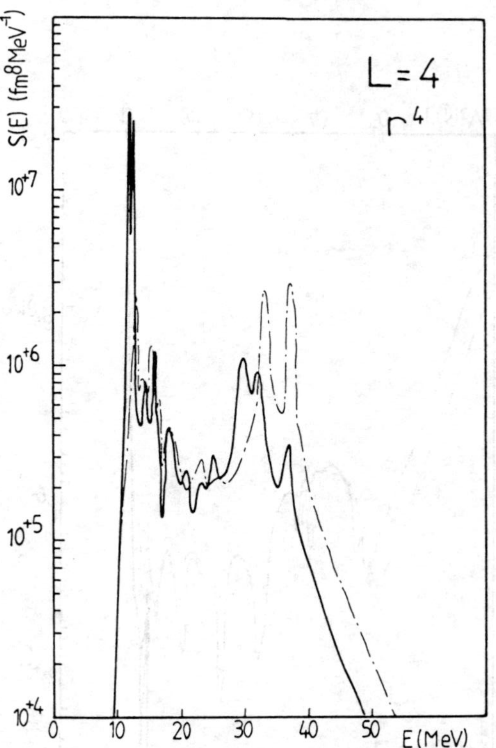
S(E) ($fm^8 MeV^{-1}$)
L = 4
r^4
10^{+7}
10^{+6}
10^{+5}
10^{+4}
0
10
20
30
40
50
E(MeV)

Fig. 7

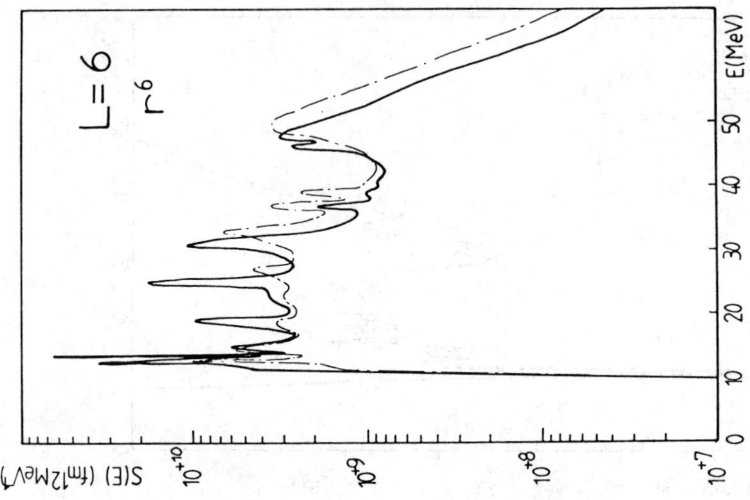
L = 6
r6
S(E) (fm12 MeV-1)
10+10
10+9
10+8
10+7
0
10
20
30
40
50
E(MeV)

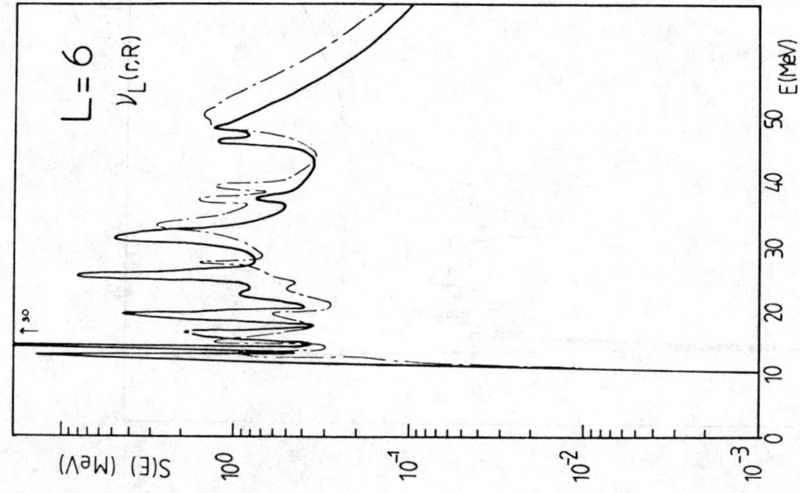
L = 6
νL(r,R)
S(E) (MeV)
↑30
100
10-1
10-2
10-3
0
10
20
30
40
50
E(MeV)

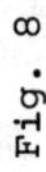
Fig. 8

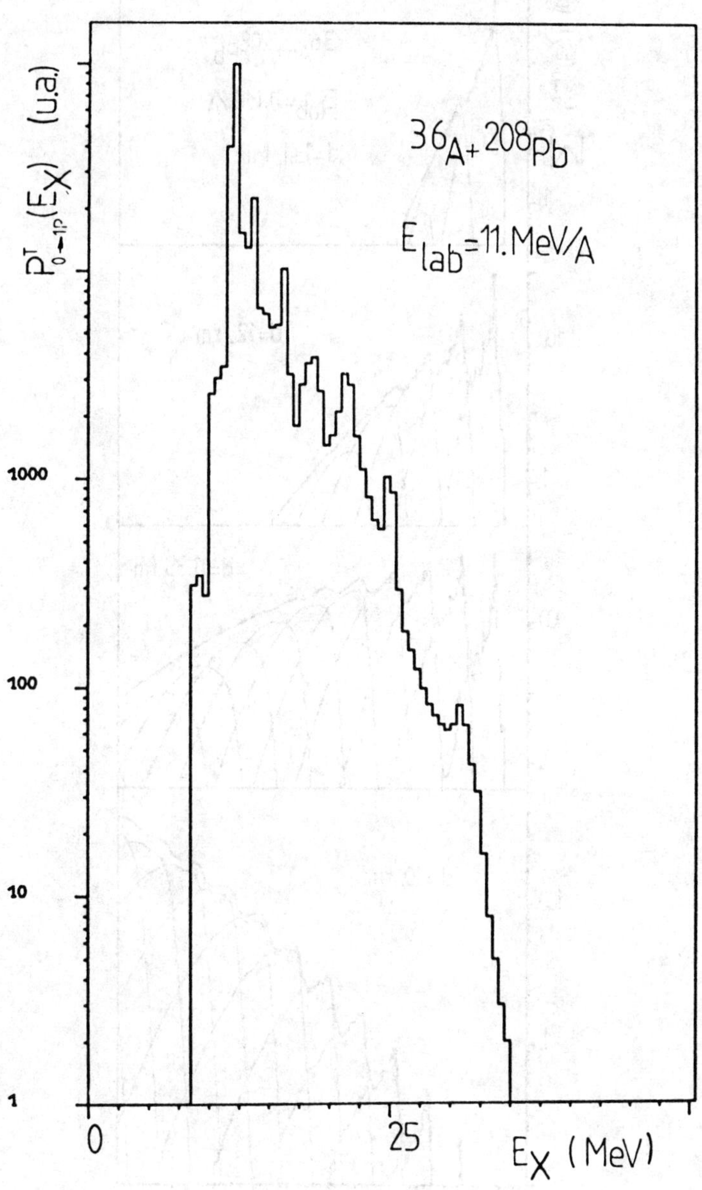

Fig. 9

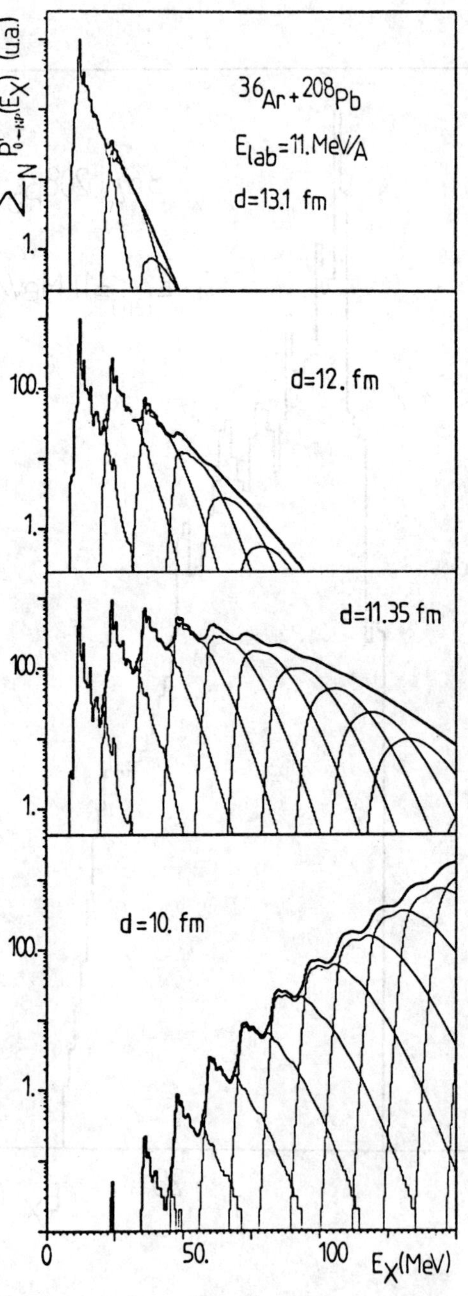
$^{36}Ar + ^{208}Pb$
E_{lab}=11.MeV/A
d=13.1 fm
d=12. fm
d=11.35 fm
d=10. fm
1.
100.
0
50.
100
E_X(MeV)

Fig. 10

SUM RULE APPROACH TO NUCLEAR COLLECTIVE MOTION

S. Stringari

Dipartimento di Fisica, Università degli Studi di Trento, 38050 POVO (TN)

ITALY

1. Introduction

In the last years the sum rule method has been extensively employed in nuclear physics in order to investigate the properties of nuclear collective excitations (see, for example, ref [1,2] and references therein). As is well known, one of the main advantages of the sum rule method is that it avoids the complete solution of the Schrödinger equation, therefore providing an alternative and direct insight on the properties of the collective motion. Of course in this series of lectures we will investigate only some aspects of the sum rule approach. In particular we will try to discuss the different kind of physics that can be learnt from the investigation of different sum rules. We will discuss in details only sum rules for electric-type excitations, the nature and the properties of magnetic transitions being less systematically known.

In sect. 2 we introduce general definitions and concepts and recall some important properties of the Hartree-Fock (HF) and random phase approximation (RPA) scheme.

In sect. 3 we discuss the inverse-energy-weighted sum rule S_{-1} and its connections with the hydrodynamic description of nuclear motion.

In sect. 3 we discuss the non-energy-weighted sum rule S_0 and show that this sum rule is particularly suited to investigate the effects of two-body correlations.

In sect. 4 we briefly recall some properties of the energy-weighted sum rule S_1 and discuss the problem of the enhancement factor for isovector excitations.

In sect. 4 we discuss the cubic-energy-weighted sum rule and its interesting link with the elastic nature of nuclear vibrations.

Finally in sect. 6 we derive sum rules for charge-exchange operators and show that also in this case the sum rule method can be successfully employed.

2. Strength function and sum rules

Let F be a general (hermitian) operator. In terms of F one can define different K-moments of the excitation strength function $S(\omega)$ associated with the operator F:

$$S_K = \int d\omega \, S(\omega) \, \omega^K \tag{1}$$

$S(\omega)$ is defined in terms of the eigenenergies ω_n and of the matrix elements of F between the ground state $|0\rangle$ and the excited states $|n\rangle$ as:

$$S(\omega) = \sum_{n \neq 0} |\langle n | F | 0 \rangle|^2 \, \delta(\omega - (\omega_n - \omega_0)) \,. \tag{2}$$

One can also write

$$S_K = \sum_{n \neq 0} |\langle n | F | 0 \rangle|^2 (\omega_n - \omega_0)^K \tag{3}$$

Evaluation of S_K is particularly interesting when the strength distribution is concentrated in a narrow region of the energy spectrum. In that case knowledge of few moments can provide useful information on important properties of the strength function $S(\omega)$, such as the peak energy, the width.....

In order to evaluate S_K one can proceed in different ways. One can evaluate explicitly the strength distribution function $S(\omega)$ starting from a given hamiltonian and solving the Schrödinger equation. Such a procedure, which gives the whole information on the excitation spectrum, is not very convenient for evaluating S_K. Alternative and more direct methods, also known as sum rule methods, have been developed to evaluate S_K. The possibility of avoiding the complete solution of the Schrödinger equations emerges naturally if one looks at eq. (3) and considers moments with $K \geq 0$. Then, by making explicit use of the closure relation $\sum_n |n\rangle\langle n| = 1$, one can find the following expressions in the case of hermitian operators (the case of non hermitian operators will be discussed in sect. 6):

$$
\begin{aligned}
S_0 &= \frac{1}{2}\langle 0|\left\{F, F\right\}|0\rangle - \left(\langle 0|F|0\rangle\right)^2 \\
S_1 &= \frac{1}{2}\langle 0|\left[F,\left[H, F\right]\right]|0\rangle \\
S_2 &= \frac{1}{2}\langle 0|\left\{\left[F, H\right], \left[H, F\right]\right\}|0\rangle \\
S_3 &= \frac{1}{2}\langle 0|\left[\left[F, H\right], \left[H, \left[H, F\right]\right]\right]|0\rangle \\
&\ldots
\end{aligned}
\tag{4}
$$

By looking at eq. (4) one observes a difference between K-odd and K-even sum rules. While the former can be completely reduced in terms of commutators, the latter involve anticommutators too. This makes evaluation of the K-even sum rules very difficult since they depend crucially on the description of the ground state $|0\rangle$. Conversely it often happens that the

K- odd sum rules are not dramatically affected by the choice of the ground state $|0\rangle$ over which one evaluates the commutators of eq. (4).

The most famous example of sum rule is known from atomic physics [3]. Consider the following hamiltonian

$$H = \sum_{i=1}^{Z} \frac{p_i^2}{2m} + \sum_{i=1}^{Z} U(r_i) + \sum_{i<j} V(r_i - r_j) \tag{5}$$

and the excitation operator

$$F = \sum_{i=1}^{Z} z_i \tag{6}$$

It is immediate to find

$$[H, F] = -\frac{i}{m} \sum_i p_i^z \tag{7}$$

and

$$S_1 = \frac{Z}{2m} \tag{8}$$

This is the well known Thomas-Reiche-Kuhn sum rule.

When K is negative it is no longer possible to reduce the moment S_K in terms of commutators or anticommutators. However some of these moments, and in particular the inverse energy weighted sum rule

$$S_{-1} = \sum_{n \neq 0} \frac{|\langle n| F |0\rangle|^2}{\omega_n - \omega_0} \tag{9}$$

can be evaluated directly, as discussed later, avoiding the calculation of $S(\omega)$.

A possible way to investigate sum rules is given by the study of the dynamic polarizability, also known as the linear response function. Let us suppose that the nucleus interacts with an external oscillating field through the interaction $-\varepsilon F \cos\omega t$ where ε is the strength of the field and F is our excitation operator. The dynamic polarizability $\alpha(\omega)$ is defined as:

$$\alpha(\omega) = \lim_{\varepsilon \to 0} \frac{\langle\psi|F|\psi\rangle - \langle 0|F|0\rangle}{\varepsilon \cos\omega t} \tag{10}$$

where $|\psi\rangle$ is the solution of the Schrödinger equation in the presence of the interaction term. $\alpha(\omega)$ is related to the excitation strength of the operator F through the relation:

$$\alpha(\omega) = 2 \sum_n \frac{\omega_n - \omega_0}{(\omega_n - \omega_0)^2 - \omega^2} |\langle n|F|0\rangle|^2 \tag{11}$$

which follows from the use of perturbation theory.

By evaluating $\alpha(\omega)$ we get information on sum rules. This follows from the expansion of eq. (11) in the limits $\omega \to 0$ and $\omega \to \infty$:

$$\alpha(\omega)_{\omega \to 0} = 2\left(S_{-1} + \omega^2 S_{-3} + \cdots\right)$$

$$\alpha(\omega)_{\omega \to \infty} = -\frac{2}{\omega^2}\left(S_1 + \frac{1}{\omega^2} S_3 + \cdots\right) \tag{12}$$

Until now we have introduced general concepts which can be applied, in principle, to very general contexts. In practice one is forced to introduce some approximation schemes in the analysis of physical problems. One of the approximation schemes that have been more extensively explored in nuclear physics in the last ten years is the HF-RPA scheme [4]. Such a scheme provides useful information on static as well as on dynamic properties of the nuclear system. The static (ground state) calculation is carried out in the Hartree-Fock approximation with the use of an effective two-body interaction. The dynamic calculation uses the RPA for which one needs the results of the HF calculation (in particular single particle energies and wave functions), in addition to particle-hole matrix elements evaluated using the same two-body interaction. Self consistency (i.e. use of the same interaction to carry out the static and the dynamic calculation) has been shown to be a crucial requirement in order to satisfy important conservation laws. An important property, which holds in the HF-RPA scheme and which essentially follows from selfconsistency, is given by eq. (11). In this case $\alpha(\omega)$ is evaluated by solving the time dependent Hartree-Fock equations while $\langle n|F|0\rangle$ and $\omega_n - \omega_0$ are matrix elements and excitation energies of the random phase approximation (F = 1-body operator) [1]:

$$\alpha_{TDHF} = \left(2 \sum_n \frac{\omega_n - \omega_0}{(\omega_n - \omega_0)^2 - \omega^2} |\langle n|F|0\rangle|^2 \right)_{RPA} \tag{13}$$

The static limit ($\omega \to 0$) of eq. (13) is particularly interesting: it clearly indicates that a static (constrained) Hartree-Fock calculation permits evaluation of the sum rule S_{-1} in the RPA approximation, thereby avoiding the explicit calculation of $S(\omega)$ [5].

In the following we will work in the framework of the HF-RPA scheme. The effective force that will be more frequently employed is the Skyrme interaction [6] . Such a force is of zero range nature and yields an expression of the type:

$$E = \int H(\vec{r})\, dv = \int \Big[\frac{\tau}{2m} + v(\rho) + a_0 (\rho\tau - \vec{J}^2) - a_1 \vec{J}_1^2 + \frac{1}{2} v_{sym}(\rho)\, \rho_1^2 + b_0 (\vec{\nabla}\rho)^2 + b_1 (\vec{\nabla}\rho_1)^2 \Big] dv \qquad (14)$$

for the energy associated with a state represented by a Slater determinant (spin orbit and Coulomb interactions as well as terms depending on τ_1 have been neglected for the sake of simplicity). The energy functional (14) depends only on a few 1-body densities:

the isoscalar and isovector densities

$$\rho(\vec{r}) = \rho_n(\vec{r}) + \rho_p(\vec{r}) = \sum_{i\sigma\tau} \varphi_i^*(\vec{r}\sigma\tau)\, \varphi_i(\vec{r}\sigma\tau)$$

$$\rho_1(\vec{r}) = \rho_n(\vec{r}) - \rho_p(\vec{r}) = \sum_{i\sigma\tau} \tau\, \varphi_i^*(\vec{r}\sigma\tau)\, \varphi_i(\vec{r}\sigma\tau) \qquad (15)$$

the kinetic energy density

$$\tau(\vec{r}) = \tau_n(\vec{r}) + \tau_p(\vec{r}) = \sum_{i\sigma\tau} \vec{\nabla}\varphi_i^*(\vec{r}\sigma\tau)\, \vec{\nabla}\varphi_i(\vec{r}\sigma\tau) \qquad (16)$$

and the isoscalar and isovector current densities

$$\vec{J}(\vec{r}) = \vec{J}_n(\vec{r}) + \vec{J}_p(\vec{r}) = \frac{1}{2i}\sum_{i\sigma\tau}\left(\psi_i^*(\vec{r}\sigma\tau)\vec{\nabla}\psi_i(\vec{r}\sigma\tau) - \vec{\nabla}\psi_i^*(\vec{r}\sigma\tau)\psi_i(\vec{r}\sigma\tau)\right)$$

$$\vec{J}_1(\vec{r}) = \vec{J}_n(\vec{r}) - \vec{J}_p(\vec{r}) = \frac{1}{2i}\sum_{i\sigma\tau}\tau\left(\psi_i^*(\vec{r}\sigma\tau)\vec{\nabla}\psi_i(\vec{r}\sigma\tau) - \vec{\nabla}\psi_i^*(\vec{r}\sigma\tau)\psi_i(\vec{r}\sigma\tau)\right) \quad (17)$$

In eq. (15)-(17) ψ_i is i-th single particle wave function with which we construct the Slater determinant. The combination $(\rho\tau - \vec{J}^2)$ in eq. (14) ensures the respect of the Galilean invariance of the force (see exercise 2).

It should be stressed that expression (14) represents a very special and particular form for the energy functional. When the effective interaction has a finite range its dependence on the 1-body density matrix $\rho^{(1)}(\vec{r}, \vec{r}') = \sum_i \psi_i^*(\vec{r}')\psi_i(\vec{r}'')$ becomes much more complicate.

From eq. (14) one can derive the HF equations for the nuclear ground state

$$H_{HF}\psi_i = \frac{\delta E}{\delta\psi_i^*} = \left(-\vec{\nabla}\frac{1}{2m^*(\vec{r})}\vec{\nabla} + U(\vec{r})\right)\psi_i = e_i\psi_i \quad (18)$$

In deriving eq. (18) we have assumed N = Z nuclei $(\rho_1 = 0)$. Equation (18) has the form of a <u>local</u> Schrödinger equation with an effective mass m^* which depends on the density only

$$\frac{1}{2m^*(\vec{r})} = \frac{1}{2m} + a_0\,\rho(\vec{r}) \quad (19)$$

whereas the potential $U(\vec{r})$ also depends on the kinetic energy density:

$$U(\vec{r}) = \frac{\partial v}{\partial \rho} + a_0 \tau - 2 b_0 \nabla^2 \rho \tag{20}$$

<u>Exercise 1.</u> Assume $F = \sum_{i=1}^{Z} f(\vec{r}_i)$ and consider hamiltonian (5). By explicitly evaluating the double commutator $[F[H F]]$, verify the result $S_1 = \frac{1}{2m} \int (\vec{\nabla} f)^2 \rho_0(\vec{r}) dv$ where ρ_0 is the (diagonal) 1-body density of the ground state $|0\rangle$.

<u>Exercise 2.</u> Consider the Slater determinant $|f\rangle = e^{i \sum_i f(\vec{r}_i)} |0\rangle$ obtained by applying the unitary transformation $e^{i \sum_i f(\vec{r}_i)}$ to the Slater determinant $|0\rangle$. By explicitly evaluating the changes in ρ, τ and $\vec{J}$ induced by such a transformation and by using eq. (14), verify that the potential energy, differently from the kinetic energy, does not change (local Galilean invariance). [Hint: the single particle wave functions φ_i^f of $|f\rangle$ are given by $\varphi_i^f(\vec{r}) = e^{i f(\vec{r})} \varphi_i^0(\vec{r})$ where $\varphi_i^0(\vec{r})$ are the single particle wave functions of $|0\rangle$] .

3. The polarizability sum rule

The sum rule

$$S_{-1} = \sum_{n\neq 0} \frac{|\langle n|F|0\rangle|^2}{\omega_n - \omega_0} \tag{21}$$

has been rather extensively investigated using the constrained Hartree-Fock (CHF) method. In practice one determines the HF ground state with the hamiltonian

$$H = H_0 - \varepsilon F \tag{22}$$

and uses the relation (see eq. (13))

$$\alpha_{CHF} = 2\, S_{-1}^{RPA} \quad . \tag{23}$$

If one uses the energy density formalism, the CHF method is equivalent to minimizing the quantity

$$E_{tot} = \int H(\vec{r})\, dv - \varepsilon \langle F\rangle \tag{24}$$

where $\langle F\rangle$ is $\int f\rho\, dv$ and $\int f\rho_1\, dv$ for isoscalar $\left(F = \sum_i f(\vec{r}_i)\right)$ and isovector $\left(F = \sum_i f(\vec{r}_i)\tau_i^3\right)$ operators respectively and $H(\vec{r})$ is defined in eq. (14).

The CHF method can be safely applied only by a proper choice of the excitation oeprator F. In fact if F excites low-lying vibrations, then evaluation of S_{-1} becomes very delicate. The CHF method has been employed for the monopole case (isoscalar [2] and isovector [7]) where, due to the nature of the excitation operator, the HF potential preserves its sphericity, and more recently for the isovector dipole excitation [8].

A method alternative to the CHF approach is given by the use of semi-classical models, by means of which one avoids the calculation of the single particle wave functions. The use of the Thomas-Fermi approximation (together with its extensions [9]) is at the basis of such approaches. The Thomas-Fermi approximation essentially fixes a relation between the kinetic energy density and the diagonal density:

$$\tau_n^{TF}(\vec{r}) = \frac{3}{5}(3\pi^2)^{\frac{2}{3}} \rho_n^{\frac{5}{3}}(\vec{r}) \quad , \quad \tau_p^{TF}(\vec{r}) = \frac{3}{5}(3\pi^2)^{\frac{2}{3}} \rho_p^{\frac{5}{3}}(\vec{r}) \quad . \tag{25}$$

Such a relation (as well as its extensions) allows us to write the kinetic energy term and the other terms of the force depending on τ, as a function of the diagonal density only. The variational calculation then can be carried out with respect to the density rather than with respect to the single particle wave functions. Of course use of approximation (25) should be justified. A considerable effort has been devoted to justify semiclassical methods in static calculations [10]. Here we recall that the validity of the Thomas-Fermi approximation in a constrained calculation can be proved in infinite systems only when the spatial modifications induced on the density are smooth (limit of long wave length). Such a result has been first derived in the framework of the Landau's theory of Fermi liquids [11]. Use of the Thomas-Fermi approximation is at the basis of the hydrodynamic description of nuclear collective motions.

One can introduce changes in the densities ρ_n and ρ_p in terms of the displacement field $\vec{u}$ as:

$$\rho_n = \rho_n^0 + \vec{\nabla}(\vec{u}_n \rho_n^0) + \frac{1}{2} \nabla^k (u_n^k \nabla^\ell (u_n^\ell \rho_n^0)) + \cdots$$

$$\rho_p = \rho_p^0 + \vec{\nabla}(\vec{u}_p \rho_p^0) + \frac{1}{2} \nabla^k (u_p^k \nabla^\ell (u_p^\ell \rho_p^0)) + \cdots \tag{26}$$

The isoscalar case is obtained by putting $\vec{u}_n = \vec{u}_p = \vec{u}$:

$$\rho = \rho_0 + \vec{\nabla}(\vec{u} \rho_0) + \frac{1}{2} \nabla^k (u^k \nabla^\ell (u^\ell \rho^0))$$

$$\rho_1 = 0 \tag{27}$$

while the isovector one is obtained by putting $\vec{u}_n = -\vec{u}_p = \vec{u}_1$:

$$\rho = \rho_0 + \frac{1}{2} \nabla^k (u_1^k \nabla^\ell (u_1^\ell \rho^0)) + \cdots$$

$$\rho_1 = \vec{\nabla}(\vec{u}_1 \rho^0) + \tag{28}$$

It is interesting to notice that the isoscalar density changes (at the second order) also during an isovector oscillation.

Assuming the Thomas-Fermi relation (25) and using equations (27) and (28) one finds the following result for the energy in the presence of the external field (see eqs. (14) and (24)):

$$\delta E = \frac{1}{18} K \int (\vec{\nabla} \vec{u})^2 \rho_0 \, dv - \varepsilon \int \vec{\nabla}(\vec{u} \rho_0) f \, dv \tag{29}$$

for the isoscalar case and

$$\delta E = \frac{1}{2} b_{vol} \int (\vec{\nabla} \vec{u}_1)^2 \rho_0 \, dv + E^v_{surf}(\vec{u}_1) - \varepsilon \int \vec{\nabla}(\vec{u}_1 \rho_0) f \, dv \tag{30}$$

for the isovector one. In deriving eqs. (29) and (30) we have neglected some surface terms which, differently from the term E^{v}_{surf} (which will be discussed later), can be ignored in the limit of large systems if one investigates compression modes.

There are some important differences between the isoscalar and the isovector cases. First at all the bulk parameters characterizing the energy variation are different in the two cases, as expected. In the isoscalar case the relevant parameter is the nuclear matter incompressibility:

$$K = 9\rho \frac{\partial}{\partial \rho}\left(\rho \frac{\partial}{\partial \rho}\left(\frac{v(\rho)}{\rho} + \frac{1}{2m^*(\rho)} \frac{\tau(\rho)}{\rho}\right)\right) . \tag{31}$$

Conversely, in the isovector case the relevant parameter is the volume symmetry energy

$$b_{vol} = v_{sym}(\rho)\rho + \frac{5}{9} \frac{1}{m^*(\rho)} \frac{\tau(\rho)}{\rho} \tag{32}$$

Another important difference between the two cases is due to the presence of the surface term in the isovector energy (30). Such a term has a particularly simple form in the case of large systems:

$$E^{v}_{surf} = \frac{1}{2}\int \left(v_{sym}(\rho) - \frac{\partial^2 v(\rho)}{\partial \rho^2}\right)\left(\vec{u}_1 \cdot \vec{\nabla}\rho\right)^2 dr \tag{33}$$

and, except for special choices of $\vec{U}_1$, gives most of the contribution to the isovector potential energy δE . In the case of Goldhaber-Teller deformations $\vec{U} = \vec{\nabla}\chi$ only this term would contribute to δE . The presence of the surface term (33), which physically gives the nuclear resistence against the formation of a neutron (proton) skin, strongly affects the solutions of the equations of motion. In the limit of large systems, the solutions with the lowest energy are found when the surface contribution (33) vanishes. This requirement is equivalent to imposing the vanishing of the radial component of the displacement field at the surface (Steinwedel-Jensen condition [12]):

$$\vec{U}_1 \cdot \vec{n} \Big|_{Surf} = 0 \qquad \text{(isovector)} \tag{34}$$

where $\vec{n}$ is the unit vector in the $\vec{r}$-direction. Using condition (34) one can write the interaction term of eq.(30) in a more convenient form

$$\varepsilon \int \vec{\nabla}(\vec{U}_1 \rho_0)\, f\, dv = \varepsilon \int \vec{\nabla}\vec{U}_1\, f\, \rho_0\, dv \tag{35}$$

Also in the isoscalar case one has to impose some constraints to the form of the displacement field. Eq. (29) has been in fact derived by neglecting some surface terms which are normally negligible, except when surface deformations of the density are involved. It then follows that eq. (29) can be used only for investigating the compression modes of the system. Such a restriction is in practice obtained by imposing that the excitation operator f vanishes at the surface [13]:

$$f(\vec{r})\big|_{surf} = 0 \qquad \text{(isoscalar)} \qquad (36)$$

Condition (36) clearly indicates that with eq. (29) we cannot explore the polarizability sum rule for operators of the type $F = \sum_i r_i^\lambda Y_{\lambda m}$. For such excitations the polarizability sum rule is very sensitive to low-lying vibrations as well as to quantum shell effects.

Because of eq. (36) also in the isoscalar case one can express the interaction term of eq. (30) in terms of $\vec{\nabla}\cdot\vec{u}$:

$$\varepsilon\int \vec{\nabla}(\vec{u}\,\rho_0)\,f\,dv = \varepsilon\int \vec{\nabla}\vec{u}\; f\,\rho_0\,dv \qquad (37)$$

Having expressed all the terms in the energy as a function of $\vec{\nabla}\cdot\vec{u}$, minimization of δE is now straightforward. One gets

$$\vec{\nabla}\vec{u} = 9\,\frac{\varepsilon}{\chi}\,f \qquad (38)$$

for the isoscalar case and

$$\vec{\nabla}\vec{u}_1 = \frac{\varepsilon}{b_{vol}}\,f \qquad (39)$$

for the isovector one. From eqs. (36) and (38) one recovers the usual boundary condition for compression modes [14] :

$$\vec{\nabla}\vec{u}\,\Big|_{surf} = 0 \qquad \text{(isoscalar)} \qquad (40)$$

In the isovector case it is possible to see that condition (34) for the displacement field implies the following condition for f:

$$\langle f \rangle = \int f \rho_0 \, dv = 0 \qquad \text{(isovector)} \qquad (41)$$

Such a condition does not imply any physical restriction for the excitation operator.

In conclusion, provided that conditions (36) and (41) are satisfied, one gets the following results for S_{-1} [13]:

$$S_{-1} = \frac{9}{2}\frac{1}{K}\int f^2 \rho_0 \, dv \qquad (42)$$

and

$$S_{-1} = \frac{1}{2}\frac{1}{b_{vol}}\int f^2 \rho_0 \, dv \qquad (43)$$

for isoscalar and isovector excitations respectively. Result (43) was first derived by Migdal [15] in the case of dipole excitations (f = z):

$$S^{M}_{-1} = \frac{1}{6}\frac{A}{b_{vol}}\langle r^2 \rangle \qquad (44)$$

Result (42) for the isoscalar monopole operator ($f = r^2 - R^2$) has been first discussed in ref. [16] .

Clearly results (42) and (43) hold only in the limit of very large systems $(A \to \infty)$. In finite nuclei surface effects cannot be neglected and can significantly change the predictions of eqs (42), (43). The importance of such effects in the evaluation of the sum rule S_{-1} has been explored both in the isoscalar [17] and in the isovector case. In the isovector case an analytical generalization of eq(43), derived in the framework of the hydrodynamic model, has been proposed in ref. [18]. Surface [19a] as well as curvature effects [19b] in the dipole case have been investigated in the framework of the liquid doplet model too. By keeping only surface corrections, generalization of eq. (44) becomes:

$$S_{-1} = S_{-1}^{M} \left(1 + \frac{5}{3} \frac{b_{surf}}{b_{vol}} A^{-\frac{1}{3}} \right) \tag{45}$$

where b_{surf} is the surface contribution to the symmetry energy coefficient entering in the semi-empirical mass formula. S_{-1} (dipole) is known experimentally from photoreaction data [20]. One then can explicitly appreciate the importance of surface effects. In table 1 we report the Migdal prediction (eq. (44)), the prediction of eq. (45) together with the experimental values ($\sigma_{-2} = 4\pi^2 e^2 S_{-1}$).

	$\sigma_{-2}A^{-5/3}$ (μb/MeV)		
	Migdal	theory	exp.
^{16}O	1.9	4.8	5.8
^{40}Ca	1.8	4.0	4.8
^{140}Ce	1.6	2.8	2.5
^{208}Pb	1.5	2.5	2.6

Table 1

As concerns the curvature contribution, in ref. [19] it has been shown that such an effect is not negligible (10∼15%) and tends to increase the value of S_{-1}, in agreement with the RPA results of ref. [8].

In conclusion we can say that evaluation of the sum rule S_{-1} provides a useful tool to investigate the properties of nuclear excitations of isovector and isoscalar compression nature. The present discussion has also indicated that hydrodynamics is not an useless theory in nuclear physics, though general considerations based on the comparison between the frequency of the collisions between particles and the frequency of giant resonances state that thermalization cannot occurr during the nuclear motion. In the case of isoscalar divergency free motions, excited by operators $F = \sum_i r_i^\lambda Y_{\lambda m}$ evaluation of S_{-1} is extremely sensitive to the low-lying part of the excitation spectrum and consequently contains only partial information on the giant state we are interested in. For these excitations the above considerations on the absence of thermalization during the nuclear oscillation have dramatic consequences and reveal the drawback of the hydrdynamic approximation as we will discuss in the next sections.

Exercise 3. Derive results (29) and (30) using expressions (27) and (28) for the density changes [hint: make use of the saturation condition for the ground state $\frac{\partial}{\partial\rho}\left(\frac{E}{A}\right) = 0$].

Exercise 4. In ref. [13] the surface term [33] has been replaced by the phenomenological expression $E^{v}_{surf} = \frac{4}{3}\frac{\rho_0}{r_0} Q \int_{surf} (\vec{U}_1 \cdot \vec{n})^2$. Using this expression derive result (45) for the dipole polarizability $\left(\frac{b_{surf}}{b_{vol}} = \frac{9}{4}\frac{J}{Q} ;\ b_{vol} = 2J \right)$

4. The non-energy weighted sum rule

The non energy weighted sum rule can be written as:

$$S_o = \int S(\omega) d\omega = \sum_{m \neq 0} |\langle m|F|0\rangle|^2 = \langle 0 F F 0\rangle - |\langle 0 F 0\rangle|^2 . \quad (46)$$

Evaluation of S_o is rather difficult because it requires knowledge of the 2-body density matrix of the ground state $|0\rangle$. The interest in evaluating S_o is clearly related to the possibility of learning something about the nature of the 2-body correlations in the ground state. In the case of electron scattering (longitudinal case) the relevant sum rule is:

$$S^c(q) = \sum_{m \neq 0} |\langle m| \sum_i e_i e^{i\vec{q}\cdot\vec{r}} |0\rangle|^2 \quad (47)$$

$S^c(q)$ is directly connected with the two body proton-proton correlation function through the relation

$$S^c(q) = \int e^{i\vec{q}\cdot\vec{s}} \rho^{(2)}_{p,p}(s) - Z^2 |F(q)|^2 + Z \quad (48)$$

where $F(q)$ is the elastic form factor $(F(0) = 1)$.

The idea of investigating $\rho^{(2)}_{pp}$ through measurements of integrated electron scattering cross sections was proposed by Gottfried a long time ago [21]. However only recently have such experiments become available [22]. Some of these experiments [22 a,b,c] have focused on the momentum transfer range q=1-2 fm^{-1}. This region is expected to be sensitive to the presence of short range as well as tensor correlations in the nuclear wave function. In the region at lower momentum transfer ($q < 1\ fm^{-1}$) the electron scattering cross section is expected to depend on the presence of long range correlations. Such correlations are responsible for the collective phenomena (giant resonances) exhibited by nuclei and their investigation is the object of these lectures.

S_o has been recently evaluated for the dipole operator $F = \sum_i z_i r_i^3$ by explicitly integrating the RPA strength [23]. Comparison of the RPA result with the uncorrelated value for S_o, obtained using the HF 2-body density, reveals the importance of dipole correlations which decrease S_o by 25 ~ 40% in good agreement with experimental data [20].

The effects of dipole correlations on the electron scattering sum rule (47) have been recently investigated by carrying out a RPA calculation in the framework of the schematic model using a dipole-dipole separable force [24]. Such a model provides a simplified inclusion of the RPA correlations and allows for a fully analytical solution of the RPA equations. The schematic hamiltonian is given by

$$H = H_o + \frac{1}{2}\chi_1 \vec{D}\cdot\vec{D} \tag{49}$$

where H_o is the harmonic oscillator hamiltonian (the value of the harmonic oscillator parameter $\alpha = \sqrt{m\omega_o}$ is fixed to reproduce the nuclear root mean square radius), $\vec{D} = \sum_i \vec{r}_i r_i^3$ is the dipole operator and χ_1 is a parameter which gives the amount of the dipole correlations. χ_1 is fixed in order to reproduce the experimental value of the dipole strength through the relation :

$$\sigma_{-1} = 60 \frac{NZ}{A} \frac{1}{\omega_c} \frac{1}{\xi} \tag{50}$$

where $\xi = \sqrt{1 + \frac{A}{m\omega_c^2} \chi_1}$. A reasonable value for ξ turns out to be $\sim$ 1.5.

$S^C(q)$ is clearly affected by the presence of dipole correlations as can be shown by the RPA result

$$S^C_{RPA}(q) = S^C_{HO}(q) + \frac{1}{4} Z \frac{q^2}{\alpha^2} |F_{HO}(q)|^2 \left(\frac{1}{\xi} - 2\right) \tag{51}$$

where S^C_{HO} is the uncorrelated harmonic oscillator prediction. In deriving eq. (51) we have taken into account also corrections due to the center of mass motion. From eq. (51) one can show that dipole correlations $(\xi > 1)$

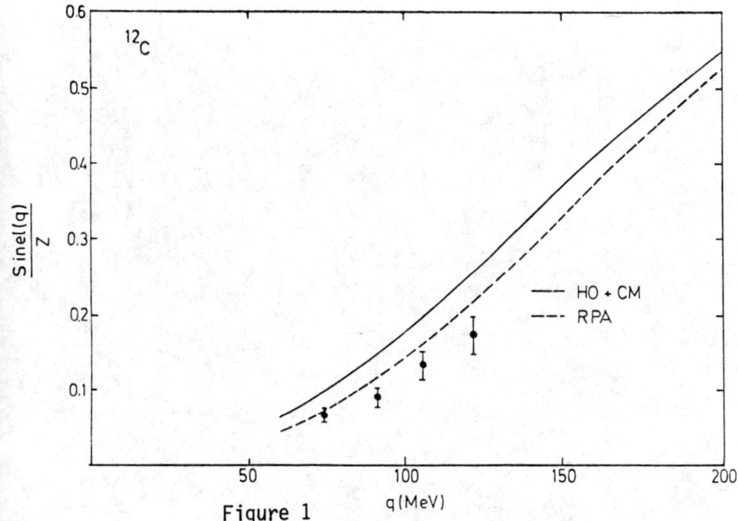

Figure 1

tend to quench the longitudinal strength. Fig. 1 shows the predictions of the RPA schematic model (including transverse corrections) together with recent experimental results for inelastic electron scattering cross sections in ^{12}C [22 d] .

Exercise 5. Result (51) has been derived taking into account center of mass corrections. This can be done by adding a term $\frac{1}{2} \chi_o \sum_i \vec{r}_i \cdot \sum_j \vec{r}_j$ to eq. (49). Determine the value of the coupling constant χ_o. Is the center of mass correction important for the evaluation of $S^c(q)$?

5. The energy-weighted sum rule

This sume rule can be evaluated in terms of a double commutator involving the nuclear hamiltonian. A theorem, due to Thouless [25], ensures that if one evaluates such a commutator on the Hartree-Fock ground state, then S_1 is evaluated with RPA accuracy. Starting from the energy functional (14) one can alternatively evaluate S_1 using the following relation:

$$S_1 = \frac{\partial^2}{\partial \nu^2} \langle \nu | H | \nu \rangle \Big|_{\nu=0} \qquad (52)$$

where $|\nu\rangle = e^{i\nu F}$. Transformation $e^{i\nu F}$ introduces changes in the densities of the HF ground state which modify the expectation value of H. If one evaluates these changes up to second order in ν one finds (see also exercise 2):

$$\begin{aligned} \vec{J}(\vec{r}) &= \nu \vec{\nabla} f \rho_0 \\ \tau(\vec{r}) &= \tau_0 + \nu^2 (\vec{\nabla} f)^2 \rho_0 \end{aligned} \qquad \text{(isoscalar)} \qquad (53)$$

and

$$\begin{aligned} \vec{J}_1(\vec{r}) &= \nu \vec{\nabla} f \\ \tau(\vec{r}) &= \tau_0 + \nu^2 (\vec{\nabla} f)^2 \rho_0 \end{aligned} \qquad \text{(isovector)} \qquad (54)$$

Inserting these results in eq. (14) and using result (52) one finally obtaines:

$$S_1 = \frac{1}{2m} \int (\vec{\nabla} f)^2 \rho_0 \, dr \qquad \text{(isoscalar)} \qquad (55)$$

$$\begin{aligned} S_1 &= \frac{1}{2m} \int (\vec{\nabla} f)^2 (1 + 2m(a_0 - a_1)\rho_0) \rho_0 \, dv = \qquad \text{(isovector)} \\ &= \frac{1}{2m} \int (\vec{\nabla} f)^2 \rho_0 \, dv \quad (1 + K) \end{aligned}$$

It is important to notice that while in the isoscalar case the local Galilean invariance ensures the vanishing of the potential contribution to S_1, in the isovector one such a contribution can be different from zero.

The value of the enhancement factor K is known experimentally in the dipole case from measurements in photonuclear reactions [20,26]. Its value is 0.5 - 0.8 when the cross section is integrated up to 140 MeV. However one should remark that an important contribution to the experimental values of K comes from the high energy region of the excitation spectrum, where short range and tensor correlations are known to play a crucial role. For this reason it is not clear what is the best value for K one should get from RPA calculations which cannot account for such correlations. A possibility is to require the RPA calculations to reproduce the dipole cross sections only in the region of the giant dipole resonance and consequently give a value of K corresponding to the integral of the cross section up to 25-30 MeV in ^{208}Pb (K = 0.2-0.3).

<u>Exercise 6</u>. Evaluate the enhancement factor K in the HF approximation (i.e. using the Hartree-Fock hamiltonian (18)). Show that one gets a wrong result .

<u>Exercise 7</u>. Evaluate the mean excitation energy for the isovector dipole and quadrupole operators $F = \sum_i z_i \tau_i^3$ and $F = \sum_i x_i y_i \tau_i^3$

as $\omega = \sqrt{S_1/S_{-1}}$ using eq. (45) for S_{-1}. Show that surface effects are important and modify the usual $A^{-1/3}$ law predicted by the Steinwedel -Jensen model (take $b_{vol}(1+K) = 70$ MeV and $\frac{b_{surf}}{b_{vol}} = 2$). The results for ω are shown in fig. 2 .

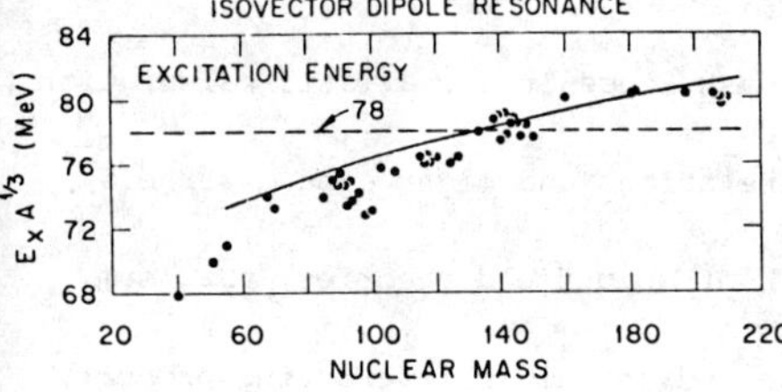

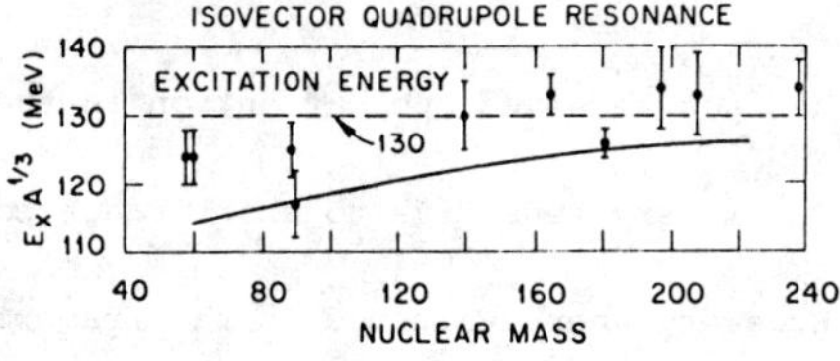

Figure 2

6. The cubic-energy-weighted sum rule

This sum rule, which can be written as:

$$S_3 = \frac{1}{2}\langle 0|\left[\left[F, H\right], \left[H, \left[H, F\right]\right]\right]|0\rangle \tag{56}$$

has not received much attention in the past since in a realistic calculation it is very sensitive to short range correlations and might even diverge. However, when evaluated in the framework of mean field theories, this sum rule has a clear physical interpretation and reveals an interesting property of Fermi systems, i.e. the elastic nature of their vibrations. To explore this point in more detail it is convenient to write S_3 in the following way [1,2] :

$$S_3 = \frac{\partial^2}{\partial\nu^2}\langle \nu | H | \nu \rangle\Big|_{\nu=0} \tag{57}$$

where

$$|\nu\rangle = e^{\nu[H, F]}|0\rangle \tag{58}$$

The sum rule S_3 can be regarded as the restoring force associated with transformation (58). Evaluation of $[H, F]$ is straightforward for isoscalar excitations, while for isovector excitations it can be evaluated in the RPA by projecting the 2-body terms on its 1p-1h components (see ref. 1). One can write:

$$[H,F] = -\frac{i}{2m} \sum_i \left(\vec{\nabla}f \cdot \vec{p}_i + h.c\right) \quad \text{(isoscalar)}$$

$$[H,F] = -\frac{i}{2m} \sum_i \left(\vec{\nabla}f\left(1+2m(a_0-a_1)\rho_0\right) \cdot \vec{p}_i + h.c.\right) \quad \text{(isovector)} \tag{59}$$

In both cases the transformation $e^{\eta[H,F]}$ looks like a scaling transformation acting on the ground state. One can construct very general scaling transformations in terms of the displacement field $\vec{u}$:

$$|\psi\rangle = e^{i\frac{1}{2}\sum_i(\vec{u}(\vec{r}_i)\cdot\vec{p}_i + h.c.)t_i}|0\rangle \tag{60}$$

where t_i is equal to 1 for isoscalar excitations and is equal to τ_i^3 for isovector excitations. Transformation (60) provides the microscopic basis for a fluid-dynamic description of nuclear motion [27-29] .

It is interesting to evaluate the energy change due to such a transformation up to terms quadratic in $\vec{u}$. In the limit of large systems ($A \to \infty$) one gets the following simplified result:

$$\delta E = \int_{vol} \left(\frac{1}{2}\lambda(\vec{\nabla}\cdot\vec{u})^2 + \frac{1}{4}\mu(\nabla_k u_l + \nabla_l u_k)^2\right) + E^v_{surf}(\vec{u}) \tag{61}$$

where E^v_{surf} (see eq. (33)) should be considered only in the isovector case. Equation (61) has the typical form of the potential energy of an elastic medium. The Lamè's coefficient λ and μ are connected to bulk properties of the nuclear system $\left(\varepsilon_F = \frac{K_F^2}{2m^*}\right)$:

$$\lambda = \begin{cases} \left(\frac{1}{9}K - \frac{4}{15}\varepsilon_F\right)\rho_0 & \text{(isoscalar)} \\ \left(b_{vol} - \frac{4}{15}\varepsilon_F\right)\rho_0 & \text{(isovector)} \end{cases} \tag{62}$$

$$\mu = \frac{2}{5}\varepsilon_F \rho_0$$

Equation (61) suggests the idea that nuclei can vibrate as an elastic medium rather than a fluid [27] . In terms of δE the sum rule S_3 can be evaluated through:

$$S_3 = \begin{cases} \delta E\left(\vec{u} = \frac{1}{m}\vec{\nabla} f\right) & \text{(isoscalar)} \\ \delta E\left(\vec{u} = \frac{1}{m}\vec{\nabla} f\,(1 + 2m(a_0 - a_1)\rho_0)\right) & \text{(isovector)} \end{cases} \tag{63}$$

The importance of the elastic effect becomes particularly evident in the case of divergency free motions excited by the operator $F = \sum_i r_i^{\lambda} Y_{\lambda m}$. In this case $\vec{\nabla}\vec{u}$ is equal to zero and the only contribution to the restoring force (61), and hence to S_3, comes from the shear term in μ. It is also interesting to notice that elasticity plays an important role also in other Fermi systems (liquid ^{3}He [30]).

Exercise 8. By choosing the usual quadrupole and octupole operators $F = \sum_i x_i y_i$ and $F = \sum_i x_i y_i z_i$, evaluate the mean excitation energy $\omega = \sqrt{S_3/S_1}$ and compare it with the experimental values $\omega_{2^+} = 65 A^{-1/3}$ MeV and $\omega_{3^-} = 108 A^{-1/3}$ MeV (use $\frac{m^*}{m} = 0.8$ and $R = 1.15 A^{1/3}$ fm).

Exercise 9. Verify that the sum rule S_3 for the dipole operator $F = \sum_i z_i r_i^3$ diverges if the ground state density is assumed to be a step function (Steinwedel - Jensen model).

Sum rules in the isospin channels

In the preceding sections we have derived sum rules for hermitian operators F. In that case we have shown that K-odd sum rules can be reduced in terms of commutators and can be evaluated in a relatively simple manner. Conversely K-even sum rules are expressed in terms of anticommutators and depend on the description of the ground state $|0\rangle$ in a more critical way. When the excitation operator is not hermitian the situation can be very different. In this section we will discuss the case of isospin dependent operators of the form

$$F^{\pm} = \sum_i f(\vec{r}_i)\tau_i^{\pm} \qquad (64)$$

where $\tau^{\pm}$ are the usual isospin operators defined by $\tau^{+}|p\rangle=\sqrt{2}|n\rangle$, $\tau^{-}|n\rangle=\sqrt{2}|p\rangle$

Operators of the form (64) enter in many physical reactions involving charge exchange $([\pi^{\pm},\pi^{0}],[p,n]\cdots$ reactions); it is therefore interesting to explore how the sum rule approach works for such excitations [31].

By denoting the third component of the nuclear isospin operator $\vec{T}$ by $T^{z}=\dfrac{N-Z}{2}$, we notice that the operators $F^{\pm}$, when applied, to a state with $T^{z}=T$, yield excited states with $T^{z}=T\pm 1$.

A schematic situation of the various excited isospin fragments is reported in fig. 3. Once the Coulomb energy is removed, the fragments with the same isospin occurr at the same energy (fig. 4):

The remarkable feature energing from the experimental data and schematically

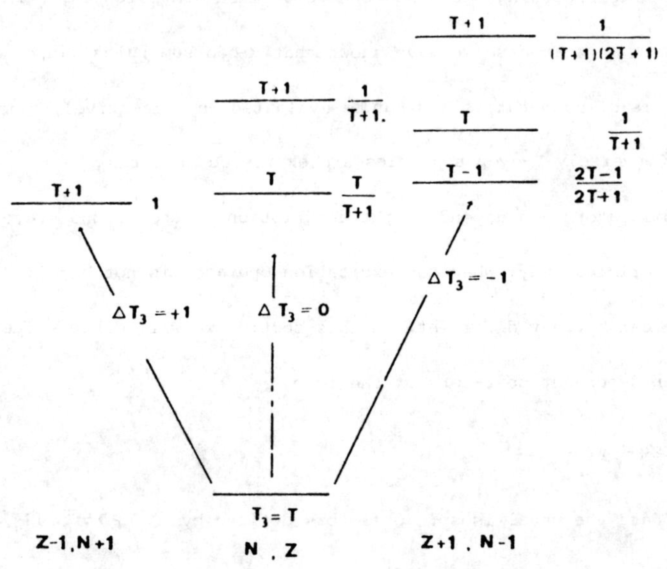

Figure 3

E_{T+1} } E_+

E_T } E_o

E_{T-1} } E_-

O

Figure 4

proposed in figs. (3) and (4), is the existence of splittings between fragments with different isospin.

One can define the strength of the operator $F^{\pm}$ in analogy to the case of hermitian operators:

$$S^{\pm}(\omega) = \sum_{n} |\langle n | F^{\pm} | 0 \rangle|^2 \delta(\omega - (\omega_n - \omega_0)) \tag{65}$$

Clearly the excited states $|n\rangle$ have $T^z = T \pm 1$. The moments of the strength are also defined in a natural way:

$$S^{\pm}_{K} \quad \int S^{\pm}(\omega)\, \omega^{K} d\omega = \sum_{n \neq 0} |\langle n | F^{\pm} | 0 \rangle|^2 (\omega_n - \omega_0)^{K} \tag{66}$$

Separate evaluation of S^{+}_{K} and S^{-}_{K} cannot be reduced in terms of commutators. However, suitable combinations of S^{+}_{K} and S^{-}_{K} admit such a reduction:

$$\begin{aligned} S^{-}_{0} - S^{+}_{0} &= \langle 0 | [F^{+}, F^{-}] | 0 \rangle \\ S^{-}_{1} + S^{+}_{1} &= \langle 0 | [F^{+} [H, F^{-}]] | 0 \rangle \\ S^{-}_{2} - S^{+}_{2} &= \langle 0 | [[F^{+}, H], [H F^{-}] | 0 \rangle \\ &\ldots \end{aligned} \tag{67}$$

Once more one can show that if $|0\rangle$ is replaced by the HF ground state, then eq (67) provides evaluation of sum rules within RPA accuracy. RPA calculations for such excitations have been recently carried out using Skyrme [32] and schematic [33] interactions.

Before evaluating the sum rules (67) we discuss how they can be employed to analyze the excitation spectrum of fig. 4. We define E_+, E_- and E as the centroids of the strength in the $\Delta T^z = +1$, -1 and 0 channels. The nature of the isogeometrical factors which weight the strength in the various T channels (and which are reported in fig. 3) are such that in the limit of large T only the state with the lowest value of T (for a fixed T^z) contributes to the strength. In this limit we are then allowed to identify E_+ E_- and E with E_{T+1}, E_{T-1} and E_T respectively.

We can conveniently introduce a "vector" and "tensor" contribution to the excitation energy as follows:

$$E_+ = E + \Delta_V + \Delta_T$$

$$E_- = E - \Delta_V + \Delta_T \tag{68}$$

from which one can also find

$$\Delta_V = \frac{1}{2}(E_+ - E_-)$$

$$\Delta_T = \frac{1}{2}(E_+ + E_- - 2E). \tag{69}$$

One expects that the leading contribution to Δ_V is a first order effect in the interaction responsible for the splitting, while the contribution to Δ_T will be a second order effect. By assuming that the strength is concentrated in a single fragment for each excitation operator $F^{\pm}$ and F, one can write:

$$S_1^{\pm} = E_{\pm} S_0^{\pm} \quad , \qquad S_1 = E\, S_0 \quad ,$$

$$S_2^{\pm} = E_{\pm}^2 S_0^{\pm} \quad , \qquad S_2 = E^2 S_0 \qquad (70)$$

We will also assume, for the sake of simplicity, that the exchange contributions to the commutator $[H,F^{\pm}]$ are the same as in $[H,F]$ and can be consequently factorized (such an assumption is reasonable only for electric operators). One can use results (70) to evaluate Δ_V and Δ_T by a suitable expansion up to second order terms. The final result is [34] :

$$\Delta_V = \frac{E^2 \langle [F^+,F^-] \rangle - \langle [[F^+,H],[H,F^-]] \rangle}{\langle [F^+[H,F^-]] \rangle} \qquad (71)$$

$$\Delta_T = \frac{1}{2} \frac{\Delta_V^2}{E} \qquad (72)$$

We now apply the above results to the dipole case $f(r)=z$.

By ignoring exchange effects, one finds:

$$\langle [F^+,F^-] \rangle = \frac{2}{3} \langle \sum_i r_i^2 \tau_i^3 \rangle$$

$$\langle [F^+,[H,F^-]] \rangle = \frac{A}{2m} \qquad (73)$$

$$\langle [[F^+,H],[H,F^-]] \rangle = \frac{1}{m^2} \frac{2}{3} \langle \sum_i p_i^2 \tau_i^3 \rangle$$

If we use harmonic oscillator wave functions for describing the ground state , we find $\langle \sum_i p_i^2 \tau_i^3 \rangle = m^2 \omega_0^2 \langle \sum_i r_i^2 \tau_i^3 \rangle$

The excitation energy E can be evaluated (consistently with the assumption of harmonic oscillator wave functions) using the schematic model of Bohr-

Mottelson which gives

$$E = \sqrt{\omega_0^2 + \frac{3}{4}\frac{V_1}{m\langle r^2\rangle}} \tag{74}$$

where V_1 is the symmetry energy potential (100 ~ 130 MeV). Inserting expression (74) in eq. (71) and using result (73) we find:

$$\Delta_V = \frac{1}{2}\frac{T}{A} V_1 \frac{\langle r_v^2\rangle}{\langle r^2\rangle} \tag{75}$$

where $\langle r_v^2\rangle = \left(N\langle r_n^2\rangle - Z\langle r_p^2\rangle\right)/(N-Z)$. These results are identical to those obtained with an exact diagonalization of the schematic model in the RPA [33]. Using V_1 = 120 MeV, $\langle r_v^2\rangle = \langle r^2\rangle\ (\langle r_n^2\rangle = \langle r_p^2\rangle$ and $E = 78\ A^{-\frac{1}{3}}$ MeV, one obtains

$$\Delta_V = 60\,\frac{T}{A}\ \text{MeV}$$
$$\Delta_T = 46\,T^2 A^{-\frac{5}{3}}\ \text{MeV} \tag{76}$$

Results (76) can be used to evaluate the isospin energy splittings though the relation [35]

$$\Delta E_+ = E_{T+1} - E_T = (T+1)\left(\frac{\Delta_V}{T}\right) + \frac{(T+1)(2T-1)}{2}\left(\frac{\Delta_T}{T^2}\right)$$
$$\Delta E_- = E_T - E_{T-1} = T\left(\frac{\Delta_V}{T}\right) - \frac{(2T+3)T}{T}\left(\frac{\Delta_T}{T^2}\right) \tag{77}$$

which take into account the correct isogeometrical dependence and can be applied also to nuclei with relatively small T.

The vector contribution Δ_V to the isospin splittings has been investigated much time ago [36]. At that time, however, rather few data on the T-1 fragment were available. Data for Δ_+ and Δ_- are now available for several nuclei [37] and are reported in table 2 together with the theoretical predictions of eqs. (76) (77). The comparison with experiments is rather encouraging and reveals that the schematic model, inspite of its simplifying assumptions, is a realistic model for investigating the excitations in the isospin channels.

	ΔE_+	ΔE_-	$\frac{A\Delta_V}{T}$	$\frac{\Delta_T A^{5/3}}{2T^2}$
^{90}Zr	3.9(4.3)	2.2 (2.9)	51 (60)	36 (23)
^{116}Sn	4.0(5.3)	2.5 (3.5	40 (60)	22 (23)
^{120}Sn	5.5(6.3)	3.6 (4.1)	53 (60)	19 (23)
^{124}Sn	6.3(7.4)	4.6 (4.6)	54 (60)	13 (23)
^{208}Pb	11.2(8.1)	4.5 (4.8)	73 (60)	23 (23)

Table 2

<u>Exercise 10</u>. Explain why in the $T^z = T+1$ channel one cannot have the excitations E_T and E_{T-1}.

<u>Exercise 11.</u> Using results (73) and eq. (70) evaluate the dipole strengths S_0^+ and S_0^-. Estimate the ratio $\frac{S_0^+}{S_0^-}$ in ^{208}Pb and explain why this ratio is very small (blocking effect).

References

1) S. Stringari, E. Lipparini, G. Orlandini, M. Traini and R. Leonardi, Nucl. Phys. A. 309 (1978) 177,189

2) O. Bohigas, A.M. Lane and J. Martorell, Phys. Rep. 51 (1979) 267

3) U. Fano and J.W. Cooper, Rev. Mod. Phys. 40 (1968) 441

4) D.J. Rowe, "Nuclear Collective Motion", Methuen, London, 1970

5) E.R. Marshalek and J. Da Providencia, Phys. Rev. C 7 (1973) 2281

6) T.H.R. Skyrme, Nucl. Phys. 9 (1959) 615;
D. Vautherin and D.M. Brink, Phys. Rev. C 5 (1972) 626

7) K. Goeke, B. Kastel and P.G. Reinhard, Nucl. Phys. A 339 (1980) 377

8) O. Bohigas, Nguyen Van Giai and D. Vautherin, Phys. Lett. B102 (1981) 105

9) C.F. Weizsäcker, Z. Phys. 96 (1935) 431

10) P. Ring and P. Schuck, "The Nuclear Many-Body Problem", Springer-Verlag, Berlin/New York, 1980

11) D. Pines and P. Nozières, "The Theory of Quantum Liquids", Benjamin, New York, 1966

12) H. Steinwedel and J.H.D. Jensen, Z. Naturforsch. 5a (1950) 413

13) S. Stringari, Ann. Phys. 151 (1983) 35

14) A. Bohr and B. Bottelson, "Nuclear Structure", vol. 2, Benjamin, New York 1975

15) A. Migdal, J. Phys. (Moscow) 8 (1944) 331

16) B.K. Jennings and A.D. Jackson, Phys. Rep. 66 (1980) 141

17) J. Treiner et al, Nucl. Phys. A 371 (1981) 253

18) E. Lipparini and S. Stringari, Phys. Lett. B 112 (1982) 421

19a) S. Stringari and E. Lipparini, Phys. Lett. B 117 (1982) 141;
J. Meyer, P. Quentin and B.K. Jennings, Nucl. Phys. A385 (1982) 269

b) E. Lipparini and S. Stringari, in Proc. of the Meeting on "Nuclear Fluid Dynamics" ICTP Trieste 1982, IAEA Vienna 1983

20) J. Ahrens et al, Nucl. Phys. A251 (1975) 479;
R. Bergére in "Electron and PHotonuclear Reactions", Lecture notes in Physics, vols. 61, 62, Springer Verlag 1977

21) K. Gottfried, Ann. Phys. 21 (1963) 29

22) a) R. Altemus et al, Phys. Rev. Lett. 44 (1980) 965
b) P. Barreau et al, Nucl. Phys. A358 (1981) 287c
c) M. Deady et al, Phys. Rev. C 28 (1983) 631
d) J.S. O'Connell et al, Phys. Rev. C 27 (1983) 2492

23) O. Bohigas in Theory and applications of moment methods in many fermion systems, ed. B.J. Dalton et al. (Plenum NY 1980) and in From collective states to quarks in nuclei, Lecture Notes in Physics, vol. 137, Springer Verlag 1981

24) S. Stringari, Phys. Rev. C, to be published

25) D.J. Thouless, Nucl. Phys. 22 (1961) 78

26) A. Lepretre, These, Orsay 1982

27) G.F. Bertsch, Ann. Phys. 86 (1974) 138; Nucl. Phys. A249 (1975) 253

28) S. Stringari, Nucl. Phys. A279 (1977) 454

29) G. Holzwarth and G. Eckart, Z. Phys. A284 (1978) 291

30) Kevin Bedell and C.J. Pethick, J. Low Temp. Phys. 49 (1982) 213

31) R. Leonardi and M. Rosa-Clot (Riv. N. Cim. 1 (1971) 1)

32) N. Auerbach and A. Klein, Nucl. Phys. A 395 (1983) 77

33) R. Leonardi, E. Lipparini and S. Stringari, Phys. Rev. C26 (1982) 2636

34) R. Leonardi in "Highly Excited States and Nuclear Structure" (Hesans 1983);
R. Leonardi, E. Lipparini and S. Stringari, to be published

35) R. Leonardi, Phys. Rec. C 14 (1976) 385

36) R. Leonardi and M. Rosa-Clot, Phys. Rev. Lett. 23 (1969) 874;
S. Fallieros and B. Goulard, Nucl. Phys. A147, (1970) 593;
R.Ö. Akyüz and S. Fallieros, Phys. Rev. Lett. 27 (1971) 1016

37) W.A. Sterremburg et al., Phys. Rev. Lett. 45 (1980) 1839

DAMPING OF COLLECTIVE MODES

P.F. Bortignon
Department of Physics, University of Padova, Padova;
and I.N.F.N., Laboratori Nazionali Legnaro, Italy

Abstract

In the first lecture, experimental information on the known giant resonances is summarized with emphasis on the damping properties. Theoretical models for the damping mechanism are presented in the second lecture, closely following ref. (1). In the third lecture, some recent calculations of giant resonance strength at high excitation energy and of the continuum background underlying the main peaks are discussed.

I. A short review of the experimental data

The giant resonances are the resonant peaks displayed by the response of the nucleus to an external field in the region of excitation energy of 10-30 MeV. Their excitation energies vary smoothly with the nuclear mass.

The dependence of the external field on the spin and isospin as well as its spatial dependence, that resolves into multipoles, determine the kind of giant vibration excited. Many kinds of nuclear vibrations have already been identified. The reader can find the bulk of the experimental data in recent review articles, cfr. (2)-(7) and references therein.

The peaks of the giant modes are in a region of very high--level density which produces the spreading of the response. Typically, the main peak has a full width half maximum (FWHM) of 3-10 MeV.

A third important experimental quantity is the fraction of the appropriate sum rule, (8), exhausted by the excited vibrations. A peak in the strength function may be considered a giant vibration if its strength exhausts most of the sum rule. We anticipate that the experimental determination of the strength in a peak is rather uncertain, depending on the assumptions about the underlying background and the peak shape.

A. The giant dipole vibration

The earliest and best known example of a nuclear vibration is the giant dipole resonance, (GDR), (2). This is a vibration of neutrons against protons, and is excited by the long-wavelength isovector field $F(\vec{r}) \sim \tau_z \vec{r}$.

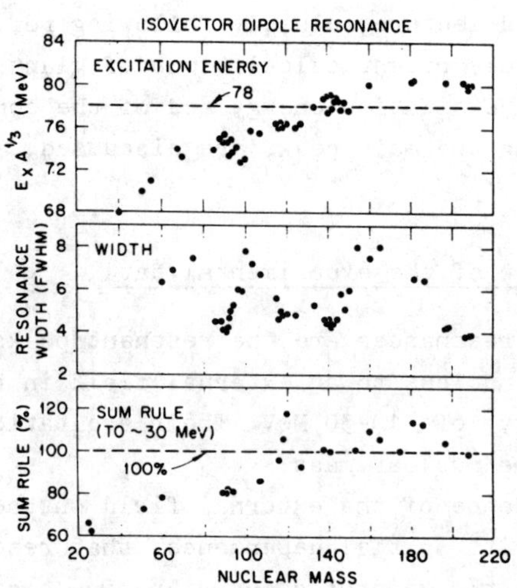

Fig. 1. Form ref. (3). Systematics of the excitation energy, width and sum rule depletion for the Giant Dipole Resonance, (2).

In fig. 1 the systematics for the excitation energy, width and strength is shown, (2). In heavy nuclei, the excitation energy varies smoothly with target mass and about 100% of the sum rule is exhausted.

Three examples of the GDR, as excited in photoneutron reaction, are shown in fig. 2.

In light nuclei, as in the He example, the peak is quite broad, owing to the escape of particles into the continuum.

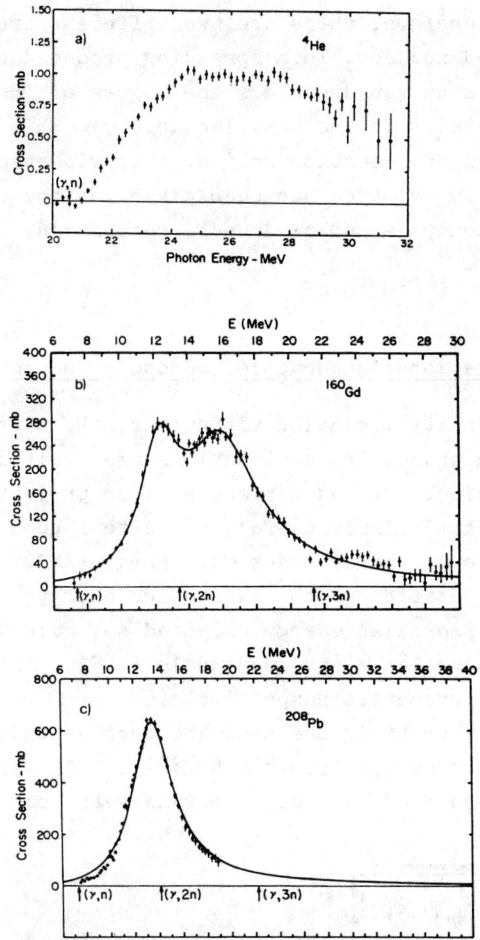

Fig. 2. The photon-neutron cross section for ^{4}He, ^{160}Gd and ^{208}Pb from ref. (2).

Experimentally, the dipole state in heavy nuclei decays primarily by emitting low-energy neutrons. The escape width $\Gamma^{\uparrow}$ is relatively small, of the order of 15% of the total width, (9), and the damping is mainly due to mixing with more complicated states, (spreading width $\Gamma^{\downarrow}$). A typical dipole response of Lorentzian shape in a heavy spherical nucleus is shown in fig. 2(c).

A double-peaked structure appears in the dipole response of deformed nuclei, as in fig. 2(b). This is understood as it follows: the dipole frequency depends on the size of the nucleus along the dipole axis. Therefore, there are two different frequencies in an axially deformed nucleus. This spreading produced by the coupling of the dipole to the quadrupole shape degree of freedom will be discussed in details in the next lecture.

I will end this section on isovector vibrations reminding you that recently evidence has been given for the observation of the isovector monopole and quadrupole resonances see refs. (3), (10),(11),(12) .

B. Density vibrations: quadrupole, monopole and octupole

In the density isoscalar vibrations, the neutrons and the protons move in phase. These vibrations are excited by inelastic scattering of electrons and various nuclear projectiles. Alpha particles are particularly useful, since to a good approximation they only excite isoscalar modes for energies $\lesssim$200 MeV. In figs 3, 4, 5 the systematics of the excitation energy, width and depletion of the isoscalar energy weighted sum rule are reported, (3), for the isoscalar giant quadrupole (GQR), monopole (GMR) and octupole (GOR) resonances respectively.

The GQR is probably the most thoroughly studied giant mode. It is seen in all nuclei from ^{16}O to ^{238}U. In fig. 6 are shown inelastic spectra from five different targets bombarded by

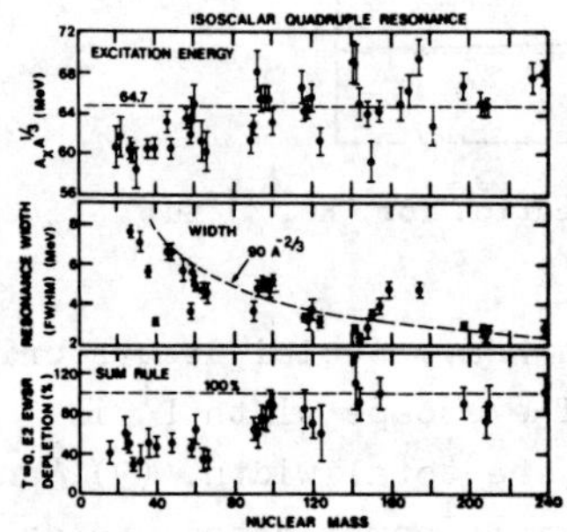

Fig. 3. From ref. (3). Systematics of the excitation energy, width and depletion of the isoscalar Energy Weighted Sum Rule (EWSR) for the Giant Quadrupole Resonance.

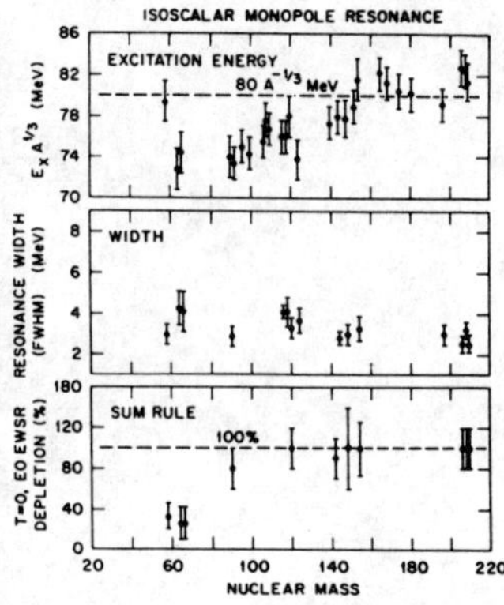

Fig. 4. From ref. (3). Systematics of the excitation energy, width and depletion of the isoscalar Energy Weighted Sum Rule (EWSR) for the Giant Monopole Resonance.

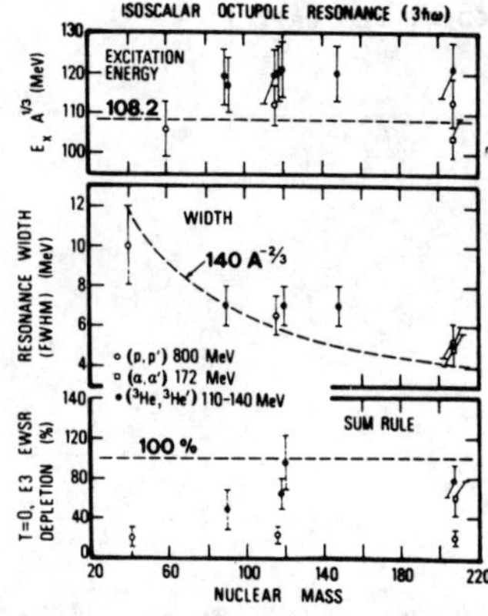

Fig. 5. From ref. (3). Systematics of the excitation energy, width and depletion of the isoscalar Energy Weighted Sum Rule (EWSR) for the Giant Octupole Resonance.

152-MeV alpha particles, (13). The broad peak is decomposed, in this experiment, in a portion marked E0 (the GMR) and in a second portion marked E2 (the GQR). The determination of the multipolarity is obtained by comparison of the measured angular distribution with those calculated in the Distorted Wave Born Approximation (DWBA) (14), as shown in figs 7 and 8. The normalization of the calculated cross section to those measured gives the strength of the resonance. There will be some uncertainty about the extracted strength associated with the model dependence of the DWBA calculations and the assumptions about the background.

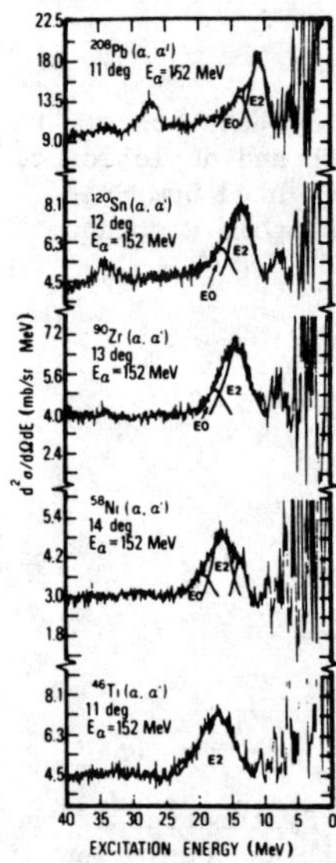

Fig. 6. Spectra from inelastic scattering of 152-MeV alpha-particles on ^{208}Pb, ^{120}Sn, ^{90}Zr, ^{58}Ni and ^{46}Ti {from ref. (3)}. The peak at higher energy in ^{208}Pb and ^{120}Sn are due to hydrogen contamination of the target.

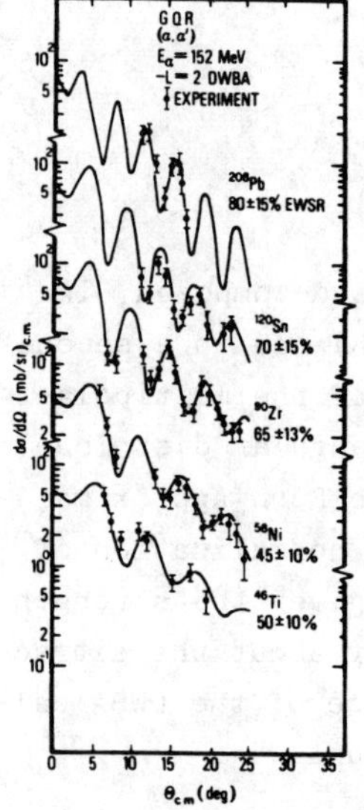

Fig. 7. Angular distribution of the E2 portion of the spectra of fig. 6, from (13). The data are compared to an L=2 DWBA calculation normalized to the indicated depletion of the EWSR.

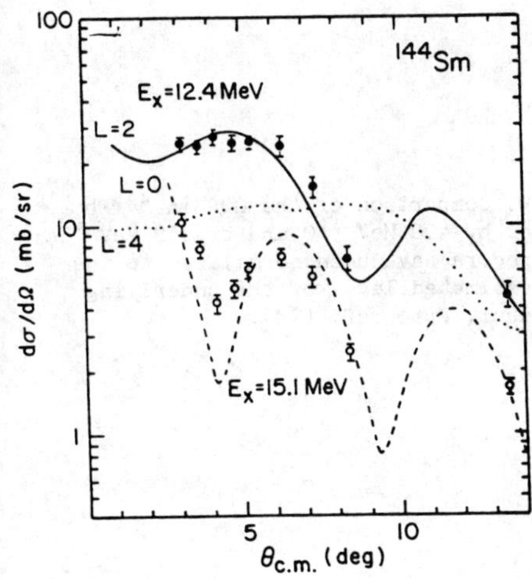

Fig. 8. Example of angular distribution of alpha particles exhibiting the forward peak characteristic of L=0 transfer, from ref. (15).

For the GQR in ^{208}Pb, there is one important difference between results of electron scattering and hadron scattering. The (e,e') data of ref. (16) indicate a very low value ($\sim$35%) for the depletion of the EWSR, much lower than the $\sim$70% of the hadron data. This is not yet understood, but needs to be resolved for a quantitative comparison with theory.

In the systematic of the quadrupole width of fig. 3, we note that the state is rather broad in light nuclei. This is partly due to the escape width $\Gamma^{\uparrow}$, (17). For example, in ^{16}O there is an important direct decay component by α-particle emission, (18). In heavy nuclei $\Gamma^{\uparrow}$ is small, but a direct neutron decay branch has been observed in ^{119}Sn, ^{92}Zr (19) and ^{208}Pb (20), via (α, α' n) experiments. Coincidence experiments are very important, providing informations on the mixing of different multipolarity in the same peak region and on small fractions of strength embedded in the continuum background, see, e.g., (21) and (22) for two recent studies on ^{40}Ca, via the reaction ^{40}Ca(α,α'α_0). In this context, coincidence experiments with high-energy heavy ions look very promising. They can profit from the large cross section for excitation of the giant resonances and from the very large peak to continuum ratio, as shown in fig. 9 from ref. (23).

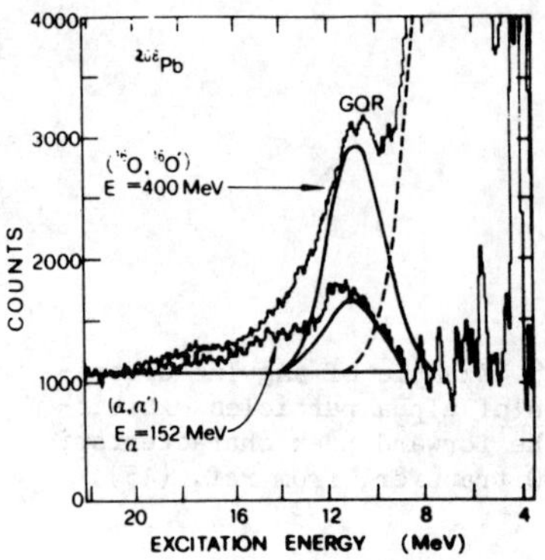

Fig. 9. Comparison of the GQR in ^{208}Pb excited by 400 MeV ^{16}O and by 152 MeV ^{4}He. The spectra have been normalized to the same estimated level of the underlying continuum. From ref. (23).

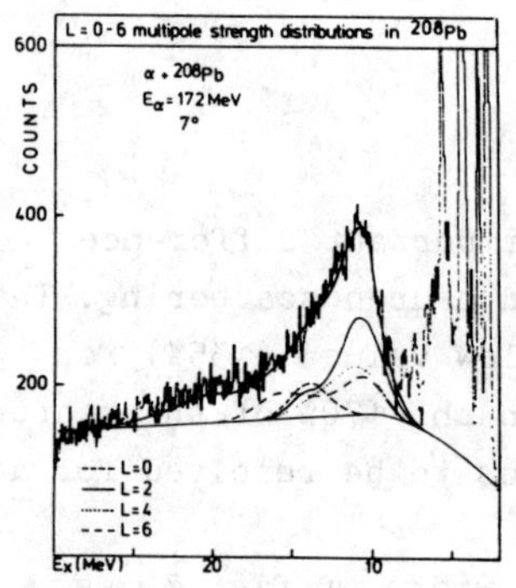

Fig. 10. From ref. (26). Relative multipole strength distributions for even multipolarities.

The GMR is best excited at small scattering angle, cfr fig. 8. It is located in energy just above the GQR, appearing in most spectra as a shoulder on the quadrupole peak. However, it is well visible in nuclei heavier than ^{40}Ca with a width of about 3 MeV, cfr also ref. (22) and (31).

Some recent experiments on ^{208}Pb, (24)-(25)-(26), have definitively indicated the presence of strength of higher even multipolarities (L=4,6) in the excitation energy region of the GQR and GMR. This is well shown in fig. 10 from ref. (26). The extracted EWSR strengths for L=4 and 6 are in the order of 30% in each case, (26).

The GOR and the giant isoscalar dipole vibration are the odd-parity giant modes recently identified at higher excitation energy, see, e.g., fig. 11 from ref. (26).

The results of few experiments for the GOR are collected in fig. 5. The presence at these high excitation energy of a very large background may explain the large discrepancies in the data, (26)-(30).

The squeezing mode, that is the isoscalar dipole compression mode, has been identified in ^{208}Pb, (26),(27),(30). The mean excitation energy is 21 MeV, the total width being about 6 MeV. The mode exhausts 70-90% of the EWSR. In the (p,p') experiment of ref. (30), E_p=201 MeV, it appears superposed on the isovector quadrupole vibration.

C. Spin vibrations

In the vibrations we have previously described, the nuclear spins were not affected and the modes carry spin S=0. In principle, we may expect isoscalar and isovector spin excitations, depending on whether the coupling to protons and neutrons is the same or not.

The recent (p,n) experiments at 100-200 MeV bombarding energy, (4), have produced a great progress in our understanding of the spin-isospin correlations in nuclei.

The simplest spin-isospin excitation is created by the Gamow-Teller beta-decay operator σt_- with no spatial dependence.

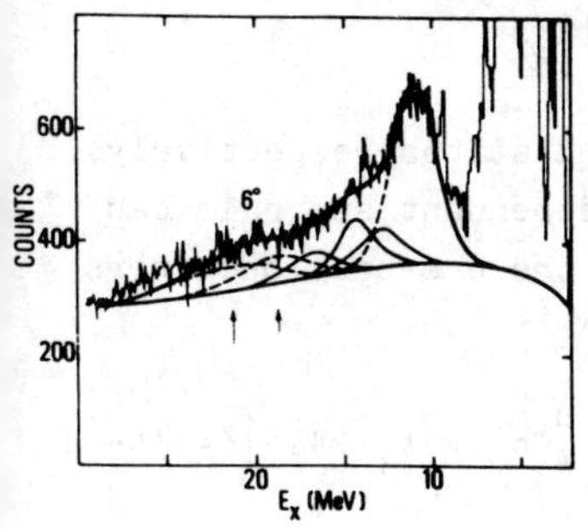

Fig. 11. Adapted from (26). Small angle spectrum of 172 MeV alpha-scattering on ^{208}Pb.

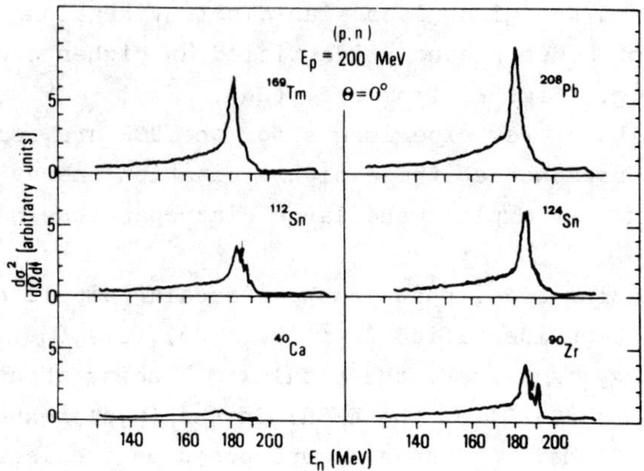

Fig. 12. Neutron time of flight spectra at 200 MeV. From ref. (4).

The angular distribution in the excitation reaction will be forward peaked, because there is no momentum transfer. This is the case for the energy spectra at zero degree shown in fig. 12. The prominent peak in the spectrum of nuclei having a neutron excess is interpreted as the Gamow-Teller collective state, (GTR), which was first discussed in ref. (32). The width of the GTR in heavy nuclei is about 4 MeV, and its excitation energy in the residual nucleus varies from about 10 MeV in lighter nuclei to 15 MeV in ^{208}Pb.

From the zero degree cross section in the impulse approximation the Gamow-Teller strength B (GT) is obtained, (33), being

$$B(GT)=\sum_{\mu} |\langle f| \sum_{k=1}^{A} \sigma_{\mu}(k)t_{-}(k)|i\rangle|^2 \tag{1}$$

where $|i\rangle$ and $|f\rangle$ are the initial and final states respectively.

For the GT operator σt_{-}, a model independent sum rule can be obtained from commutator relations for the one-body operators t_{-} and t_{+}, (34). The sum rule is

$$S_{\beta^-}-S_{\beta^+}=\sum_{f}\sum_{\mu}|\langle f| \sum_{k=1}^{A} \sigma_{\mu}(k)t_{-}(k)|i\rangle|^2-\sum_{f'}\sum_{\mu}|\langle f'| \sum_{k=1}^{A} \sigma_{\mu}(k)t_{+}(k)i\rangle|^2=3(N-Z) \tag{2}$$

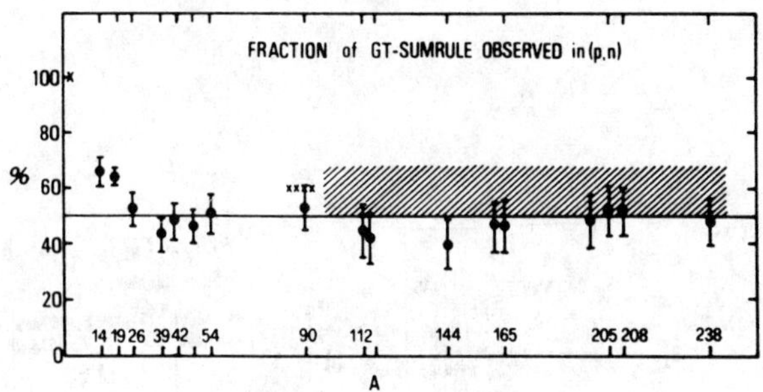

Fig. 13. From ref. (4). Fraction of the GT-sum rule strength observed in (p,n) reactions for E_p=160 MeV. The cross hatched region corresponds to the uncertainty in defining the background, the limit including everything as GT-strength.

In the evaluation, no internal degrees of freedom for the nucleon are considered.

In fig. 13, it is shown that only ∿50% of the GT-sum rule is observed in (p,n) reactions, at least in the E_x<30 MeV part of the spectra, cfr fig. 12.

Two possible quenching mechanisms will be discussed in the third lecture. They are the coupling to the Δ_{33}(E=300 MeV, s=3/2, t=3/2) resonance and the mixing with high-lying configurations, via a residual interaction.

In fig. 14(a), the broad structure appearing at angles different from zero (cfr the spectrum at 4.5 degrees) is interpreted as the spin dipole resonance excited by the operator $r(Y_1\sigma)_J t_-$, being $J^\pi=0^-,1^-,2^-$, (35). So far, the different spin components have not been resolved, (35). In fig. 14(b), examples of angular distributions with transfer of orbital angular momentum L=0,1,2 are shown.

In fig. 15(a), spectra are displayed for (p,n) reactions on ^{90}Zr at E_p=120 and 200 MeV, to emphasize the energy dependence of the cross section for the excitation of spin transfer (GT)- and non spin transfer (IAS)- modes. This is understood in terms of energy dependence of the t-matrix, (36)-(37), as it is shown in figs 15(b) and 15(c).

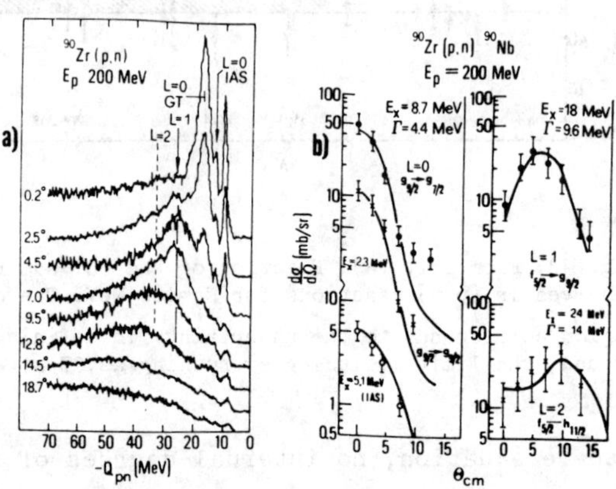

Fig. 14. From ref. (4). In (a), neutron spectra are shown at different angles for the $^{90}Zr(p,n)^{90}Nb$ reaction at E_p=200 MeV. In (b), examples of angular distributions from the same reaction. See text.

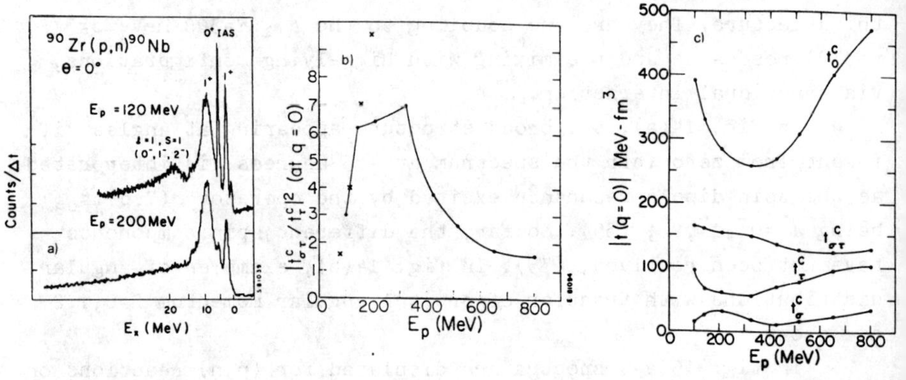

Fig. 15. From ref. (4). In (a), neutron spectra for the reaction $^{90}Zr(p,n)^{90}N$ at E_p=120 and 200 MeV. In (b) and (c), energy dependence of t-matrix elements for the central part of the interactions in the different channels, (37). The crosses in (b) are experimental values, (38).

Electron scattering, (5), and the (p,p') reaction, (6), give information on the response of the nucleus to another spin excitation operator, the field σt_z, which is close to the electromagnetic operator M1.

In fig. 16, backward-angle inelastic electron scattering spectra are displayed, the arrows pointing to strong $J^\pi=1^+$ states. We note the strongly excited M1 giant resonance in ^{48}Ca, the large fragmentation of the mode in ^{44}Ca and the excitation of M1 states in ^{40}Ca, indicating strong ground state correlations, (40). (No M1 transition exists in ^{40}Ca in the independent particle shell model, see the third lecture). In fig. 17, the example of the fragmented M1 giant mode in ^{50}Ti is reported, (41).

From the spectra of fig. 16, the strength of the M1 transitions have been deduced, (39). By comparison to the results of calculations performed in the full fp-shell model space, (42), it is found that only half of the predicted strength is seen experimentally, (5).

In fig. 18, the quenching factor defined as (5)

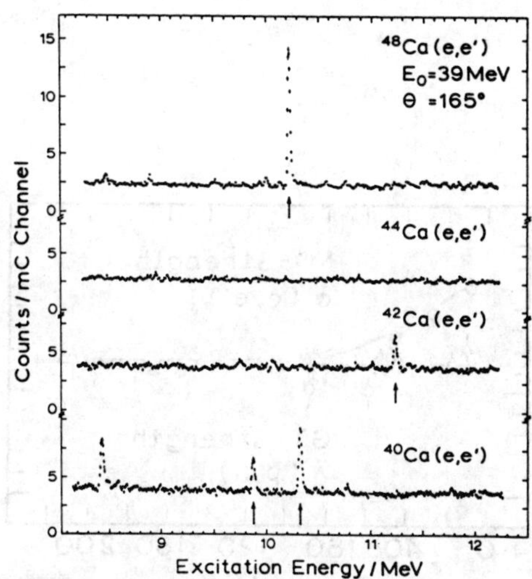

Fig. 16. From refs (5) and (39). Inelastic electron scattering spectra on the even-even Ca isotopes. The arrows point to strong $J^\pi=1^+$ states. In ^{90}Ca, the peak at ∿9.9 MeV is a doublet containing only a weakly excited 1^+ state.

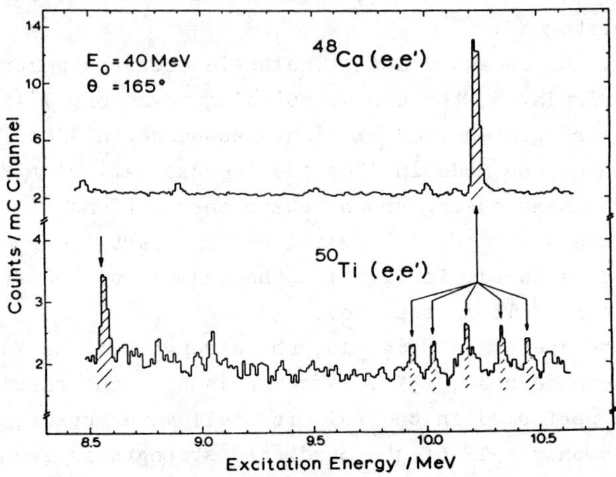

Fig. 17. From ref. (5) and (41). Comparison of (e,e') spectra in ^{48}Ca and ^{50}Ti, the hatched areas indicating 1^+ states.

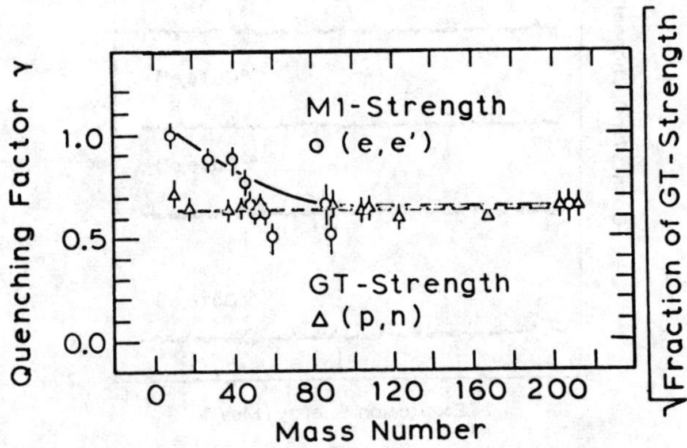

Fig. 18. From ref. (5). Mass dependence of quenching factors for M1 transitions (open circles) and Gamow-Teller transitions (open triangles). The solid and dashed lines are to guide the eye.

$$\gamma = \left[\sum B(M1)_{exp.} / \sum B(M1)_{th.} \right]^{\frac{1}{2}} \quad (3)$$

is reported as function of the nuclear mass. It is noted that no model independent sum rule exists for M1 transitions and the quenching factor may only be obtained by comparison with model calculations.

The figs. 19 and 20 show examples of the results obtained in the study of the M1 giant resonance using the (p,p') reaction at $E_p \cong 200$ MeV, cfr fig. 15(c).

In fig. 21 (table 1), a summary of the obtained data is presented. The attenuation (quenching) factor Q is defined as the ratio of experimental to theoretical, predicted cross section. It appears to be almost constant with mass number. For ^{48}Ca, the value obtained from fig. 20(b) is 0.30, about the same as in the more complex cases in fig. 19 (and table 1). The excitation energy of the M1 giant mode in, e.g., ^{90}Zr is not very different from

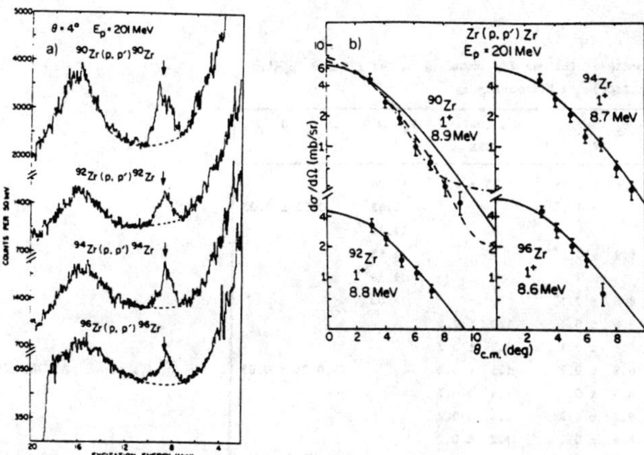

Fig. 19. From refs (6). (a), Spectra of protons inelastically scattered from Zr isotopes at 4°. The arrows indicate the centroids of the M1 resonance. (b) Angular distributions for the M1 states in (a). The points are the measured values, the dotted-dashed curve is from the (p,n) reaction at 200 MeV, (4), the solid and dashed curves are calculations described in refs (6).

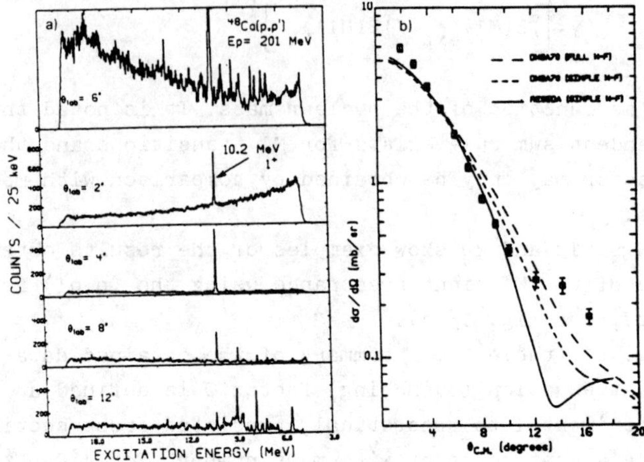

Fig. 20. From refs (6) and (43). (a) Spectra of protons inelastically scattered from ^{48}Ca. The uppermost spectrum has been scaled to show the weakly excited states.(b) Angular distributions for the 1^+ state at 10.2 MeV in (a). The points are the measured values, the curves calculations discussed in ref. (43).

TABLE 1 : Summary of the results obtained in the 200 MeV (p,p') excitation of M1 resonances.

Nucleus	E_x (MeV) T_0	Γ (MeV) FWHM	E_x (MeV) T_0+1	Q
^{51}V	10.15 ± 0.15	1.35 ± 0.1	13.08	
^{58}Ni	8.5 ± 0.1[a)]		10.65	0.23 ± 0.03
			11.36[b)]	
^{60}Ni	8.9 ± 0.1[a)]		11.85	
			12.58[b)]	
^{62}Ni	8.8 ± 0.1[a)]		14.03	
^{68}Zn	9.6 ± 0.1	1.0 ± 0.1		
	8.6 ± 0.1	0.9 ± 0.2		
^{90}Zr	8.9 ± 0.2	1.5 ± 0.2		0.26 ± 0.03
^{92}Zr	8.8 ± 0.2	1.4 ± 0.2		
^{94}Zr	8.7 ± 0.2	1.4 ± 0.2		
^{96}Zr	8.6 ± 0.2	1.2 ± 0.2		
^{92}Mo	9.0 ± 0.1	1.1 ± 0.1		0.34 ± 0.05
	7.95 ± 0.1	0.70 ± 0.05		
^{94}Mo	8.6 ± 0.15	2.35 ± 0.15		
^{96}Mo	8.4 ± 0.15	2.3 ± 0.15		
^{98}Mo	8.5 ± 0.15	2.2 ± 0.2		
^{100}Mo	8.5 ± 0.15	2.8 ± 0.2		
^{120}Sn	8.4 ± 0.15			0.27 ± 0.05
^{124}Sn	8.7 ± 0.2			
^{140}Ce	8.6 ± 0.2			0.25 ± 0.07

Q is the attenuation factor defined in the text
a) Centroid energy
b) Centroid energy when high energy structures are included

Fig. 21. From refs (6).

the GT state, but the width is much narrower. One reason for this is the low-level density in the even-even nuclei.

In the (p,p') reaction on ^{28}Si, one isoscalar 1^+ state has been identified, cfr fig. 22. The obtained value for the quenching factor Q is 0.25. The result is important, because only configuration mixing can produce sizeable isoscalar quenching, see the third lecture.

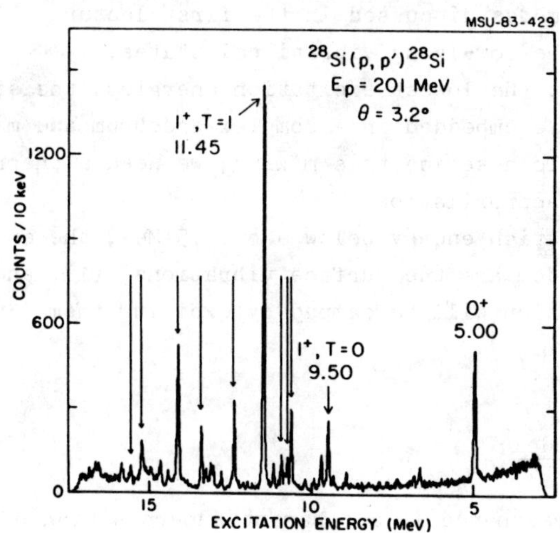

Fig. 22. From ref. (44). Spectrum of protons inelastically scattered from ^{28}Si. The arrows indicate the T=0 and nine T=1 1^+ states.

II. Theoretical models of vibrational damping

The properties of the nuclear excitations discussed above ar conveniently expressed in terms of their strength functions, defined as (1)

$$S(E) = \sum_f \langle f|F|0\rangle^2 \delta(E-E_f) \qquad (4)$$

Here $\langle f|F|0\rangle$ is the matrix element of the operator creating the nuclear excitation, $|0\rangle$ is the initial states and $|f\rangle$ the eigenstates of the nucleus.

At the lowest excitation energies, the theory of the strength function can be understood in terms of mean-field theory, as discussed in the lectures of C.Mahaux (36) and N. Van Giai (45).

The single-particle motion is described by the Hartree-Fock (HF) field (36). Treating this single-particle potential as a dynamic quantity, the vibrations of the nucleus around the ground state are described in the random-phase approximation (RPA) theory (45). The RPA provides a good description both of the mean energies of the various modes discussed in the first lecture and of the properties of the low-lying vibrational states.

At all but the lowest excitation energies, the simple modes of excitation are embedded in a complex spectrum and mix with the nearby states. To describe this mixing, we need a theory beyond the mean-field approximation.

For excitation energy below about 15 MeV, the only important degrees of freedom are the surface vibrations, (1), and the single-particle motion will be damped by exciting them (36), (46).

A. The naive model

A giant resonance is a coherent superposition of particle--hole excitations (45).

In a simple model, we can assume that the particle and hole decay independently and write, for a single-component wave function, the spreading width $\Gamma^{\downarrow}_{ph}$ as

(5) $$\Gamma^{\downarrow}_{ph}=\Gamma^{\downarrow}_{p}+\Gamma^{\downarrow}_{h} .$$

After the necessary refinements because of the energy dependence of the spreading widths of the particle ($\Gamma^{\downarrow}_{p}$) and hole ($\Gamma^{\downarrow}_{h}$) states (1)(36), and because of the many particle-hole configurations involved, it has been shown that a too large (about a factor 2) $\Gamma^{\downarrow}_{v}$ is obtained for the giant vibrations (1),(47).

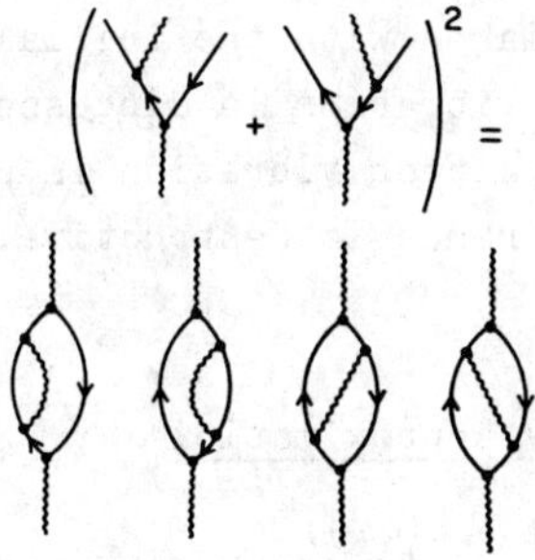

Fig. 23. From ref. (1). Perturbation graphs for the self-energy of a vibration.

B. A better theory

In writing eq. 5), we disregarded the coherence between particle and hole in the collective giant modes. This coherence can strongly reduce the coupling of the collective states to 2-particle-2-hole configurations. This happens, e.g., for the plasmon excitation of a metal ($\Gamma_v^{\downarrow} \cong 0.2$ eV at E≅10 eV) and this reduction of the coupling is built into the quantum collision integral in the Landau theory of collective vibrations (1).

The graphs of fig. 23 show the two amplitudes describing the coupling of the giant modes to the doorway states containing 1 particle-1 hole configurations and a surface vibration. The coherent sum produces four terms. The two graphs on the left arise from the independent damping of the particle and the hole, evaluated at the giant vibration energy, as in eq. 5). The other two diagrams represent vertex corrections and they typically interfere destructively. This happens because the particle and the hole usually couple to other degrees of freedom with opposite signs. The sign of the interference terms for all possible values of the spin and isospin quantum numbers associated with the initial vibration and with the intermediate doorway vibration is discussed in detail in ref. (47). In particular, if either vibration is purely scalar (density vibration), the interference is destructive.

C. The strength function in the doorway approximation

The strength function defined by eq. 4) can be expressed formally in terms of the resolvent of the Hamiltonian

$$S(E)=\frac{1}{\pi}\,\mathrm{Im}\langle 0|F(H-E-i\eta)^{-1}F|0\rangle \tag{6}$$

The simple state $|a\rangle$ (that is, e.g., the RPA giant vibrations) is produced from the ground state by the operator F,

$$|a\rangle=\frac{F|0\rangle}{(\langle 0|FF|0\rangle)^{\frac{1}{2}}} \tag{7}$$

and the Hamiltonian is separated into parts that act on $|a\rangle$ and the rest

$$H=E_a|a\rangle\langle a|+\sum_\alpha\Big[V_{a\alpha}(|\alpha\rangle\langle a|+|a\rangle\langle\alpha|)+E_\alpha|\alpha\rangle\langle\alpha|\Big]+H' \tag{8}$$

The states $|\alpha\rangle$ with $V_{a\alpha}\neq 0$ are the doorway states, in this case the selected 2p-2h states in fig. 23.

If H' is neglected, the resolvent can be evaluated algebraically, cfr App. A of (1) and references therein. It is convenient to average the strength function over energy with a weight function,

$$\rho(E-E')=\frac{1}{\pi}\,\frac{I}{(E-E')^2+I^2} \tag{9}$$

The averaging parameter I takes account of the coupling to more complicated configurations not included in the model space. The averaging with the function 9) is equivalent to evaluating S at the complex energy E+iI.

Then, the averaged strength function can be expressed in the form

(10) $$\overline{S}(E)=\frac{<0|FF|0>}{\pi}\frac{\Gamma(E)/2+I}{[E_a-E-\mathrm{Re}\Sigma(E)]^2+[\Gamma(E)/2+I]^2}$$

$$\mathrm{Re}\Sigma(E)=\frac{V_{a\alpha}^2(E_\alpha-E)}{(E_\alpha-E)^2+I^2}$$

$$\Gamma(E)=2\sum_\alpha\frac{V_{a\alpha}^2 I}{(E_\alpha-E)^2+I^2}=2\mathrm{Im}\Sigma(E)$$

where $\Sigma(E)$ is the self-energy of the mode $|a>$. Four contributions to the $\Sigma(E)$ of the giant vibrations are shown in fig. 23.

D. Results of microscopic calculations

Many authors have recently studied the gross features of the giant resonances in the framework of the strength function discussed above, (48)-(50). I will briefly discuss the calculations in (48), more details may be found in the quoted references.

To calculate the diagrams of fig. 23 in ref. (48) the single--particle energies and eigenfunctions are obtained in the Hartree-Fock approximation, (HF), with the Skyrme III interaction (51). An effective mass $m^*/m=0.76$ (for nuclear matter) is associated to this force. The coupling will eventually increase the effective mass to the value $m^*/m\cong 1$ around the Fermi surface, as discussed in (36).

The giant vibrations λ, e.g. the GQR, and the surface vibrations λ', which are coupled to the single-particle and single-hole states, are calculated in the RPA using a schematic separable interaction with matrix elements

(11) $$V_{php'h'}(\lambda)=K<j_p||r\frac{\partial U}{\partial r}Y_\lambda||j_h><j_{p'}||r\frac{\partial U}{\partial r}Y_\lambda||j_{h'}>.$$

The constant K is determined by the condition that the spurious 1^- state is at zero energy. The properties of the vibrations are in overall agreement with the experimental data. This treatment leads to similar results for collective surface modes as those one would obtain using as residual interaction the density derivative of the HF field (45). The RPA properties of the λ' vibrations fully determine (52) the matrix elements of the particle-vibration coupling Hamiltonian δU appearing in the vertices of the graphs in fig. 23, being

$$\delta U(\vec{r}) \equiv \langle g.s.|V(\vec{r}-\vec{r}\,')|\lambda'\rangle . \tag{12}$$

The interaction $V(\vec{r}-\vec{r}\,')$ is the residual interaction used in the RPA. For spin and isospin vibrations, like the GTR, a residual interaction of the type

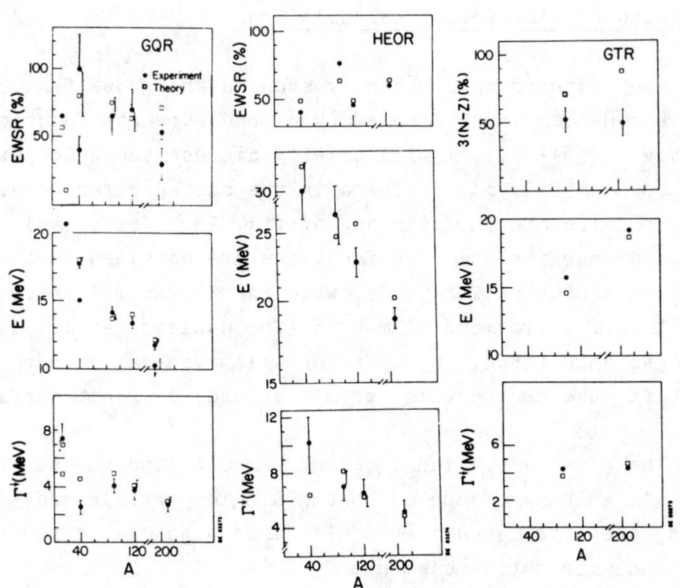

Fig. 24. Comparison of the experimental data, (3)(4), with the theoretical results (48) for the centroid energy, the spreading width and the fraction of sum rule exhausted by the Giant Quadrupole Resonance, the Giant Octupole Resonance and the Gamow-Teller Resonance of some closed-shell nuclei. Adapted from (53).

(13) $$V(\vec{r}_1-\vec{r}_2)=V_{\sigma\tau}\delta(\vec{r}_1-\vec{r}_2)(\vec{\sigma}_1\cdot\vec{\sigma}_2)(\vec{\tau}_1\cdot\vec{\tau}_2)$$

replaces the force (11) with the strength $V_{\sigma\tau}$=200 MeV fm^3.

In fig. 24, the results of refs (48) for the centroid energy, the spreading width $\Gamma^{\downarrow}$ and the fraction of sum rule exhausted by the GQR, the GOR and the GTR are compared with experimental data. The value $\Gamma^{\downarrow}$ is obtained from the standard deviation σ of the strength function as

(14) $$\Gamma^{\downarrow}=\sqrt{8\ln 2}\ \sigma$$

The eq. (14) is valid for a Gaussian distribution, while the calculated strength functions may have a very different shape, as shown in fig. 25 for the GQR and in fig. 26 for the GTR in ^{208}Pb. A more significant comparison with the experimental data can be performed in terms of spectra (54), as it is displayed in fig. 27 for the α-scattering experiment from ^{208}Pb of ref. (13).

It is worthwhile to outline two features of the calculations (48).

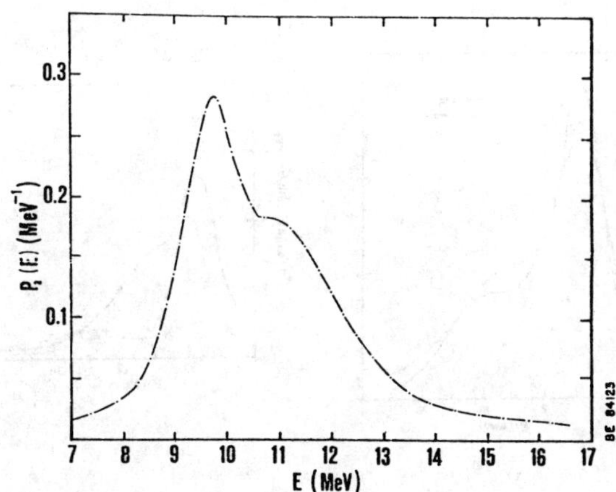

Fig. 25. Strength function (normalized to unity) for the GQR in the ^{208}Pb nucleus. Adapted from (48).

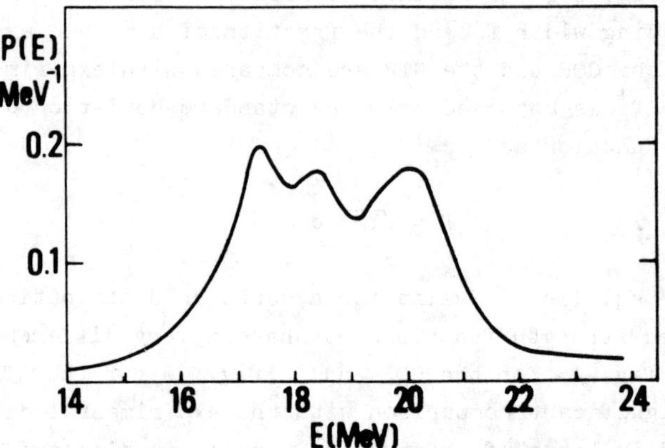

Fig. 26. Strength function (normalized to unity) for the Gamow-Teller giant resonance in the ^{208}Pb. Adapted from the second of ref. (48).

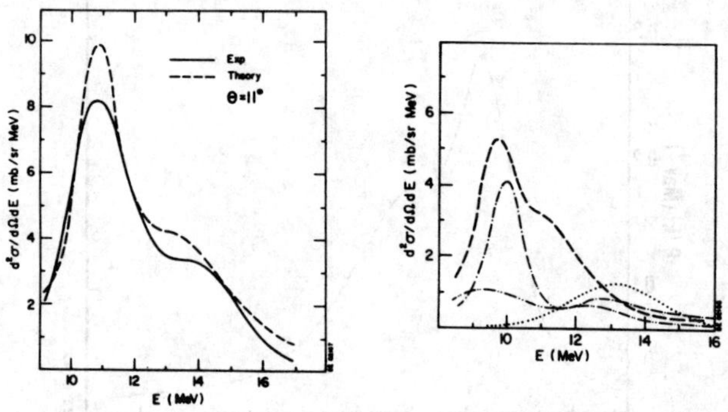

Fig. 27. On the left side the experimental (13) and theoretical (54) spectra at Θ=11° for 152 MeV α-scattering from ^{208}Pb are displayed. On the right side the predicted contribution of the GQR (---), GMR (....) and of the L=4 (-·-·-) and L=6 (-·-·-·-) multipolarities are shown.

In fig. 28 are collected few examples of the strong destructive interference discussed in sect. IIB between the contributions to the matrix element $V_{a\alpha}$ coupling the GQR, the GDR and the GTR of ^{208}Pb to the doorway states. This cancellation (roughly a factor of 2) helps explain why the giant vibrations are seen as distinct peaks up to much higher excitation energy than the single-particle states, (36).

The largest contributions (∿50%) to the damping are connected with the presence of the low-lying octupole vibration in the doorway states, (48)(49). However, a number of authors uses in the calculation of the strength functions uncorrelated 2p-2h states (55)-(59).

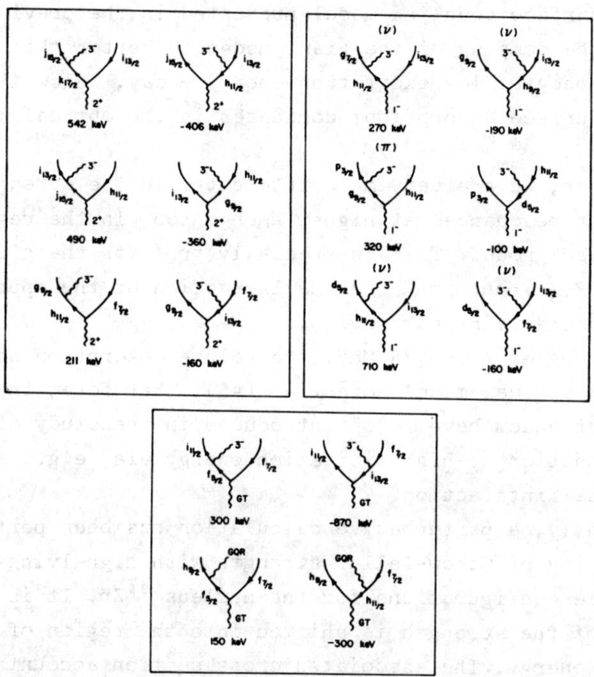

Fig. 28. Examples of cancellation between the contributions of fig. 23 to the matrix elements $V_{a\alpha}$ coupling the GQR, GDR and GTR of ^{208}Pb to a "doorway" 2p-2h state containing a collective surface vibration, (47).

For giant vibrations at relatively low energy, the spreading width of the main peak evaluated in this framework is smaller than the value for $\Gamma_v^{\downarrow}$ obtained including collective vibrations in the doorways. This is not surprising (60), because the density of uncorrelated 2p-2h states is lower (it may be a factor of two lower in the region of the GQR, (49)) and the coupling in eq. (12) smaller, being proportional to the transition moment $\beta_{\lambda'}$ of the vibration λ' in the doorway states.

III. Giant resonance strength at high excitation energy

The surface coupling model presented in the previous section describes the damping of the giant modes to better than a factor of two accuracy at low excitation energy - say, below 15-20 MeV - where the surface absorption dominates in the optical model potential (46).

However, we are very much interested in the strength function of the giant resonances at higher energy too, in the region of the continuum background. This is specially true for the giant modes, like the GTR, exhausting a small fraction of the appropriate sum rule in the main-peak region.

At energies above 25 MeV, the volume absorption dominates the nucleon optical-model potential (46). Therefore, the volume degrees of freedom have to be introduced in the study of the strength function at high excitation energy via, e.g., a G-matrix (36) residual interaction.

Recently, a perturbative calculation has been performed (61) for the mixing of Gamow-Teller strength with high-lying two-particle two-hole configurations for the nucleus ^{90}Zr. It is found that about 50% of the strength is shifted into the region of 15-50 MeV excitation energy. The associated cross section accounts for a substantial part (about 30%) of the continuum background in the spectrum of fig. 12, (61)(62). The interaction used in ref. (61) is a Yukawa fit to a G-matrix based on the Reid soft-core potential and with a long-range part identical to a one-pion-exchange potential (63). In this calculation, the contribution of the tensor part of the interaction to the strength is roughly the 30%. Of the same order is the total contribution of the ground state correlations, which tend to add incoherently to the final state correlations.

A similar calculation has been performed for the nucleus ^{40}Ca (53), using the same procedure and residual interaction of ref. (61). In this case, any strength associated to the GT operator σt_- comes from the fround state correlations. From the lowest order 2 particle-2 hole contributions, the strength function of fig. 29 is obtained. The GT strength associated to the lowest peak in fig. 29 is B(GT)=0.26. Experimentally, a 1^+ state at E_x=2.7±0.1 MeV in ^{40}Sc has been excited in the ^{40}Ca(p,n)^{40}Sc reaction at E_p=160 MeV (64), the associated B(GT) being equal to 0.21±0.04.

The agreement is encouraging, because the quenching of the σt_- operator produced by the coupling to the Δ_{33} resonance (65) has not been considered in the calculation.

In fig. 30, a preliminary comparison with the experimental

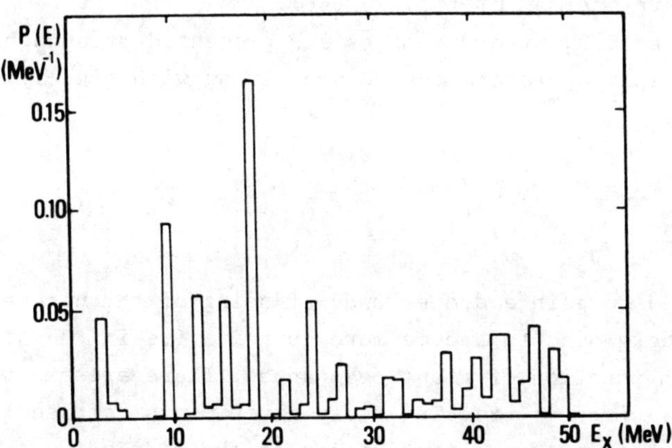

Fig. 29. Strength function for the operator σt_- in ^{40}Ca. The energy scale is with respect to the ground state of ^{40}Sc (53).

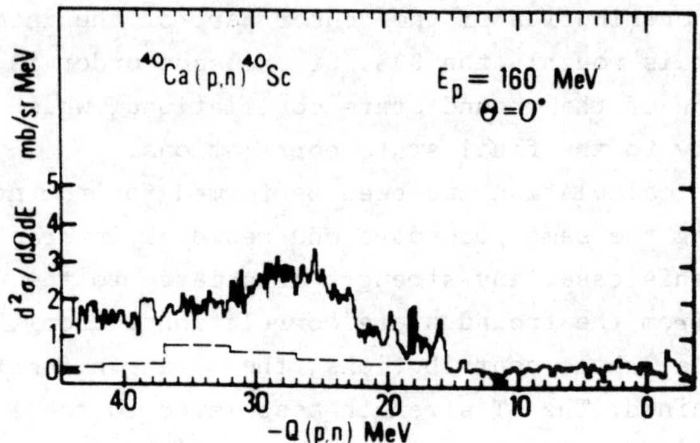

Fig. 30. Adapted from refs (53)(66). The neutron spectrum for the reaction $^{40}Ca(p,n)^{40}Sc$ at E_p=160 MeV and at zero degrees (64) is compared with the theoretical results (---) for the L=0 Gamow Teller transitions calculated as discussed in the text.

spectrum is displayed up to very high excitation energy, where the Q-value dependence reduces the cross section associated to the strength distribution of fig. 29 (66).

In refs (67), other examples are presented of quenching of the spin-isospin operators due to the mixing with high-lying configurations.

Perspectives

We need to gain a deeper understanding of the nature of the continuum background to become more quantitative in the study of the strength function of giant resonances. There are exciting theoretical developments (68). The decay experiments, cfr sect. II, are necessary to test the predictions of the models.

Acknowledgements

Discussions with G.F.Bertsch, R.A.Broglia and Xia Ke-ding are gratefully acknowledged.

References

(1) G.F. Bertsch, P.F. Bortignon and R.A. Broglia, Rev. Mod. Phys. 55 (1983) 287

(2) B.L. Berman and S.C. Fultz, Rev. Mod. Phys. 47 (1975) 713

(3) F.R. Bertrand, Nucl. Phys. A354 (1981) 129

(4) C. Gaarde, Nucl. Phys. A396 (1983) 127

(5) A. Richter, Phys. Scr. T5 (1983) 63
A. Richter, Proceedings of the Int. Conf. on Nucl. Phys., Florence (1983), eds P. Blasi and R.A. Ricci, p. 189

(6) C. Djalali et al., Nucl. Phys. A388 (1982) 1
G.M. Crawley et al. preprint IPNO DRE 83-15 Orsay, 1983

(7) J. Speth and A. van der Woude, Rep. Prog. Phys. 44 (1981) 719

(8) S. Stringari, these Proceedings and references therein

(9) F.T. Kuchnir et al., Phys. Rev. 161 (1967) 1236

(10) J.D. Bowman et al., Phys. Rev. Lett. 50 (1983) 1195

(11) D.M. Drake et al., Phys. Rev. Lett. 47 (1981) 1581

(12) R. Leicht et al., Nucl. Phys. A362 (1981) 111

(13) F.E. Bertrand et al., Phys. Rev. C22 (1980) 1832

(14) See e.g. G.R. Satchler, Proc. Int. School of Physics E. Fermi, Course LXIX, eds A. Bohr and R.A. Broglia, 1977, p. 271

(15) D.C. Youngblood et al., Phys. Rev. Lett. 39 (1977) 1188

(16) G. Kühner et al., Phys. Lett. 104B (1981) 189

(17) G.T. Wagner, in Nuclear Physics, eds C. Dasso, R.A. Broglia and A. Winther, 1982, p. 385

(18) K.T. Knöpfle et al., Phys. Lett. 74B (1978) 191

(19) K. Okada et al., Phys. Rev. Lett. 48 (1982) 1382

(20) W. Eyrich et al., Phys. Rev. C29 (1984) 418

(21) F. Zwarts et al., Phys. Lett. 125B (1983) 123

(22) S. Brandenburg et al., Phys. Lett. 130B (1983) 9

(23) F. Plasil, Nucl. Phys. A400 (1983) 417
F.E. Bertrand et al., Journ. de Phys. C4 (1984) 209

(24) J.R. Tinsley et al., Phys. Rev. C28 (1983) 1417

(25) T. Yamagata et al., Phys. Lett. 123B (1983) 169

(26) H.P. Morsch et al., Phys. Rev. C28 (1983) 1947

(27) H.P. Morsch et al., Phys. Rev. Lett. 45 (1980) 337

(28) T.A. Carey et al., Phys. Rev. Lett. 49 (1980) 239

(29) T.S. Yamagata et al., Phys. Rev. C23 (1981) 937

(30) C. Djalali et al., Nucl. Phys. A380 (1982) 42

(31) M. Buenerd, in Nuclear Physics, eds C.H. Dasso, R.A. Broglia and A. Winther, 1982, pag. 361

(32) K. Ikeda et al., Phys. Lett. 3 (1963) 271

(33) C.D. Goodman et al., Phys. Rev. Lett. 44 (1980) 1755

(34) C. Gaarde et al., Nucl. Phys. A334 (1980) 248

(35) C. Gaarde et al., Nucl. Phys. A369 (1981) 258

(36) C. Mahaux, these Proceedings

(37) W.G. Love and M.A. Franey, Phys. Rev. C24 (1981) 1073

(38) T.N. Taddeucci et al., Phys. Rev. C25 (1982) 1094

(39) W. Steffen et al., Phys. Lett. 95B (1980) 23

(40) W. Grass et al., Phys. Lett. 84B (1979) 296

(41) G. Eulenberg et al., Phys. Lett. 116B (1982) 113

(42) J.B. McGrory and B.H. Wildenthal, Phys. Lett. 103B (1981) 173

(43) G.M. Crawley et al., Phys. Lett. 127B (1983) 322

(44) N. Anantaraman et al., MSUCL-444 preprint, 1983

(45) N. van Giai, these proceedings

(46) J. Rapaport, Phys. Rep. 87 (1982) 25

(47) P.F. Bortignon, R.A. Broglia and C.H. Dasso, Nucl. Phys. A398 (1983) 221

(48) P.F. Bortignon and R.A. Broglia, Nucl. Phys. A371 (1981) 405
P.F. Bortignon, R.A. Broglia and F. Zardi; Proceedings of the Conference on Spin Vibration in Nuclei, Telluride 1982, in press

(49) J. Wambach, V.K. Mishra and Li Chu-Hsia, Nucl. Phys. A380 (1982) 285
H.R. Fiebig and J. Wambach, Nucl. Phys. A386 (1982) 381

(50) V.G. Soloviev, Ch. Stoyanov and V.V. Voronov, Nucl. Phys. A399 (1983) 141
J.S. Dehesa et al., Phys. Rev. C15 (1977) 1858

(51) M. Beiner et al., Nucl. Phys. A238 (1975) 29

(52) A. Bohr and B.R. Mottelson, Nuclear Structure, vol. II (Benjamin, Reading, 1975)
P.F. Bortignon, R.A. Broglia, D.R. Bes and R. Liotta, Phys. Rep. 30C (1977) 305

(53) P.F. Bortignon, R.A. Broglia and Xia Ke-ding, Journ. de Phys. C4 (1984) 209

(54) F.E.Serr, P.F. Bortignon and R.A. Broglia, Nucl. Phys. A393 (1983) 109

(55) M. Damos and W. Greiner, Phys. Rev. 138 (1965) 876B

(56) T. Hoshino and A. Arima, Phys. Rev. Lett. 37 (1976) 266
W. Knüpfer, Lect. Notes in Phys. 158 (1982) 225
W. Knüpfer and M.G. Huber, Z. Phys. A276 (1976) 99

(57) S. Adachi and S. Yoshida, Nucl. Phys. A306 (1978) 53
S. Adachi, Phys. Lett. 125B (1983) 5
S. Adachi and N. Auerbach, Phys. Lett. 131B (1983) 11
S. Adachi, these Proceedings

(58) K. Muto et al., Phys. Lett. 118B (1982) 261 and Phys. Lett. 127B (1983) 291

(59) B. Schwensinger and J. Wambach, Phys. Lett. 134B (1984) and Phys. Rev. C29 (1984) 663

(60) G.F. Bertsch et al., Phys. Lett. 80B (1979) 161

(61) G.F. Bertsch and I. Hamamoto, Phys. Rev. C26 (1982) 1323

(62) O. Scholten, G.F. Bertsch and H. Toki, Phys. Rev. C27 (1983) 2975

(63) G.F. Bertsch et al., Nucl. Phys. A284 (1977) 399

(64) T.N. Taddeucci et al., Phys. Rev. C28 (1983) 2511

(65) W. Weise, Nucl. Phys. A396 (1983) 373 and references therein
K. Nakayama et al., Phys. Rev. Lett. 52 (1984) 500
T. Chaon, K. Shimizu and A. Arima, Phys. Lett. 138B (1984) 345

(66) C. Gaarde, private communication

(67) I.S. Towner and F.C. Khanna, Nucl. Phys. A399 (1983) 335 and references therein
W. Bentz et al., Nucl. Phys. A412 (1984) 481 and references therein

(68) G.F. Bertsch and O. Scholten, Phys. Rev. C25 (1982) 804
H. Esbensen and G.F. Bertsch, preprint MSUCL-445 (1984)
F. Osterfeld and A. Schulte, Phys. Lett. 138B (1984) 23

II. NUCLEAR DYNAMICS AT LOW ENERGY

QUASI ELASTIC HEAVY ION REACTIONS

D. M. Brink

Department of Theoretical Physics
University of Oxford
1 Keble Road
Oxford, OX1 3NP
U.K.

Lectures 1 and 2: Semi-Classical Scattering.

1. Elastic Scattering

Lectures 1 and 2 will concentrate on the application of path integral methods to derive semi-classical approximations to scattering problems.

Scattering problems are usually formulated in terms of the scattering wave function. This satisfies the Schrödinger equation and the boundary condition

$$\psi(\underset{\sim}{r}) = e^{ikz} + \frac{1}{r} e^{ikr} f(\theta) \tag{1}$$

when r is large. In (1) the incident wave is parallel to the z-axis, θ is the scattering angle and the incident energy is $E_0 = \hbar^2 k^2/2m$. Feynman's propagator $K(\underset{\sim}{r}, \underset{\sim}{p}_0, T)$ gives the amplitude for a final position $\underset{\sim}{r}$ at time T when the initial momentum is $\underset{\sim}{p}_0$. We want to relate this propagator to the scattering wave function.

This can be done by noting that the unitary time development operator between a final state $|\underset{\sim}{r}\rangle$ in the coordinate representation and an initial state $|\underset{\sim}{p}_0\rangle$ in the momentum representation

$$K(\underset{\sim}{r}, \underset{\sim}{p}_0, T) = \langle \underset{\sim}{r} | \exp(-iHT/\hbar) | \underset{\sim}{p}_0 \rangle \tag{2}$$

In (2) $H = H_0 + V_0$ where $H_0 = \underset{\sim}{p}^2/2m$ is the kinetic energy operator and $V_0(r)$ is the potential causing the scattering. The state $|\underset{\sim}{p}_0\rangle$ is an eigenstate of H_0 with energy E_0. The initial and final states in (2) are normalized so that

$$\langle \underset{\sim}{r} | \underset{\sim}{p}_0 \rangle = \exp(i \underset{\sim}{p}_0 \cdot \underset{\sim}{r}/\hbar) \tag{3}$$

A scattering wave with incident momentum $\underset{\sim}{p}_0$ is given by a matrix element of the Moller wave operator (Taylor [1])

$$\psi(\underset{\sim}{r}, \underset{\sim}{p}_0) = \langle \underset{\sim}{r} | \Omega_+ | \underset{\sim}{p}_0 \rangle \tag{4}$$

which is defined by

$$\Omega_+ = \lim_{T\to\infty} \exp(-iHT/\hbar)\exp(iH_0T/\hbar) \tag{5}$$

We obtain a relation between the scattering wave function and the propagator (2) by using (4) and (5),

$$\psi(\underset{\sim}{r}) = \lim_{T\to\infty} K(\underset{\sim}{r}, \underset{\sim}{p}_0, \underset{\sim}{T})\exp(iE_0T/\hbar) \tag{6}$$

Both the asymptotic form (1) and the relation (6) have to be modified in the presence of a Coulomb potential because of the long range character of the field. The generalization of the asymptotic form (1) is

$$\psi(\underset{\sim}{r}) \sim e^{i(kz + n\ln k(r-z))} + \frac{1}{r} f(\theta) e^{i(kr - n\ln(2kr))} \tag{7}$$

where $n = Z_1 Z_2 e^2/\hbar v$ is the Sommerfeld parameter. Equation (5) is modified using a method given by Dollard [2] and equation (6) is replaced by

$$\psi(\underset{\sim}{r}, \underset{\sim}{p}_0) = \lim_{T\to\infty} K(\underset{\sim}{r}, \underset{\sim}{p}_0, T)\exp i\phi_c(E_0, T) \tag{8}$$

$$\phi_c(E_0, T) = E_0T/\hbar + n\ln(4E_0T/\hbar)$$

2. A Path Integral Formula for the Propagator

We consider the scattering of a particle with mass m by a potential V (r). The propagator K in equation (2) can be written as a path integral

$$K(\underset{\sim}{r}, \underset{\sim}{p}_0, T) = \int D[\underset{\sim}{r}(t)]\exp\left(\frac{i}{\hbar} S[\underset{\sim}{r}(t)]\right) \tag{9}$$

$$S[\underset{\sim}{r}(t)] = \int_0^T L(\underset{\sim}{r}, \dot{\underset{\sim}{r}})\, dt + \underset{\sim}{p}_o \cdot \underset{\sim}{r}_o \, . \tag{10}$$

In equation (10) $L = m\dot{r}^2/2 - V_o(r)$ is the Lagrangian. The integral (9) is taken over all paths satisfying the boundary conditions

$$m\,\dot{\underset{\sim}{r}}(0) = \underset{\sim}{p}_o \quad , \quad \underset{\sim}{r}(T) = \underset{\sim}{r} \, . \tag{11}$$

The most common form for the propagator (Feynman and Hibbs [3] Schulmann [4]) corresponds to a transition from an initial position $\underset{\sim}{r}_o$ to a final position $\underset{\sim}{r}$. This form is related to (9) by aFourier transform

$$K(\underset{\sim}{r}, \underset{\sim}{p}_o, T) = \int d\underset{\sim}{r}_o \, K(\underset{\sim}{r}, \underset{\sim}{r}_o, T)\, exp(i\, \underset{\sim}{p}_o \cdot \underset{\sim}{r}_o / \hbar)$$

The boundary conditions in (9) are more appropriate for a scattering problem. Koeling and Malfliet [5] have studied a path integral formulation of scattering theory based on the S-matrix rather than the Møller wave operator. They consider a propagator from an initial momentum $\underset{\sim}{p}_o$ to a final momentum $\underset{\sim}{p}$.

3. The Semi-Classical Wave Function

A semi-classical expression for the propagator is obtained by evaluation of the path integral in (9) with the stationary phase approximation. A stationary path satisfies the condition

$$\delta S[\underset{\sim}{r}(t)] = \int_0^T dt \left(\frac{\partial L}{\partial \underset{\sim}{r}} - \frac{d}{dt} \frac{\partial L}{\partial \dot{\underset{\sim}{r}}} \right) \cdot \delta \underset{\sim}{r}(t) + \underset{\sim}{p} \cdot \delta \underset{\sim}{r} + \underset{\sim}{r}_o \cdot \delta \underset{\sim}{p}_o \tag{12}$$

$$= 0$$

for small variations $\delta \underset{\sim}{r}(t)$ about the path $\underset{\sim}{r}(t)$. Paths determined by (12) satisfy the classical equations of motion with the boundary conditions (11). The action $S(\underset{\sim}{r}, \underset{\sim}{p}_o, T)$ for a stationary path is a function of the final position $\underset{\sim}{r}$, the initial momentum $\underset{\sim}{p}_o$ and the time interval T. Equation (12) shows that the final momentum $\underset{\sim}{p}$ and the initial position $\underset{\sim}{r}_o$ and the energy E are given by

$$\underset{\sim}{p} = \frac{\partial S}{\partial \underset{\sim}{r}} \quad , \quad \underset{\sim}{r}_o = \frac{\partial S}{\partial \underset{\sim}{p}_o} \quad , \quad E = -\frac{\partial S}{\partial T} \tag{13}$$

The stationary phase evaluation of the path integral (9) gives

$$K(\underset{\sim}{r}, \underset{\sim}{p}_o, T) = \sum_s K_s(\underset{\sim}{r}, \underset{\sim}{p}_o, T)$$
$$K_s(\underset{\sim}{r}, \underset{\sim}{p}_o, T) = A_s(\underset{\sim}{r}, \underset{\sim}{p}_o) \exp\left(\frac{i}{\hbar} S(\underset{\sim}{r}, \underset{\sim}{p}_o, T)\right) \tag{14}$$

where the sum s is over all stationary paths $r_s(t)$ satisfying the boundary conditions (11). The amplitude

$$A_s(\underset{\sim}{r}, \underset{\sim}{p}_o) = \left[D_s(S)\right]^{\frac{1}{2}} e^{-\frac{1}{2} i \nu_s} \tag{15}$$

where ν_s is a Maslov index and $D_s(S)$ is a VanVleck [6] determinant. This can be written in three equivalent ways by using (13)

$$D(S) = \det\left(\frac{\partial^2 S}{\partial r_i \partial p_{oj}}\right) = \det\left(\frac{\partial p_i}{\partial p_{oj}}\right) = \det\left(\frac{\partial r_{oj}}{\partial r_i}\right) \tag{16}$$

The pre-exponential factor A in (15) can be obtained by using the methods of Levit and Smilansky [7] . Formula (16) is given by Gutzwiller [8] and others. Miller [9] has given a simple derivation of the magnitude of A based on a unitary argument.

A semi-classical formula for the scattering wave function is obtained by substituting (16) into equations (6) or (8). It is

$$\psi(\underset{\sim}{r}, \underset{\sim}{p}_o) = \sum_s A_s(\underset{\sim}{r}, \underset{\sim}{p}_o) \exp\left(\frac{i}{\hbar} W_s(\underset{\sim}{r}, \underset{\sim}{p}_o)\right) \tag{17}$$

$$W_s(\underset{\sim}{r}, \underset{\sim}{p}_o) = \lim_{T\to\infty} \left(S_s(\underset{\sim}{r}, \underset{\sim}{p}_o, T) + \phi_c(E_o, T)\right)$$
$$= \int_{\underset{\sim}{r}_o}^{\underset{\sim}{r}} p \cdot d\underset{\sim}{r} - \hbar(kr_o - n \ln 2kr_o) \tag{18}$$

The integral in (18) is taken along a classical path with initial momentum $\underset{\sim}{p}_o$ and final position $\underset{\sim}{r}$. When the initial position $\underset{\sim}{r}_o$ is far from the scatterer the $\underset{\sim}{r}_o$ dependence in the integral in (18) is cancelled by the r_o dependence in the second term.

When the final position $\underset{\sim}{r}$ is far from the scatterer the sum in (17) contains two kinds of terms. One is from a path which passes far from the target and which is hardly influenced at all by $V_0(r)$. This term gives rise to the first part of the asymptotic form of the wave function (7). The other terms correspond to trajectories which pass close to the target and are deflected by it. The action integral along this path can be evaluated in polar coordinates and equation (18) for W becomes

$$W(r, p_o)/\hbar = kr - n \ln 2kr + 2\delta(\lambda) - \lambda\Theta(\lambda). \tag{19}$$

In (19) λ is the semi-classical angular momentum. It is related to the impact parameter b by $\lambda = kb$ and to the partial wave quantum number ℓ by $\lambda = \ell + \frac{1}{2}$. The function $\Theta(\lambda)$ is the classical deflection angle for an orbit with angular momentum λ and $\delta(\lambda)$ is the semi-classical phase shift. These two quantities are related by

$$2\frac{d\delta(\lambda)}{d\lambda} = \Theta(\lambda) \tag{20}$$

The semi-classical phase is given by the asymptotic form of the integral

$$\int_{r_t}^{r} p_r\, dr \sim \hbar\left(kr - n \ln 2kr + \delta(\lambda) - \tfrac{1}{2}\pi\lambda\right) \tag{21}$$

when r is large. In (21) p_r is the radial momentum

$$p_r(r) = \left[2m\left(E - V_0(r) - \hbar^2\lambda^2/2mr^2\right]^{\frac{1}{2}} \tag{22}$$

and r_t is the classical turning point which is found by solving $p_r(r_t) = 0$. When $V_o(r)$ is the Coulomb potential of a point charge then $\delta(\lambda)$ is the semi-classical Coulomb phase shift

$$\sigma(\lambda) = \tfrac{1}{2} n \ln(n^2+\lambda^2) - n + \lambda \tan^{-1}(n/\lambda)$$

where n is the Sommerfeld parameter. The pre-exponential factor A corresponding to (19) is given by

$$|A|^2 = \sigma_{cL}(\lambda)/r^2 \tag{23}$$

where

$$\sigma_{cL}(\lambda) = \lambda(d\lambda/d\Theta)/k^2 \sin(\Theta(\lambda)) \tag{24}$$

is the classical cross-section associated with angular momentum λ. Combining (23), (19), (17) and (7) gives a semi-classical formula for the scattering amplitude

$$\begin{aligned} f(\theta) &= \Sigma_s f_s(\theta) \\ f_s(\theta) &= \sqrt{\sigma_{cL}(\lambda_s)}\, \exp i\left(2\delta(\lambda_s) - \lambda_s \Theta(\lambda_s) - \tfrac{1}{2}\pi\nu_s\right) \end{aligned} \tag{25}$$

In (25) the sum is over all classical orbits with the same angle of observation θ ; that is the deflection angle $\Theta(\lambda_s)$ correspon-ding to angular momentum λ_s should be related to θ by

$$\Theta(\lambda_s) = \pm\theta - 2m\pi \tag{26}$$

where m is an integer. The relation (26) allows for trajectories which orbit around the target m times before emerging. Equation (25) takes into account the interference between the amplitudes associated with different orbits which have the same scattering angle. The integer ν_s is a Maslov or Morse index and is important for fixing the phase of the interference.

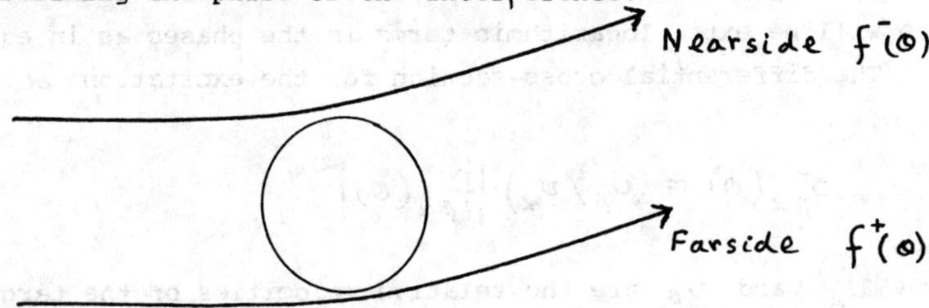

Fig. 1. Illustration of nearside and farside orbits.

In many cases terms in (26) with $m = 0$ give the dominant contribution to the cross-section. The positive sign in (26) is a nearside orbit while the negative sign corresponds to a farside orbit. These orbits are illustrated in fig. 1.

4. Inelastic Scattering.

We now show how the arguments used in the first part of this lecture can be generalized to apply to inelastic scattering. We suppose that the target nucleus can be excited but do not consider excitation of the projectile. The initial state of the target is denoted by $|\alpha\rangle$ and has internal energy $\mathcal{E}_\alpha$. The final state is $|\beta\rangle$ with internal energy $\mathcal{E}_\beta$. The relative momentum in the initial state is $\underset{\sim}{p}_\alpha$ and the kinetic energy is $E_\alpha = p_\alpha^2/2m$. The corresponding quantities in the final state are $\underset{\sim}{p}_\beta$ and E_β. Conservation of energy gives

$$E_\alpha + \mathcal{E}_\alpha = E + \mathcal{E}_\beta = E$$

where E is the total energy. The asymptotic form of the total scattering wave function in the channel $|\beta\rangle$ is

$$\begin{aligned}\psi &\sim |\beta\rangle \frac{1}{r} f_{\beta\alpha}(\theta)\, e^{ik_\beta r} \quad ; \quad \beta \neq \alpha \\ &\sim |\alpha\rangle \left(e^{ik_\alpha z} + \frac{1}{r} e^{ik_\alpha r} f_{\alpha\alpha}(\theta)\right)\end{aligned} \tag{27}$$

Here $f_{\beta\alpha}(\theta)$ is the scattering amplitude for scattering from the state $|\alpha\rangle$ to the state $|\beta\rangle$. The form (27) neglects the effects of the long range Coulomb field. In the presence of a Coulomb field there will be extra logarithmic terms in the phases as in equation (7). The differential cross-section for the excitation α to β is

$$\sigma_{\beta\alpha}(\theta) = (v_\beta/v_\alpha)\, |f_{\beta\alpha}(\theta)|^2 \tag{28}$$

where v_α and v_β are the relative velocities of the target and projectile in the initial and final states.

The hamiltonian for the interactinb system is

$$H = p^2/2m + V_0(r) + h(r, \xi)$$
$$h(r, \xi) = H_0(\xi) + V(r, \xi) \tag{29}$$

where $\underset{\sim}{r}$ is the relative coordinate of the target and projectile and $\underset{\sim}{p}$ is the conjugate momentum. The internal coordinates of the target and projectile are denoted by ξ, $H_0(\xi)$ is the internal hamiltonian and $V(\underset{\sim}{r}, \xi)$ is the coupling which produces excitation. The central potential $V_0(r)$ contains the monopole part of the Coulomb interaction whereas the central part of the optical potential can be included either in $V_0(r)$ or in $V(\underset{\sim}{r}, \xi)$. If the scattering problem is solved exactly the final results are identical. If approximate methods are used different ways of dividing the interaction between $V_0(r)$ and $V(\underset{\sim}{r}, \xi)$ yield different approximations.

Pechukas [10] writes a path integral for the inelastic propagator $K_{\beta\alpha}(\underset{\sim}{r}, t_1, \underset{\sim}{p}_0\ t_0)$ which gives the amplitude for the system to move from a relative momentum $\underset{\sim}{p}_0$ at $t = t_0$ to a final position $\underset{\sim}{r}$ at $t = t_1$, while the internal state changes from α to β.

$$K_{\beta\alpha}(\underset{\sim}{r}\, t_1, \underset{\sim}{p}_0 t_0) = \int D[\underset{\sim}{r}(t)]\, T_{\beta\alpha}[\underset{\sim}{r}(t)]\, exp\left(\tfrac{i}{\hbar} S[\underset{\sim}{r}(t)]\right) \tag{30}$$

The action $S[\underset{\sim}{r}(t)]$ is given by

$$S[\underset{\sim}{r}(t)] = \int_{t_0}^{t_1} \left(\tfrac{1}{2} m \dot{\underset{\sim}{r}}^2 - V_0(r)\right) dt + \underset{\sim}{p}_0 \cdot \underset{\sim}{r}_0 \tag{31}$$

and $T_{\beta\alpha}[r(t)]$ is the amplitude for a transition between the initial state α and the final state β as the relative motion passes along $\underset{\sim}{r}(t)$. The integral (30) is over all paths satisfying the boundary conditions

$$m\dot{\underset{\sim}{r}}(t_0) = \underset{\sim}{p}_0 \quad , \quad \underset{\sim}{r}(t_1) = \underset{\sim}{r} \tag{32}$$

The amplitude $T_{\beta\alpha}[\underset{\sim}{r}(t)]$ is given by

$$T_{\beta\alpha}[\underset{\sim}{r}(t)] = \langle \beta | U[\underset{\sim}{r}(t), t_1, t_0] | \alpha \rangle \tag{33}$$

where U is the unitary operator describing the time development of the internal state when the relative motion is $\underset{\sim}{r}(t)$. It is a solution of the Schrødinger equation

$$i\hbar \frac{\partial U}{\partial t} = h(\underset{\sim}{r}(t), \xi)\, U \tag{34}$$

with initial condition $U = 1$ when $t = t_o$. The path integral (30) is a direct generalisation of (9). It was derived by Pechukas [10] with different boundary conditions. He specified the initial position rather than the initial momentum, but the present choice (32) is more suitable for use in scattering problems.

The path integral (30) is more complicated than the corresponding integral (9) for the elastic propagator and is more difficult to evaluate. Pechukas [10] expands about a path which makes the phase of the integrand in (30) stationary, that is

$$\delta S + \delta(\mathrm{Im}\ln T_{\beta\alpha}). \tag{35}$$

This method has the disadvantage that the equation for the stationary path $\underset{\sim}{r}(t)$ is complicated and the pre-exponential factor which is the generalisation of A in equation (15) cannot be evaluated accurately (Sukumar [11]). A simpler approach is to assume that $T_{\beta\alpha}[\underset{\sim}{r}(t)]$ is a slowly varying function of the path $\underset{\sim}{r}(t)$ and to evaluate it for the stationary path $\underset{\sim}{r}_s(t)$ determined by the condition $\delta S = 0$. Then

$$K_{\beta\alpha}(\underset{\sim}{r}\,t_1, \underset{\sim}{p}_o\,t_o) \simeq \sum_s T_{\beta\alpha}[r_s(t)]\, K_s(\underset{\sim}{r}\,t_1, \underset{\sim}{p}_o\,t_o) \tag{36}$$

where K_s is the elastic propagator in equation (14). The sum ia over all stationary paths $r_s(t)$ satisfying the boundary condition (32). Equation (36) for the propagator leads to an approximate formula for the inelastic amplitude

$$f_{\beta\alpha}(\theta) \simeq \sum_s t_{\beta\alpha}(\lambda_s)\, f_s(\theta) \tag{37}$$

In equation (37) $f_s(\theta)$ is the semi-classical elastic scattering amplitude (25) and $t_{\beta\alpha}(\lambda_s)$ is the transition amplitude calculated for the orbit $\underset{\sim}{r}_s(t)$ in the interaction representation

$$T_{\beta\alpha}\left[\underset{\sim}{r}_s(t)\right] = t_{\beta\alpha}(\lambda_s)\exp\left(\frac{i}{\hbar}(\varepsilon_\alpha t_0 - \varepsilon_\beta t_1)\right) \tag{38}$$

It is useful to choose the orbit $\underset{\sim}{r}_s(t)$ so that $t = 0$ correspondsng to the distance of closest approach. Then $t_{\beta\alpha}[r_s(t)]$ is independent of the final position r and $t_1(r)$ is the time from the point of closest approach to r. Dos Aidos and Brink [12] show that the phase depending on t_1 in (38) gives the modification of the phase of the scattered wave function required by energy conservation to first order in the excitation energy

If $V_0(r)$ in (29) is the monopole part of the Coulomb interaction the stationary orbit $\underset{\sim}{r}_s(t)$ is a Coulomb orbit and there is just one term in the sum (37)

$$f_{\beta\alpha}(\theta) \simeq f_c(\theta)\, t_{\beta\alpha}(\lambda_\theta) \tag{39}$$

where $f_c(\theta)$ is the Coulomb elastic scattering amplitude. Then $t_{\beta\alpha}$ is calculated for a Coulomb orbit with angular momentum $\lambda_\theta = n \cot \frac{1}{2}\theta$ Equation (39) gives a cross-section

$$\sigma_{\beta\alpha}(\theta) = (v_\beta / v_\alpha)\, \sigma_c(\theta)\, \left| t_{\beta\alpha}(\lambda_\theta)\right|^2 \tag{40}$$

This kind of formula is used in the Alder-Winther [13] theory of Coulomb excitation.

5. A Uniform Approximation for Elastic Scattering.

The stationary phase formula (17) for the scattering wave function breaks down in the neighbourhood of caustics. These are points, lines or surfaces where $A_s(\underset{\sim}{r}, \underset{\sim}{p}_0)$ becomes infinite. In these circumstances it is often possible to obtain good results by using a uniform approximation to the path integral.

In the following we give a very simplified account of uniform approximations. We refer to Berry [14], Levit and Smilansky [7] or Knoll and Schaeffer [15] for a more complete treatment.

Suppose I and J are two multi-dimensional integrals which both depend on m parameters $a_1 \ldots\ldots a_m$. The integral J is a uniform approximation to I if there is a correspondence between the stationar points of I and J as a function of the parameters a_i and if the stationary phase evaluations of both integrals give the same results for all parameter sets in some region. In particular, the caustics of I and J must map onto each other. Away from the caustics the stationary phase method gives a good approximation to I for these parameter values. The idea is that I can still be a good approximati to J near the caustics where the stationary phase methods fails. Som mathematical conditions for this hope to be realized have been reviewed by Duistermaat [16]. We shall assume that the appropriate conditions are satisfied.

Uniform approximations have many nice features. The integrals I and J can have a different number of dimensions. In particular, I can be a path integral and J a one- or two-dimensional integral. This extension is possible because the method depends on the type of caustic and not on the number of dimensions. There is no unique uniform approximation to a path integral. Different approximations must be asymptotically equal as $\hbar \to 0$.

We now give a diffraction integral which is a useful uniform approximation to the elastic scattering amplitude $f(\theta)$ for forward and intermediate scattering angles. It can be written as an integral over a two-dimensional angular momentum vector $\underset{\sim}{\lambda} = k\underset{\sim}{b}$ which is proportional to the impact parameter vector $\underset{\sim}{b}$. The integra is

$$f(\theta) = \frac{1}{2\pi i k}\left(\frac{\theta}{\sin\theta}\right)^{\frac{1}{2}} \sum_{m=-\infty}^{\infty} (-1)^m \int_0^{\infty}\int_0^{2\pi} \lambda\, d\lambda\, d\phi\; e^{i F_m(\lambda,\phi;\theta)} \tag{41}$$

where λ and ϕ are the components of $\underset{\sim}{\lambda}$ in cylindrical coordinates and

$$F_m(\lambda,\phi;\theta) = 2\delta(\lambda) - \lambda\theta\cos\phi + 2m\pi\lambda \tag{42}$$

It is a simple matter to check that the stationary phase evaluation of (41) gives the same result as the stationary phase formula (25) obtained from the path integral. Different values of m allow for deflection angles $\Theta(\lambda) = \pm\theta - 2m\pi$ which orbit around the target several times before emerging as in equation (26). The m = 0 term in (41) gives the dominant contribution to $f(\theta)$ in most applications. The integral over ϕ in (41) can be evaluated to give a Bessel function $J_0(\lambda\theta)$ and if only the m = 0 term is retained

$$f(\theta) = \frac{1}{ik}\left(\frac{\theta}{\sin\theta}\right)^{\frac{1}{2}}\int_0^\infty \lambda d\lambda\, J_0(\lambda\theta) e^{2i\delta(\lambda)} \tag{43}$$

Formula (43) can be obtained by other methods (cf. Frahn [17]) and it can also be regarded as a generalization of the eikonal approximation.

Equation (41) is a valid uniform approximation for forward and intermediate angles but it is singular at $\theta = \pi$ and cannot be used for backward scattering. An alternative form which is valid for θ near π is best written in terms of $\vartheta = \pi - \theta$

$$f(\theta) = \frac{1}{2\pi ik}\left(\frac{\vartheta}{\sin\vartheta}\right)^{\frac{1}{2}}\sum_{n=-\infty}^{\infty}(-1)^n\int_0^\infty\int_0^{2\pi}\lambda d\lambda\, d\phi\, e^{i\bar{F}_n(\lambda,\phi;\theta)}$$

$$\bar{F}_n(\lambda,\phi;\vartheta) = 2\delta(\lambda) - \lambda\vartheta\cos\phi + (2n-1)\pi\lambda$$

6. A Diffraction Formula for Inelastic Scattering.

Equations (37) and (39) are stationary phase formulae for the inelastic scattering amplitude. Equation (37) includes interference effects but not diffraction and both break down on the caustics at $\theta = 0$ and $\theta = \pi$. They also fail near a rainbow caustic, In this section we give a uniform approximation which is valid for all except backward angles and which is a direct generalization of the diffraction integral (41). Equation (41) contains a sum over m but normally only the m = 0 term is important and we omit the other terms in the discussion in this section. They can be added easily if required. The generalization of (41) is

$$f_{\beta\alpha}(\Theta)=\frac{1}{2\pi i k}\left(\frac{\Theta}{\sin\Theta}\right)^{\frac{1}{2}}\int_0^{\infty}\int_0^{2\pi}\lambda\,d\lambda\,d\phi\,t_{\beta\alpha}(\lambda,\phi)\,e^{i(2\delta(\lambda)-\lambda\Theta\cos\phi)} \tag{44}$$

The amplitude $t_{\beta\alpha}(\lambda,\phi)$ is calculated for an orbit in the potential $V_0(r)$ with angular momentum λ. The plane of the orbit makes an angle ϕ with the (x,y)-plane. Here we use axes with the incident beam direction $\underset{\sim}{k}_\alpha$ parallel to the z-axis and the y-axis in the direction of $\underset{\sim}{k}_\alpha \times \underset{\sim}{k}_\beta$ where $\underset{\sim}{k}_\beta$ is the wave-vector of the scattered particle. These are standard axes for the distorted wave Born approximation and we call them DWBA-axes. If the integral (44) is evaluated by the stationary phase approximation treating $t_{\beta\alpha}(\lambda,\phi)$ as a slowly varying function of λ and ϕ then (44) reduces to (37), so (44) can be regarded as a uniform approximation to the path integral for $f_{\beta\alpha}(\Theta)$.

When the initial and final states α and β are angular momentum eigenstates with quantum numbers (J_α, M_α) and (J_β, M_β) the ϕ dependence of $t_{\beta\alpha}$ is given by

$$t_{\beta\alpha}(\lambda,\phi) = t_{\beta\alpha}(\lambda)\,e^{i(M_\alpha - M_\beta)\phi} \tag{45}$$

This is because the total z-components of angular momentum is conserved and the change in the z-component of orbital angular momentum is $M_\beta - M_\alpha$. When (45) is substituted into (44) the ϕ-integral can be evaluated to give a Bessel function and

$$f_{\beta\alpha}(\Theta)=\frac{i^{M-1}}{k}\left(\frac{\Theta}{\sin\Theta}\right)^{\frac{1}{2}}\int_0^{\infty}\lambda\,d\lambda\,t_{\beta\alpha}(\lambda)\,e^{2i\delta(\lambda)}\,J_M(\lambda\Theta) \tag{46}$$

where $M = M_\beta - M_\alpha$. Formulae equivalent to (46) have been obtained by Potgieter and Frahn [18], Hahne [19], Frahn [20] and others. Using conventional methods starting from the DWBA. The arguments here show that the same results can be derived from the path integral. The path integral formula (46) is even more general because the $t_{\beta\alpha}(\lambda)$ can include multiple excitation effects.

In the following we restrict the discussion to the case where the initial state α has zero angular momentum and the final state β has angular momentum quantum numbers (L, M). We write the amplitude

$f_{\beta\alpha}$ as f_{LM}. When the scattering angle Θ is large the integral over ϕ in (44) can be calculated by the stationary phase approximation. There are two stationary points at $\phi = 0$ and $\phi = \pi$. Using the stationary phase formula we get

$$f_{LM}(\Theta) = f^{-}_{LM}(\Theta) + f^{+}_{LM}(\Theta) \tag{47}$$

$$f^{\pm}_{LM}(\Theta) = \frac{1}{ik}\int_0^\infty \left(\frac{\lambda}{2\pi \sin\Theta}\right)^{\frac{1}{2}} d\lambda\, t_{LM}(\lambda)\, e^{i(2\delta(\lambda) \pm \lambda\Theta \mp \frac{1}{4}\pi)}$$

The nearside amplitude (fig. 1) f^{-}_{LM} comes from the stationary point $\phi = 0$ while the farside amplitude is the contribution from $\phi = \pi$. The factor $(-1)^M$ comes from the term $\exp(-iM\phi)$ in (45) evaluated at $\phi = \pi$.

It is often convenient to write (47) with respect to Basel axes. The x-axis is parallel to the incident beam direction $\underset{\sim}{k}_\alpha$, while the z-axis is perpendicular to the reaction plane direction $\underset{\sim}{k}_\alpha \times \underset{\sim}{k}_\beta$. The scattering amplitude $\mathcal{F}_{LM}(\Theta)$ in the new coordinate system is related to f_{LM} by a rotation

$$\mathcal{F}_{LM}(\Theta) = (-1)^M \sum_K i^K d^L_{KM}(\tfrac{1}{2}\pi)\, f_{LK}(\Theta) \tag{48}$$

(Frahn [20]) . The $d^L_{KM}(\frac{1}{2}\pi)$ are rotation matrices. In the new axes (47) takes the form

$$\mathcal{F}_{LM}(\Theta) = \mathcal{F}^{-}_{LM}(\Theta) + (-1)^L \mathcal{F}^{+}_{LM}(\Theta) \tag{49}$$

$$\mathcal{F}^{\pm}_{LM}(\Theta) = \frac{1}{ik}\int_0^\infty \left(\frac{\lambda}{2\pi \sin\Theta}\right)^{\frac{1}{2}} d\lambda\, B_{L\mp M}(\lambda)\, e^{i(2\delta(\lambda) \pm \lambda\Theta \mp \frac{1}{4}\pi)}$$

The transition amplitudes $B_{LM}(\lambda)$ in the Basel axes are obtained from the $t_{LM}(\lambda)$ in (47) by a transformation like (48). The structure of the farside term in (49) can be understood by noting that it corresponds to $\phi = \pi$ and is obtained from $\phi = 0$ by a rotation through π about the x-axis in the Basel coordinate system. For such a rotation

$$B_{LM}(\lambda) \rightarrow (-1)^L B_{L-M}(\lambda) \tag{50}$$

Formulae equivalent to (47) and (49) have recently been used by Dean and Rowley [21] .

Lecture 3: Angular Correlations.

1. An Angular Correlation Experiment.

In this lecture an angular correlation experiment made recently by Bhowmik and Rae [22] is discussed. They study the α -decay of excited states produced by inelastic scattering. An example is

$$^{12}C + {}^{18}O \rightarrow {}^{12}C + {}^{18}O^* \rightarrow {}^{12}C + ({}^{14}C + \alpha)$$

at a laboratory energy of 82 MeV. A systematic dependence of the correlations on the scattering angle in the center-of-mass frame is observed, which allows the transferred angular momentum to be determined. It is possible to extract information about the spins of excited states from the experimental data.

2. The Angular Distribution.

Following the argument in section 6 of lecture 2 the inelastic scattering amplitude for exciting a state with angular momentum (L,M) from the ground state in the Basel axes is decomposed into nearside and farside components

$$\mathcal{F}_{LM}(\theta) = \mathcal{F}^{-}_{LM}(\theta) + (-1)^L \mathcal{F}^{+}_{LM}(\theta) \tag{51}$$

In order to get a qualitative feeling for the form of $\mathcal{F}_{LM}(\theta)$ we write the phase $\delta(\lambda) = \delta_n(\lambda) + \sigma(\lambda)$ as a sum of a nuclear part $\delta_n(\lambda)$ and a Coulomb part $\sigma(\lambda)$. Then we put

$$B_{LM}(\lambda)\, exp(2i\,\delta_n(\lambda)) = B_{LM}\, g_M(\lambda - \lambda_g) \tag{52}$$

Here $g_M(\lambda-\lambda_g)$ is peaked around the grazing angular momentum λ_g. It decreases for $\lambda \ll \lambda_g$ because Im $\delta_n(\lambda)$ becomes large due to absorption by the imaginary part of the optical potential. For $\lambda \gg \lambda_g$, g_M becomes small because the interaction is too weak to cause excitation, that is, $B_{LM}(\lambda)$ becomes small. The simplest possibility is to assume that $g_M(\lambda-\lambda_g)$ is independent of M and we shall do this in the present section. A more accurate treatment would take the effects of energy and angular momentum into account. Following Strutinsky [23] we approximate the integral in (47) by putting $\sqrt{\lambda} = \sqrt{\lambda_g}$ and expanding the Coulomb phase about the grazing angular momentum λ_g to first order

$$2\sigma(\lambda) \simeq 2\sigma(\lambda_g) + (\lambda-\lambda_g)\Theta_g \tag{53}$$

where $\Theta_g = \Theta(\lambda_g)$ is the Coulomb grazing angle. It is related to λ_g by the Rutherford formula. Then the nearside and farside components of the amplitude can be written in the form

$$\mathcal{F}^{\pm}_{LM}(\theta) \simeq B_{L\mp M}\, C^{\pm}(\theta)\,\tilde{g}(\theta_g \pm \theta)$$

$$C^{\pm}(\theta) = \frac{1}{ik}\left(\frac{\lambda}{2\pi\sin\theta}\right)^{\frac{1}{2}} e^{i(2\sigma(\lambda_g) \pm \theta\lambda_g \mp \frac{1}{4}\pi)} \tag{54}$$

where $\tilde{g}$ is the Fourier transform of g

$$\tilde{g}(x) = \int_{-\infty}^{\infty} e^{ix\mu}\, g(\mu)\, d\mu \tag{55}$$

Strutinsky pointed out that the nearside amplitude $\mathcal{F}^-$ gives the dominant contribution to the cross-section for reactions with a large Sommerfeld parameter. Note that the nearside and farside amplitudes have phases which depend on Θ in the following way

$$\mathcal{F}^-_{LM}(\theta) \propto e^{-i\lambda_g\theta} \quad , \quad \mathcal{F}^+_{LM}(\theta) \propto e^{i\lambda_g\theta} \tag{56}$$

Consider the excitation of a state with L = 0. Using (51) and (54) the inelastic cross-section is

$$\sigma(\theta) = \sigma^-(\theta) + \sigma^+(\theta) + 2\left(\sigma^-(\theta)\sigma^+(\theta)\right)^{\frac{1}{2}}\cos\left(2\lambda_g\theta - \tfrac{1}{2}\pi + 2\eta\right) \tag{57}$$

where $\sigma^-(\theta)$ and $\sigma^+(\theta)$ are the cross-sections calculated from the nearside amplitude $\mathcal{F}^-$ and the farside amplitude $\mathcal{F}^+$. The third term in (57) is due to nearside-farside interference $\eta = \arg[\tilde{g}(\theta_g+\theta)/\tilde{g}(\theta_g-\theta)]$. Equations (57) predicts nearside-farside interference oscillations with spacing $\Delta\theta \simeq \pi/\lambda_g$. When state with $L \neq 0$ is excited equation (57) is replaced by

$$\sigma_L(\theta) = \sum_M \Big[\sigma^-_{LM}(\theta) + \sigma^+_{LM}(\theta) + 2\left(\sigma^-_{LM}(\theta)\,\sigma^+_{LM}(\theta)\right)^{\frac{1}{2}}\cos\left(2\lambda_g\theta + L - \tfrac{1}{2}\pi + 2\eta_M\right)\Big] \tag{58}$$

In principle the cross-section $\sigma_L(\theta)$ in (58) can still show interference oscillations. In practice inelastic scattering tends to polariZe the spin perpendicular to the reaction plane and this damps out the oscillations. From (54)

$$\sigma^-_M(\theta) \propto |B_{LM}|^2 \quad , \quad \sigma^+_M(\theta) \propto |B_{L-M}|^2$$

The average of the magnetic quantum number M along the z-axis perpendicular to the reaction plane is (Hahne [19])

$$\begin{array}{ll} \text{Nearside} & \langle M\rangle \simeq QR/\hbar v < 0 \\ \text{Farside} & \langle M\rangle \simeq -QR/\hbar v > 0 \end{array} \tag{59}$$

where Q is the reaction Q-value which is negative for inelastic scattering, R is the strong absorption radius and v is the related velocity at the point of closest approach. The nearside and farside polarizations have opposite directions. The nearside cross-section σ^-_M will be large for $M<0$ and small for $M > 0$ if $Q \ll 0$. The farside cross-section will have the oppos ite M-dependence. As a result the interference terms in (58) will be small and the cross-section $\sigma_L(\theta)$ will be smooth especially if L is large.

3. The Experiment.

Bhowmik and Rae [22] measure the double differential cross-section $d^2\sigma/d\Omega_\theta\, d\Omega_\psi$ where the solid angle Ω_θ refers to the scattering angle of the ^{12}C in the center-of-mass systems and Ω_ψ to the solid angle associated with the direction of the α -particle. Both θ and ψ are measured relative to the direction of the incident beam. The angular distribution $d\sigma/d\Omega_\theta$ is a smooth function of angle. This is in line with the argument in section 2, but Bhowmik and Rae find that the double differential cross-section has a lot of structure.

DWBA calculations show this same structure. The double differential cross-section has a pattern of ridges which is correlated with the spin of the state excited inelastic scattering. In particular there is a definite correlation between the slope

$$\nu = d\theta/d\psi \qquad (60)$$

of the ridges and the angular momentum of the excited state. This slope depends on the system and the incident energy but is not very sensitive to the Q-value provided Q is reasonably large. Table 1 shows values of ν extracted from DWBA calculations and Table 2 shows ν obtained from the experimental data. The numbers are from the paper of Bhowmik and Rae [22]. It is clear that ν can be used to make spin assignments.

Table 1. Values of ν from DWBA

2^+	3^-	4^+	5^-	6^+
0.14	0.20	0.27	0.34	0.42

Table 2. Values of ν from Experiment

E(MeV)	7.10	7.83	8.21	10.30	11.59
L^π	4^+	5^-	2^+	4^+	5^-
ν	0.21 ± .03	0.30 ±.03	0.13 ± .01	0.23 ± .02	0.34 ± .03

4. Theory of the Angular Correlation

The angular correlation between the direction ψ of emission of the alpha particles and the scattering angle θ is determined by the amplitude

$$\mathcal{F}(\theta,\psi) = \Sigma_M \mathcal{F}_{LM}(\theta)\, Y_{LM}(\tfrac{1}{2}\pi, -\psi) \tag{61}$$

The angle $\frac{1}{2}\pi$ appears in (61) because the alpha particles are measured in the reaction plane. The right hand side contains $-\psi$ because ψ and θ are measured on opposite sides of the incident beam direction. Equation (61) cannot be used if θ is too small because equation (49) is based on the stationary phase evaluation of the ϕ integral in (44).

The amplitude (61) has a simple dependence on θ and ψ if only one polarization state contributes. Suppose that the nearside amplitude $\mathcal{F}^-_{LM} \neq 0$ only for M = -K and the farside amplitude $\mathcal{F}^+_{LM} \neq 0$ only for M = K. Then

$$\mathcal{F}(\theta,\psi) = Y_{L-K}(\tfrac{1}{2}\pi, 0)\left[\mathcal{F}^-_{L-K}(\theta)\, e^{iK\psi} + (-1)^{L-K} \mathcal{F}^+_{LK}(\theta)\, e^{-iK\psi}\right] \tag{62}$$

If $\mathcal{F}(\theta,\psi)$ is separated into its near and farside components (56) and (62) show that

$$\mathcal{F}^-(\theta,\psi) \sim e^{i(K\psi - \lambda_g \theta)}$$

$$\mathcal{F}^+(\theta,\psi) \sim e^{-i(K\psi - \lambda_g \theta)}$$

It follows that the ridges in the interference pattern in the double differential cross-section

$$d^2\sigma / d\Omega_\theta \, d\Omega_\psi \sim |\mathcal{F}(\theta,\psi)|^2 \tag{63}$$

correspond to lines with

$$\lambda_g \theta - K\psi = \text{const.}$$

This relation predicts that the slope ν of the ridges in (63) is given by $\nu = K/\lambda_g$. The above result has been derived by Bhowmik and Rae [22] starting with the semi-classical approximation to DWBA [18, 19, 20] .

Lecture 4: Nucleon Transfer.

1. Semi-classical Nucleon Transfer.

A typical example of a nucleon transfer reaction is

$$^{16}O + {}^{208}Pb \rightarrow {}^{15}O + {}^{209}Pb$$

in which a neutron is tranferred from the oxygen nucleus into the lead nucleus. It is possible to obtain a very simple approximate formula for the transfer cross-se ction by assuming that the projectile passes the target on a Rutherford orbit and that the transfer occurs when the projectile and target are close together. When the Sommerfeld parameter is large the cross-section would have the form given in equation (40)

$$d\sigma/d\Omega \simeq (d\sigma/d\Omega)_R \; P_t(\theta) \tag{64}$$

and is the product of the Rutherford cross-section and a transfer probability $P_t(\theta)$ calculated for the Rutherford orbit with scattering angle θ .

$$P_t(\theta) = |A_t(\theta)|^2 \exp\left[-\frac{2}{\hbar}\int_{-\infty}^{\infty} W(\underset{\sim}{r}(t))\,dt\right] \tag{65}$$

In equation (65) $A(\theta)$ is a transfer amplitude and the second factor is the attenuation due to the imaginary part of the optical potential $W(r)$. The integral involving W in (65) is taken along a Rutherford orbit with scattering angle θ . The first factor in (43) is the probability that a single nucleon transfer occurs while the second is the probability that no more complicated process happens. The amplitude $A_t(\theta)$ is calculated from the perturbation formula

$$A_t(\theta) = -\frac{i}{\hbar}\int_{-\infty}^{\infty} \langle \psi_f | V_1 | \psi_i \rangle \tag{66}$$

Equation (64) can be replaced by a uniform approximation like (44) for lighter systems where diffraction effects are important.

The remainder of this lecture will present a simple formula for calculating the transfer amplitude $A_t(\theta)$. The method has been developed by Lo Monaco and Brink [24] and is based on work of Hasan and Brink [25].

We consider the case of neutron transfer. The neutron is initially in a state ψ_i bound in a single particle potential V_1 which represents the shell model potential of the first nucleus. It is transferred into a single particle state ψ_f in the final potential V_2. The potential V_1 moves past V_2 during the transfer and the relative motion is described by a Rutherford orbit. Hence the potentials $V_1(\underset{\sim}{r},t)$ and $V_2(\underset{\sim}{r},t)$ are functions of the neutron coordinate $\underset{\sim}{r}$ and the time.

The initial state $\psi_i(r,t)$ satisfies the time-dependent Schrödinger equation for a neutrom in the potential V_1

$$i\hbar \frac{\partial \psi_i}{\partial t} = (T + V_1)\psi_i \tag{67}$$

where T is the kinetic energy operator ($T = -\hbar^2\nabla^2/2m$). Similarly

$$i\hbar \frac{\partial \psi_f}{\partial t} = (T + V_2)\psi_f \tag{68}$$

In order to derive (66) we note that the neutron is affected by both the potentials V_1 and V_2 and its wave function satisfies

$$i\hbar \frac{\partial \psi}{\partial t} = (T + V_1 + V_2)\psi \tag{69}$$

The amplitude (66) is obtained by solving (69) by perturbation theory with the initial condition

$$\psi(r,t) \to \psi_i(r,t) \quad , \quad t \to -\infty$$

The amplitude $A_t(0)$ is the overlap of ψ_f with ψ when t is large

$$A_t(0) = \langle \psi_f | \psi \rangle \quad , \quad t \to \infty \tag{70}$$

Using (68) and (69) we obtain

$$i\hbar \frac{\partial}{\partial t}\langle \psi_f | \psi \rangle = \langle \psi_f | T + V_1 + V_2 | \psi \rangle - \langle \psi_f | T + V_2 | \psi \rangle$$

$$= \langle \psi_f | V_1 | \psi \rangle \tag{71}$$

$$\simeq \langle \psi_f | V_1 | \psi_i \rangle \tag{72}$$

Equation (71) is exact while (72) is obtained by making the perturbation approximation $\psi \simeq \psi_i$. Equation (66) is obtained from (72) by integrating both sides with respect to t.

Equation (72) was used by Hasan and Brink [25] but it is difficult to calculate. A much simpler result can be obtained in the case of a peripheral collision when the potentials V_1 and V_2

do not overlap during the collision. If this is the case it is possible to draw a surface Σ between the potentials so that $V_1(r,t) = 0$ for all t in the region R_2 on one side of Σ while $V_2(r,t) = 0$ for all t in the region R_1 on the other side of Σ. The calculation proceeds as follows:

$$\langle\psi_f|V_1|\psi_i\rangle = \int \psi_f^*(\underset{\sim}{r},t)\, V_1(\underset{\sim}{r},t)\, \psi_i(\underset{\sim}{r},t)\, d^3\underset{\sim}{r}$$

$$= \int_{R_1} \psi_f^*(\underset{\sim}{r},t)\, V_1(\underset{\sim}{r},t)\, \psi_i(\underset{\sim}{r},t)\, d^3\underset{\sim}{r} \qquad (73)$$

Equation (73) holds because $V_1(r,t) = 0$ in R_2. Now we use (67) and write

$$\langle\psi_f|V_1|\psi_i\rangle = \int_{R_1} \psi_f^*(\underset{\sim}{r},t)\left(i\hbar\frac{\partial\psi_i}{\partial t} + \frac{\hbar^2}{2m}\nabla^2\psi_i\right)d^3\underset{\sim}{r} \qquad (74)$$

Integrating (74) by parts gives

$$\langle\psi_f|V_1|\psi_i\rangle = \frac{\hbar^2}{2m}\int_\Sigma d^2\underset{\sim}{S}\cdot\left(\psi_f^*\nabla\psi_i - (\nabla\psi_f^*)\psi_i\right) \qquad (75)$$

$$+ \int_{R_1}\left(i\hbar\frac{\partial\psi_f}{\partial t} + \frac{\hbar^2}{2m}\nabla^2\psi_f\right)^*\psi_i\, d^3\underset{\sim}{r}$$

$$+ i\hbar\frac{\partial}{\partial t}\int\psi_f^*\psi_i\, d^3\underset{\sim}{r}$$

Using (68) the second term in equation (75) becomes

$$\int_{R_1}\psi_f^* V_2\psi_i\, d^3\underset{\sim}{r} = 0$$

because $V_2(r,t) = 0$ in the region R_1. If (75) is integrated over time the third term vanishes because ψ_i and ψ_f have no overlap as $t \to \pm\infty$. This is because ψ_f and ψ_i are bound states in the potentials V_2 and V_1 which are far away from each other as $t \to \pm\infty$. Hence (66) reduces to

$$A_t(\theta) = \frac{\hbar}{2mi}\int dt\int_\Sigma d^2\underset{\sim}{S}\cdot\left(\psi_f^*\nabla\psi_i - (\nabla\psi_f^*)\psi_i\right) \qquad (76)$$

Equation (76) involves a surface integral over the surface Σ separating V_1 and V_2. It is exactly equivalent to (66) if the two potentials V_1 and V_2 do not overlap during the collision. If they do have an overlap there are corrections. Equation (76) has some very nice features. It is symmetric between the initial and final states whereas (66) looks asymmetric because it contains V_1. The evaluation of (76) requires the wave functions ψ_i and ψ_f only on the surface Σ. As Σ is outside V_1 and V_2 the wave functions ψ_i and ψ_f can be replaced by their asymptotic forms. It is interesting to note that the integrand in (76) resembles a quantal probability current

$$j(r) = \frac{\hbar}{2mi}\left(\psi^* \nabla\psi - \psi\nabla\psi^*\right)$$

The amplitude $A_t(\theta)$ in (76) has been evaluated by Lo Monaco and Brink [25] with the symplifying assumption that the Rutherford orbit of relative motion can be replaced by a constant velocity orbit tangential to the Rutherford orbit at the point of closest approach between V_1 and V_2. This is a reasonable approximation for small angle scattering because the transfer takes place near the point of closest approach.

The result given in [25] for transfer from an initial single particle state ψ_i with angular momentum quantum numbers (ℓ_1, m_1) to a final state (ℓ_2, m_2) is

$$A(\ell_2 m_2, \ell_1 m_1) = -4\pi i \frac{\hbar}{mv} C_1 C_2^* (-1)^{m_1} \times Y^*_{\ell_2 m_2}(\beta_2)\, Y_{\ell_1 m_1}(\beta_1)\, K_{m_1 - m_2}(\eta d) \tag{77}$$

In this formula C_1 and C_2 are normalization constants determined by the asymptotic forms of the initial and final bound states

$$\phi_i(r) \simeq C_1 \gamma_1 k_{\ell_1}(\gamma_1 r)\, Y_{\ell_1 m_1}(\theta,\phi) \tag{78}$$

$$\phi_f(r) \simeq C_2 \gamma_2 k_{\ell_2}(\gamma_2 r)\, Y_{\ell_2 m_2}(\theta,\phi) \tag{79}$$

Here k_ℓ is a spherical Bessel function $(\gamma k_\ell(\gamma r) \sim e^{-\gamma r}/r$ when $r\gamma$ is large) and γ_1 and γ_2 are related to the bound state energies ε_1 and ε_2 of the initial and final states by

$$\varepsilon_1 = -(\hbar^2/2m)\gamma_1^2 \quad , \quad \varepsilon_2 = -(\hbar^2/2m)\gamma_2^2$$

The function $K_{m_1-m_2}(\eta d)$ in (77) is a Bessel function. It has the asymptotic form

$$K_{m_1-m_2}(\eta d) \sim (\pi/2\eta d)^{\frac{1}{2}} e^{-\eta d}$$

The constant η is given by

$$\eta^2 = -(2m/\hbar^2)\,\overline{\varepsilon}$$

$$\overline{\varepsilon} = \tfrac{1}{2}(\varepsilon_1 + \varepsilon_2) - \tfrac{1}{4}\left[\frac{(\varepsilon_1 - \varepsilon_2)^2}{\frac{1}{2}mv^2} + \tfrac{1}{2}mv^2\right] \tag{80}$$

where $\overline{\varepsilon}$ us a kind of average bound state energy. It depends on the velocity v of relative motion. The angles β_1 and β_2 are complex angles given by

$$\tan\beta_1 = ik_1/\gamma_1 \quad , \quad \tan\beta_2 = ik_2/\gamma_2$$

where $Q = \varepsilon_1 - \varepsilon_2$ is the reaction Q-value and

$$\hbar k_1 = -(Q + \tfrac{1}{2}mv^2)/v \quad , \quad \hbar k_2 = -(Q - \tfrac{1}{2}mv^2)/v$$

References

[1] Taylor, J. R. (1972). 'Scattering Theory'. New York: John Wiley.

[2] Dollard, J. D. (1964). J. Math. Phys. 5, 729.

[3] Feynman, R. P. & Hibbs, A. R. (1965). 'Quantum Mechanics and Path Integrals'. New York: McGraw Hill.

[4] Schulman, L. S. (1981). 'Techniques and Applications of Path Integrals'. New York: John Wiley.

[5] Koeling, T. & Malfliet, R. A. (1975). Phys. Rep. C31, 159.

[6] Van Vleck, J. H. (1928). Proc. Nat. Acad. Sci. 14, 178.

[7] Levit, S. & Smilansky, U. (1977). Ann. Phys. 103, 198, (1977). Ann. Phys. 108, 165.

[8] Gutzwiller, M. C. (1967). J. Math. Phys. 8, 1929, (1969). J. Math. Phys. 10, 1004.

[9] Miller, W. H. (1970). J. Chem. Phys. 53, 1949.

[10] Pechukas, P. (1969). Phys. Rev. 181, 166, 174.

[11] Sukumar, C. V. (1984). J. Phys. G10, 81.

[12] Dos Aidos, F. & Brink, D. M. to be published.

[13] Alder, K. & Winther, A. (1966). 'Coulomb Excitation'. New York: Academic Press.

[14] Berry, M. V. (1976). Adv. in Phys. 25, 1.

[15] Knoll, J. & Schaeffer, R. (1976). Ann. Phys. 97, 307, (1977). Phys. Rep. C31, 159.

[16] Duistermaat, J. J. (1974). Communs. Pure Appl. Maths, 27, 207.

[17] Frahn, W. E. (1984). Vol. 1, 'Heavy-Ion Science'. Ed. D. A. Bromley, New York: Plenum Press.

[18] Potgieter, J. M. & Frahn, W. E. (1967), Nucl. Phys. A92, 84.

[19] Hahne, F. J. W. (1967). Nucl. Phys. A104, 545.

[20] Frahn, W. E. (1980). Phys. Rev. C21, 1820.

[21] Dean, D. R. & Rowley, N. (1984). J. Phys. G10, 493.

[22] Rae, W. D. M. & Bhowmik, R. K. (1984). Nucl. Phys. A420, 320.

[23] Strutinsky, V. M. (1964). JETP 19, 1401. (1973). Phys. Lett. 44B, 245.

[24] Lo Monaco, L. & Brink, D. M. to be published.

[25] Hasan, H. & Brink, D. M. (1978). J. Phys. G4, 1573, (1979). J. Phys. G7, 1501.

Lecture I

Basics and Heavy Ion Scattering in Time Dependent Hartree Fock Theory

Morton S. Weiss

Lawrence Livermore National Laboratory
University of California
Livermore, CA 94550

Work performed under the auspices of the U.S. Department of Energy by the Lawrence Livermore National Laboratory under contract number W-7405-ENG-48.

Time Dependent Hartree-Fock theory, TDHF, is the most sophisticated, microscopic approach to nuclear dynamics yet practiced. Although, as you will see, it is far from a description of nature it does allow us to examine multiply interactive many-body systems "semi quantum mechanically" and to visualize otherwise covert processes. In two lectures I can at best adumbrate an elementary introduction of what has been in recent years a very prolific activity. I hope those of you who are intrigued by specific areas will use the informal discussions to elucidate details.

In this first lecture I will state some of the properties of the TDHF equations leaving the interested reader to one of several excellent review articles[100] for the derivations. Then with a brief nod to technique, I will describe some of the applications to the collision of heavy ions. I will then take the last quarter of this lecture to literally interpret visualize with a 15 minute color movie of a heavy ion reaction which will illustrate many of the previous points. The second lecture will be less amusing and emphasize special applications of this theory.

One of the most seductive aspects of TDHF is that it is a no free parameter theory. While this does indeed provide those of us who have practiced it with a feeling of righteousness when talking to our more macroscopic collegues with their invented bulk and surface energies and various forms of anthropromorphic dissipation, what does this mean? Clearly we haven't started from nucleon-nucleon scattering data, nor QCD and calculated heavy ion reactions. What has, of course, happended is that in some distant past we have introduced lots of free parameters; generations of sophisticated scientists have varied them to fit other phenomena and now we can ignore our disreputable origins with a "no parameter theory". TDHF is only the dynamical extension of ordinary, static, Hartree-Fock (HF) theory.

Hartree-Fock is used to study atomic nuclei through the introduction of an effective 2 body potential of which a typical example is

$$v(\vec{k}_1, \vec{k}_2) = t_o (1+\chi_o P_\sigma) + \frac{1}{2} t_1 (k_1^2 + k_2^2) + t_2 \vec{k}_1 \cdot \vec{k}_2$$

$$+ i W_o (\sigma_1 + \sigma_2) \cdot (\vec{k}_1 x \vec{k}_2)$$

The t's and W_o and x's are free parameters invented to force HF theory to adequately mimic certain choosen properties of some specially selected nuclei. This is enormously successful! Figure 100 shows you a comparison of the cross section acquired from precise measurements of elastic electron scattering on ^{208}Pb to the theoretical quantity[101] derived from the HF density.

In static HF theory the nuclear wave function is approximated by a slater determinant of independent particle orbits which obey

$$(t + \nu) \varphi_\alpha = \varepsilon_\alpha \varphi_\alpha$$

The φ_α and ν are, as you know from previous lectures, found not to be independent in this approximation and the equations solved iteratively. Physically each orbit is in the potential derived from the nucleur material of all the other orbits, subject to the constrant of orthonormality imposed by the Pauli-principle.

In extending this to dynamics the same physics is retained. The wave function for the system to be evolved in time must be choosen to be a slater determinant. For a collision of a nucleus with A_1 paticles onto a nucleus with A_2 particles the systems wave function is a slater determinant of $(A_1 + A_2)$ orbitals. The equations they obey are the obvious extension of the static HF equations;

$$\frac{\partial}{\partial t} \varphi_\alpha (\vec{r},t) = (t + \nu) \varphi_\alpha (\vec{r},t)$$

an added complication in solving these equations is that now the ν depends implicitly upon the time because it is a function of the wave functions (or at least the density). These equations can be derived in a variety of ways, perhaps most insightfully from a variational principle

$$\frac{\delta}{\delta \varphi^*_\alpha(\vec{r},t)} \int_{t_1}^{tz} dt \langle \psi(\vec{r},t) \mid i \hbar \frac{\partial}{\partial t} - H \mid \psi(\vec{r},t) \rangle = 0$$

where $\psi (\vec{r},t) = \frac{1}{\sqrt{A!}} \det \varphi_\alpha (\vec{r}_i,t)$ and $\vec{r} = \vec{r}_1, \vec{r}, \ldots \vec{r}_{A_1} + A_2$

These equations have the pleasant properties that

$$\frac{d}{dt} \langle \varphi_\alpha(\vec{r},t) \mid \varphi_\beta(\vec{r},t) \rangle = 0 \tag{a}$$

which conserves orthonormality.

$$\frac{d}{dt} \langle \psi(\vec{r},t) \mid H \mid \psi(\vec{r},t) \rangle = 0 \tag{b}$$

which conserves energy

$$\frac{d}{dt} \langle \psi(\vec{r}_1,t) \mid \begin{matrix} \vec{P} \\ \vec{L} \end{matrix} \mid \psi(\vec{r},t) \rangle = 0 \tag{c}$$

where $\vec{P}$ and $\vec{L}$ are, respectively, the linear and angular momentum. These conserved quantities can be very useful tests of any numerical schemes. In addition the invariance under time reversal can also be used to test the accuracy of a proceedure.

The necessity of specifying an initial condition leads us to the last invariance, gallilean. A TDHF calculation is started by solving the static HF equations and then placing the target and projectile onto a spatial grid. The individual static solutions are then boosted in velocity and direction so as to specify the initial (relative) angular momentum and energy. Figure 101 from Flocard et al,[102] shows a deeply inelastic scattering for ^{16}O onto ^{16}O at E lab 105 MeV and an initial angular momentum of 5 $\hbar$. We can extract the scattering angle and the energy lost to internal excitation. For the same energy but an initial angular momentum of 13 $\hbar$, Fig. 102 shows a different phenomena: fusion. Although in this case the calculation was persued only to $2.8 . 10^{-21}$ seconds, other calculations have been carried an order of magnitude longer and the system remained fused.

The properties of this system can be summarized in Fig. 103 which show a variety of trajctories labelled by their initial conditions. Figure 104 shows the relative distance of two ^{16}O as a function of time for a nearly central collision at 32 MeV in the lab. This behaviour of a converging radius substantiates our interpretation of this type of behaviour as fusion. Our information for $^{16}O + {}^{16}O$ at E lab = 105 MeV can be summarized in Fig. 105. Here we see both scattering angle and energy lost into excitation as a function of the initial angular momentum. This can be compared to experiment by using a classical interpretation. For example the fusion cross section is

$$\sigma_{fus} [E \text{ lab}] = \frac{\pi\hbar^2}{\mu E_{lab}} \sum_{\ell<}^{\ell>} (2\ell+1)$$

where μ is the reduced mass and l>< are the lowest and highest angular momentum at which fusion starts and stops, respectively. The interpretation we make of Fig. 105 is that from central collisions (0 $\hbar$ to 13 $\hbar$) the system underwent deeply inelastic scattering, then had 0.8 barns of fusion and then more deeply inelastic scattering tapering to peripheral scattering and then plain old coulomb scattering.

If this is your first exposure to TDHF and if it is not, you really shouldn't be reading this lecture, you are appropriately horrified at the fact that there appears to be calculated fusion at 15 $\hbar$ but not at 0 $\hbar$. Much effort has been spent on this point. Figure 106 shows O^{16} onto ^{40}ca at E lab 250 MeV and 20 $\hbar$ initial angular momentum.[103] It is quite clear the offending projectile punches through the target even though at intermediate times it looks as if a true compound system has been formed. Figure 107 from Ref. 104 shows the structure of the fusion cross section as a function of energy and initial angular momentum. Clearly the fall in the cross section for fusion at increasing energy is due, in this calculation, to the opening of the low l window. This is reflected in Fig. 108 which compares predicted and calculated cross section for fusion. In spite of the heroic efforts of many excellent experimenters the existence of the low l window has been neither verified nor disproved.[105]

Now all of the calculations I have shown you were performed in three dimensional coordinate space with a very simple form of the HF potential.[100] In addition one ignored both intinsic spin and charge, thus reducing the number of orbits to be propagated by a factor of 4. To study much larger nuclei some of these simplifications must be removed to make the physics believable. However, something else must happen or the calculations become intractable in terms of computer time and memory. A compromise has been made in the form of the frozen approximation.[106]

Here the complexity of the three dimensional TDHF calculation is enormously simplified by assuming the single particle orbitals can be factored into a part that depends upon the coordinates in the reaction plane and the time and a part dependings only upon the direction normal to the reaction plane and time independent.

$$\psi_i\ (\vec{r},t) = \varphi_i\ (X,Y,t)\ \chi_i\ (Z)$$

The $\chi_i(Z)$ are choosen at the beginning of the problem to be one dimensional harmonic oscillator waves functions whose oscillator parameter is adjusted to minimize the total energy of the static. HF solutions with this choice the TDHF equations become two dimensional:

$$i\hbar \frac{\partial\varphi}{\partial t} i\,[(X,Y,t)] = \left(-\frac{\hbar^2}{2m}\left(\frac{\partial^2}{\partial x2} + \frac{\partial^2}{\partial Z^2}\right) + (T_Z)_i + W_i(X,Y)\right)[\phi_i(X,Y,t)]$$

where $W_i\ (X,Y) = \int dZ|X_i(Z)|^2\ W(X,Y,Z)$

$$(T_Z)_i = \frac{h^2}{2m}\int dt\ \left|\frac{d\chi i}{dZ}\right|^2$$

This reduces the time of a calculation by nearly an order of magnitude while retaining the full three dimensional kinematics. Clearly we are suppressing the possibility of energy going into this normal direction. As can be seen in Ref. 106, extensive testing of this approximation was performed both statically and dynamicaly by comparing the frozen approximation against the full three deminsional calculation. For time periods of interest and energies up to several MeV/nucleon the frozen approximation was in excellent agreement with the full calculation.

One more approximation must be discussed before we can discuss the collision of large nuclei; the filling approximation. As you know the ground states of very many nuclei are deformed. In principle we could choose a spectrum of orientations for each of our initial states. This is impossible. Instead, when we solve the static HF equations, the nucleons outside the last filled shell are forced have normalization less than one so that the last unfilled shell is uniformly occuppied. This spherizes the static HF solution. Unfortunately this alters total and relative binding energies and makes mass transfers suspect.

With these caveats, the full Skyrme III potential (without spin orbit) was used to calculate ^{86}Kr onto ^{139}La at E lab = 505 MeV[107] and Figs. 109, 110 shows the projected density for 5 $\hbar$ (fusion) and, 84 $\hbar$ (deeply inelastic scattering). Even with all of the approximations above, each impact parameter took about one hour on a Cray I computer and involved the evaluation of 146 wave functions. Table I shows the results for the 13 initial conditions studied. These were calculated with a time step of $9.0 \cdot 10^{-24}$ sec on a spatial grid of 1.2 fm. The interaction time in Table I is a subjective decision on the time the compound system existed. Z_{LF} is the charge on the light fragment well after separation.

Experiments[108] have measured the fusion and deeply inelastic cross section for this system at this energy. The measured fusion cross section, 170 ± 50 m.b. is consistent with all angular momentum from 0 $\hbar$ to 66 ± 10 $\hbar$ fusing. The TDHF calculation has $60 < \ell$ fusion < 80 and is consistent with experiment.

The measured deeply inelastic scattering cross section is

$$\sigma_{exp} \text{ (d.i.)} = 1020 \pm 200 \text{ m.b}$$

and the σ_{calc} (d.i.) = 987 m.b.

Figure 111 compares experiment with inelasticity as a function of scattering angle.

Mass and charge transfers are also measured as are the distribution in mass and charge of the final fragments. Time does not permit a discussion of that type of calculation. Suffice that while the initial TDHF slater determinant is factored into its target and projectile, the final one is not. Off block diagonal terms proliferate, which means that the final fragment is not an eigenvalue of the number operator. Nevertheless the calculated distributions are much smaller than experiment, in part due to the slater det. limitation and possibly also to the "soliton nature" of the TDHF solutions (assuming they are logically separate). The inadequarcy of the calculation to get the appropriate mass transfer would be more puzzeling if we were not making the filling approximation. Because of it, the relative binding energy of both the initial nuclei and final fragments is mis-represented and it is reasonable that mass and charge flow would be incorrectly calculated.

I will conclude this lecture with a movie of the $^{86}Kr + {}^{139}La$ calculation. I remind you that you are watching the projected density taken directly from the calculation. There is no artistic interpretation. I suggest you pay particular attention both to the properties of the neck both as to its structure and time dependence. This type of behaviour is not included in macrosopic calculations. In addition you may find the periphery or surface most intriuging. Lastly you will notice that the exterior shape is very distorted in the final fragments regardless of the level of violence in the collision whereas the interior is much more sensitive to the impact parameter.

At best this is an abbreviated menu of what has been done in this area. Many calculations with a multitude of approximations exist and if you have the dubious taste to find that interesting I hope this will provide an entry to the literature.

References

100. S. E. Koonin, Progress in Particle and Nuclear Physics, 4, 1979, D. H. Wilkinson, ed. Pergamon Press, N. Y.

J. W. Negle, Reviews of Modern Physics 54, 913, (1982).

101. J. L. Friar, J. Heisenberg and J. W. Negle, Proc. Workshop in Intermediate Energy Electromagnetic Interactions with Nuclei, MIT, 1977, A. M. Bernstein, ed.

102. H. Flocard, S. E. Koonin and M. S. Weiss, Phys. Rev. C17, 1682, (1978).

103. M. S. Weiss, private communication.

104. P. Bonche, et al., Phys. Rev. C20, 641 (1979).

105. J. Huizenga, private communication.

106. K. R. Sandhya Devi and M. R. Strayer, J. Phys. G4, K97 (1978) and Phys. Lett. 77B, 135 (1978). S. E. Koonin, et al., Phys. Lett. 77B, 13 (1978).

107. P. Bonche, et al., (no longer in preparation); results and discussion may be found in M. S. Weiss, Dynamics of H. I. Collision, N. Cindro, ed, (1981).

108. R. Vandenbosch et al., Phys. Rev. C17, 1672 (1978).

Figure Captions

Figure 100 Elastic scattering cross section of ^{208}Pb compared to that calculated from HF density. From Ref. 101.

Figure 101 ^{16}O onto ^{16}O at E lab = 105 MeV, initial angular momentum = 5 $\hbar$. The curves are isocontours of density projected onto the reaction plane. The scattering angle and inelasticity can be extracted. Fig. Ref. 102.

Figure 102 As above but for initial angular momentum = 13 $\hbar$. For obvious reasons this result is assumed to represent fusion. From Ref. 102.

Figure 103 The radius vector between centers of mass for ^{16}O onto ^{16}O at E lab = 105 MeV labelled by their initial angular momentum. From Ref. 102.

Figure 104 As above for E ebo = 32 MeV. It is this type of damping with time that makes us confident this type of event represents fusion. From Ref. 102.

Figure 105 Scattering angle and energy loss for ^{16}O + ^{16}O at E lab = 105 MeV as a function of initial angular momentum, Ref. 102.

Figure 106 ^{16}O onto ^{40}Ca at E lab = 250 MeV for a nearly central collision. From Ref. 103.

Figure 107 The structure in initial angular momentum of the fusion cross section for ^{16}O + ^{40}Ca as a function of energy. From Ref. 104.

Figure 108 Experimental and calculated fusion cross section for ^{16}O + ^{40}Ca. From Ref. 104.

Figure 109 ^{86}Kr + ^{139}La at E lab = 505 MeV for initial angular momentum 5 ℏ. Each color shows approx. a 10% change in projected density.

Figure 110 As above for 84 ℏ.

Figure 111 Experimental results for ^{86}Kr + ^{139}La at E lab = 505 MeV on which are superimposed results of TDHF calculations.

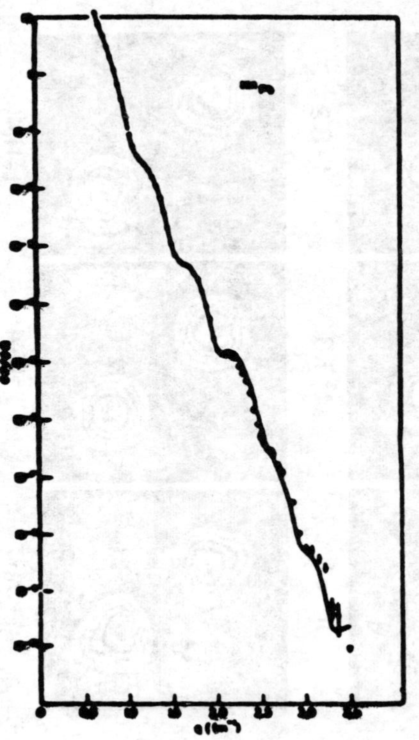

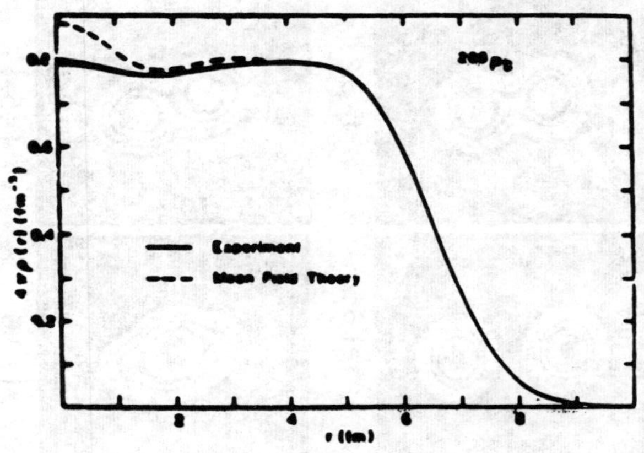

FIG. 100

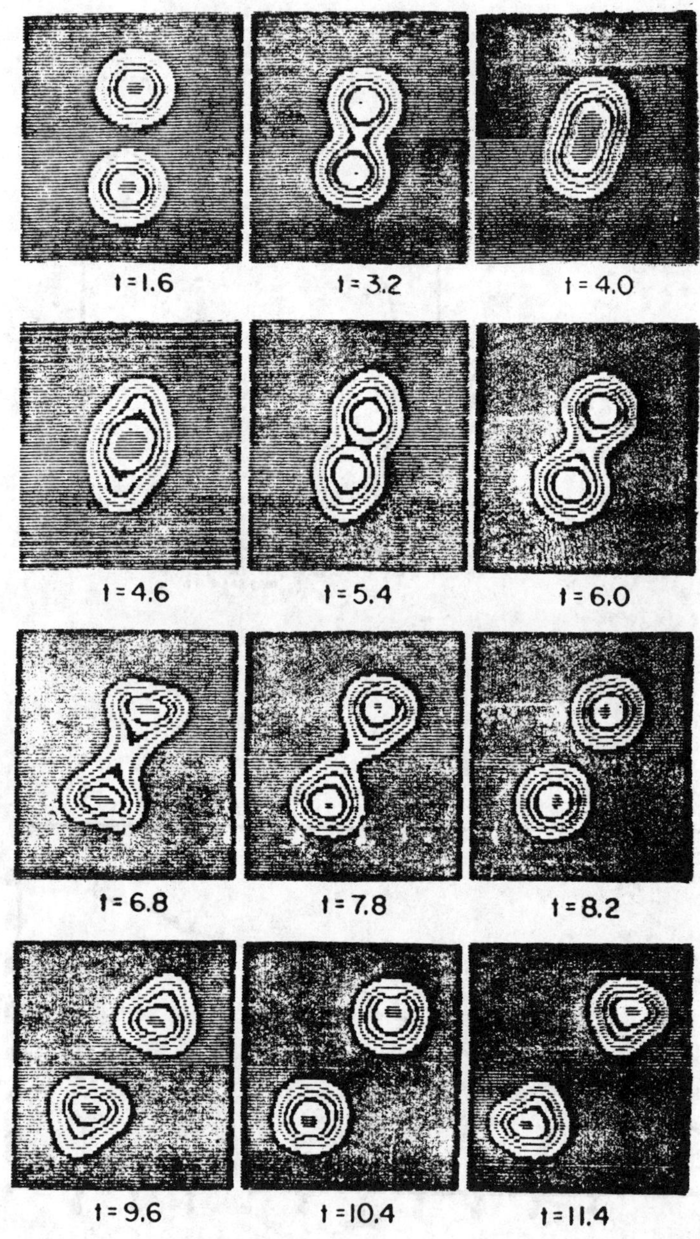
t=1.6
t=3.2
t=4.0
t=4.6
t=5.4
t=6.0
t=6.8
t=7.8
t=8.2
t=9.6
t=10.4
t=11.4
XBL 777-1530

FIG. 101

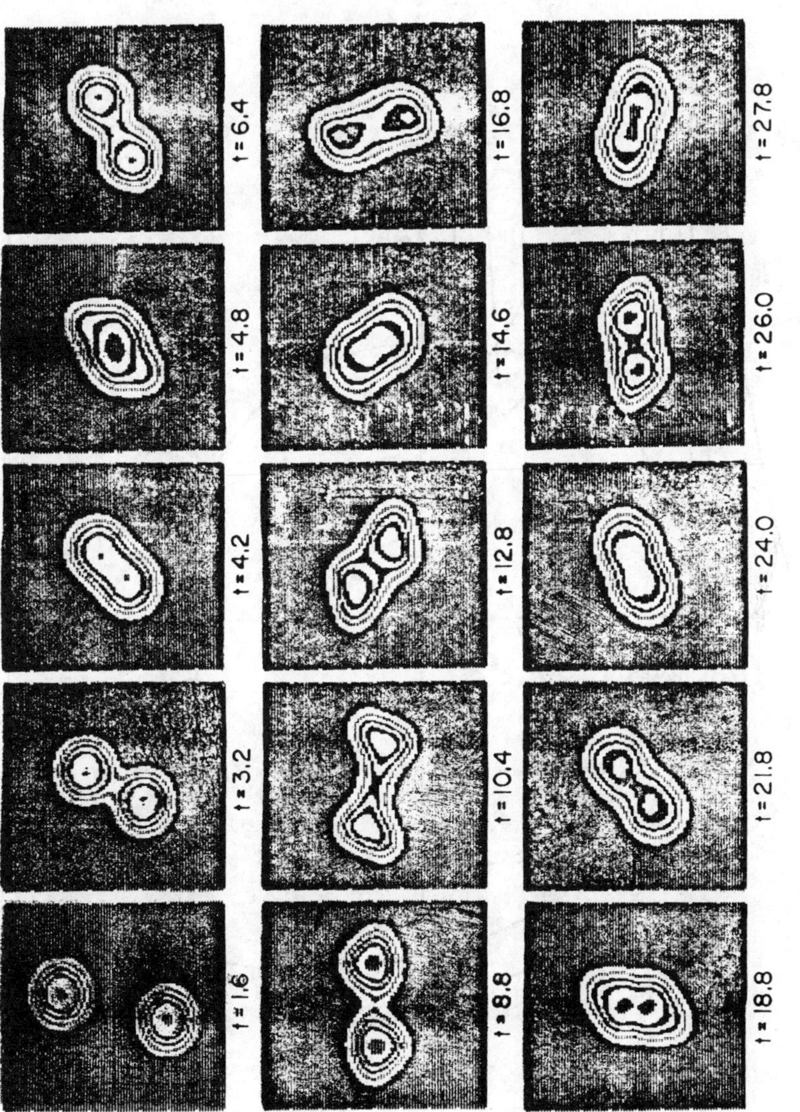
t = 1.6
t = 3.2
t = 4.2
t = 4.8
t = 6.4
t = 8.8
t = 10.4
t = 12.8
t = 14.6
t = 16.8
t = 18.8
t = 21.8
t = 24.0
t = 26.0
t = 27.8
XBL 777-1527

FIG. 102

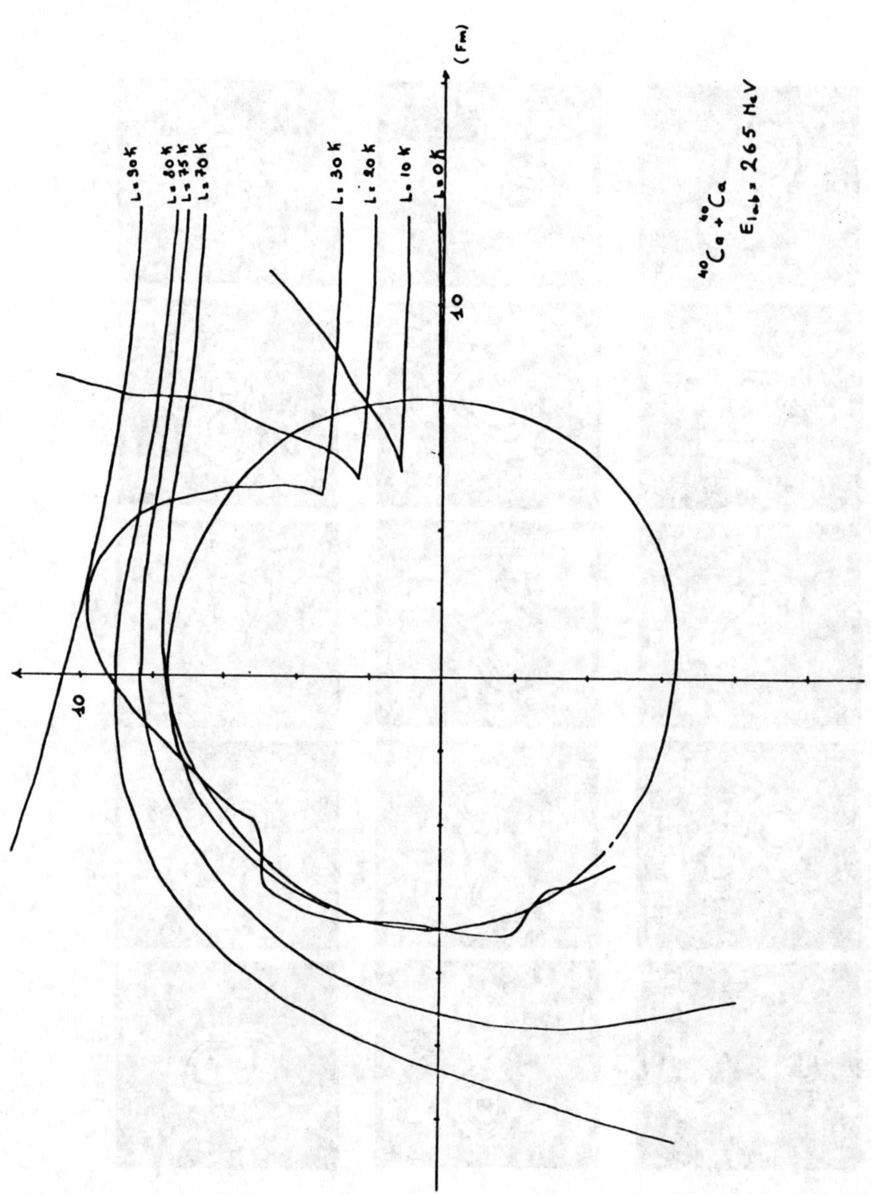
L=90ħ
L=80ħ
L=75ħ
L=70ħ
L=30ħ
L=20ħ
L=10ħ
L=0ħ
(Fm)
10
10
40Ca + 40Ca
Elab = 265 MeV

FIG. 103

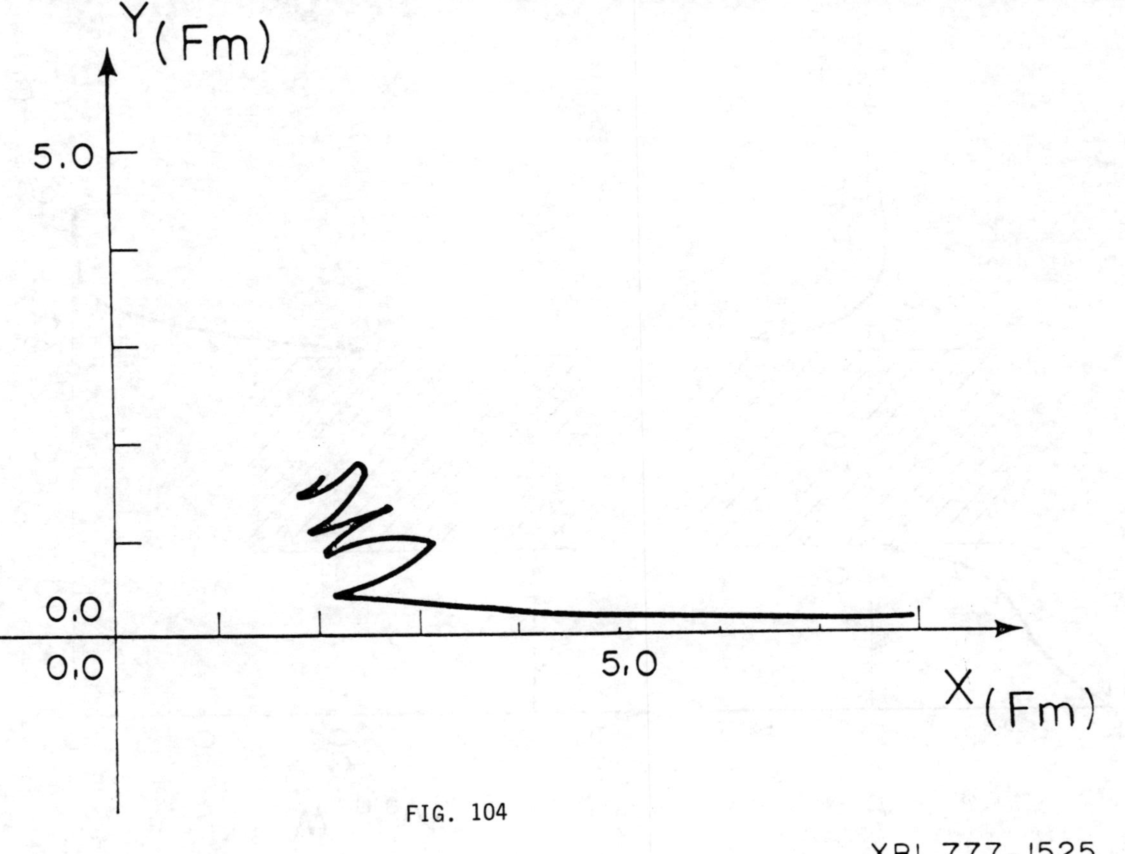

FIG. 104

XBL 777-1525

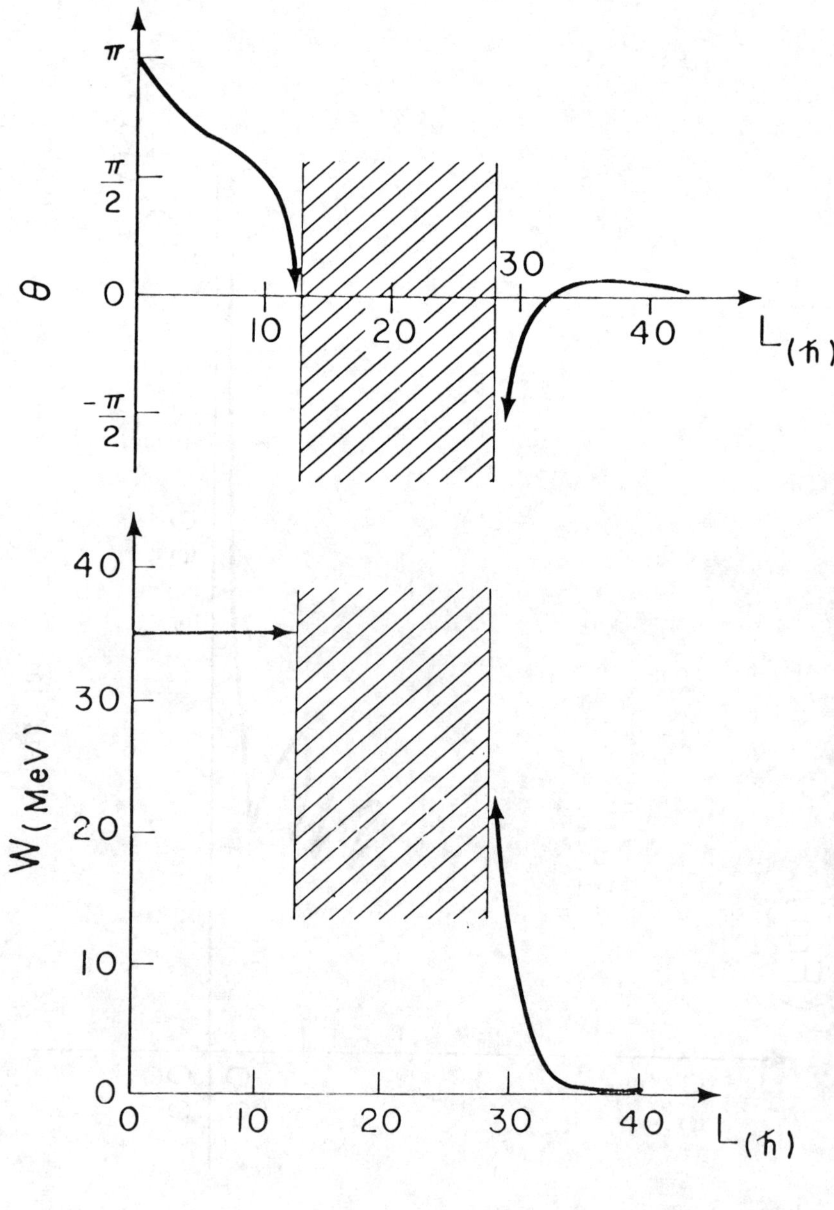

π
π/2
θ
0
-π/2
10
20
30
40
L(ħ)
W(MeV)
0
10
20
30
40
0
10
20
30
40
L(ħ)
XBL 777-1524

FIG. 105

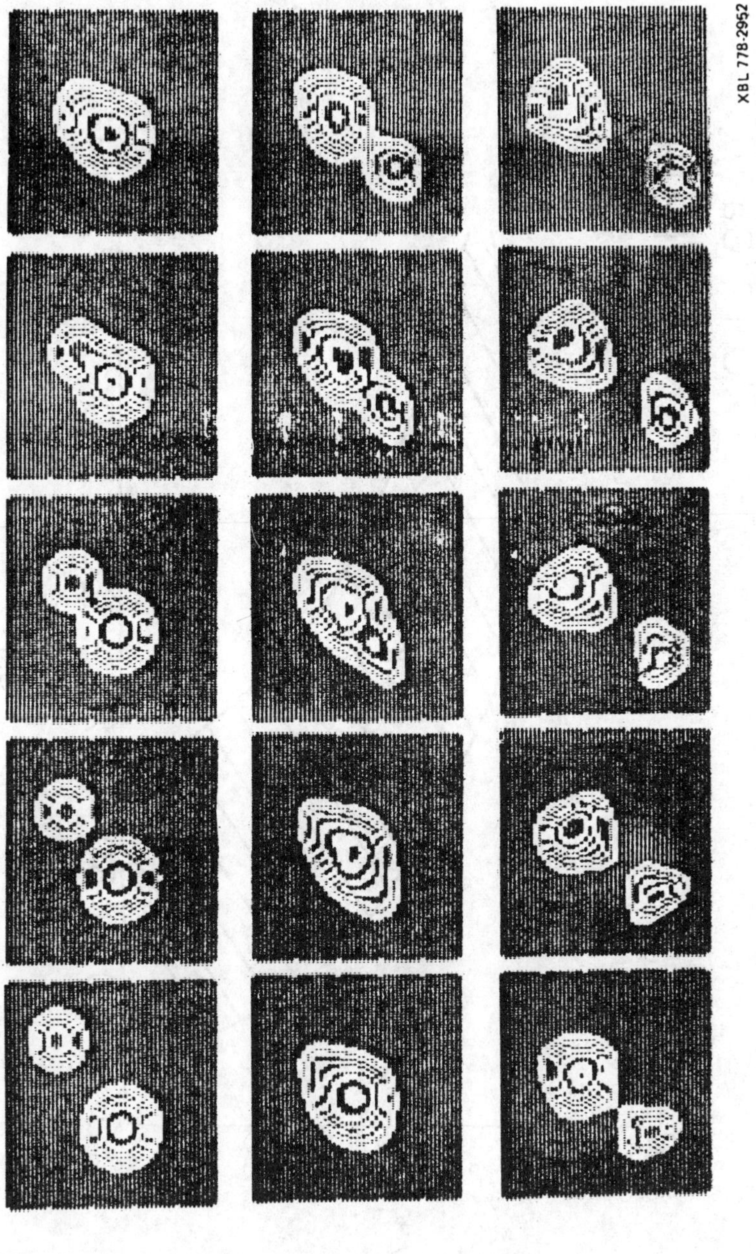

FIG. 106

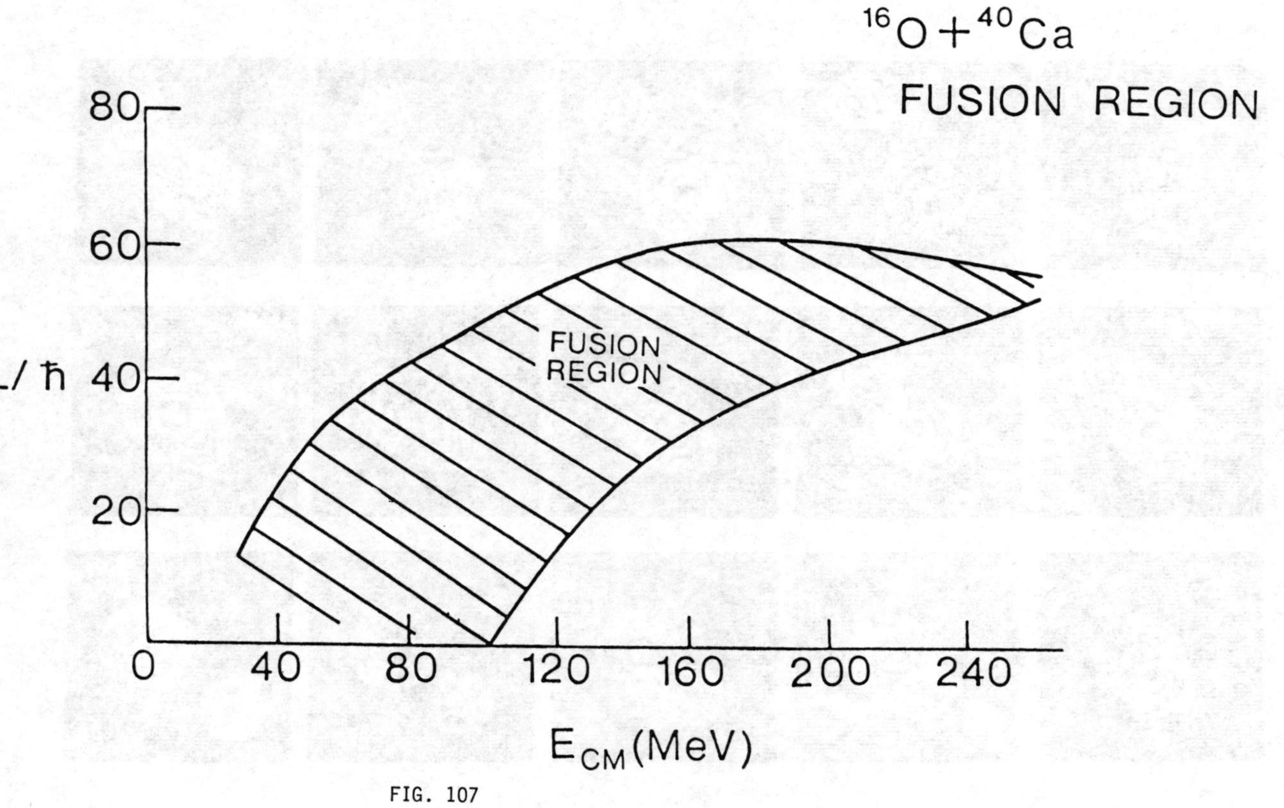

$^{16}O + ^{40}Ca$
FUSION REGION
80
60
L/ħ 40
20
0
40
80
120
160
200
240
FUSION REGION
E_{CM}(MeV)

FIG. 107

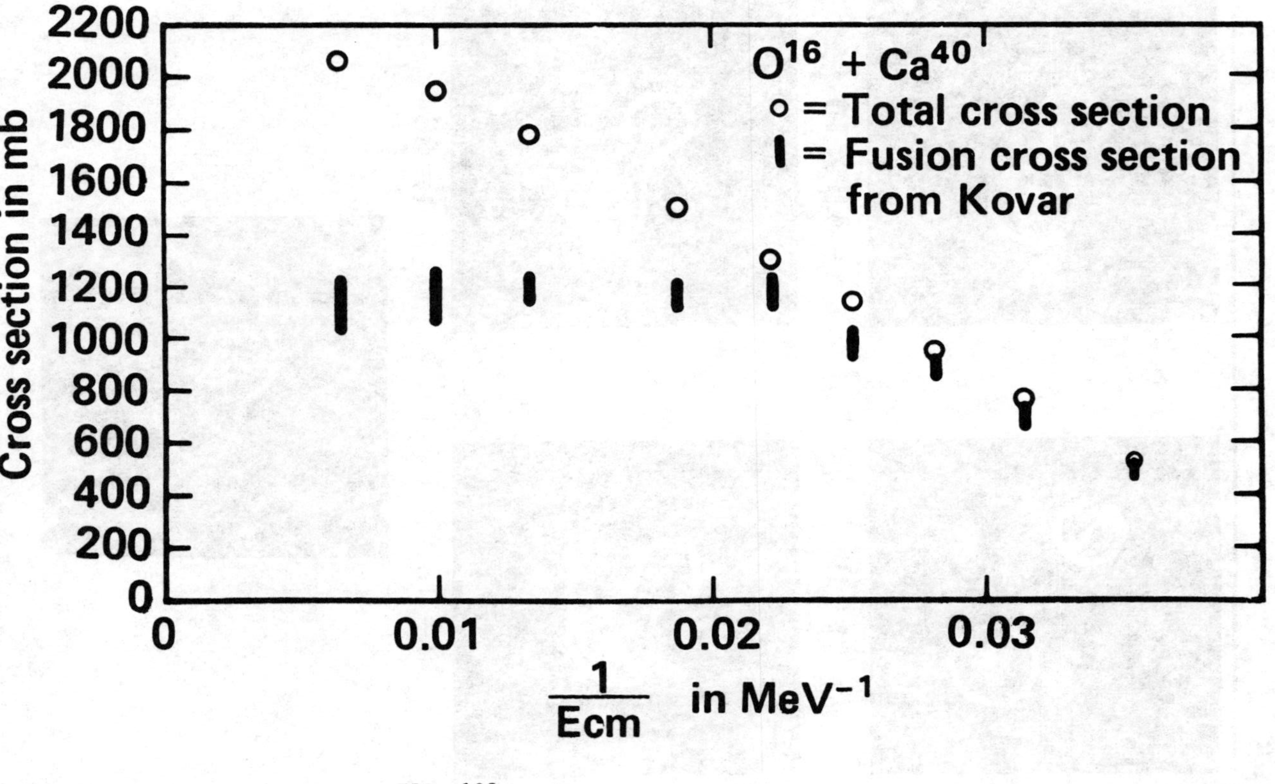

O^{16} + Ca^{40}
○ = Total cross section
▮ = Fusion cross section
from Kovar
Cross section in mb
2200
2000
1800
1600
1400
1200
1000
800
600
400
200
0
0
0.01
0.02
0.03
$\frac{1}{Ecm}$ in MeV^{-1}

FIG. 108

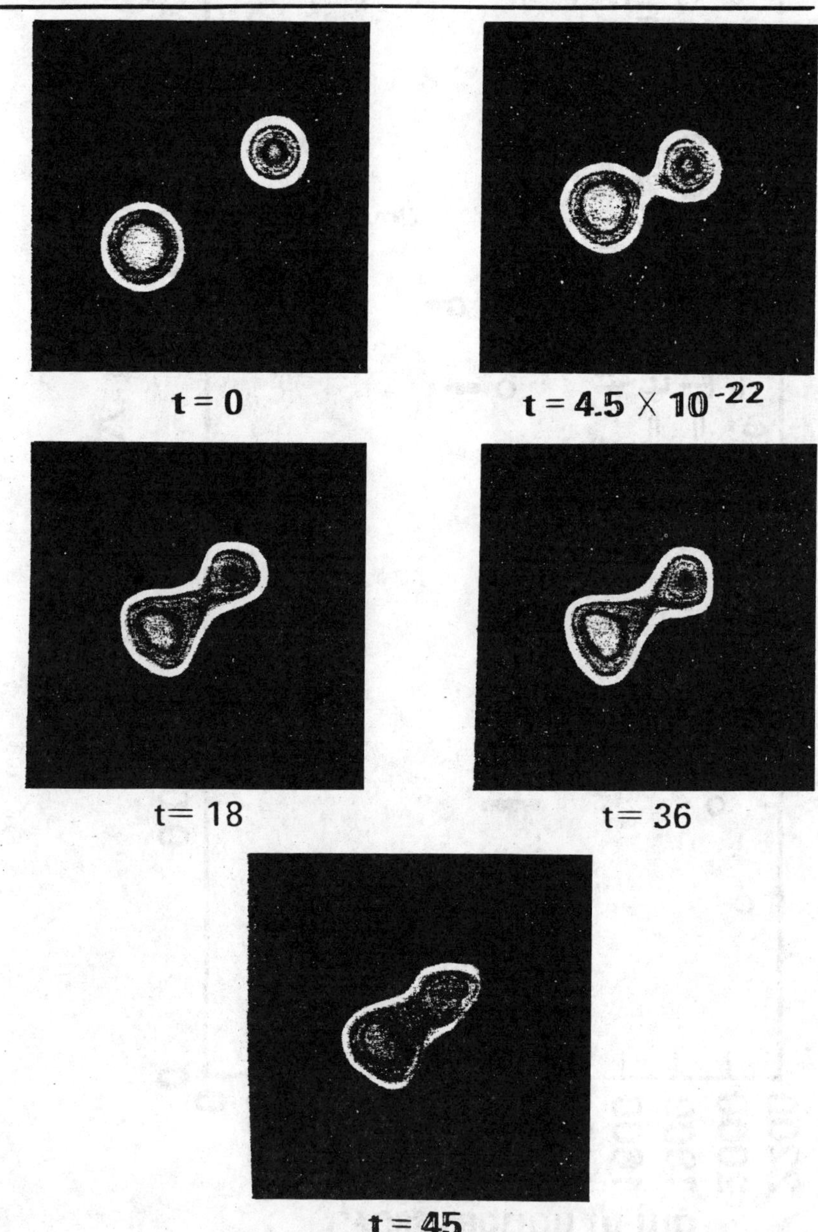

FIG. 109

$\ell = 84$

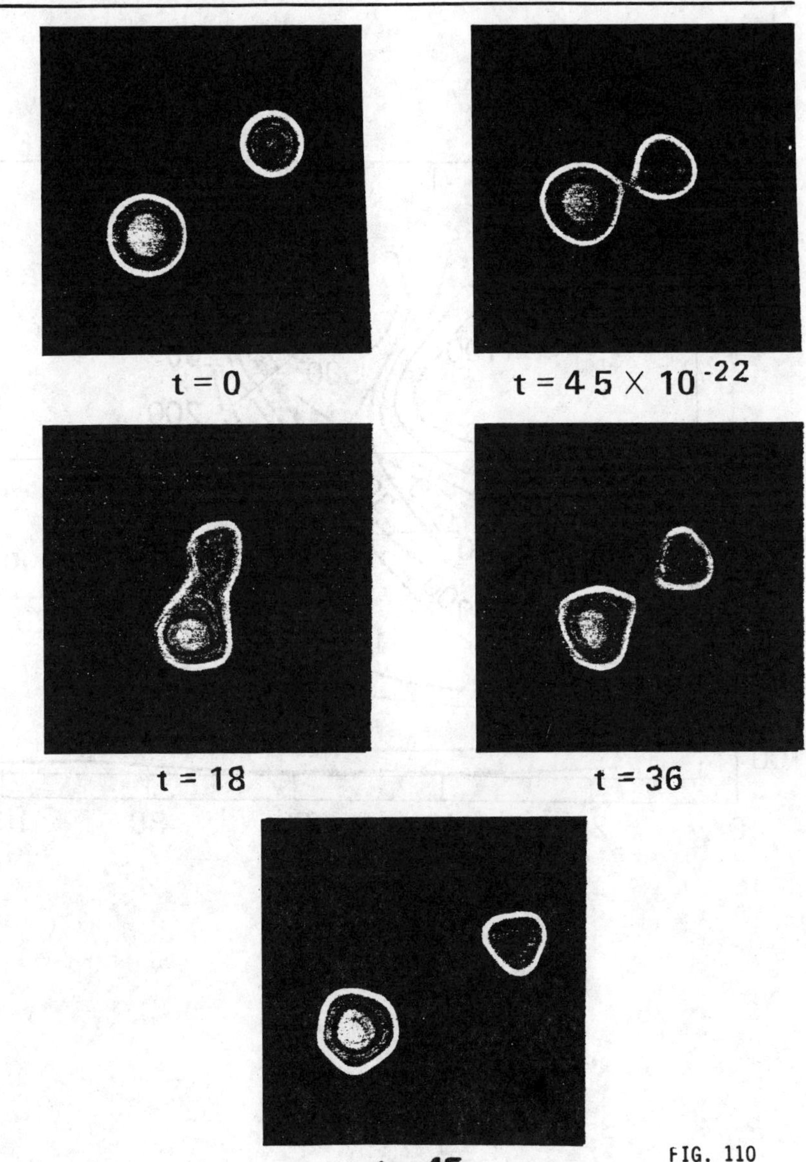

FIG. 110

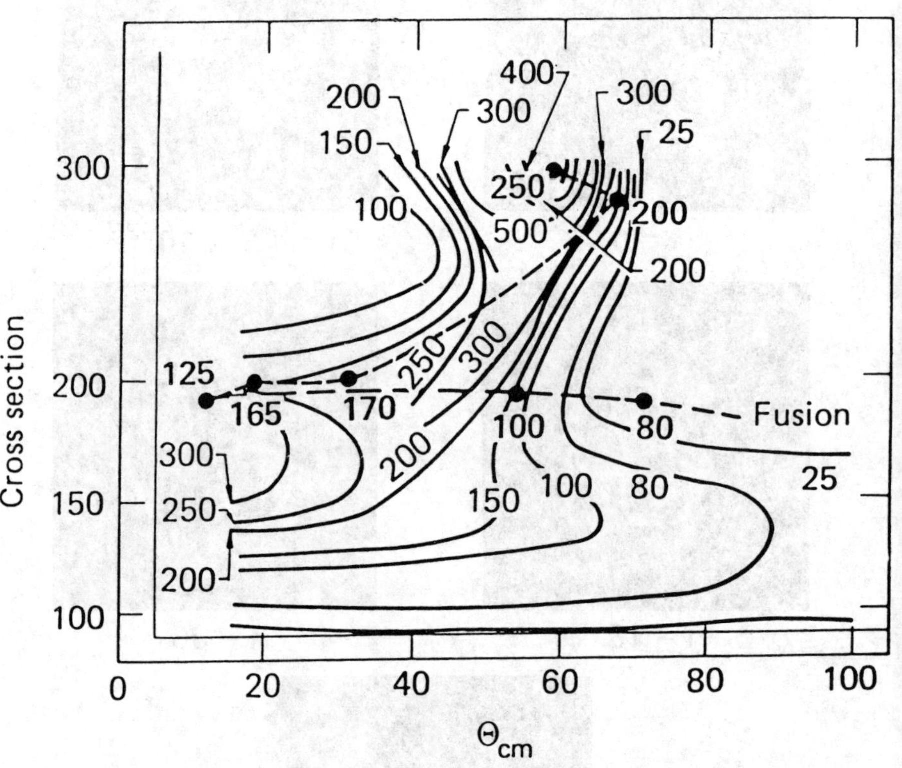
Cross section
300
200
150
100
0
20
40
60
80
100
Θ_{cm}
400
200
300
300
25
150
250
100
200
500
200
125
250
300
165
170
100
80
Fusion
300
200
250
100
80
25
150
200

FIG. 111

LECTURE II

Time Dependent Hartree-Fock Theory:

Special Applications

M. S. Weiss*
Lawrence Livermore National Laboratory
Livermore, CA 94550

*Work performed under the auspices of the U. S. Department of Energy by the Lawrence Livermore National Laboratory under contract No. W-7405-ENG-48.

In this second lecture I will address some TDHF applications that are somewhat subtler than just crashing nuclei together and looking at large chunks of nuclear material. I will also attempt to address some of the technical problems of doing TDHF should any of you have the bad taste to want to do so and also point out some areas where work remains to be done.

A. Promptly Emitted Particles (PEP)

We will restrict our discussion to what is called Fermi jetting. Epistemologically this is a simple concept. If two nuclei collide at an energy sufficiently high above the coulomb barrier it seems reasonable that some of the nucleons will have velocity components made up of their Fermi motion added to the relative motion which will be so high that they won't be captured into the compound system or even remain in the deeply inelastically excited fragments. Although theorists can isolate processes, in real nuclei lots of things happen at once and fast-forward nucleons can come not only from Fermi jets but also from peripheral reactions and pre-compound emission.[200]

With these caveats in mind let us examine the TDHF calculation of Devi[201] et al. In this work ^{16}O was impacted upon ^{93}Nb at E_{lab} = 204 MeV, for initial angular momenta from 0 to 50 $\hbar$. In figure 201, the top half shows the results for a central collision. The left side is the density in the reaction plain greater than 10% of normal nuclear density (nnd), the right side 1 to 10% of nnd. The little puff of nuclear material is the TDHF candidate for a Fermi jet. It is a candidate with excellent credentials because it is moving with a velocity slightly higher than the Fermi velocity. It is emitted in a time comparable to the transit time of an uninhibited projectile nucleon traversing the target. Averaged over impact parameters (see bottom half fig. 201: the qualitative behavior is the same as for a central collision), one has

$$N_{\nu}(\text{TDHF}) = 0.19 \text{ neutrons in the jet}$$

and

$$N_{\nu}(\text{expt}) = 0.15 \pm 0.05.$$

Some of the technical difficulties, however, which must be understood, make many of the calculated numbers uncertain to a factor of 2. There are numerical problems of accuracy which make calculations of the velocity, particularly large ones, inaccurate by 50%. The number of nucleons in the jet is uncertain because the ^{16}O punches through the ^{93}Nb on the tail of the jet and the separation into projectile and jet is arbitrary to perhaps a factor of two. It is also amusing to find that the TDHF results do not differ markedly from a simple Fermi gas model. This comparison is shown in Fig. 202 for the number of particles, scattering angle and mean velocity as a function of incident angular momentum. In the Fermi gas model, the distribution of boosted projectile velocities are compared to a fiducial v_t choosen so that particles with velocity greater that v_t had kinetic energies bigger than the average nuclear potential energy. These numbers can be corrected for window effects[202] without qualitative changes. The bottom curve of Fig. 202 shows the effect of numerical inaccuracy in that the jet velocity increases with incident angular momentum which is unrealistic.

Another calculation of PEP with TDHF was performed by Stocker[203] et al. at three to six times the energy per nucleon in the lab than the calculation of Devi et al.[201] They studied ^{12}C onto ^{197}Au at 360 MeV to 1020 MeV. In contrast to the previous calculation carried out in coordinate space and finite grid this work was done in momentum space, with the potential calculated in coordinate space and then fast-Fourier transformed. Fig. 203 illustrates their results which one sees are qualitatively similar to the work of Devi et al.[201] The momentum space calculation is in principle more accurate at higher energies because the TDHF propagation is gallilean invariant. In practice one must be very careful about not choosing too low an upper limit to the momentum and in connecting the potential between representations. Both calculations suffer from the problem of nucleon radiation. Being performed in a finite box, the emitted nucleons will reflect from the walls and eventually re-enter the compound system, which is unphysical. Therefore in performing these calculations one must be very careful to follow even small quantities of density in time and terminate the calculation before it becomes invalid. Alternatively one can add absorbing walls if one wants to follow late time behavior. To my knowledge no one has carried out such a program.

B. Large Amplitude RPA

TDHF has been used to study the radiation, and associated damping, of a nuclear giant resonance in the extremely interesting study of D. Vauterin and S. Stringari.[204] Here a simple Hartree-Fock potential was used to construct the self-consistent of solution to ^{16}O. The ^{16}O was then placed on a one dimensional radial grid in an external radial potential

$$V_{ext} = e^{-\gamma r^2}$$

and then evolved in time by the TDHF equations. The external potential raised the total energy of the system by approximately 1 MeV and it began to perform (monopole) oscillations. The top graph in Fig. 204 shows the number of particles at a distance 18 to 20 fm from the center of the excited nucleus. The calculation becomes meaningless for t $> 3 \cdot 10^{-22}$ sec because radiated particles have had time to reflect off the spherical cavity wall and to be re-absorbed by the nucleus. The bottom curve shows an arbitrary representation of the radius of the radiating nucleus as a function of time. It is arbitrary because the radius at which one says this point is nucleus and that point is radiated nucleon is subjective. The choice made by the authors is to define r_c as the root mean square radius of all the nuclear density between 0 and 5 fm.

One sees that in $2 \cdot 10^{-22}$ seconds about .04 nucleons have been radiated. Therefore in 10^{-18}-10^{-16} seconds, a compound nucleus lifetime, the nucleus would radiate too much. We are observing a direct process, that is $\Gamma\uparrow$. The oscillation in r_c damps very quickly, which considering the small amount of nuclear material evaporated I find very surprising. The authors extract a damping time [$1.5 \cdot 10^{-22}$ sec] and find it is consistent with the RPA width, calculated in a very different way. It is also interesting but not shocking that putting more initial energy into the "cold" nucleus (by increasing λ) also increases the number of particles radiated, and a larger decrease of the eschatological r_c.

In Figure 205, the same technique is applied to ^{40}Ca. One sees an order of magnitude increase in the number of nucleons radiated and a large decrease in the damping time [also consistent with RPA calculations]. Although this study by Vauterin and Strinjari[204] is limited to monopole resonances it is nevertheless the first study of a macroscopic amplitude giant resonance calculation and the only one of particle radiation.

C. Exciting Giant Resonances in H.I. Collisions

The previous study of a giant resonance leads us to examine another application of TDHF to the possibility that not only are giant resonances excited in heavy-ion collisions but also that some of them are at unusually high energy. The motivation for this work was the very provocative experiments of Frascia et al.[205] Fig. 206 purports to show the nuclear excitation for the scattering of ^{40}Ca onto ^{40}Ca at E_{lab} = 400 MeV. One sees structure in the cross section for several different final nuclei. The labels indicate excitation energy in the residual nucleus.

The application to TDHF was obvious.[206] Using the techniques described in the first lecture, the energy and impact parameter were choosen to reproduce the experimental situation (i.e. the scattering angle). After the nuclei were well separated, the coulomb interaction between the two nuclei was turned off. Fig. 207 shows the density profiles of the excited nucleus. Then in a coordinate system moving with the center of mass of the final nucleus, the cartesian moments

$$M_{pqr} = \iiint d\tau\, \rho\,(x, j, z, t)\, x^p y^q\, Z^r$$

were calculated for $2 \cdot 10^{-21}$ sec. Then after transforming the moments to one rotating with the principal axes of the excited nucleus, we fourier analyzed the M's. The time $2 \cdot 10^{-21}$ sec was choosen to reduce the spread due to finite time to 2 MeV and to keep that year's computer cost over-run manageable.

Fig. 208 shows the r.m.s. radius as a function of time and Fig. 209 the fourier transform of one of the quadrupole moments. Fig. 210 shows a mix of even and odd moments. It is provacative that the calculation mimics the experimental dependence on inelasticity. Frascaria et al.[205] find the high energy structure most pronounced if the collision is peripheral. If it is more central, therefore more inelastic, the amplitude of the high energy structure is reduced by an order of magnitude. The same is true for the calculations.

There are many questions of interpretation in these results. The calculation has no way of discriminating between high energy vibrations and "multiple photons." Having calculated the moments in a cartesian representation we cannot make contact with the conventional multipole folk-lore. This can be eventually repaired although perhaps some of you will be provoked to do this in a timely fashion. In addition the model for ^{40}Ca was made simple for calculational reasons and does not have the richness of a "real nucleus." Nevertheless, it would appear that the TDHF calculation mimics this exotic experimental structure, namely collective motion at energies too high to seem appropriate. However, heavy ions emphasize a property that your conventional giant resonance exciters [protons, electrons] don't have, namely the interaction need not, indeed cannot except at the periphery, act as a one body operator. It could be that the high energy collective excitations are multiple multipole giant resonances. It is not difficult to think up amusing ways of using TDHF to test that hypothesis.[207]

The last topic I will address is the application, not truly successful as yet, of TDHF to study nuclear fission. The only published work which I know of is by Negele et al.,[208] and the only unpublished with which I am familiar is by Weiss.[209] Discussing this will allow us also to reillustrate the importance of symmetries in TDHF.

In principle TDHF could offer much insight into the fission process. Fission is clearly the grossest example of nuclear collective phenomena. Because pure TDHF is valid only for positive total energies it cannot be used to study tunneling in real time. Instead one must choose an initial condition from which fission can proceed through positive energies. That is, a position on the "far" side of the saddle point must be taken. One can take advantage of the presumption that in induced fission near the barrier the collective velocity will be negligible near the saddle to simply start the TDHF solution with the static HF solution at zero collective velocity on the static barrier below the saddle.

The questions TDHF calculations can address are the time scale of the process, the nature of the "viscosity" or more sharply the interplay between fragment kinetic and potential energy. Not independent is the question of fragment shape along the path and at scission. If one believed TDHF was sufficiently free of restrictions to adequately represent the application of that method to this problem then one could test the mean field approximation. Experimental data abound on fragment distribution (including assymmetry) and mean translational kinetic energies.

The formidable technical problems associated with this type of work led the authors of ref. 207 to consider ^{236}U with axial symmetry, reflection symmetry, constant gap pairing and no spin orbit interaction. Of these the axial symmetry is the most serious. As the nucleus moves along the fission path its initial set of occupied m projection levels will not remain the lowest (fig. 211). Were it to remain fixed with the initial choice it could never fission. The authors of ref. 207 "cured" this pathology by introducing pairing not for its effect on the energy surface, which was already unrealistic because of the lack of spin orbit, but to allow changes in m projection of the occupied orbits at crossings. The strength of the pairing gap was varied from .72 to 6.0 MeV. Of these only the first is realistic for ^{236}U.

Not surprisingly, the results depend very strongly on the value of the pairing gap.

Pairing Gap MeV	Time to Fission 10^{-21} ec	Kinetic Energy MeV
0.7	5.0	166
2.0	3.4	142
6.0	2.2	---
Experiment		168

Figure 212 shows the static barrier as a function of quadrapole moment. The calculations are initiated 1 MeV below the top of the barrier. Fig. 213 shows the shape as a function of time and compares it to various macroscopic calculations.

What one learns from this paper is that a) the authors have worked very hard because this is a very difficult calculation, b) maybe you can use TDHF to quantitatively study fission but you will either have to remove the symmetries or learn how to calculate the pairing, c) TDHF can be used to qualitatively, literally, look at fission and provide provocative comparisons with macroscopic theories.

In the spirit of removing symmetries, to see if that would provide a pairing free method of changing orbits during the fission process, Weiss used a simple model of ^{80}Zr with only one symmetry; reflection in the plane perpendicular to the fission direction. There was no pairing to mix orbits, which were characterized only by the parity perpendicular to the reaction plane. Unfortunately, the static fission barrier was not very carefully mapped. Instead somewhat arbitrary quadrupole constraints in x, y, and z were used to find a "high point" but which was assumed to be higher than the saddle. When used as an initial starting point the nucleus had the excellent sense to make an enormous initial excursion in quadrupole moment towards fission then slow down and very expensively begin to slosh back and forth. Whether this was due to inept mappings of the barrier or the reluctance of the single particle levels to change remains a great mystery. In principle we could afford to break the last symmetry to answer this question but like so many other things in this cornucopia, they have not yet been done.

Very likely a combination of breaking symmetry and pairing will both be required for a realistic study of fission. Having this available would be enormously exciting. TDHF bypasses the vexing problem of calculation a dynamic mass. Indeed in the region between level crossings (assuming some symmetry remains) the computational measurement of collective velocity tests the concept of a collective (or cranking) mass. A variety of exotic phenomena could be studied. In principle a muonic atom of ^{238}U could be studied in radiationless cascade leading to fission. Probabilities of excitation in the second well, which fragment the muon ends up on, etc. could be learned.

While it is probable that the initial enthusiasm for using TDHF in heavy ion scattering has appropriately been exhausted, a variety of interesting unsolved problems still remain within the context of ordinary, real time, TDHF. The introduction of more efficient computer facilities and new computers should permit solution of even those that are most intransigent.

References

200 M. Blann, Nuclear Physics A235, 211 (1974) and Phys. Rev. C23, 205 (1981).

201 K. R. S. Devi et al., Phys. Rev. C24, 2521 (1981).

202 K. Mohring, W. J. Swiatecki and M. Zielinska-Pfabe, Nuclear Science Div. Report, Lawrence Berkeley National Laboratory, Berkeley, CA (1983).

203 H. Stocker et al., Phys. Letts. 101B, 379 (1981).

204 D. Vauterin and Stringari, Proceedings of TDHF Conference, Orsay, 1978.

205 N. Frascaria et al., Z. Phys. A299, 73 (1981).

206 H. Flocard and M. S. Weiss, Phys. Letts. 105B, 14 (1981).

207 Proc. HVAR Conference on Heavy Ion Collisions, 1982, (Ed. N. Cindro, North Holland).

208 J. W. Negele et al., Phys. Rev. C17, 1098 (1978).

209 M. S. Weiss, private communication.

Figure Captions

Fig. 201 Density profile for ^{16}O onto ^{93}Nb at 204 MeV in the Lab. Left side is for densities greater than 10% of normal nuclear density (nnd). Right side for density less than 10% for nnd. From Ref. 201.

Fig. 202 Same collision as Fig. 201. Number of particles in the jet, scattering angle and mean velocity in units of Fermi velocity vs incident angular momentum. From Ref. 201.

Fig. 203 Left two columns are continuous of constant density for collision of ^{12}C onto ^{197}Au at 360 MeV in Lab., for impact parameters of 1 and 8 fm. Right column similarly but Lab energy of 102 MeV and impart parameter of 9.4 fm. From Ref. 203.

Fig. 204 Number of particles radiation and radius of nucleus vs time for monopole excitation of ^{16}O. From Ref. 204.

Fig. 205 Similar to above for ^{40}Ca. From Ref. 204.

Fig. 206 Excited spectrum subsequent to heavy ion collision of $^{40}Ca + {}^{46}Ca$ at E_{lab} = 400 MeV. From Ref. 205.

Fig. 207 Density contours of ^{40}Ca subsequent to TDHF collision at E_{lab} = 400 MeV and center of mass scattering angle of 3°. From Ref. 206.

Fig. 208 RMS radius vs time for nucleus in Fig. 207. From Ref. 206.

Fig. 209 Fourier transform of y^2 moment of nucleus in Fig. 207. From Ref. 206.

Fig. 210 Fourier transform of various moments similar to Fig. 209. From Ref. 206.

Fig. 211 Standard Nillson diagram of single particle energy levels vs deformation. Note crossings where lowest levels change m projection.

Fig. 212 Lower right curve is HF energy vs quadrapole moment. TDHF calculation started 1 MeV down on right from top of barrier. From Ref. 208.

Fig. 213 Left column shows time history for shapes of fissioning nucleus in TDHF. Other shapes are from various macroscopic models. From Ref. 208.

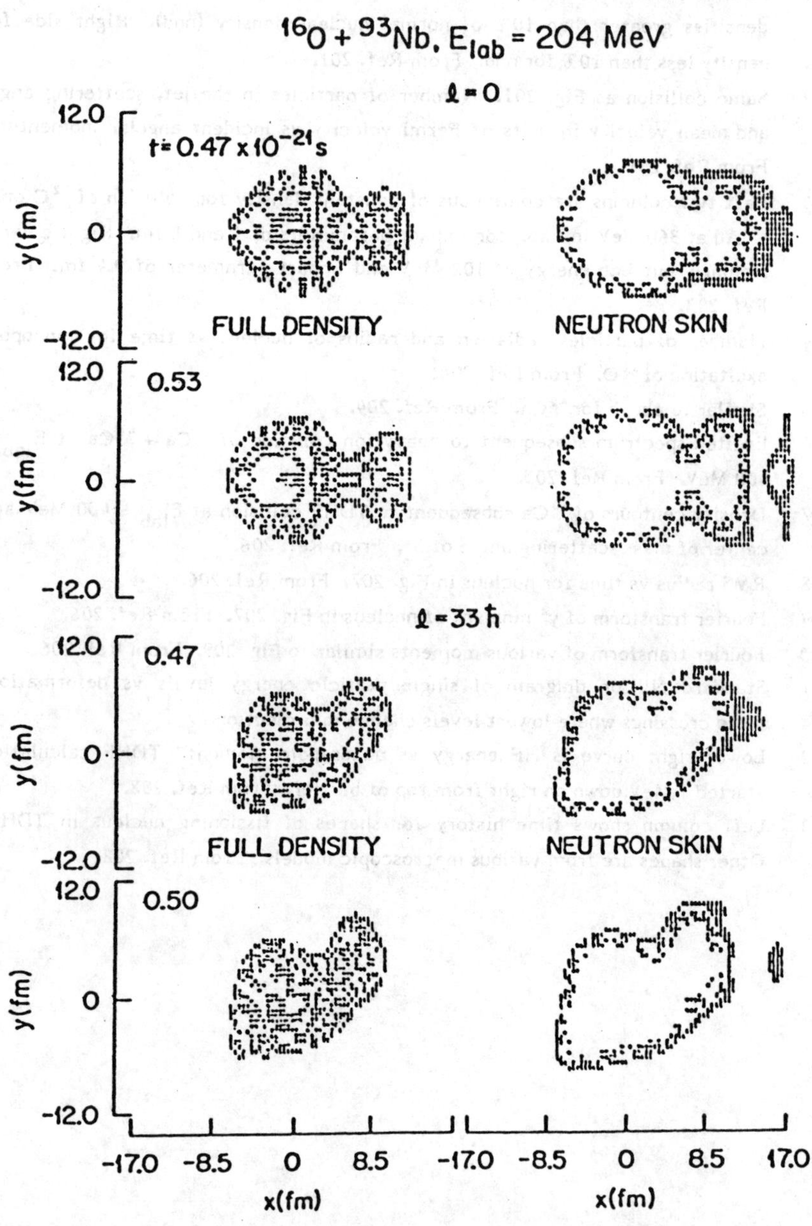

16O + 93Nb, Elab = 204 MeV
ℓ=0
t=0.47 x 10-21 s
FULL DENSITY
NEUTRON SKIN
0.53
ℓ=33ħ
0.47
FULL DENSITY
NEUTRON SKIN
0.50
y(fm)
12.0
0
-12.0
-17.0
-8.5
0
8.5
17.0
x(fm)

FIG. 201

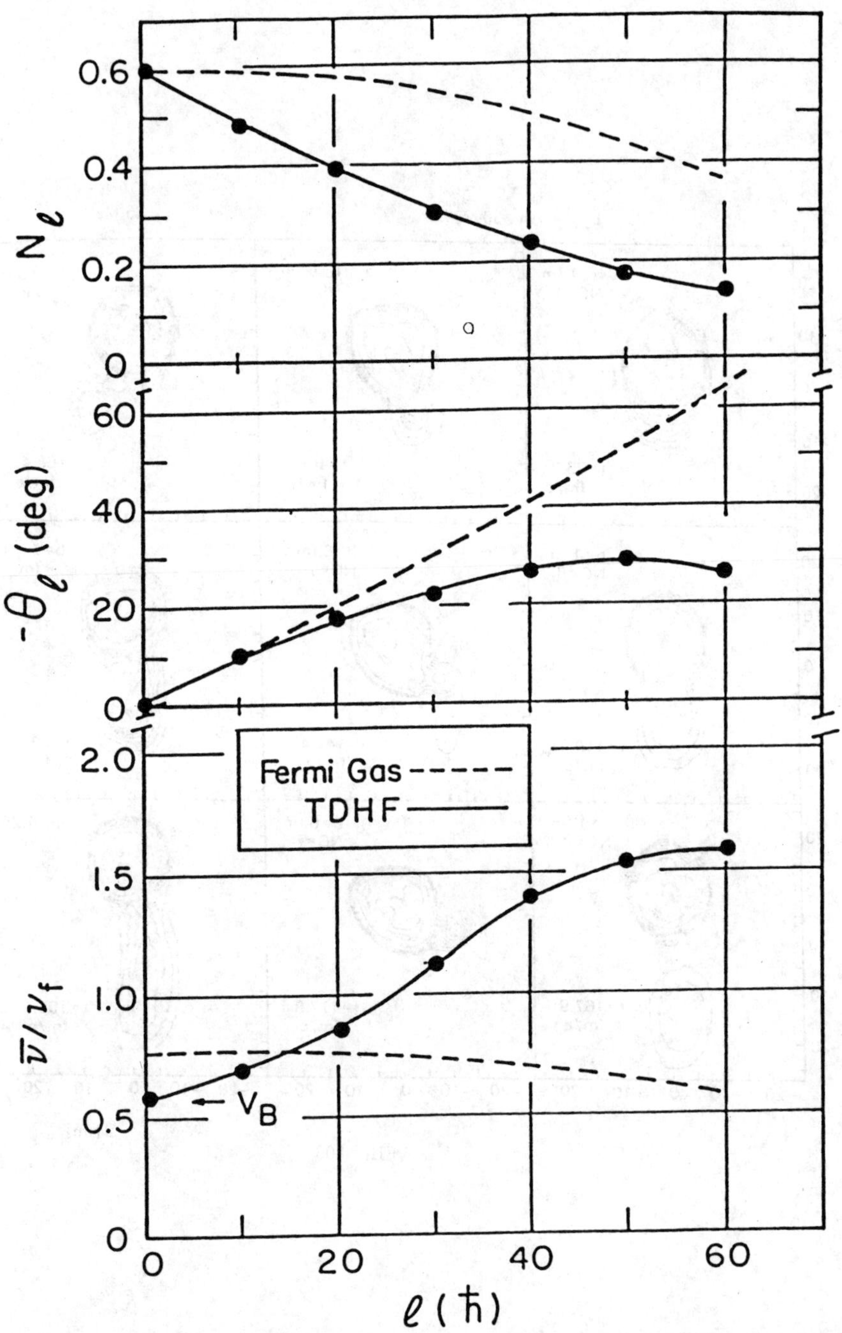

0.6
0.4
0.2
0
N_ℓ
60
40
20
0
$-\theta_\ell$ (deg)
2.0
1.5
1.0
0.5
0
$\bar{v}/v_f$
Fermi Gas
TDHF
V_B
0
20
40
60
$\ell(\hbar)$

FIG. 202

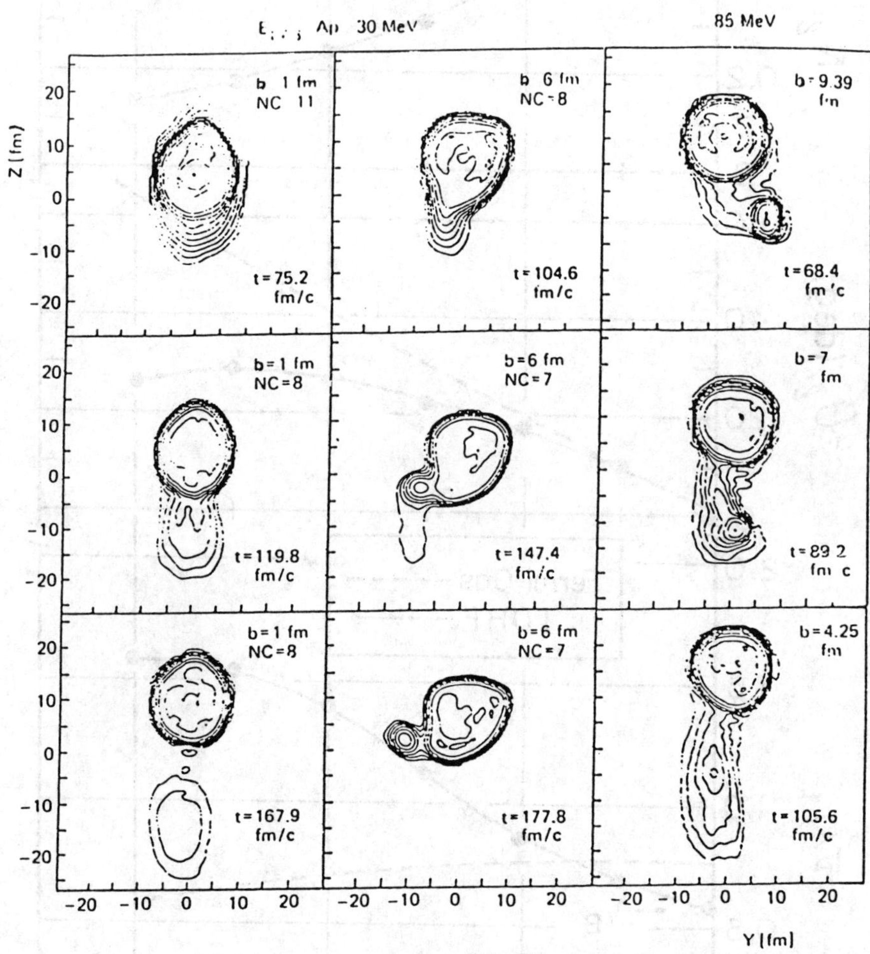
30 MeV
85 MeV
Z (fm)
b=1 fm
NC=8
t=75.2 fm/c
t=104.6 fm/c
t=68.4 fm/c
b=6 fm
NC=7
b=7 fm
t=119.8 fm/c
t=147.4 fm/c
t=89.2 fm/c
b=4.25 fm
t=167.9 fm/c
t=177.8 fm/c
t=105.6 fm/c
b=9.39 fm
Y (fm)

FIG. 203

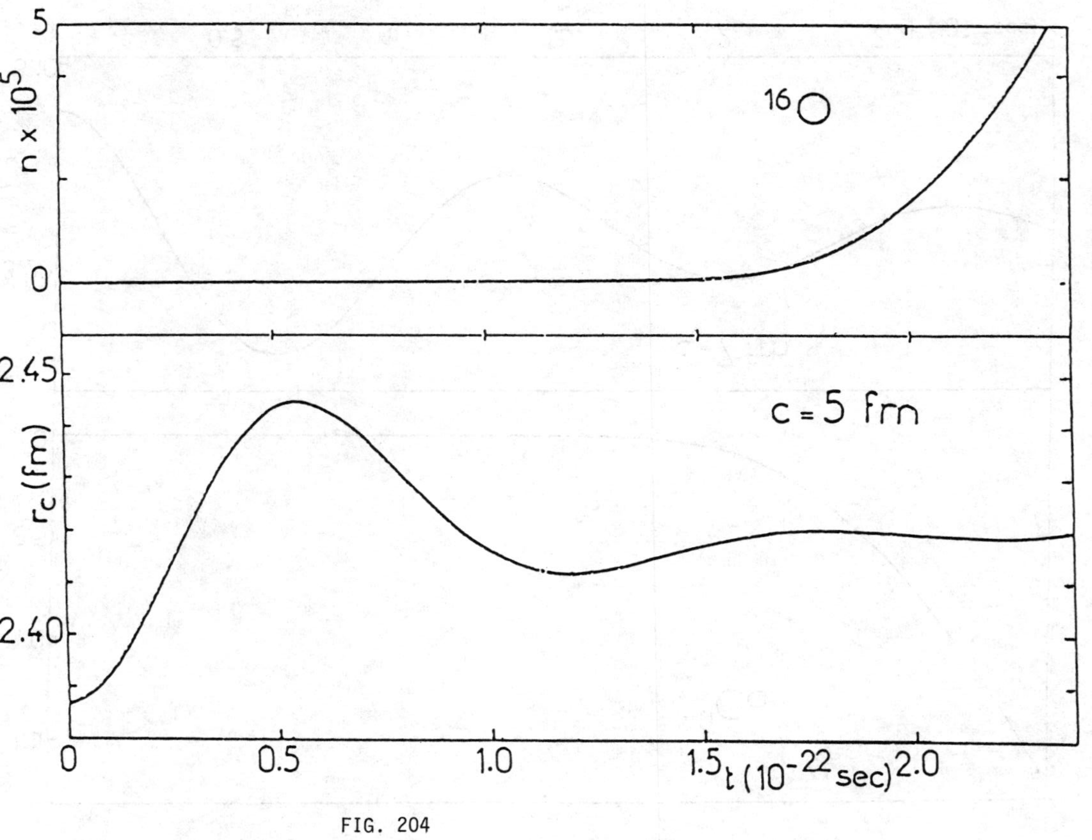

FIG. 204

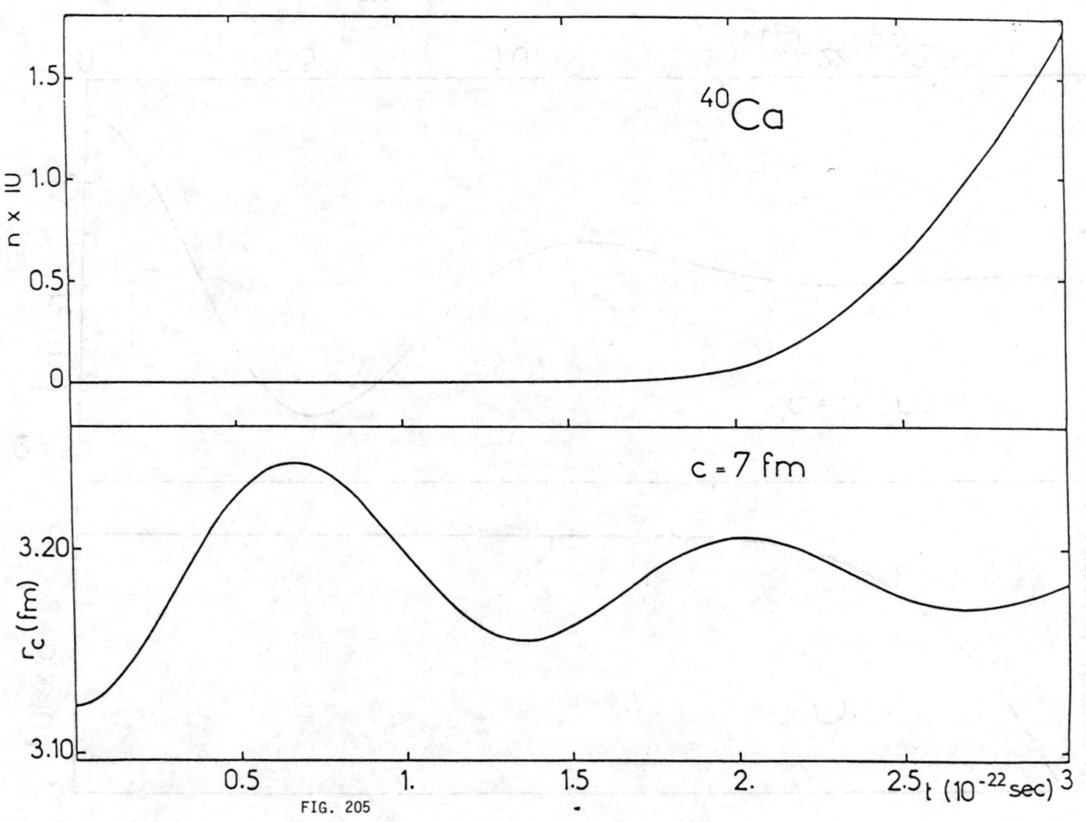

FIG. 205

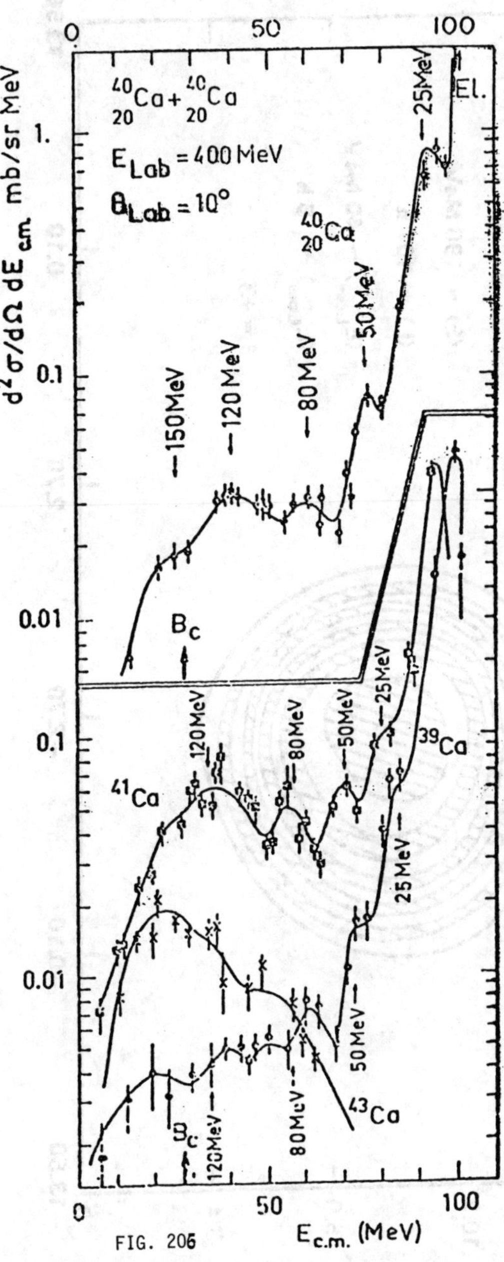

$^{40}_{20}Ca + ^{40}_{20}Ca$
$E_{Lab} = 400$ MeV
$\theta_{Lab} = 10°$
$^{40}_{20}Ca$
El.
25MeV
50MeV
80MeV
120MeV
150MeV
B_c
^{41}Ca
^{39}Ca
^{43}Ca
$d^2\sigma/d\Omega\, dE_{c.m.}$ mb/sr MeV
$E_{c.m.}$ (MeV)

FIG. 206

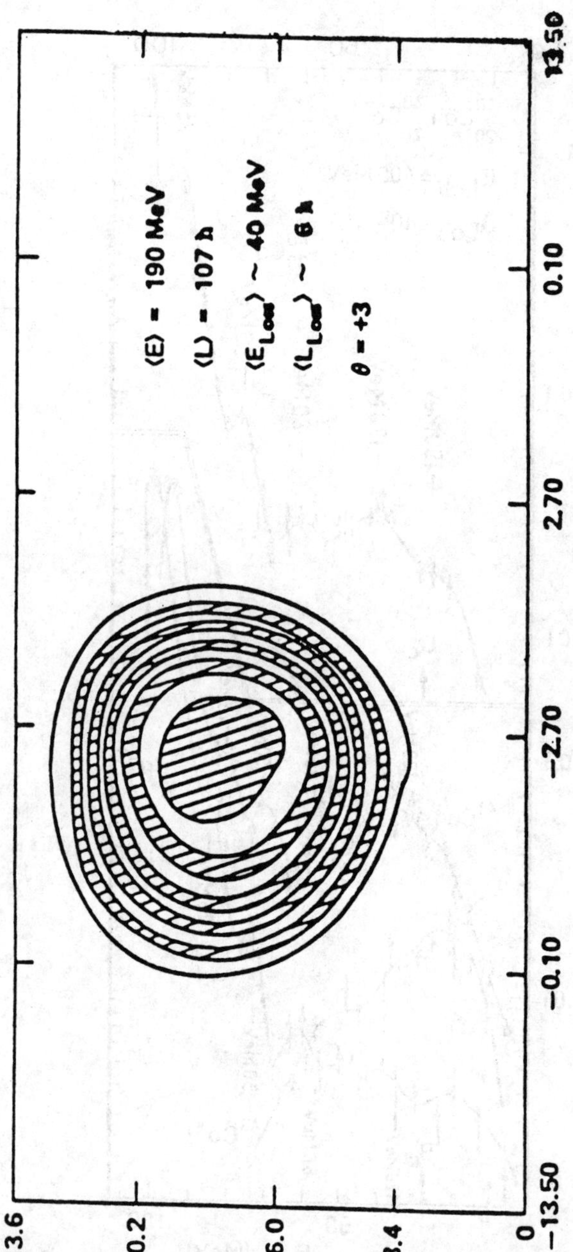
⟨E⟩ = 190 MeV
⟨L⟩ = 107 ħ
⟨E_Loss⟩ ~ 40 MeV
⟨L_Loss⟩ ~ 6 ħ
θ = +3
13.6
10.2
6.0
3.4
0
−13.50
−0.10
−2.70
2.70
0.10
13.50

FIG. 207

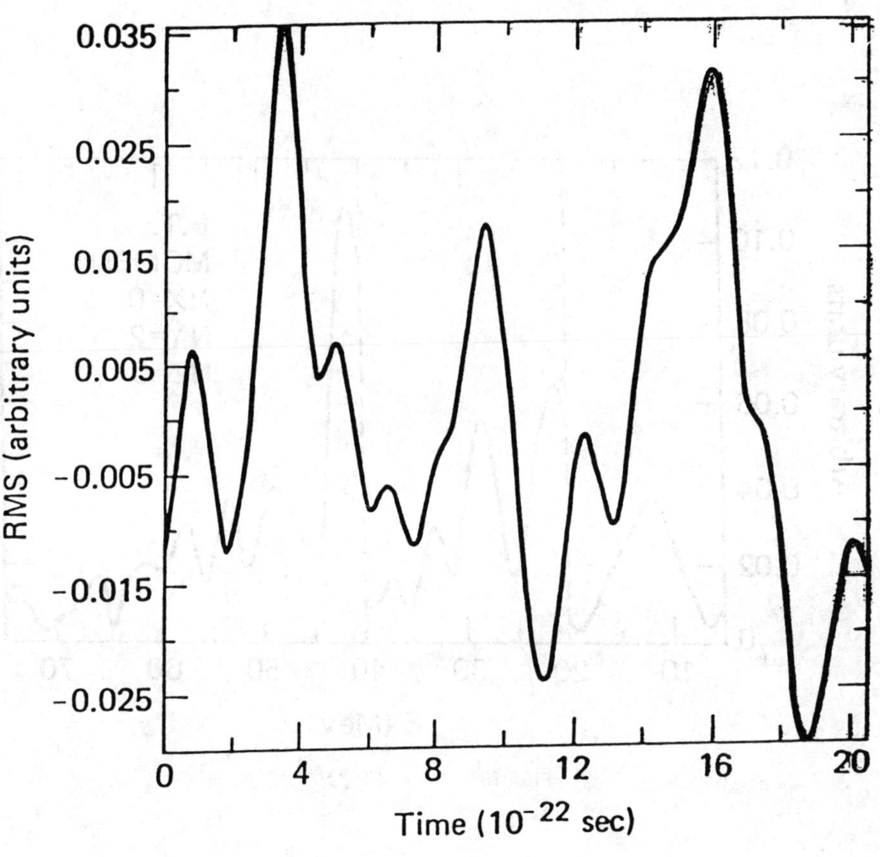
0.035
0.025
0.015
0.005
-0.005
-0.015
-0.025
RMS (arbitrary units)
0
4
8
12
16
20
Time (10^{-22} sec)

FIG. 208

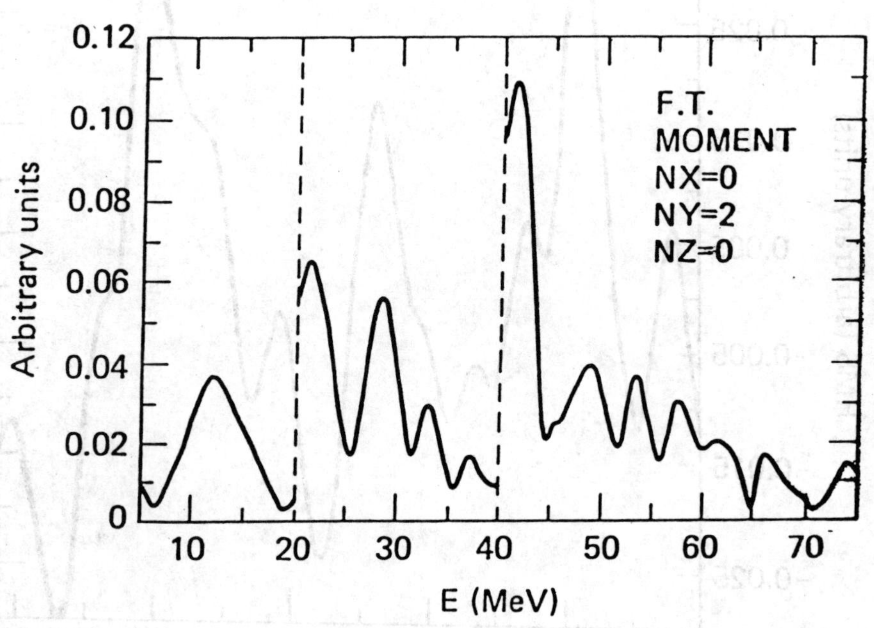
0.12
0.10
0.08
0.06
0.04
0.02
0
Arbitrary units
F.T.
MOMENT
NX=0
NY=2
NZ=0
10
20
30
40
50
60
70
E (MeV)

FIG. 209

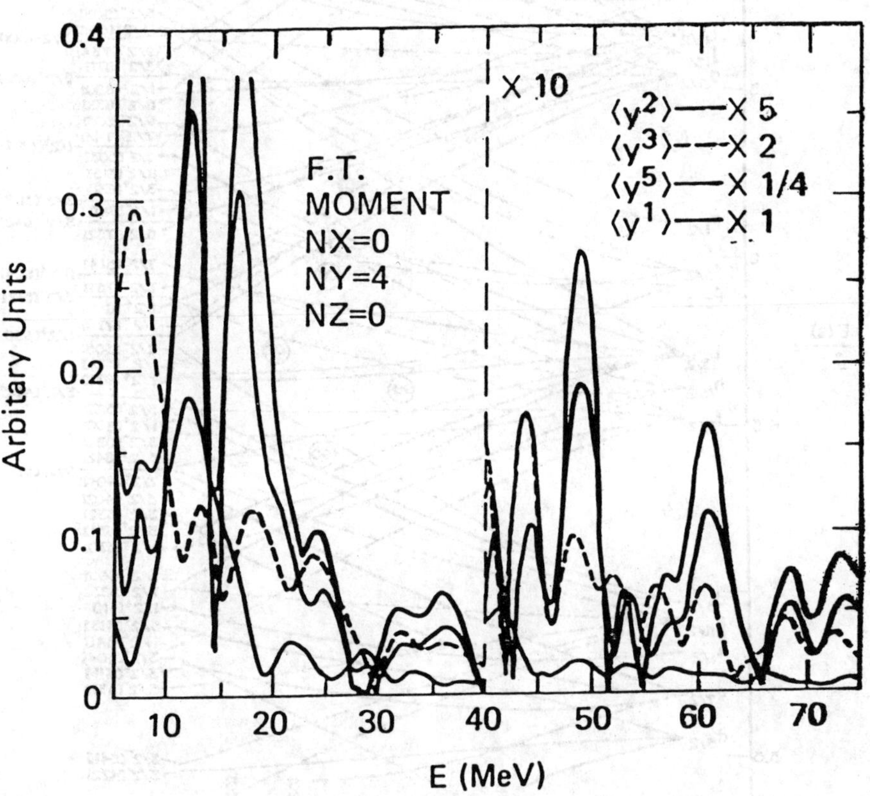
0.4
0.3
0.2
0.1
0
Arbitary Units
10
20
30
40
50
60
70
E (MeV)
X 10
F.T.
MOMENT
NX=0
NY=4
NZ=0
⟨y²⟩——X 5
⟨y³⟩- - -X 2
⟨y⁵⟩——X 1/4
⟨y¹⟩——X 1

FIG. 210

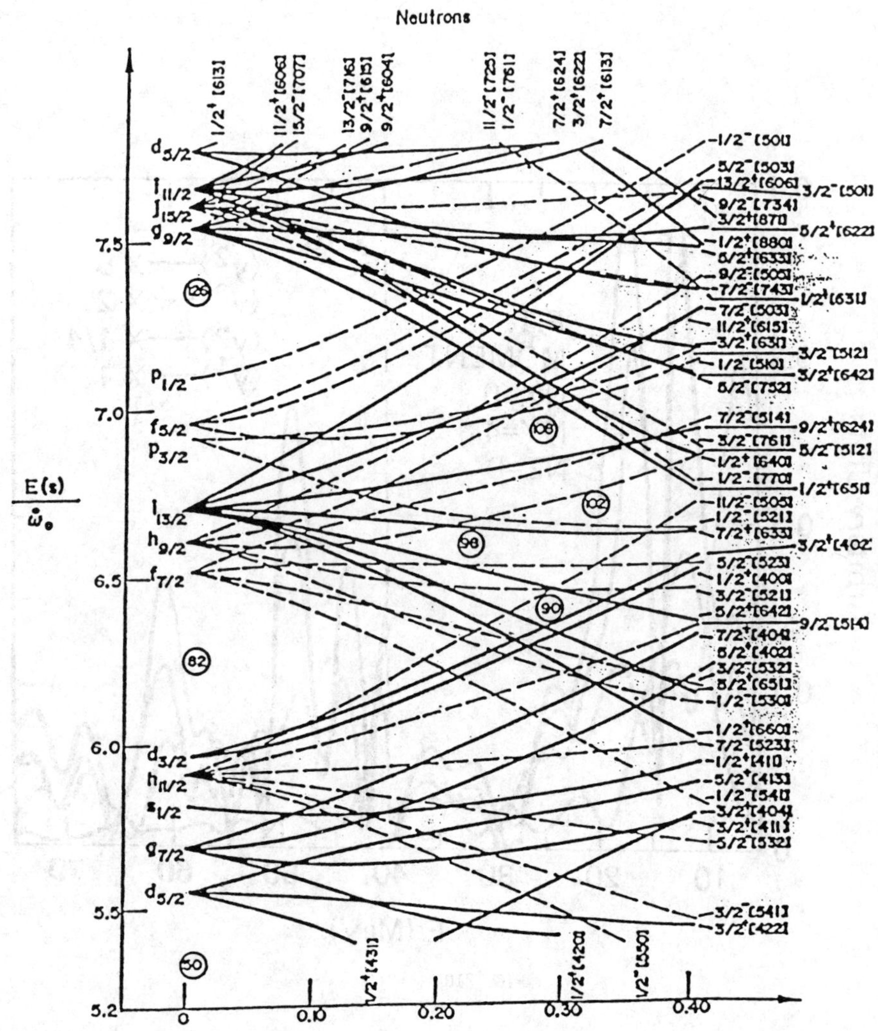

Neutrons
E(ε)/ω̊₀
d5/2
i11/2
j15/2
g9/2
p1/2
f5/2
p3/2
i13/2
h9/2
f7/2
d3/2
h11/2
s1/2
g7/2
d5/2
126
108
102
98
90
82
50
7.5
7.0
6.5
6.0
5.5
5.2
0
0.10
0.20
0.30
0.40
1/2+ [613]
11/2+ [606]
15/2- [707]
13/2- [716]
9/2+ [615]
9/2+ [604]
11/2- [725]
1/2- [761]
7/2+ [624]
3/2+ [622]
7/2+ [613]
1/2- [501]
5/2- [503]
13/2+ [606]
3/2- [501]
9/2- [734]
3/2+ [871]
5/2+ [622]
1/2+ [880]
5/2+ [633]
9/2- [505]
7/2- [743]
1/2+ [631]
7/2- [503]
11/2+ [615]
3/2+ [631]
3/2- [512]
1/2- [510]
3/2+ [642]
5/2- [752]
7/2- [514]
9/2+ [624]
3/2- [761]
5/2- [512]
1/2+ [640]
1/2- [770]
1/2+ [651]
11/2- [505]
1/2- [521]
7/2+ [633]
3/2+ [402]
5/2- [523]
1/2+ [400]
3/2- [521]
5/2+ [642]
9/2- [514]
7/2+ [404]
5/2+ [402]
3/2- [532]
5/2+ [651]
1/2- [530]
1/2+ [660]
7/2- [523]
1/2+ [410]
5/2+ [413]
1/2- [541]
3/2+ [404]
3/2+ [411]
5/2- [532]
3/2- [541]
3/2+ [422]
1/2+ [431]
1/2+ [420]
1/2- [550]

FIG. 211

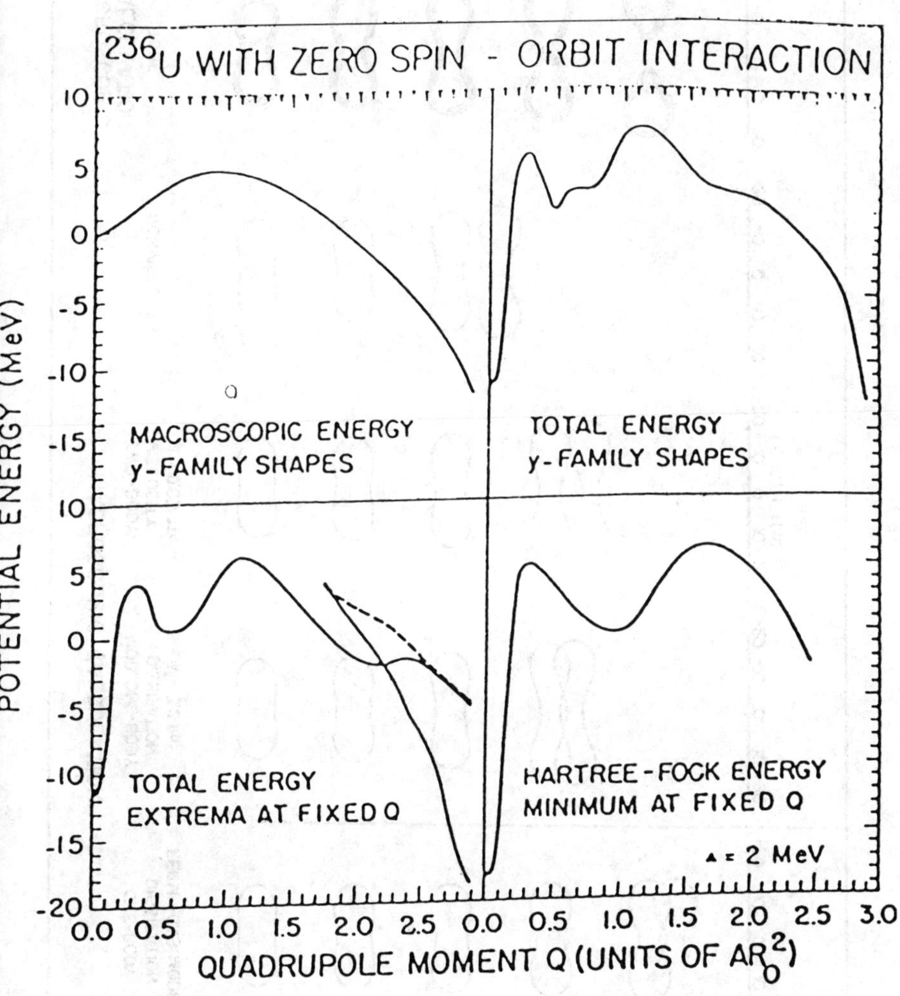

236U WITH ZERO SPIN - ORBIT INTERACTION
POTENTIAL ENERGY (MeV)
MACROSCOPIC ENERGY
y-FAMILY SHAPES
TOTAL ENERGY
y-FAMILY SHAPES
TOTAL ENERGY
EXTREMA AT FIXED Q
HARTREE-FOCK ENERGY
MINIMUM AT FIXED Q
▲ = 2 MeV
QUADRUPOLE MOMENT Q (UNITS OF AR0^2)

FIG. 212

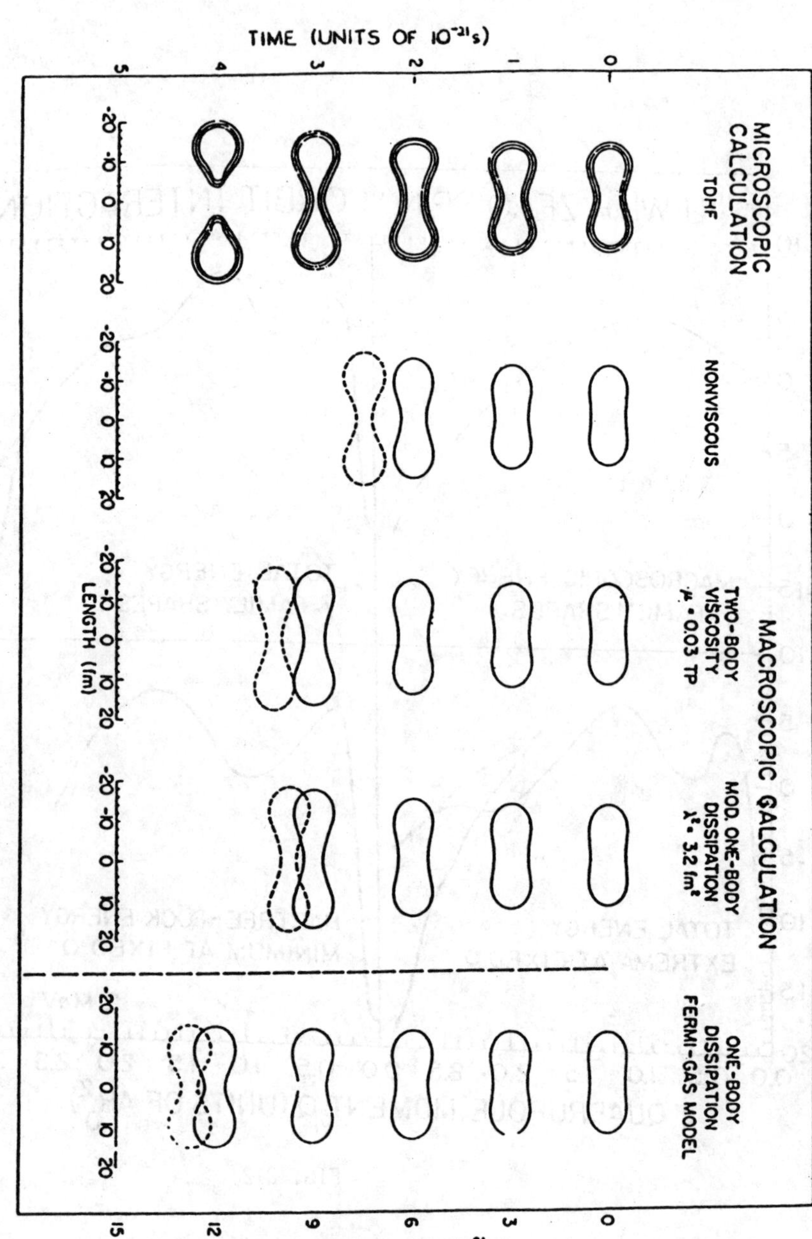

TIME (UNITS OF 10^{-21}s)
MICROSCOPIC CALCULATION TDHF
MACROSCOPIC CALCULATION
NONVISCOUS
TWO-BODY VISCOSITY μ = 0.03 TP
MOD. ONE-BODY DISSIPATION k_s = 3.2 fm
ONE-BODY DISSIPATION FERMI-GAS MODEL
LENGTH (fm)
TIME (UNITS OF 10^{-21} s)

FIG. 213

SURFACE MODES AND DEEP INELASTIC REACTIONS

C.H. Dasso
Nordita
Blegdamsvej 17
DK-2100 Copenhagen

The contents of the lectures delivered by C.H. Dasso can be found in the following references:

C.H. Dasso: "Introduction to Coherent Surface Excitations", Proceedings of the Nuclear Physics Workshop, ICTP, Trieste, Miramare, 5 - 30 Oct., 1981, p. 565, North Holland (1982)

R.A. Broglia, C.H. Dasso and A. Winther, Varenna (1979): "Nuclear Structure and Heavy-Ion Collisions", LXXVII Corso Int. School E. Fermi, Academic Press (1981), p. 327

"RANDOM MATRIX THEORY OF NUCLEAR DYNAMICS"

H.A. Weidenmüller
Max-Planck-Institut für Kernphysik
Heidelberg
F.R.G.

The contents of the lectures delivered by H.A. Weidenmüller can be found in the following references:

H.A. Weidenmüller: "Transport Theories of Heavy Ion Reactions", Progr. in Particle and Nucl. Phys., Vol. 3, 49 (1981)

J. Verbaarschot, H.A. Weidenmüller and M. Zirnbauer: "Evaluation of Ensemble Averages for Simple Hamiltonians Perturbed by a GOE Interactions Ann. of Physics 153, 367 (1984)

J. Verbaarschot, H.A. Weidenmüller and M. Zirnbauer: "Statistical Nuclear Theory as an Anderson Model of Dimensionality Zero", Phys. Rev. Lett. 52, 1597 (1984)

H.A. Weidenmüller: "Statistical Theory of Nuclera Reactions as Gaussian Orthogonal Ensembles", Ann. of Phys. (in press)

O. Bohigas, M.J. Giannoni and C. Schmit: "Characterisation of Chaotic Quantum Spectra and Universality of Level Fluctuation Laws", Phys. Rev. Lett. 52, 1 (1984)

Classical Description of Heavy Ion Collisions

D. H. E. Gross

Bereich Kern- und Strahlenphysik des Hahn-Meitner-Instituts Berlin
und Fachbereich Physik der Freien Universität Berlin
Glienicker Straße 100, D-1000 Berlin 39

Abstract: The usefulness of classical concepts in the theory of heavy ion collisions is demonstrated in three examples. First we pass over classical Coulomb-scattering and explain some basic features of elastic scattering, then we treat low-energy deep inelastic collisions in detail and discuss various concepts of statistical dissipation phenomena. In the last chapter we discuss the statistical fragmentation of target nuclei after ultra relativistic proton bombardments.

Prologue

I was asked to give these lectures on 'Classical Description of Heavy Ion Collisions' as an introduction to the physical principles of heavy ion reactions. Certainly, and this must be said very clearly, the rich phenomena of HIR cannot be understood in classical terms alone. The classical view is necessarily an over-simplified view. Many quite interesting wave-mechanical effects exist. Often it is possible to describe them by a slight extension of classical mechanics by using the semiclassical approximation. An early introduction of the latter into the theory of HIR, which allows for the treatment of several coupled degrees of freedom, typical for HIR, was given in (7).

Here, in these lectures, I take the opportunity to illustrate some basic physical mechanisms of HIR by abstracting from any wave-mechanical complications.

I. Introduction

The classical description of collisions of heavy-nuclei has proved to be very successful in the past. Why is a classical description reasonable?

The first argument is that the de Broglie-wave length λ of heavy ions is very small compared to characteristic lengths in nuclei. Take the example of ^{40}Ar at a bombarding energy of 400 MeV. Asymptotically we have for λ:

$$\lambda_{\infty} = \frac{2\pi\hbar}{P_{\infty}} \simeq 0.2 \text{ fm} \tag{1.1}$$

Typical nuclear radii are 4 to 7 fm, the surface thickness is about 2 fm. However, at the point of closest approach the radial momentum of relative motion becomes zero and one may worry, but the width of the first maximum of the radial wave functions is normally quite small also. E.g. for ^{16}O+^{208}Pb at ℓ=o and relative energy of 65 MeV it is about 1 fm.

The second argument which forces us to use a classical description of relative motion is the following: In heavy-ion collisions a large amount of kinetic energy is transferred (dissipated) from relative motion into heating the nuclei. It is not possible to write down a Schrödinger equation which includes dissipative terms. The reason for this is that dissipation comes from the coupling to the many intrinsic background degrees of freedom. In general the motion of these d.f. destroys the phase-information contained in the relative motion. This latter argument certainly is the main reason, why classical models have been applied so often for heavy-ion collisions.

The most simple classical motion of two heavy nuclei is the motion in their Coulomb field. The effective potential is

$$V_{eff}\ (r) = \frac{L^2}{2\mu r^2} + \frac{Z_1 Z_2 e^2}{r} \tag{1.2}$$

where L is the angular momentum, μ the reduced mass and Z_i the proton number of the two nuclei. The total energy is conserved:

$$E = \frac{1}{2}\ \mu\ \dot{r}^2 + V_{eff}\ (r) \tag{1.3}$$

Then the velocity r at each point of the trajectory is

$$\dot{r} = \sqrt{\frac{2}{\mu}\ (\ E - V_{eff}\ (r)\ )} \tag{1.4}$$

with $dt = \frac{dr}{\dot{r}}$ and $L = \mu r^2 \dot{\phi}$ we can calculate the deflection angle ϕ easily (c.f. figure 1.1)

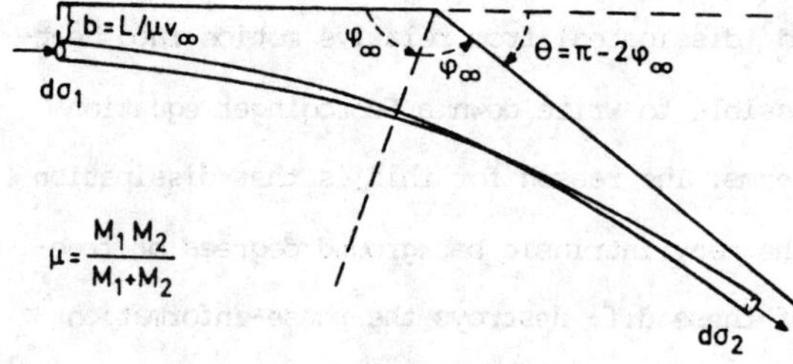

Fig. 1.1. Rutherford scattering

$$\phi(r) = \int_{r_{min}}^{r} \dot{\phi}\, dt = \int_{r_{min}}^{r} \frac{L}{\mu r^2} \frac{dr}{\dot{r}} = \text{arc cos}\left\{ \frac{1 + \frac{1}{r}\frac{L^2}{\mu_1 Z_1 Z_2 e^2}}{\sqrt{\frac{2\, EL^2}{\mu(Z_1 Z_2 e^2)^2} + 1}} \right\} \tag{1.5}$$

$$\text{or} \quad \frac{P}{r} = -1 + \varepsilon \cos \phi \tag{1.6}$$

$$\text{with } P = \frac{L^2}{\mu Z_1 Z_2 e^2} \quad , \qquad \varepsilon = \sqrt{1 + \frac{2EL^2}{\mu(Z_1 Z_2 e^2)^2}} \tag{1.7}$$

The scattering angle θ (figure 1.1) is

$$\theta = \pi - 2\phi(r = \infty) \tag{1.8}$$

The differential cross-section is defined by that area $d\sigma_1$ of initial conditions from which the trajectories lead to the final solid angle $d\Omega$ of observation. Due to flux conservation the measured number dN_2 of particles coming out into the counter window $d\sigma_2 = r^2 d\Omega$ with flux j_2 is the same as the number coming in with flux j_1 at the entrance area $d\sigma_1$ at impact parameter $b(\theta)$

$$dN_2 = j_2\, r^2 d\Omega_2 = j_1\, d\sigma_1 \tag{1.9}$$

i.e. the measured quantity

$$\frac{1}{j_1}\frac{dN_2}{d\Omega_2} = \frac{d\sigma_1}{d\Omega_2} = \frac{b\, db\, d\phi}{\sin\theta\, d\theta\, d\phi} = \frac{L}{P_\infty^2}\ \frac{1}{\sin\theta\ \frac{d\theta}{dL}} \tag{1.10}$$

or for pure Coulomb-scattering from eq. (1.5), we get the Rutherford formula:

$$\frac{d\sigma_1}{d\Omega} = \left(\frac{Z_1 Z_2 e^2}{2\mu v_\infty^2} \right)^2 \frac{1}{\sin^4 \theta/2} \tag{1.11}$$

In reality the nuclei do not move on pure Coulomb trajectories. At small impact parameters the nuclei may touch and start to feel the nuclear attraction which will reduce the scattering angle. In some cases it may even pull the trajectories around to the "other side" i.e. to negative angles.

We expect the scattering anlge θ(L), the deflection function, not to be monotonically rising with decreasing L but to bend down again (c.f. figure 1.2)

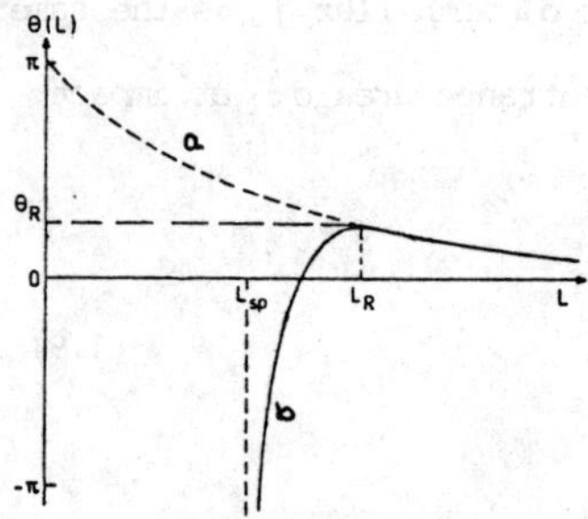

Fig. 1.2. Classical deflection function θ(L), a: pure Rutherford scattering, b: with short-range nuclear attraction. L_R, θ_R = rainbow, $L_{sp.}$ = orbiting angular momentum

Moreover, if the nuclei touch they will excite one another and loose kinetic energy (we call this dissipation of energy or friction). These trajectories will not contribute to elastic scattering they are "absorbed".

If one plots the differential cross-section $\frac{d\sigma}{d\Omega} / \frac{d\sigma\,\mathrm{Ruth}}{d\Omega}$ versus the scattering angle, we expect at small angle (i.e. large impact parameters) the ratio above to be one. At larger scattering angle (smaller distances the trajectories are "absorbed" out of the elastic channel. Due to the finite range of the nuclear force this absorbtion is smooth. The deflection of the trajectories by the nuclear attraction leads to smaller angles and some of the missing flux appears at small angles. Of course some part of this rise is also due to wave mechanical Fresnel diffraction [7].

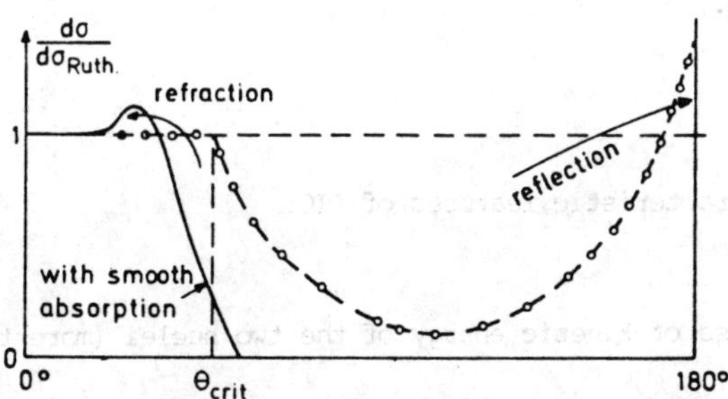

Fig. 1.3. Qualitative features of elastic scattering

II. Deep inelastic collisions (DIC)

Before I go into some details let me just show typical experimental features of DIC for the case of the collision of Xe with Bi at 1422 MeV.

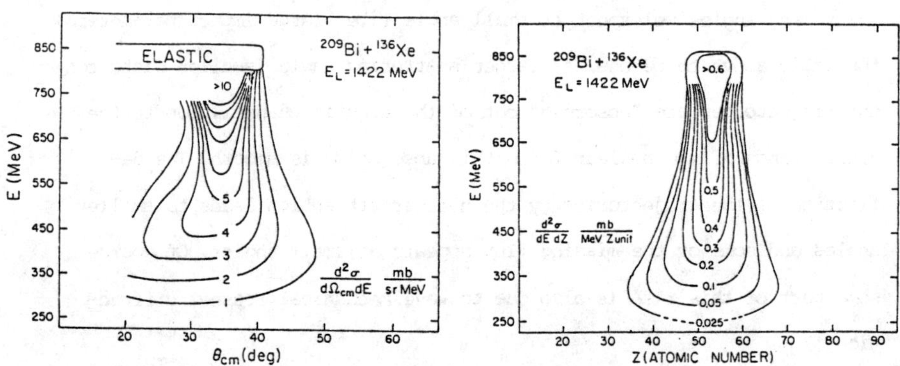

Fig. 2.1. and 2.2. Experimental $\frac{d^2\sigma}{d\Omega dE}$ and $\frac{d^2\sigma}{dEdz}$ for $^{209}Bi+^{136}Xe$ at 1422 MeV [1].

We find the following characteristic features of DIC:

a) There is an extreme loss of kinetic energy of the two nuclei (more than 500 MeV)

b) The deep-inelastic cross-section is about 80% of the total reaction cross-section

c) Nearly for all of these processes the projectile is scattered into an angle of 35° in the CM-system within a narrow angular range of $\Delta\theta=10°$. Of course other systems show different angular distributions.

d) The change of the mass of the collision partners is astonishingly small inspite of the huge excitations. Consequently the two nuclei seem not to overlap strongly. The damping (friction) must be a surface phenomenon (c.f. ch. 2.5.).

Apparantly, many back-ground degrees of freedom participate to produce the huge inelasticity. However, <u>one</u> d.f., the relative motion $\vec{R},\vec{P}$, dominates the process and does not equilibrize. Otherwise we would not see such a focussing of the scattering angle. All others d.f. are highly involved but, for the first glance, no one attains to particular importance (See, however, the discussion about deformations later on). These d.f. should be treated as a background (heat-bath).
We should, therefore, try to find some effective equation of motion for $\vec{R}(t)$, $\vec{P}(t)$ with parameters (forces) depending smoothly on the background. No Schrödinger description exists for dissipative motion, because the motion of the background d.f. destroys the phase-coherences. Therefore, only a classical equation of motion can be found for the $\vec{R},\vec{P}$ d.f.. There are only a few exceptions from this rule: E.g. harmonic motion in $\vec{R},\vec{P}$ coupled to harmonic bath d.f. However, this is no real exception because here quantum and classical dynamics coincide.

2.1 Quantum versus classical dynamics

Let me discuss this case in some more detail because many essential features of dissipative motion can be learned from it. First we consider the quantal motion of a single degree of freedom. The Hamiltonian is

$$\hat{H} = \frac{\hat{p}^2}{2m} + V(\hat{x}) \tag{2.1.1}$$

The density operator $\hat{\rho}(x)$ obeys the von Neumann equation:

$$\dot{\hat{\rho}} = \frac{1}{i\hbar} [\hat{H}, \hat{\rho}] \tag{2.1.2}$$

We take the Wigner-transform of $\hat{\rho}$ and study its equation of motion:

$$\rho (x,p,t) \overset{\text{def}}{=} \int ds \langle x + \frac{s}{2} | \hat{\rho} | x - \frac{s}{2} \rangle e^{- isp/\hbar} \tag{2.1.3}$$

$$= \int \frac{d q}{2\pi\hbar} \langle p + \frac{q}{2} | \hat{\rho} | p - \frac{q}{2} \rangle e^{iqx/\hbar} \tag{2.1.4}$$

which is a scalar function of both x and p. The states $|x\rangle$ or $|p\rangle$ are time independent states. Now we find:

$$\langle p + \frac{q}{2} | [\frac{\hat{p}^2}{2m}, \hat{\rho}] | p - \frac{q}{2} \rangle = \frac{p q}{m} \langle p + \frac{q}{2} | \hat{\rho} | p - \frac{q}{2} \rangle \tag{2.1.5}$$

and

$$\langle x + \frac{s}{2} | [V(\hat{x}), \hat{\rho}] | x - \frac{s}{2} \rangle$$

$$= \left(V(x+\frac{s}{2}) - V(x- \frac{s}{2}) \right) \langle x + \frac{s}{2} | \hat{\rho} | x - \frac{s}{2} \rangle$$

$$= \sum_k \frac{2}{(2k + 1)!} \left(\frac{s}{2} \right)^{2k+1} \left(\frac{d^{2k+1}}{dx^{2k+1}} V (x) \right) \langle x + \frac{s}{2} | \hat{\rho} | x - \frac{s}{2} \rangle \tag{2.1.6}$$

This leads to

$$\dot{\rho}\ (x,p,t) = \sum_k \frac{1}{(2k+1)} \left(\frac{i\hbar}{2} \right)^{2k} \frac{d^{2k+1}}{dx^{2k+1}} V(x) \frac{\partial^{2k+1}}{\partial p^{2k+1}} \rho(x,p,t)$$

$$- \frac{p}{m} \frac{\partial}{\partial x} \rho(x,p,t) \tag{2.1.7}$$

or if V(x) is of second order in x:

$$\dot{\rho}(x,p,t) = \frac{1}{i\hbar} [\hat{H},\hat{\rho}]_{x,p} = \frac{\partial V}{\partial x}\frac{\partial \rho}{\partial p} - \frac{p}{m}\frac{\partial \rho}{\partial x} = \{H,\rho\}_{Poisson} \tag{2.1.8}$$

where $\{\ \ \}_{Poisson}$ is the classical Poisson bracket $\{A,B\}$

$$\{A,B\}_{Poisson} = \frac{\partial A}{\partial x}\frac{\partial B}{\partial p} - \frac{\partial A}{\partial p}\frac{\partial B}{\partial x} \tag{2.1.9}$$

which is shown to coincide exactly with the quantal commutator $\frac{1}{i\hbar}[\hat{H},\hat{\rho}]$ for a harmonic $\hat{H}$. Consequently, the dynamics of the quantal and the classical motion is identical for this case. Quantum systems may differ only in the initial distribtions $\rho(x,p,t{=}o)$ which must fulfill Heisenbergs uncertainty, of course. – The generalization of this result to arbitrary many d.f. is straight forward.

2.2 Harmonic Brownian particle coupled a harmonic bath

This simple case will introduce us in a most transparant way to dissipative dynamics and especially to Brownian motion. A macroscopic particle

(Brownian particle) R,P, and mass M is coupled to very many bath oscillators q_i, p_i, m_i, ω_i:

$$H = \frac{P^2}{2M} + V(R) + \sum_i \left\{ \frac{p_i^2}{2m_i} + \frac{m_i\omega_i^2}{2} [q_i + f_i(R)]^2 \right\} \tag{2.2.1}$$

The equation of motion for the Brownian particle is

$$M\ddot{R} + \frac{d}{dR} V + \sum_i m_i\omega_i^2 [q_i + f_i(R)] \frac{df_i}{dR} = 0 \tag{2.2.2}$$

and for the bath oscillators

$$\ddot{q}_i + \omega_i^2 [q_i + f_i(R)] = 0 \tag{2.2.3}$$

If we would know R(t) we could solve (2.2.3) for $q_i(t)$:

$$q_i(t) = \hat{q}_i(t) - \int_0^t ds\, f_i[R(s)]\, \omega_i \sin[\omega_i(t-s)] \tag{2.2.4}$$

where $\hat{q}_i(t)$ is the motion of the free bath-oscillator.

$$\hat{q}_i(t) = q_i(0) \cos \omega_i t + \frac{p_i(0)}{m_i\omega_i} \sin \omega_i t \tag{2.2.5}$$

This, substituted into eq. (2.2.2), gives an effective equation of motion for the Brownian-particle alone:

$$M\ddot{R} + \frac{dV}{dR} + \int_o^t ds \sum_i m_i \omega_i^2 f_i'(t)\cos\left[\omega_i(t-s)\right] f_i'(s)\dot{R}(s) = F^+(t) \qquad (2.2.6)$$

with

$$f_i'(t) = \frac{d\, f_i\left[R(t)\right]}{d\, R(t)} \quad , \quad f_i\left[R(o)\right] = o \qquad (2.2.7)$$

and the "random" force $F^+(t)$:

$$F^+(t) = -\sum_i m_i \omega_i^2 f_i'(t)\, \hat{q}(t) \qquad (2.2.8)$$

We may interpret eq. (2.2.6) as a non-Markovian Langevin equation for our Brownian particle:

$$M\ddot{R} + \frac{dV}{dR} + \int^t ds\ \gamma(t-s)\dot{R}(s) = F^+(t) \qquad (2.2.9)$$

with the (retarded) friction kernel

$$\gamma(t-s) = \sum_i m_i \omega_i^2 f_i'(t) \cos\left[\omega_i(t-s)\right] f_i'(s) \qquad (2.2.10)$$

and intimately related to it, the "random" force $F^+(t)$, which is the only quantity in eq. (2.2.6) that depends explicitly on the (in general unknown) initial conditions $q_i(o)$, $p_i(o)$ of the bath oscillators.

For the case of linear coupling $f_i(R) = \alpha_i \cdot R$ and a harmonic potential $V(R) = \frac{1}{2} M\Omega^2 R^2$ the Langevin-equation becomes linear in R(t) and $q_i(o)$, $p_i(o)$.

Then, $\langle R(t)\rangle$ averaged over the initial conditions $q_i(o)$, $p_i(o)$ of the bath obeys the equation with friction:

$$\frac{d^2}{dt^2} \langle R(t)\rangle + \langle M\Omega^2 R(t)\rangle + \int_o^t ds\ \gamma(t-s) \frac{d}{ds} \langle R(s)\rangle = o \tag{2.2.11}$$

Moreover, the relation between $\gamma(t-s)$ and $F^+(t)$ can be put into the form:

$$\langle F^+(t)\ F^+(s)\rangle = \langle e(t{=}o)\rangle\ \gamma(t-s), \tag{2.2.12}$$

where we assumed that the initial conditions are uncorrelated and all bath-oscillators have initially the same average energy $\langle e(t{=}o)\rangle$ (equipartition law):

$$\langle q_i(o)\ p_j(o)\rangle = o \tag{2.2.13}$$

$$\langle q_i(o)\ q_j(o)\rangle = \delta_{ij} \frac{\langle e(o)\rangle}{m_i \omega_i^2} \tag{2.2.14}$$

$$\langle p_i(o)\ p_j(o)\rangle = \delta_{ij}\ m_i\ \langle e(o)\rangle \tag{2.2.15}$$

Equation (2.2.12) between the autocorrelation function of $F^+(t)$ and the friction kernel is the famous fluctuation-dissipation therorem. The memory-time, which is the width in t-s, is intimately related to the freqency sprectrum of the bath-oscillators. If they have a dense ω_i-distribution

$$\rho(\omega_i) = \alpha_i^2/(2m_i\omega_i^2\pi) \tag{2.2.16}$$

we find

$$\langle F^+(t)F^+(s)\rangle = \langle e(o)\rangle\ \delta(t-s) \tag{2.2.17}$$

and call F^+ a proper stochastic force with a white-noise spectrum. The Langevin equation (2.2.6) is a proper Markovian Langevin equation then.

$$M\ddot{R}(t) + M\Omega^2R(t) + \beta\dot{R}(t) = F^+(t) \tag{2.2.18}$$

with the friction coefficient (tensor) β given by:

$$\beta_{\alpha\beta} = \frac{1}{\langle e(o)\rangle} \int ds \langle F_\alpha(t)F_\beta(s)\rangle \tag{2.2.19}$$

The initial energy $\langle e(o)\rangle$ of each bath oscillator is, of course, the initial temperature T of the bath.

The main lesson to be learned from this simple example is:

a) A macroscopic degree of freedom R,P coupled to a bath of many d.f. whose initial conditions are known with some probability only, performs a complicated motion according to a non-Markovian Langevin equation like eq. (2.2.6). If the bath d.f. have a dense spectrum this motion may become random and damped by a friction force $\beta\dot{R}$. Then the random force $F^+(t)$ is a "white noise" i.e. its autocorrelation function is proportional to $\delta(t-s)$ and closely related to the friction tensor β by the fluctuation-dissipation theorem (eq. (2.2.19).

b) For the harmonic case no perturbation expansion (in powers of the coupling α_i) is needed to obtain this result. I.e. it does not matter whether we have "weak" or "strong" coupling. The main difficulties come from anharmonicities which are discussed in detail in ref. [2]. The classical evolution coincides with the quantal one (c.f.ch. 2.1).

2.3 Piston model and window formula for friction

a) Piston Model

What is the physical origin of friction? Let us consider a large container filled with an ideal gas of density ρ. Inside the container a small piston with the area A moves with velocity $\dot{R}$. A particle being reflected on the left side transfers a momentum to the piston of

$$\Delta P = 2m(\dot{r}-\dot{R}) \tag{2.3.1}$$

$A\rho\,|\,\dot{r}-\dot{R}\,|$ particles hit the piston per unit time. The force acting on the piston from the left is

$$F_{\ell} = A\rho 2m(\dot{r}-\dot{R})^{2} \qquad (2.3.2)$$

Similarly the collision from the right side leads to a force acting from the right onto the piston of

$$F_{r} = A\rho 2m(\dot{r}+\dot{R})^{2} \qquad (2.3.3)$$

The total force acting on the piston from the right amounts to

$$F = F_{r} - F_{\ell} = 8A\rho m|\dot{r}||\dot{R}| \qquad (2.3.4)$$

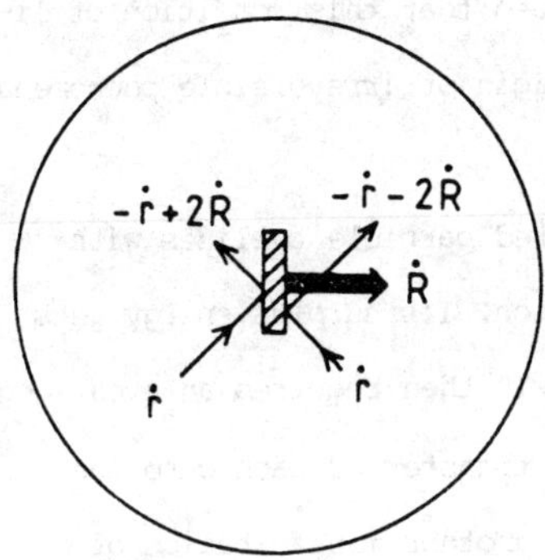

Fig. 2.3. Piston having a velocity $\dot{R}$ moving in an ideal gas

It is a proper frictional force proportional to the velocity of the piston. Two conditions must be satisfied for this conclusion to be correct:

a) The mean free path of the gas particles must be larger than the geometrical dimensions of the piston.

b) No particle should hit the piston a second time.
   Otherwise some of the energy transfered to the particles in a first collision may be given back to the piston later on.

The second condition of "never come back" can be obtained in various ways:

a) Either the container is sufficiently large or very irregularly shaped that during the time t of observation the particles do not return. I.e. in order to obtain dissipation a <u>separation of time scales</u> is required. The recurrence time T must be large compared to the observation time t.

This is a very fundamental condition for irreversible processes, which we shall frequently meet later on. It is important to note clearly that <u>no assumption of a stochastic process</u> was necessary. In many text books of statistical mechanics [3] it is argued that this condition of limited observation-time is the basic origin of irreversible phenomena.

b) Another possibility is that the scattered particle collides with another particle far away from the piston. Its surplus energy gets distributed among all other particles. It then requires an even longer time T for the dissipated energy to be transferred back onto the piston. It is evident that (b) is just another manifestation of a separation of time scales, $T \gg t$.

In heavy-ion collisions the time of contact of the two ions is quite short $\sim 10^{-21}$ sec so that we can expect that many processes, which are strictly speaking reversible, act in an irreversible manner.

Later the piston model was used by Swiatecki [4] for the damping of surface vibrations in nuclei (wall formula). Here, of course, it is the self-consistent nuclear potential-surface that acts as the piston. It is not clear, whether one can treat it as moving with its own dynamics and whether the "never come back" condition holds in that case. We will see in the next chapter another more clear application of the piston model, which does not suffer from this criticism.

b) Window friction

Swiatecki [5] argues that due to the exchange of nucleons between the two nuclei relative momentum is dissipated into intrinsic excitation. The main idea is as follows

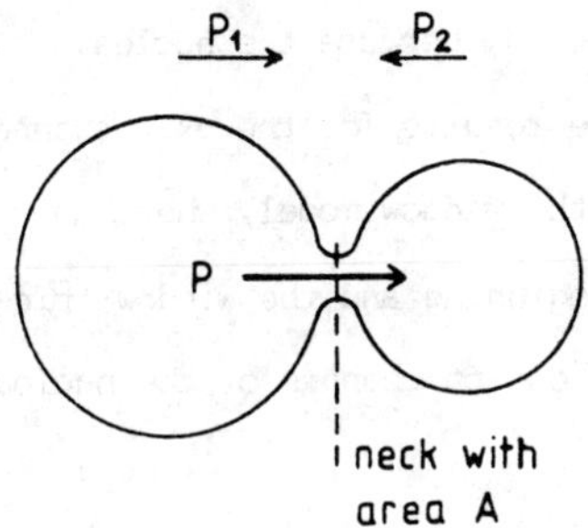

Fig. 2.4. Exchange of nucleons through a window

If in the approach phase a nucleon jumps from nucleus 1 to nucleus 2 with momentum $\vec{p}$ the relative momentum $\vec{P}_{rel}$ of the two nuclei is slowed down:

$$\vec{p}_1+\vec{p}_2 = o \qquad (2.3.6)$$

$$\vec{P}_{rel} = \frac{m_2}{M}\vec{p}_1 - \frac{m_1}{M}\vec{p}_2 \qquad (2.3.7)$$

$$\vec{P}'_{rel} = \frac{m_2+m}{M}(\vec{p}_1-\vec{p}) - \frac{m_1-m}{M}(\vec{p}_2+\vec{p}) = \vec{P}_{rel} - \vec{p} \qquad (2.3.8)$$

A jump from $2 \rightarrow 1$ induces a change

$$\vec{P}'_{rel} = \frac{m_2 - m}{M}(\vec{p}_1 + \vec{p}) - \frac{m_1 + m}{M}(\vec{p}_2 - \vec{p}) = \vec{P}_{rel} + \vec{p} \qquad (2.3.9)$$

however, here $\vec{p}$ points from $2 \rightarrow 1$ and is opposite to $\vec{P}_{rel}$.

Therefore in both cases the relative motion is slowed down in the entrance channel. In the exit channel the situation is more complicated and our simple model gives an accelleration. Some authors [6] obtain an accelleration similarly, whereas Swiatecki [5] argues that the static force due to the expansion of the system should be subtracted, because the nuclear density does not change. Then a damping force results for the exit channel also. However, an unambiguous derivation of the window model, where nuclear saturation is taken care of from the beginning and the window-friction-force is not mixed with spurious forces due to changes of the nuclear volume has not yet been given.

I, personally, have difficulties in understanding the "wall" as well as the "window" model because of the self consistency problem.

In the following we want to consider another effective damping mechanism which may even be more important for the heavy-ion damping.

2.4 Perturbation-theoretical approach to nuclear friction

If the two colliding nuclei do not overlap strongly we have a situation which is very similar to the piston model. The tail of the single particle potential of one nucleus moves across the tail of the density distribution of the other nucleus thereby exciting it. For weakly overlapping nuclei the coupling V can be treated in the lowest order. If we work in the two-center potential basis this is also possible in the limit of slow motion, i.e. if the adiabatic coupling $\dot{V}$ is small. It is evident that in this limit we have just the "never comeback" situation. Recurrences, i.e. effects of higher order are neglegible. The separation of time scales is assumed explicitly.

As was shown on several occasions [7,8,9,10] one obtains for the dissipative rate of energy loss in a straight forward manner the expression:

$$\dot{E}(t) = - \int_{-\infty}^{0} d\tau \sum_{\substack{n>\varepsilon_F \\ m<\varepsilon_F}} \{ \langle m|\dot{R}(t+\tau)\nabla V(t+\tau)|n\rangle$$

$$* \frac{e^{-i(\varepsilon_n-\varepsilon_m)\tau/\hbar}}{\varepsilon_n-\varepsilon_m} \langle n|\dot{R}(t)\nabla V(t)|m\rangle + c.c.\} \qquad (2.4.1)$$

where $|n\rangle$, ε_n are the single particle states and energies in one of the nuclei and V(t) is the s.p. potential of the other nucleus at the distance R(t). Here, there are <u>no problems with self consistency</u>, because V does not refer to the same nucleus as the states $|n\rangle$ and $|m\rangle$.

In ref. [8] I showed that the same result follows for a weakly interacting piston inside a Fermi gas. There it gives a proper dissipative rate of energy-loss $\propto \dot{R}^2$.

Exactly the same result is obtained in the linear response theory of Siemens and Hofmann [11]. However, the time dependence of the first matrix element $V(t+\tau)$ in (2.4.1) does not appear in ref. [11] because V was assumed to separate in the relative and internal coordinates. This is very unrealistic and leads to a vanishing friction force whenever the spectrum ϵ_n is discrete. I discussed that problem in ref. [12] and showed how the τ-dependence of the first matrix element (i.e. the "memory") is essential for obtaining a finite friction. This idea was investigated in detail in a recent numerical calculation by Pal and Ganguli [13,14]. Their result confirms our previous estimates and phenomenological findings. It turned out that the long memory T in the one-body friction is the essential reason for its great strength [14]. In that respect the one-body friction is qualitatively different from the two-body friction which being more Markovian, is also considerably weaker at low energies.

In a recent paper Nemes and Weidenmüller [15] studied the damping of quadrupole vibrations α in nuclei at somewhat higher temperatures. They reached at a similar conclusion about the importance of the memory time T. In their case the friction changed from an $\dot{\alpha}$ to an $\dot{\alpha}^3$ dependence. In our case of the friction in the relative motion at the beginning of contact of the two nuclei care must be taken when one wants to expand the friction in powers of $\dot{R}$. This problem may be illuminated by the following transparent example:

If our piston moves in a Fermi gas having a gap Δ in it's spectrum, the energy rate turns out to be proportional to $\dot{R}\ \theta(|\dot{R}|-\Delta/2P_F)$, where P_F is the Fermi-momentum and $\theta(x)$ the unit step function. The latter comes from the fact that the recoil energy transferred to a particle in a reflection by the piston is $2\dot{R}P_F$ and must be larger than the gap Δ. This result for the energy-rate is <u>not analytic at $\dot{R}$=0</u>. It may be meaningless to look for a mathematical expansion in powers of $\dot{R}$ at $\dot{R}$=0, when the spectrum has a gap. However, this condition is fulfilled already at energies of $\frac{m}{2}\dot{R}^2 > \Delta^2/16\,\epsilon_F \sim 2\cdot 10^{-3}\Delta^2$[MeV] per nucleon. Thus, one may perfectly neglect this condition in any estimate which is relevant for heavy-ion collisions [c.f. a similar discussion in ref. [14]].

2.5 Surface-friction model for deep inelastic collisions and capture [16-22]

After having obtained a qualitative insight into the nature of nuclear friction we try to put our ideas into a simple model and compare it with the data. We neglect all quantal effects and assume the two nuclei to move along classical trajectories.

We also allow the nuclei to deform dynamically (this was first discussed in [18]). Figure (2.5) shows the degrees of freedom we want to consider explicitely:

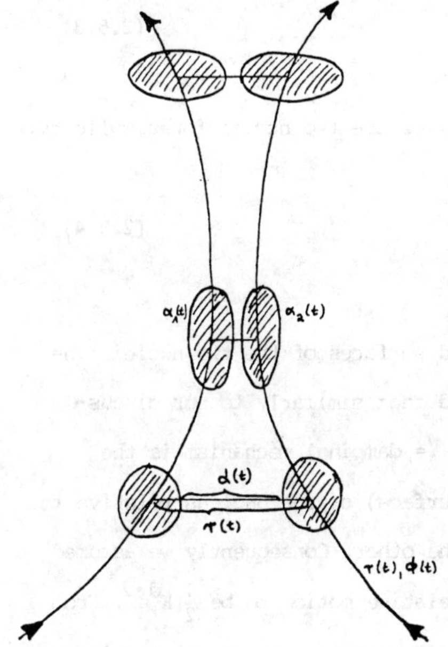

Fig. 2.5 Degrees of Freedom considered in the surface-friction model

The surface-friction model has been studied recently in great detail by P. Fröbrich [19-22]. We will follow his parametrization and discuss his results.

The dynamics is controlled by the Lagrangian

$$\mathcal{L} = \frac{P_r^{\,2}}{2\mu} + \frac{\ell^2}{2\mu r^2} + \sum_{i=1}^{2} \left\{ \frac{\pi_i^{\,2}}{2D_i} - V(r,\alpha_i) - \frac{1}{2} C_i \alpha_i^{\,2} \right\} \tag{2.5.1}$$

where $\alpha_1 . \alpha_2, \pi_1, \pi_2$ are the shape vibrations (quadrupole deformations α_i and conjugate momenta π_i and the D_i, C_i are from liquid-drop model) of the two nuclei, and the Rayleigh dissipation function:

$$\mathcal{R} = \frac{1}{2} K^d \dot{d}^2 + \frac{1}{2} \sum_i K^i \dot{\alpha}_i^{\,2} + K^\phi r^2 (\dot{\phi} - \dot{\phi}_s)^2 \tag{2.5.2}$$

$$d = r - \sum_i R_i \{ 1 + \alpha_i Y_{20}(\theta = o, \phi = o) \} \tag{2.5.3}$$

i.e. we assume a "nose to nose" symmetry of the two nuclei for simplicity.

$$\dot{d} = \dot{r} - \sum_i R_i \dot{\alpha}_i Y_{20} \tag{2.5.4}$$

Here d is the distance of the juxtaposed surfaces of the two nuclei. The distance of the centres is r. We assumed that similarly to our discussion in chapter 2.4 the main excitation (= damping) mechanism is the motion of the s.p. potential profile (surface) of one nucleus relative to the density distribution (surface) of the other. Consequently we assumed the rate of energy dissipation of the relative motion to be $\frac{1}{2} K^d \dot{d}^2$. From $\mathcal{L}$ and $\mathcal{R}$ we obtain the coupled equations of motion in the usual fashion:

$$\frac{d}{dt}\frac{\partial \mathcal{L}}{\partial \dot{q}} - \frac{\partial \mathcal{L}}{\partial q} + \frac{\partial \mathcal{R}}{\partial \dot{q}} = o \tag{2.5.5}$$

i.e.

$$\dot{r} = P_r/\mu \tag{2.5.6}$$

$$\dot{P}_r = -\frac{\partial V(r,\alpha_i)}{\partial r} - \frac{\ell^2}{\mu r^3} - \frac{\partial \mathcal{R}}{\partial \dot{d}}\frac{\partial \dot{d}}{\partial \dot{r}} \tag{2.5.7}$$

$$\dot{\phi} = \ell/(\mu \dot{r})^2 \tag{2.5.8}$$

$$\dot{\ell} = -K_\phi(\ell-\ell_s)/\mu \quad , \; \ell_s = \ell_{sticking} \tag{2.5.9}$$

$$\dot{\alpha}_i = \pi_i/D_i$$

$$\dot{\pi}_i = -\frac{\partial V(r,\alpha_i)}{\partial \alpha_i} - \frac{\partial \mathcal{R}}{\partial \dot{d}}\frac{\partial \dot{d}}{\partial \dot{\alpha}_i} - K^i\dot{\alpha}_i \tag{2.5.10}$$

Explicitely, the surface-friction terms are:

$$\frac{\partial \mathcal{R}}{\partial \dot{d}}\frac{\partial \dot{d}}{\partial \dot{r}} = K^d\left(\dot{r} - \sum_j R_j \alpha_j Y_{20}\right) = K_r^r\dot{r} + K_r^j\dot{\alpha}_i \tag{2.5.11}$$

$$\frac{\partial \mathcal{R}}{\partial \dot{d}}\frac{\partial \dot{d}}{\partial \dot{\alpha}_i} = K^d\left(\dot{r} - \sum_j R_j \dot{\alpha}_j Y_{20}\right)R_i Y_{20} = K_i^r\dot{r} + K_i^j\dot{\alpha}_j \tag{2.5.12}$$

Here the nondiagonal term K^i_r, K^r_i are dissipate couplings between the relative motion and the shape degrees of freedom. They play a decisive role in the dynamics as will be discussed later.

The dynamics can be illuminated by following the fate of trajectories starting with different initial angular-momenta ℓ. Fist we neglect the deformations for simplicity [23].

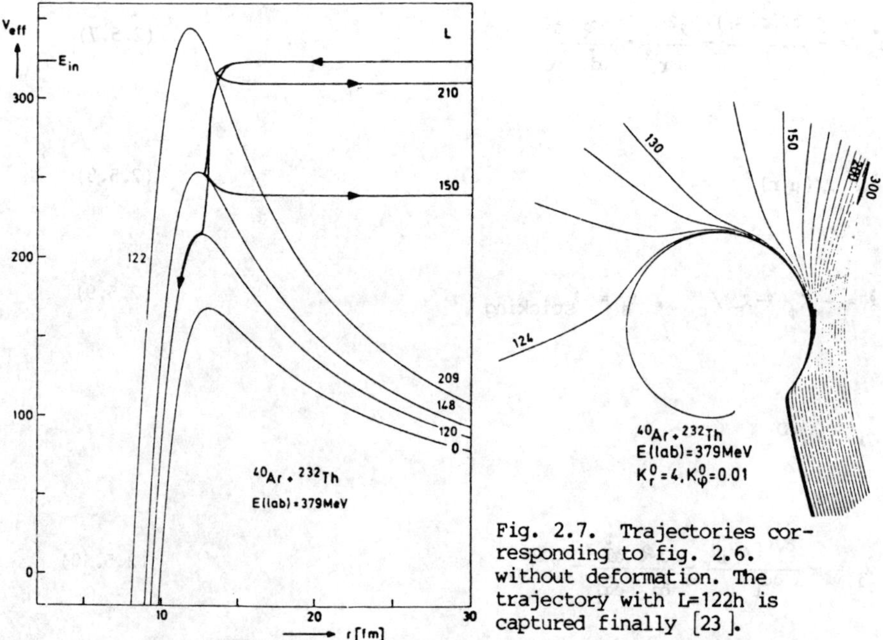

Fig. 2.7. Trajectories corresponding to fig. 2.6. without deformation. The trajectory with L=122h is captured finally [23].

Fig. 2.6. Effective potentials (including the centrifugal potential) and trajectories starting with various angular momenta L. Without deformation [23].

Due to the short range of the friction eq. (2.5.17) the energy dissipation is limited to distances below ~14 fm in the present example. Trajectories starting with high ℓ=210 h do not approach very closely. Thus, they suffer only a small loss of energy. A trajectory starting with ℓ=150 h penetrates significantly into the dissipative region, looses a substantial amount of energy but bounces off the effective potential barrier and comes out finally as a deep-inelastic event. Any trajectory with ℓ less than or equal to 122 h reaches inside the effective barrier and is finally captured. Recently, the energy-loss before capture was discussed by Swiatecki in the frame of his extra push model [24]. An additional energy shift (here due to friction) is a natural ingredient of our model also (table 3.1 ref.[25]).

However, the energy-loss obtained in these spherical model calculations is still too small. In the exit channel the Coulomb-repulsion is too large and the nuclei separate with more kinetic energy than observed experimentally. Here the deformations and especially the nondiagonal dissipative coupling between relative motion and deformations mentioned above are of crucial importance. During the approach of the two nuclei the relative motion d(t) of the two juxtaposed surfaces is slowed down by the friction. I.e., when the centres of the two nuclei approach with relative volocity r, the nuclei become oblate deformed due to the damping of d. Correspondingly in the exit channel the nuclei will be pulled to long prolate shapes because r increases faster than d. This mechanism of dissipative coupling is the clue for allowing a long contact of the two nuclei in the exit channel and for much more dissipation of energy below the spherical Coulomb-barrier.

Very similar equations of motion are proposed by the Copenhagen group [26]. Of course they include more deformation degrees of freedom than the low lying quadrupole mode as we do in the surface-friction model. The latter ones are the most important ones, however. There is one important difference between the Copenhagen model and the surface friction model. The dissipative coupling term $K_i^r r$ is absent in their equation (4.8) which corresponds to our eqs. (2.5.10./12) for the intrinsic deformation (The corresponding term $K_r^i\dot{\alpha}_i$ of (2.5.7./11) is included in their eq. 4.2, however). $K_i^r\dot{r}$ comes from the symmetry of the Rayleigh-dissipation function and has severe consequences for the deformations. As discussed above this coupling term drives the deformations and is mainly responsible for the long prolate deformations in the exit channel.

The nuclear potential $V_n(r,\alpha_i)$ is taken as

$$V_n(\vec{r},\alpha_1,\alpha_2) = \frac{1}{2}\left\{ \int U_1(\vec{r}-\vec{s},\alpha_1)\ \rho_2(\vec{s},\alpha_2)\ d^3s \right.$$

$$\left. + \int U_2(\vec{r}-\vec{s},\alpha_2)\ \rho_1(\vec{s},\alpha_1)\ d^3s \right\} \qquad (2.5.13)$$

where $U_i(r,\vec{\alpha}_i)$ is the s.p. potential of nucleus i:

$$U_i(\vec{r},\alpha_i) = \frac{Up}{1 + \exp\left[\dfrac{r - R_p(\alpha_i,\hat{r})}{a_p}\right]} \qquad (2.5.14)$$

$$R_p(\alpha_i,\hat{r}) = \{ 1 + \alpha_i Y_{20}(\hat{r}) \}\ 1.25\ A_i^{1/3}[\text{fm}] \qquad (2.5.15)$$

$$\rho_i(\vec{r},\alpha_i) = \frac{\rho_o}{1 + \exp\left[\dfrac{r-R_d(\alpha_i,\hat{r})}{a_d}\right]}$$

$$\rho_o = 0.17\ \text{fm}^{-3} \qquad (2.5.16)$$

$$R_d(\alpha_i,\hat{r}) = \{ 1 + \alpha_i Y_{20}(\hat{r}) \} [1.12A_i^{1/3} - 0.86A_i^{-1/3}] \text{ [fm]}$$

$$a_d = 0.54 \text{ fm}$$

and the remaining parameters of the s.p. potential U are taken corresponding to the experimental trends [17,21]:

	p shell nuclei	s-d shell nuclei	others
U_p	-30 MeV	-40 MeV	-50 MeV
a_p	0.35 fm	0.45 fm	0.65 fm

The friction parameters are chosen as [21]:

$$K^d = [\frac{\partial V_n(r,\alpha_i)}{\partial r}]^2 * 3.5 \cdot 10^{-23} \text{ [sec/MeV]} \quad (2.5.17)$$

$$K^i = (C_i D_i)^{1/2} * 20 \quad (2.5.18)$$

$$K^\phi = [\frac{\partial V_n(r,\alpha_i)}{\partial r}]^2 * 0.01 \cdot 10^{-23} \text{ [sec/MeV]} \quad (2.5.19)$$

This form of the friction coefficient K_d is taken according to our previous experience [23] and to rough similarity to the force-force correlation function (eq. 2.4.1).

Let us discuss the results for some examples. Fig. 2.8 shows the close similarity of the potential V(r) for $^{16}O+^{16}O$ with potentials obtained from microscopic calculations.

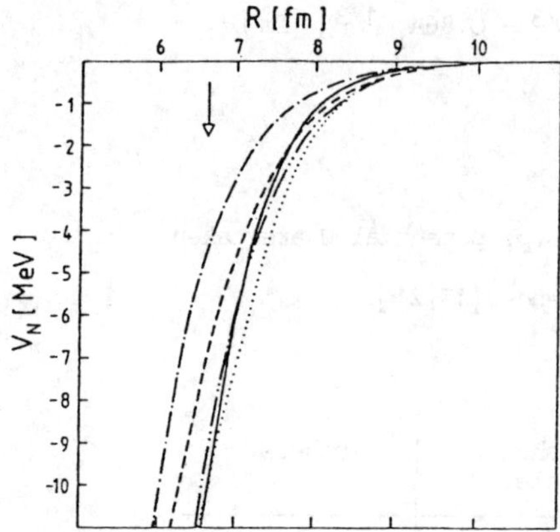

Fig. 2.8. Comparison of different nuclear potentials for the $^{16}O+^{16}O$ reaction [21]:
dashed-dotted line: proximity model [H. Krappe]
solid line: surface friction single folding potential (eq. 2.5.13)
dashed: Krappe-Nix-Sierk-potential [H. Krappe]
dashed-double-dots: double-folding potential
(Satchler) [H. Krappe]
dotted: ATDHF-potential [K. Goeke]
The arrow indicates the distance of closest approach for a trajectory at the energy (E_{cm} = 50 MeV) close to fusion (ℓ = 23).

The potential calculated according to eq. (2.5.13) agrees quite well with the microscopic potential from ATDHF and the double-folding potential whereas the Krappe-Nix-Sierk potential and much more the proximity-potential are considerably weaker at the point of closest approach.

In Fig. 2.9 we display the deformations of the nuclei as a function of the distance along a trajectory with ℓ=180 h [17]. Here, the deformations α_1 and α_2 have been forced to be the same. Here α is the ratio of the two principal axes. One sees clearly the oblate deformation α >1 (upper curve) on the way in. At the point of closest approach the oblate deformation is maximal. On the way out (lower curve) the nuclei are pulled to long prolate (α<1) deformations until their nuclear contact ceases at about r=22 fm.

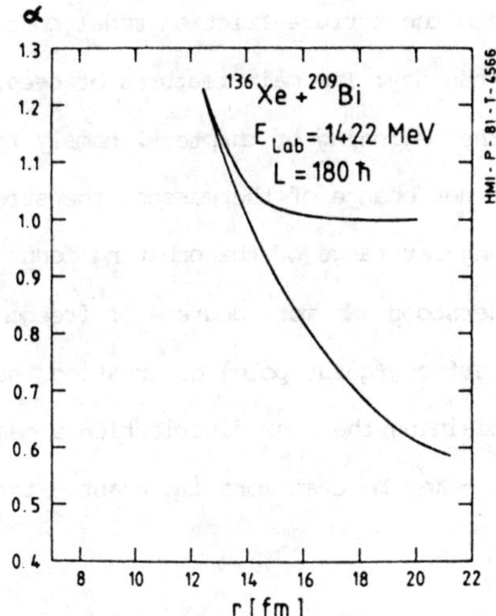

Fig. 2.9 Deformation α = the ratio of principal axes along a trajectory with ℓ=180 h from [17].

The result of the calculation of the Wilczynski pattern when also fluctuations of relative motion and deformation are included is shown in figs. (2.10) and (2.11) (from ref. [22]). In fig. (2.10) we can see by how much the energy-loss is increased when deformations are allowed.

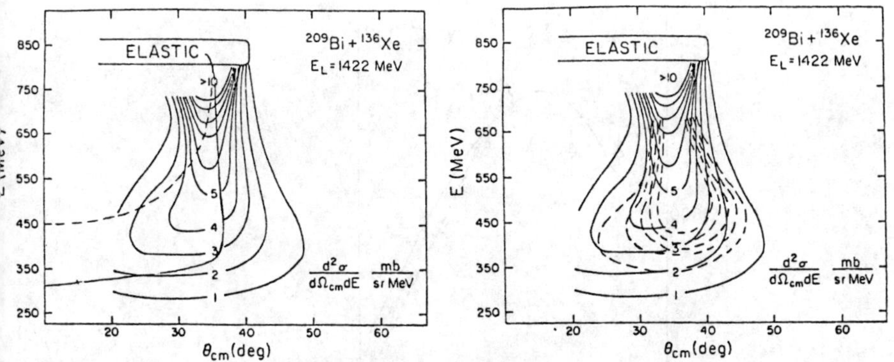

Fig. 2.10 Ridge-line of the energy-angle correlation calculated [22] from the surface-friction model and compared with the data [1] (dashed line: without deformation; solid line: with deformation)

Fig. 2.11 Comparison of the theoretical [22] and experimental [1] Wilczynski-pattern when fluctuations in relative motion are allowed in the theory.

Summarizing our experience with the surface-friction model concerning deep-inelastic collisions we can say: The main features of deep inelastic collision data mentioned at the beginning of chapter 2 namely the large energy loss at relative small net change of the masses, the strong focussing of $d^2\sigma/dEd\Omega$ to a narrow angular range or the orbiting found for other systems can be reasonably understood. No more degrees of freedom than relative motion and low-lying surface-(quadrupole) deformations have to be followed explicitely. For explaining the mass distribution a mass transport has to be introduced and - may be even more important - the necking in the exit chananel.

The most beautiful success of the surface friction model is the consistent discription of the excitation functions for capture above the barrier. We show some representative cases in figures (2.12) to (2.16). A systematic survey over more than 100 different systems can be found in [21].

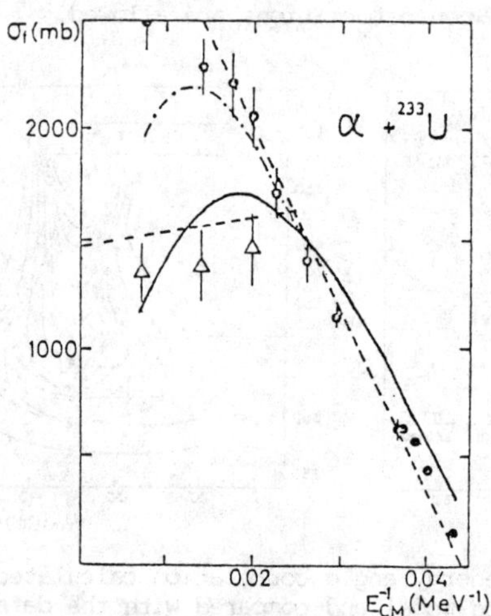

Fig. 2.12. The experimental complete (open triangles) and incomplete (solid circles) fusion cross sections for the $\alpha + {}^{233}U$ reaction [33] are compared with the results of the surface friction model (solid line), the critical distance model (straight lines) of ref. [34] and the proximity model [33] (dashed-dotted lines).

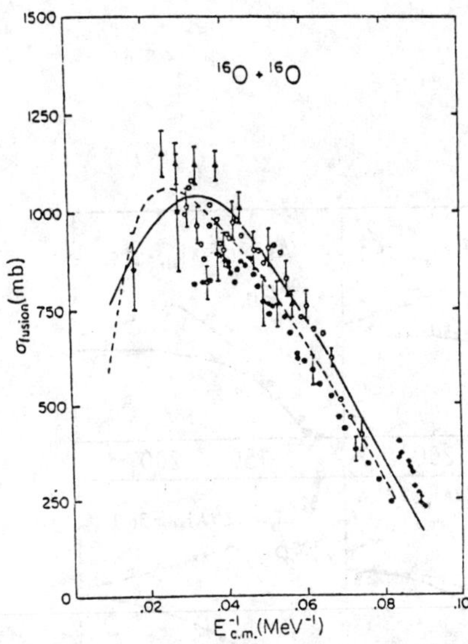

Fig. 2.13. The measured fusion cross section for the $^{16}O + ^{16}O$ collision (diamonds: ref. [35], dots: ref. [36], triangles: ref. [37], other data: ref. [38-40]) is compared with the surface-friction (solid line) and the spherical proximity (dashed line) calculations.

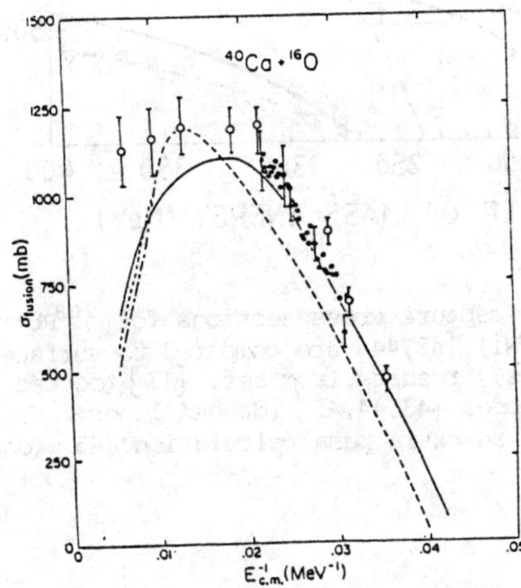

Fig. 2.14. Same as fig. 2.13. except for the $^{16}O + ^{40}Ca$ collision. Data are from refs. [41]: dots and [42]: open circles.

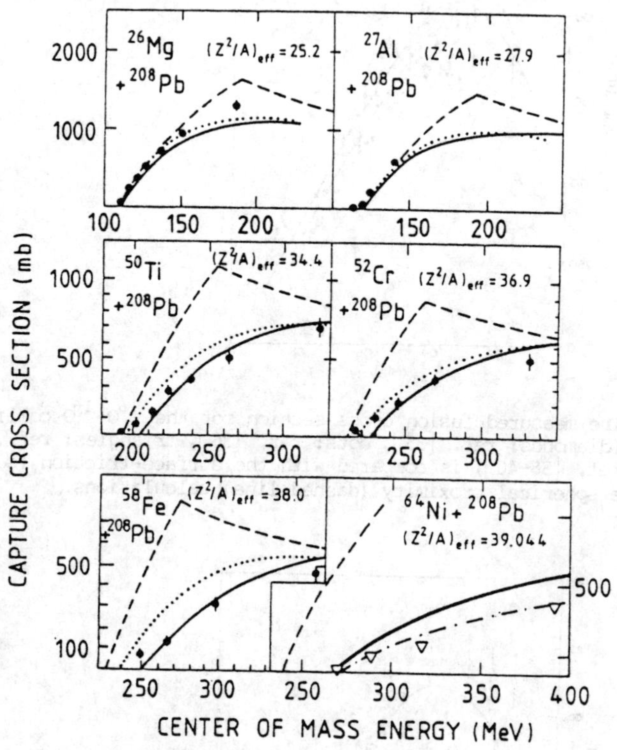

Fig. 2.15. Experimental capture cross sections for ^{208}Pb+(^{26}Mg, ^{27}Al, ^{50}Ti, ^{52}Cr, ^{58}Fe, ^{64}Ni) [43,44] are compared to surface friction calculations (solid lines), results from ref. [17] (dotted lines) and the conventional proximity model [43,44,45] (dashed lines). For the ^{208}Pb+ ^{64}Ni system a comparison with an extra push calculation [45] (dashed-dotted line) is also shown.

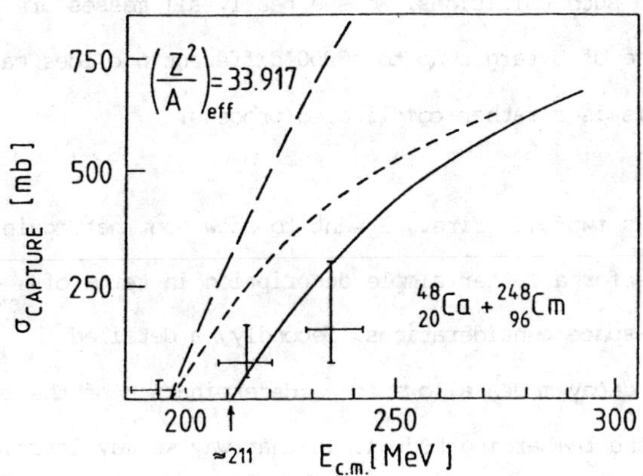

Fig. 2.16. The experimental symmetric fragmentation cross section for $^{48}Ca + ^{248}Cm$ [46] is compared with a surface friction model calculation (solid line), the proximity model (long-dashed line) and the extra push model (short-dashed line).

III. Classical statistical theory for the fragmentation of heavy nuclei after ultra-relativistic proton bombardment

In very high energetic (E>10 GeV) proton-nucleus collisions most of the incoming protons just "go through" the target nucleus. Some nucleons may be shot out of the target but most of the rest, the "rest target", remains only moderately excited. It may expand somewhat and finally break into several pieces of various size. Figure 3.3 shows a typical mass-yield of fragments produced in such collisions. We see nearly all masses are produced. For the case of U-target up to ~3000 different nuclides can be observed. Clearly this is a rather complicated process.

My interest in this is twofold. First, I want to show you that again just the complexity allows for a rather simple description in terms of statistical or phase-space considerations. Secondly, a detailed understanding of the decay modes allows for a determination of the total energy deposited by the bombarding proton. In that way we may learn about the energy dissipation, - friction? -, of ultra relativistic nucleons in nuclear matter. This is a natural extension of our previous studies at low energies.

A huge phase-space is available for the various decay-channels of the RT. It is limited only by the constraints due to the four global conservation laws of baryon number, charge, energy and momentum. The most unbiased assumption is that all decay channels which are allowed by these are equally probable. I.e. the decay of the RT is governed by the volume of the open phase space, the partition sum:

$$\zeta(A,Z,E,\vec{P}) = \sum_{\{n_i\}} \delta\big(E-\sum_i n_iE_i-C\{n_i\}\big)\, \delta(A-\sum_i n_iA_i)$$

$$* \delta(Z-\sum_i n_iZ_i)\, \delta^3(\vec{P}-\sum_i n_i\vec{P}_i) \tag{3.1}$$

where we labelled any decay channel α by the set of numbers $n^{\alpha}_i = 0,1,2,..$ counting the numbers of fragments occuring in the channel α. Here each fragment is distinguished by its baryon number A_i, its charge Z_i, its intrinsic state of excitation $\varepsilon_i{}^*$ and its momentum $\vec{P}_i$ and position $\vec{r}_i$. The energy of each decay channel α is

$$E_\alpha = \sum_i n_i{}^\alpha E_i - C\,\{n_i^\alpha\}$$

where $E_i=-B_i+\varepsilon_i{}^* + P_i{}^2/2m_i$ is the energy of the fragment i with binding energy B_i and

$$C\,\{n_i{}^\alpha\} = \frac{1}{2}\sum_{ik} \frac{Z_iZ_k}{|\vec{r}_i-\vec{r}_k|}\, n_i{}^\alpha n_k{}^\alpha \tag{3.3}$$

is the Coulomb interaction of the fragments in channel α. Here we took care of the fact that at the moment where the fragments become decoupled the nuclear interaction is zero and only the Coulomb repulsion remains. The sum in eq. (3.1) runs over all different sets $\{n_i\}$. For the excited states ε_i^* we allow only the stable ones. States above the lowest threshold for particle decay are counted under the label of the daughter nuclei.

The probability of occurence of a single decay channel α is simply

$$P_\alpha = 1/\zeta \tag{3.4}$$

I.e. knowing the partition sum ζ the expectation value of any observable is easily calculated. For instance, the mean number of fragments of a specific kind i is given by

$$\langle n_i \rangle = \frac{1}{\zeta} \sum_{\{n_i\}} n_i^\alpha \delta(E-\ldots)\,\delta(A-\ldots)\,\delta(Z-\ldots)\,\delta^3(\vec{P}-\ldots) \tag{3.5}$$

It is very convenient to consider the Laplace transform of ζ :

$$\zeta(A,Z,T,\vec{P}) = \int_0^\infty dE e^{-E/T}\ \zeta(A,Z,E,\vec{P})$$

which is called the canonical distribution and represents the available phase space if not the energy but the temperature T is given. In the case of macroscopic systems and also for the decay of heavy target nuclei the density of states at a given energy (we consider excitation energies of several 100 MeV) is huge and is fast rising with energy and there is no essential difference between $\zeta(T)$ and $\zeta(E)$.

Performing the Laplace transform on eq. (3.1) we find the much simpler expression

$$\zeta(A,Z,T,\vec{P}) = \sum_{\{n_i\}} e^{-(\sum n_i E_i - C\{n_i\})/T} \delta(A..)\,\delta(Z..)\,\delta^3(\vec{P}..) \tag{3.6}$$

and for $\langle n_i \rangle$:

$$\langle n_i \rangle = - T \frac{\partial}{\partial E_i} \ln \zeta(A,Z,T,\vec{P}) \tag{3.7}$$

The partition sum $\zeta(A,Z,T,\vec{P})$ is now to be calculated. The most severe problem is imposed by the Coulomb-interaction. At the moment of disassembly our system corresponds to a real gas with short-range nuclear interaction, which are responsible for the condensation into stable fragments and a long-range Coulomb interaction. Various methods are developed for solving this problem [27]. Here we cannot go into the details. Let me only discuss some of the results in order to demonstrate how powerful such a simple classical theory can be. I should mention that this theory is a natural extension of the conventional theory for the decay of a low excited compound nucleus. Here the temperatures are significantly higher such that the RT-nucleus does not decay sequentially into binary decay modes but decays simultanously into many (also heavy) fragments.

Let me first show what happens to the mass yield when we rise the temperature of the system:

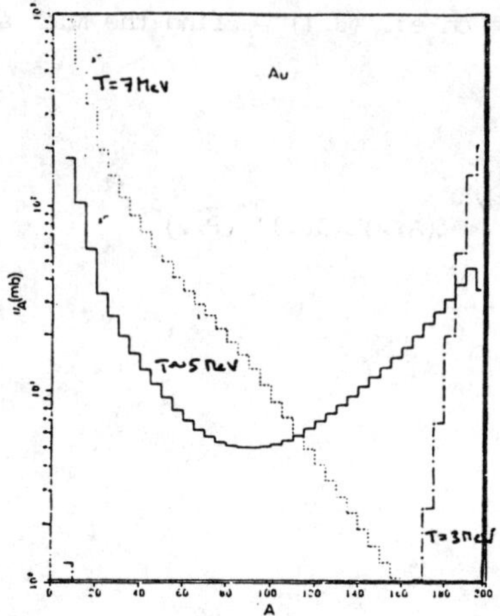

Fig. 3.1 Mass yield of the fragmentation of Au at three different temperatures

At low temperatures, T<3 MeV, the nucleus evaporates a few nucleons and possible α-particles but survives as a heavy residue. At "high" temperatures, T~7 MeV the nucleus has no chance to survive but decays into many small fragments. This must be so, as the nucleons are bound by about 7 MeV only (binding fraction).

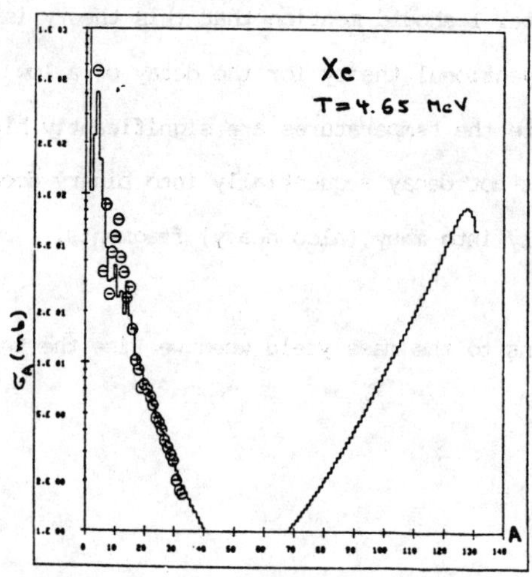

Fig. 3.2. Mass yield for the fragmentation of Xe. Data points from [28]

At intermediate temperatures, T~5 meV, we obtain the U-shaped mass yield which reproduces the data quite well (fig. 3.2-4).

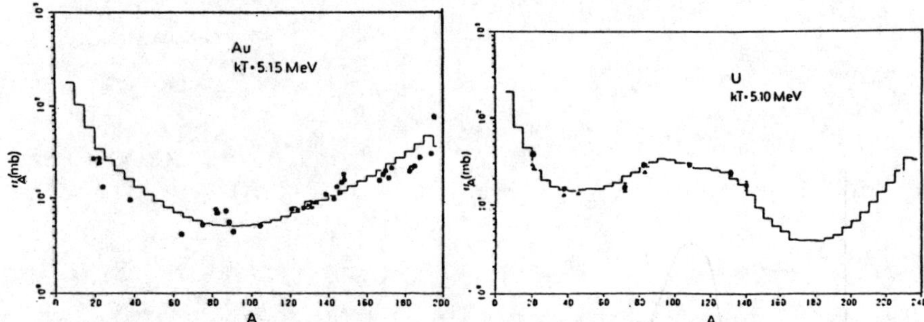

Fig. 3.3. Mass yield for the fragmentation of Au. Data points from [29]

Fig. 3.4. Mass yield for the fragmentation of U
Data points from [30]. Newer data [31] show an additional peak at A~190. This is presumably due to shell-effects, which are not included in our calculations. We use liquid-drop binding energies only.

One may wonder, why the decay happens in this narrow temperature band T~5-6 MeV, only. A possible hint for an explanation is the fact that the systems are all close to a phase transition. It is not a phase transition of the usual macroscopic kind like gas ↔ liquid, but it is a specific transition which exists only in finite (and small) highly charged systems. Figure (3.1) shows what we mean by this. Varying the temperature from T~3 MeV to T~7 MeV, the mass-distribution changes from a deeply U-shaped (eventually with a fission peak in the middle) to one where only small fragments are observed (multifragmentation). This transition is crucially controlled by the Coulomb interaction and the finite charge of the system. No infinite systems can have distributions which are U-shaped or peaked in the "middle". It is quite interesting to see that the specific heat has a clear maximum at the transition temperature of T=6 MeV. Here the temperature is very insensitive to substantial changes of the ex-

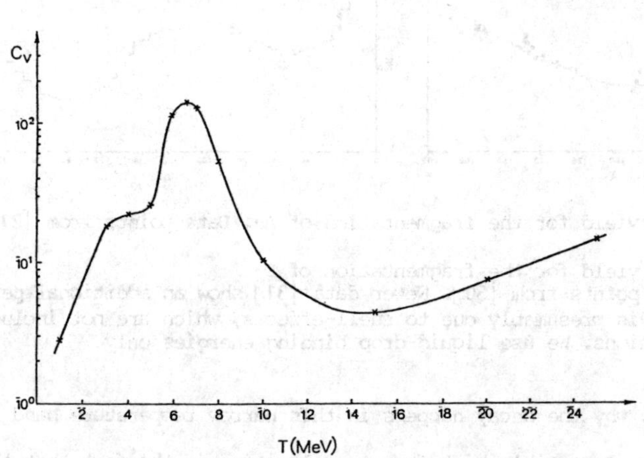

Fig. 3.5 Heat capacity c_V as a function of the temperature T [32]
At T~6 the sharp peak demonstrates the phase transition

citation energy (The same is observed in daily-life with a pot of boiling water). May be this is the reason for the fact that all fragmenting systems decay at roughly the same temperature.

I can not present more about this highly interesting topic. More details can be found in the literature [27,32].

References

1) H.J. Wollersheim, W.W. Wilcke, J.R. Birkelund, J.R. Huizenga, W.U. Schröder, H. Freiesleben and D. Hielscher, Phys. Rev. C24 (1981) 2114

2) D.H.E. Gross, Z. Physik A291 (1979) 145
D.H.E. Gross, Proc. Symp. Berlin, Oct. 23-25, 1979 ed. W.v. Oertzen, Lecture Notes in Physics 117, Springer (1980), p. 81

3) R. Becker, Theorie der Wärme, Berlin, Heidelberg, New York, Springer (1966)

4) J. Randrup and W.J. Swiatecki, Ann. Phys. 124 (1980) 193

5) J. Blocki, T. Boneh, J.R. Nix, J. Randrup, M. Robel, A.J. Sierk and W.J. Swiatecki, Ann. Phys. 113 (1978) 338

6) J.N. De, S.K. Kataria, S.S. Kapoor and V.S. Ramamurthy, Nucl. Phys. in print

7) D.H.E. Gross, in "Heavy-Ion, High-Spin States and Nuclear Structure" IAEA-SMR-1416, Vienna (1975) Vol.I p. 27

8) D.H.E. Gross, Nucl. Phys. A240 (1975) 472

9) D.H.E. Gross, Lectures given at the IPN Orsay (1976), IPNO/TH 77-10 (1977)

10) H.A. Weidenmüller, Progress in Particle and Nuclear Physics, Vol. 3 (1980) 49

11) H. Hofmann and P.J. Siemens, Nucl. Phys. A275 (1977) 464

12) D.H.E. Gross, Phys. Lett. 68B (1977) 412

13) S. Pal and N.K. Ganguli, Nucl. Phys. A370 (1981) 175, Phys. Lett. in print

14) S. Pal, Thesis Univ. of Calcutta (1983) and Nucl. Phys. A in print

15) M.C. Nemes and H.A. Weidenmüller, Phys. Rev. C24 (1981) 944

16) D.H.E. Gross, R.C. Nayak and L. Satpathy, Z. Phys. A299 (1981) 63

17) D.H.E. Gross and L. Satpathy, in "Nuclear Physics" edit. by C.H. Dasso, R.A. Broglia and A. Winther, North Holland 1982 and D.H.E. Gross and L. Satpathy, Phys. Lett. 110B (1982) 31

18) H.H. Deubler and K. Dietrich, Nucl. Phys. A277 (1977) 493

19) P. Fröbrich, B. Strack and M. Durand, Nucl. Phys. A406 (1983) 557

20) P. Fröbrich, Phys. Lett. 122B (1983) 338

21) P. Fröbrich, Preprint HMI-P 83/17Th (1983) submitted to Phys. Rep.

22) P. Fröbrich, Proc. of Workshop on Semiclassical Methods in Nuclear Physics Grenoble, 5-8.3.1984, to be published in J. Physique-Colloque

23) D.H.E. Gross, H. Kalinowski, Phys. Rep. 45 (1981) 175

24) W.J. Swiatecki, Physica Scripta Vol. 24 (1981) 113

25) D.H.E. Gross, H. Kalinowski and J.N. De, Proc. Symp. Classical and Quantum Mechanical Aspects of Heavy Ion Collisions, Heidelberg (1974), Lecture Notes in Physics Vol. 33, (Springer, Heidelberg) p. 194

26) R.A. Broglia, C.H. Dasso and A. Winther,Varenna (1979), Nuclear Structure and Heavy-Ion Collisions (1981) LXXVII Corso Soc. Italiana di Fisica-Bologna-Italy, p. 327

27) D.H.E. Gross, L. Satpathy, Meng Ta-chung and M. Satpathy, Z.f.Phys. A309 (1982) 41

D.H.E. Gross, Physica Scripta T5 (1983) 213

Meng Ta-chung, in Proc. Asia Pacific Phys. Conf. Singapore (1983) in press

D.H.E.Gross "Multifragment production in high-energy proton-nucleus collisions". Preprint HMI-Berlin (1984)

28) A.S. Hirsch, A. Bujak, J.E.Finn, L.J. Gutay, R.W. Minich, N.T. Porile, R.P. Scharenberg,B.C. Stringfellow and J. Turkot, preprint Purdue and Fermi Lab (1983)

J.E. Finn, S. Agarwal, A. Bujak, J. Chuang, L.J. Gutay, A.S. Hirsch, R.W. Minich, N.T. Porile, R.P. Scharenberg and B.C. Stringfellow, Phys. Rev. Lett 49 (1982) 1321

29) S.B. Kaufmann, M.W. Weisfield, E.P. Steinberg, B.D. Wilkins and D. Henderson, Phys. Rev. C14 (1976) 1121

30) G. Friedlander, Proc. Symp. Phys. and Chemistry of Fission, Salzburg (1965) p. 265, IAEA, Vienna

J. Hudis, T. Kirsten, R.W. Stoenner and O.A. Schaeffer, Phys. Rev. C2 (1970) 2019

31) B.V. Jacak, W. Loveland, D.J. Morrissey, P.L. McGanghey and G.T. Seaborg, preprint Berkeley (1982)

32) Sa Ban-hao, D.H.E. Gross and Meng Ta-chung, preprint HMI Berlin (1983) and submited to Nucl. Phys.

33) W.G. Meyer, V.E. Viola,jr., R.G. Clark, S.M. Read and R.B. Theus, Phys. Rev. C20 (1979) 1716

34) T. Matsuse, A. Arima and S.M. Lee, Phys. Rev. C26 (1982) 2338

35) H. Spinka and H. Winkla, Nucl. Phys. A233 (1974) 456

36) J.J. Kolata, R.C. Fuller,R.M. Freeman, F. Haas, B. Heusch and A. Gallman, Phys. Rev. C16 (1977) 891

37) B. Fernandez, C.Gaarde, J.S. Larsen, S. Pontopiddan and F. Videbaek, Nucl. Phys. A306 (1978) 259

38) A. Weidinger, F. Busch, G. Gaul, W. Trautmann and W. Zipper, Nucl. Phys. A263 (1976) 511

39) M. Conjeaud, S. Harar, F. Saint-Lawrent, J.M. Loiseaux, J. Menet and J.B. Viano, Proc. Symp. on Macroscopic Features of Heavy-Ion Collisions and Pre-equilibrium Processes, Hakone, Japan, 1977

40) I. Tserruya, Y. Eisen, D. Pelte, A. Gavron, H. Oeschler, D. Berndt and H.L. Harney, Phys. Rev. C18 (1978) 1688

41) M.N. Namboodiri, E.T. Chulick and J.B. Natowitz, Nucl. Phys. A263 (1976) 491

42) S.E. Vigdor, D.G. Kovar, P. Sperr, J. Mahoney, A. Menaca-Rocha, C. Olmer and M.S. Zisman, Phys. Rev. C20 (1979) 2147

43) H. Sann, R. Bock, Y.T. Chu, A. Gobbi, A. Olmi, U. Lynen, W. Müller, S. Bjørnholm and H. Esbensen, Phys. Rev. Lett. 47 (1981) 1248

44) R. Bock, Y.T. Chu, M. Dakowski, A. Gobbi, E. Grosse, A. Olmi, H. Sann, D. Schwalm, U. Lynen, W. Müller, S. Bjørnholm, H. Esbensen, W. Wölfli and E. Morenzoni, Nucl. Phys. A388 (1982) 334

45) W.J. Swiatecki, Nucl. Phys. A376 (1982) 275

46) H. Gäggeler (private communication)

FAST FISSION PHENOMENON AND ITS IMPLICATIONS ON FUSION

Ngô Christian
Service de Physique Nucléaire - Métrologie Fondamentale
CEN Saclay, F91191 Gif-sur-Yvette Cedex, France

Abstract

We review recent experimental data on fusion which call for a better understanding. A macroscopic dynamical model allowing to understand these data is described. In this approach a phenomenon intermediate between deep inelastic reactions and compound nucleus formation appear : fast fission. The conditions under which fast fission can be observed, as well as its properties are discussed. We make a comparison with another approach of the same problems : the extrapush model. Finally we review a simple dynamical model of fusion which reproduces pretty well a large number of experimental fusion excitation functions.

Fast fission is a mechanism which has been proposed[1-3]) to explain experimental results which were hard to understand in the standard description of dissipative heavy ion collisions. Although there is, up to now, no direct evidence of this mechanism, we have got a consistent picture of dissipative heavy ion reactions. Most of the unexplained experimental observations are now understood without changing our old understanding of more classical experimental observations in heavy ion collisions.

In these lectures I would like to review this subject and make the connection with other new ideas related to the comprehension of the fusion process. We will treat the following items :

We will first briefly recall the experimental problems which have called for new concepts. Then we shall describe the properties of the fast fission mechanism as it appears in a dynamical approach to heavy ion reactions. In particular we shall investigate under which conditions it can be observed. We shall discuss the extension of such a dynamical approach to heavier systems and see how the fast fission process is modified. At this stage a connection with the extrapush model of Swiatecki[4]) will be done. Finally we shall describe the notion of dynamical fusion barrier which plays the same role as the extrapush in understanding fusion.

This paper is somehow a synthetic summary of the oral presentation which uses other materials from ref.[5]). For a complete survey of the question we refer the reader to refs.[5-9]) where much more details can be found.

I. EXPERIMENTAL DATA CALLING FOR NEW CONCEPTS

I.1 Critical angular momentum

The fusion cross section, σ_F, is experimentally defined as the evaporation residues plus the fission-like cross sections. This quantity is often parametrized by a critical angular momentum, ℓ_{cr}, related to σ_F by :

$$\sigma_F = \frac{\pi}{k^2} (\ell_{cr} + 1)^2 \qquad (1)$$

where k is the wave number. Equation (1) has been obtained assuming that the lowest impact parameters give fusion and that the sharp cut off approximation is valid. With these assumptions the critical angular momentum is the largest ℓ value giving fusion.

One of the basic question we have to address concerns the identity between fusion and compound nucleus formation. It is closely related to how much angular momentum a compound nucleus can carry. Liquid drop studies[10]) show, for instance, that the effective barrier against fission decreases when the angular momentum of the compound nucleus increases. For a value, denoted here by ℓ_{B_f}, this fission barrier disappears. Since some amount of time is necessary to form a compound nucleus in the real sense, it is reasonable to think that it is not possible to form a compound nucleus when $\ell \geq \ell_{Bf}$. In the case where fusion would be identical to compound nucleus formation, ℓ_{cr} should always be smaller than ℓ_{Bf}. Compilation of the existing data[7]) shows that this is not true and several measurements do show that ℓ_{cr} can notably exceed ℓ_{Bf}. One possibility would be to say that σ_F not only contains complete fusion but that there is also a contribution of incomplete fusion. However, measurements of light fast particles[11]), which are usually associated with this last process, show a too small multiplicity and cannot account for the difference between ℓ_{cr} and ℓ_{Bf}. Therefore, we have to conclude that fusion cannot be identified with compound nucleus formation. Then what happens when $\ell_{Bf} \leq \ell \leq \ell_{cr}$ for which there is fusion but not compound nucleus formation?

I.2 Widths of fission-like mass distributions of the Ar + Ho system at high bombarding energies

The starting point of fast fission was a series of experiments[1]) on the Ar + Ho system done at different bombarding energies. In fig. 1 are shown (dots) the full width half maximum (FWHM) of the mass distribution of the fission-like products as a function of the excitation energy of the "compound nucleus". The FWHM increases with the excitation energy E*. Because the temperature, T, rises with increasing E*, we expect the mass distribution to broaden when T increases due to statistical fluctuations. An estimation of this effect[1]) gives however a too small increase compared to the experimental observation. The results of these experiments might tell us that the stiffness of the potential energy surface along the mass asymmetry coordinate could be strongly angular momentum dependent. Another possibility would be that the fission-like products associated to the largest partial waves are coming from another mechanism different from fission following compound nucleus formation. This mechanism would only contribute when $\ell_{B_f} \leqslant \ell \leqslant \ell_{cr}$. In fig. 1, the excitation energy corresponding to ℓ_{cr} larger than ℓ_{B_f} is at about 80 MeV and is indicated in the figure. It is precisely in this region that one might guess a particularly large increase of the FWHM. This last preliminary conclusion is supported by investigations of heavier systems (Cl + Au in ref.[11]) and Cl + U in ref.[12])). Indeed, as the system becomes heavier ℓ_{B_f} decreases and for the Cl + U system, for instance, we have mainly to deal with ℓ values larger than ℓ_{B_f}. In this case the FWHM turns out to be almost constant over the investigated bombarding energy range (240-350 MeV) whereas it varies for the Cl + Au system (204-317 MeV) but to a smaller extend than for the Ar + Ho combination. These observations suggest that when $\ell_{B_f} \leqslant \ell \leqslant \ell_{cr}$ there could be a contribution of a mechanism different

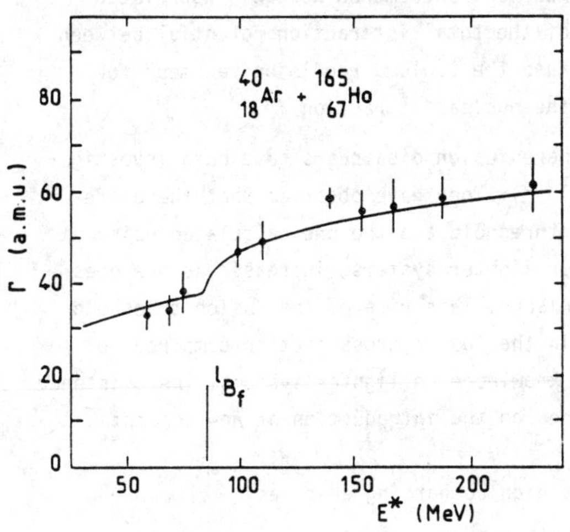

Fig. 1 – Full width half maximum, Γ, of the fission like mass distribution, as a function of the excitation energy of the fused system, for Ar + Ho. The dots are the experimental points in ref.[1]). The full curve is the result of the calculation of ref.[7]).

from ordinary fission. This process would give larger fission-like mass distributions.

I.3 Disappearance of fusion for very heavy systems

When the projectile and the target become very heavy it becomes no longer possible to make them fuse together[13]. This occurs when the product, Z_1Z_2, of the atomic numbers of the two ions becomes larger than about 2500-3000. The disappearance of this important phenomenon has been correlated with the vanishing of the pocket of the total interaction potential between two hèavy nuclei[14]. It is so because the Coulomb repulsion becomes, for all overlaps, much stronger than the nuclear attraction.

Systems close to the limit where fusion disappears have been investigated in great details by Bock et al.[15]. They have observed that the difference between the measured fusion threshold and the one calculated using prescriptions working very well for lighter systems, increases as one goes towards the limit where fusion vanishes. This rise of the fusion threshold leads, of course, to a reduction in the fusion cross section compared to what can be extrapolated from our knowledge on lighter systems. The existence of such a difference has also pushed on the introduction of new concepts.

I.4 Fusion cross section defect at high bombarding energies

At bombarding energies just above the fusion threshold the fusion cross section increases almost linearly as a function of $1/E_{CM}$, the inverse of the center of mass bombarding energy. This can be explained in a simple classical picture by looking if it is possible, for a given impact parameter, to overcome the associated fusion barrier. Then σ_F is simply given by :

$$\sigma_F = \pi R_{12}^2 \left(1 - \frac{V_{12}}{E_{CM}}\right) \qquad (2)$$

where R_{12} and V_{12} are respectively the location and the height of the fusion barrier for a head-on collision.

It is a well known experimental fact that at higher bombarding energies the experimental σ_F is smaller than the value calculated with eq.(2). An attractive idea proposed to understand this fusion cross section defect was the notion of critical distance proposed in refs.[16] and [17]. For medium systems this method was rather successful but could not explain why very heavy systems do not fuse. Furthermore a theoretical justification of this notion was missing.

It should be noted that this point is correlated to the preceding one since, in both cases, a fusion cross section defect is noticed.

II. THEORETICAL INVESTIGATION OF FAST FISSION

During a heavy ion collision, the two nuclei remain unchanged before they reach the interaction region. There, if the overlap is large enough, dissipative phenomena can take place. A part of the kinetic energy in the relative motion can be transformed into intrinsic excitation of the total system. This loss of energy can be such that the system remains trapped in the interaction region. In this case we say that there is fusion.

II.1 Limits of fusion

In order to fuse the total interaction potential of the system, plotted versus the interdistance separating the two nuclei, should have a pocket. Using the sudden approximation and the energy density formalism[18]) we can show[5]) that, for a head-on collision, the pocket disappears when :

$$\left(\frac{Z^2}{A}\right)_{eff} = \frac{4\ Z_1 Z_2}{A_1^{1/3} A_2^{1/3} (A_1^{1/3} + A_2^{1/3})} \lesssim 48 \qquad (3)$$

This condition expresses the fact that the nuclear force, at the distance where it is maximum, is larger than the modulus of the Coulomb force.

II.2 The dynamical fast fission model

For large overlaps, the two nuclei, initially supposed to be spherical, become deformed. Various shape degrees of freedom are excited and, for instance, a neck appears between the two ions creating in this way a single composite system with two centers. Investigation of the future evolution of the system needs a good description of these shape deformations. Since these excitations transform a potential landscape where the two nuclei are spherical (sudden potential) to one where some of the shape degrees of freedom have relaxed (adiabatic potential) it is tempting to describe this transition in a phenomenological way. This was done in ref.[3]) where a dynamical transition between a sudden potential[18]) in the entrance channel and an adiabatic one[19]) in the exit channel was done. The degree of completeness of the transition depends upon the overlap between the two ions.

The collision of the two nuclei is described by means of four collective degrees of freedom : the distance, R, separating the center of mass of the two nuclei, the corresponding polar angle, the mass asymmetry of the system and the neutron excess of one of the fragments. The dynamical evolution of the collision is followed by means of a transport equation derived by Hofmann and Siemens[20]).

This model allows to describe deep inelastic properties as well as fusion. For some systems it shows new features which we shall discuss now.

II.3 The appearance of fast fission

When conditions are fulfilled, the model reveals the existence of a mechanisms intermediate between deep inelastic reactions and compound nucleus formation. This can be seen, for instance, for the 340 MeV Ar + Ho system. In fig. 2 typical mean trajectories are shown versus mass asymmetry and radial distance.

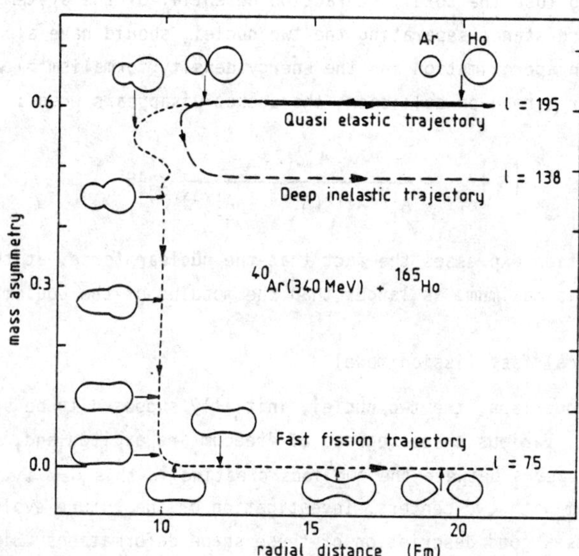

Fig. 2 - Few mean trajectories for various initial values of the orbital angular momentum, ℓ, plotted in the plane radial distance-mass asymmetry. Three kinds of mechanism are illustrated in this plot : 1) quasi-elastic process for ℓ=195, 2) deep inelastic collision for ℓ=138 and 3) fast fission phenomenon for ℓ=75. For $\ell < \ell_{B_f}$ = 72, a compound nucleus is formed. This figure has been extracted from ref.[3]).

-ℓ =195 corresponds to a quasi elastic reaction : we have little mass and energy exchanged during the interaction.

- ℓ=138 represents typically a deep inelastic trajectory : a lot of kinetic energy is lost in the relative motion and a non negligeable mass transfer occurs for this particular ℓ value.

- ℓ=75 shows a new kind of phenomenon. The system is trapped into the pocket of the entrance potential. Mass asymmetry relaxes to equilibrium and, simultaneously, the sudden potential changes to the adiabatic one. However, for such a large value of the angular momentum the fission barrier of the compound nucleus does not exist any more. Consequently the system cannot remain caught any longer in the interaction region and separates in two fragments. We have a phenomenon similar to fission following compound nucleus formation except that we start from a two-center system. This is schematically illustrated in fig. 3. The interaction time for such a collision is smaller than the one corresponding to compound nucleus formation followed by fission. It is of the order of 10^{-20}s which is nevertheless larger than those involved in deep inelastic reactions.

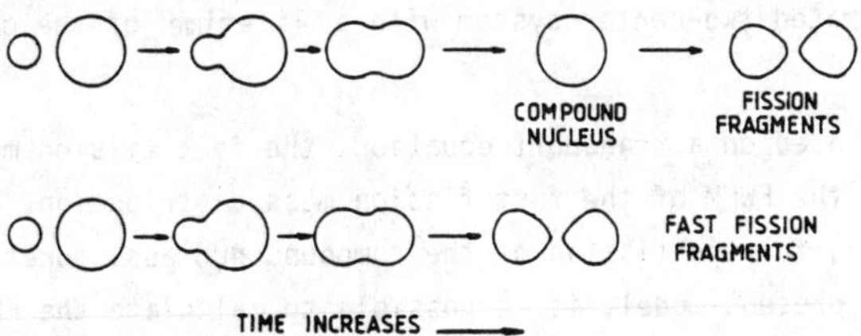

Fig. 3 - Schematic picture of compound nucleus fission and fast fission.

- For this particular system ℓ_{Bf} = 72. When $\ell < 72$ the system which is trapped in the entrance channel remains trapped in the adiabatic potential because the fission barrier still exists and has a configuration less compact than the one of the pocket.

In conclusion, for a system like Ar + Ho, fast fission only occurs when $\ell_{Bf} \leqslant \ell \leqslant \ell_{cr}$.

II.4 Fast fission phenomena for heavy systems : quasifission

When the fissility parameter, $\frac{Z^2}{A}$, increases the saddle configuration becomes more and more compact. For big nuclei it can become less elongated than the pocket configuration. For a symmetric system this occurs when the following condition is fulfilled :

$$\frac{Z^2}{A} \gtrsim 38.5 \qquad (4)$$

where Z and A are the atomic and mass numbers of the compound nucleus. In this case, even if $\ell < \ell_{Bf}$, the system which is trapped into the pocket of the sudden potential cannot remain trapped when the adiabatic landscape is reached, because it has an elongation larger than the one associated to the pocket : it reseparates in two fragments. This special type of fast fission, which occurs for $0 \leq \ell \leq \ell_{cr}$, has been called quasifission by Swiatecki[4]) who was the first to point out this possibility.

II.5 Properties of fast fission

According to the macroscopic model described above, fast fission should have very similar properties to those of fission following compound nucleus formation. This makes difficult to get unambiguous proofs of its existence. At the present stage it is the simplest possibility to understand the data but that is all. The most advanced experimental proof might be found in experiments done by Ho et al.[21]) who have found evidence of light particles emitted by an equilibrated two-center system with a lifetime of the order of 10^{-20}s.

Since it is based on a transport equation, the fast fission model permits to calculate the FWHM of the fast fission mass distribution. Using the results of ref.[22]), for the fission of the compound nucleus, together with the output of the present model, it is possible to calculate the FWHM of the fission-like products. This is shown by the full line in fig. 1.

When $\frac{Z^2}{A} \lesssim 38.5$, the fast fission threshold is larger than the one corresponding to compound nucleus formation. This is illustrated in fig. 4 which shows the fusion excitation function of the Ar + Ho system. The full curve is the calculation of ref.[3]) and the dots are the experimental data of ref.[1]). The full curve can be decomposed in two terms : compound nucleus formation and fast fission.

II.6 The four classes of dissipative heavy ion collisions

We summarize schematically in fig. 5 the four classes of dissipative heavy ion collisions which can be predicted by the macroscopic dynamical model of ref.[3]). In this figure the entrance sudden and the adiabatic potentials are represented as a function of R. This one dimensional plot is just to get a feeling of the real collision which occurs in fact in multi-dimensional space.

In fig. 6 we show the range of ℓ values, or impact parameters, to which these dissipative phenomena are associated and fig. 7 summarizes the conditions under which fusion, fast fission and quasifission occur.

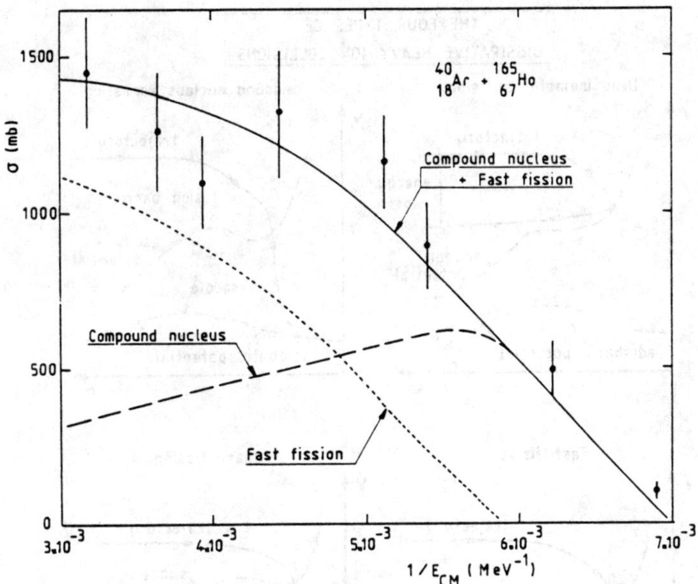

Fig. 4 - Experimental fusion cross section (dots) from ref.[1]) plotted as a function of $1/E_{CM}$, the inverse of the center of mass bombarding energy. It is compared with the calculated fusion cross section of ref.[3]) (full curve). The fusion cross section is the sum of the compound nucleus and of the fast fission cross sections. Their corresponding excitation functions are also shown in the figure. This figure is extracted from ref.[3]).

III. COMPARISON WITH THE EXTRAPUSH MODEL

III.1 The dynamical extrapush model

Swiatecki[4]) has developed a dynamical model for head-on collisions of two heavy ions. There are three macroscopic variables : one associated to the distance separating the two fragments, one connected with mass asymmetry and a last one related to the size of the neck connecting the two nuclei. The dynamical evolution of the macroscopic variables, is followed by Newton equations with friction forces proportional to the velocities and given by the one body approach[23]). Except for the interdistance of the two separated nuclei, inertial forces were neglected (overdamped motion in the sense of Kramers[24])).

Three particular configurations play an important role in this approach : 1) the contact one (it is at this point that the neck degree of freedom is unfrozen), 2) the conditional saddle (saddle point calculated under the condition that the mass asymmetry degree of freedom remains frozen to its initial value), 3) the unconditional saddle (usual saddle point). To each of

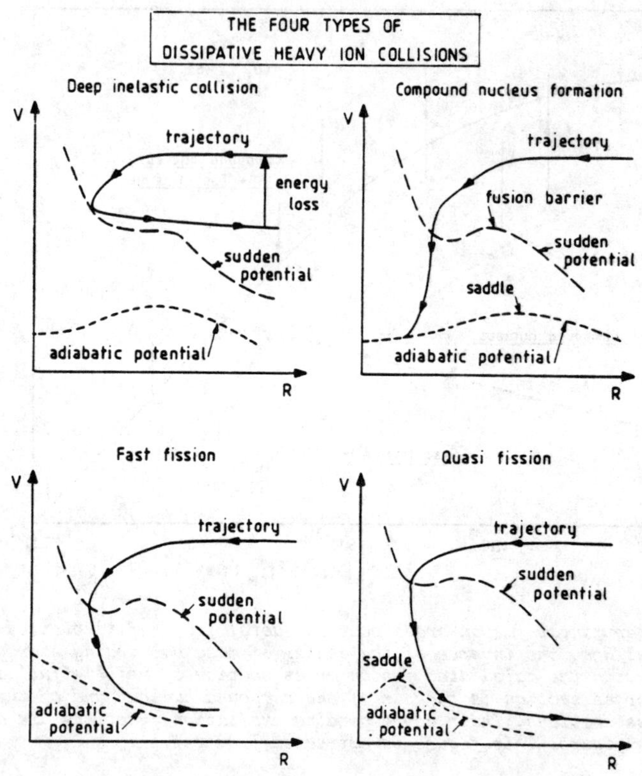

Fig. 5 - Typical illustration of the four dissipative mechanisms occuring in a heavy ion reaction : Top left *: the system is not trapped but it looses a lot of kinetic energy in the relative motion : we have a deep inelastic collision.* Top right *: the system is trapped in the entrance channel. The sudden potential goes to the adiabatic one but the saddle configuration is elongated enough to keep the system trapped : we have compound nucleus formation.* Bottom left *: the system is trapped but the fission barrier of the compound nucleus has vanished due to angular momentum. Therefore it desintegrates in two almost equal fragments because mass asymmetry had time to reach equilibrium : we have fast fission.* Bottom right *: the compound nucleus has a fission barrier but the saddle configuration is too compact to keep the trapped system : we have also fast fission or quasi-fission.*

these configurations three thresholds are associated ; they are shown in fig. 8 borrowed from ref.[25]). Fusion occurs only if the conditional saddle is reached. For heavy systems this configuration is more compact than the contact one (case of fig. 8) and some extra energy, over the contact point, is needed to reach it and to compensate the energy loss due to friction : it is the extra-push. For light systems this extrapush is zero because the contact configuration is more elongated than the conditional saddle. The above considerations explain why the fusion threshold becomes larger than expected for heavy systems[15]).

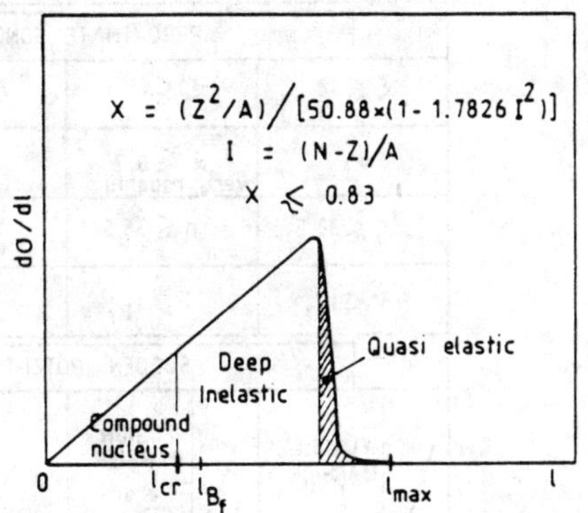

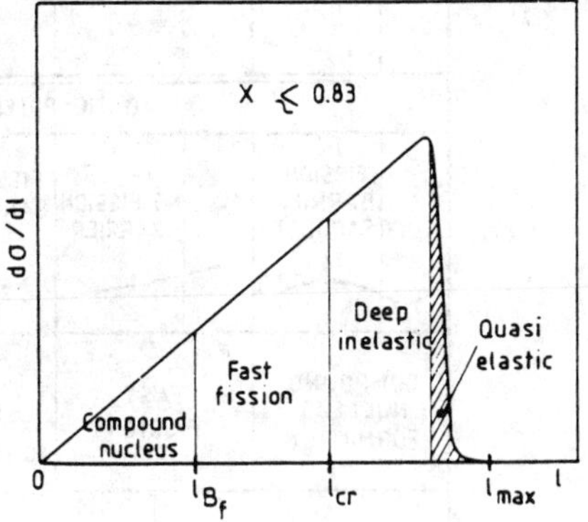

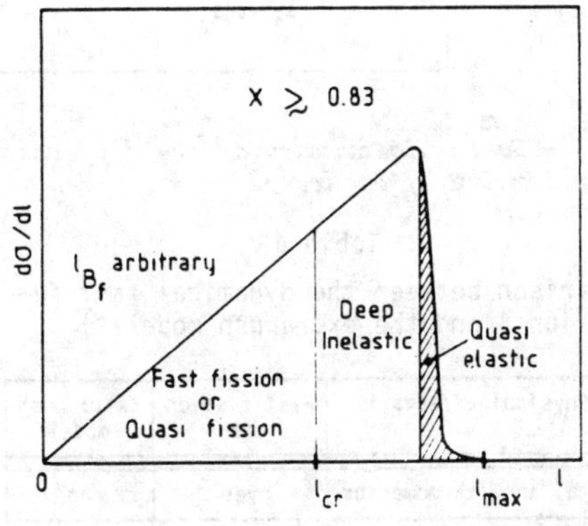

Fig. 6 - Schematic representation of the different ranges of ℓ values associated to the four dissipative mechanisms which can be observed in heavy ion reactions.

Fusion does not mean necessarily compound nucleus formation. To form a real compound nucleus the system should reach the unconditional saddle point. The extra energy above the contact configuration necessary to do that is called the extra extrapush.

III.2 Comparison with the dynamical fast fission model

It is interesting to make a brief comparison between the dynamical extra push and the fast fission models. Indeed both aim to give a good overall understanding of fusion. In table I are summarized the main points in which they differ. In addition to that we should add one more difference : in the extrapush model the two nuclei remain unchanged until they reach the contact configuration. At this step the neck degree of freedom is unfrozen and relaxes very fast to equilibrium ($\sim 10^{-22}$s). At variance, in the fast fission model the transition between the sudden and the adiabatic potential occurs more slowly ($\sim 10^{-21}$s) but starts before the two nuclei are in close contact.

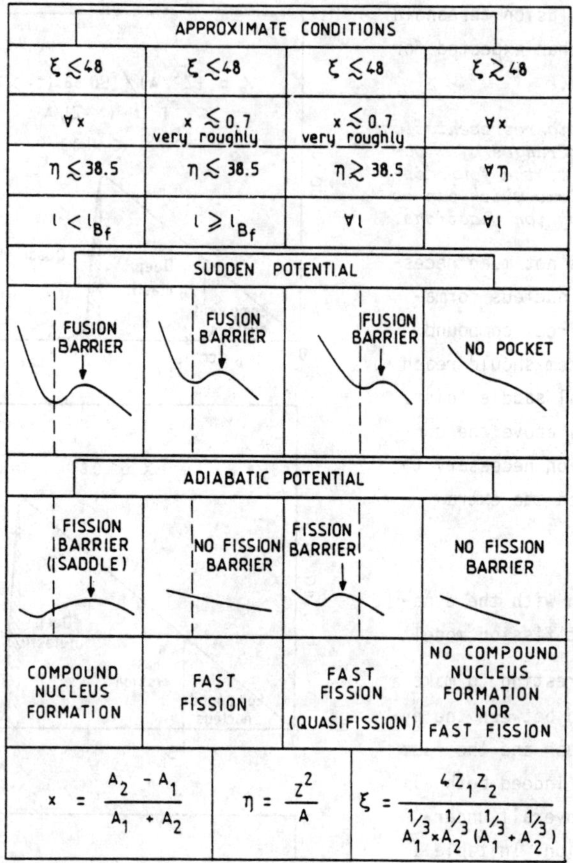

Fig.7 - Schematic summary of the different mechanisms following fusion and their domains of occurence.

Table 1

Comparison between the dynamical fast fusion[3]) and the extrapush models[4])

Physical effects	Fast fission model	Extra push model
Orbital angular momentum	yes	no
Statistical fluctuations	yes	no
Deformations	simulated	yes
Overdamped approximation	no	yes

III.3 The static extrapush model

Swiatecki[4,25]) has parametrized some of the results obtained for a head-on collision by simple analytical formulas. In particular the extra push ΔE can be expressed as a function of the effective fissility parameter $(Z^2/A)_{eff}$ defined in eq.(3) : it is zero below a certain threshold and goes like a parabola

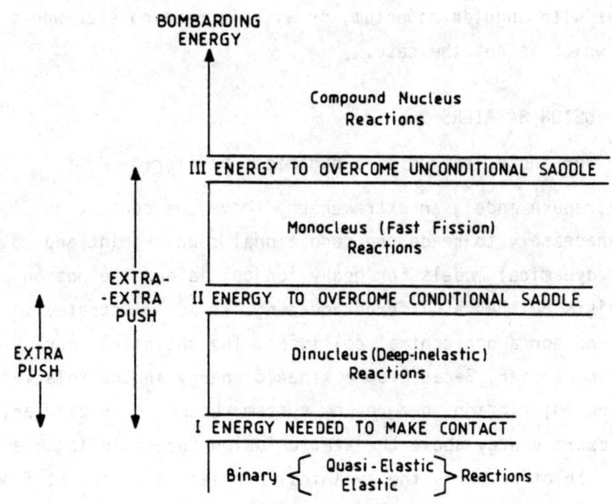

Fig. 8 - Schematic illustration of the different mechanisms which are obtained in the dynamical model of Swiatecki[4]) for head-on collisions. If the bombarding energy is not sufficient to reach the contact configuration (I) we get elastic or quasi elastic processes. If the energy is sufficient to make contact but not to reach the conditional saddle configuration we have to deal with deep inelastic reactions. To reach the conditional saddle from the contact configuration an extra energy is needed : it is the extrapush. If this point is reached there is fusion but not necessarily compound nucleus formation. To reach this latter situation more extra energy is needed (extra extrapush) otherwise there is quasi-fission (fast fission). This figure is taken from ref.[25]).

above. This formula can be extended to non central collision by generalizing $(Z^2/A)_{eff}$ to non zero orbital angular momentum[3]) :

$$\left(\frac{Z^2}{A}\right)_{eff}(\ell) = \left(\frac{Z^2}{A}\right)_{eff} + g(f\ell)^2 \frac{A_1 + A_2}{A_1^{1/3}A_2^{1/3}(A_1^{1/3} + A_2^{1/3})} \tag{5}$$

To a constant factor, $(Z^2/A)_{eff}(\ell)$ is the sum of the Coulomb and centrifugal forces over the nuclear one at the point where this last quantity is maximum. f is the proportion of angular momentum which remains in the relative motion after tangential friction has acted. We would like to stress that a real problem exists as far as the choice of this quantity is concerned. Several prescriptions can be used, for instance rolling or sticking, but the good choice is still not clear. For example, for the systems investigated at GSI in ref.[15]), the rolling seems to be more appropriate whereas, for the Ar + Ho system, the sticking is better[5]). In fact it should be noted that eq.(5) just simulates angular momentum by an equivalent Coulomb force. However we know that it is only a very rough approximation otherwise the evolution of the fission barrier with angular momentum, or with increasing size would be closely related, which is not the case.

IV. DYNAMICAL FUSION BARRIERS

IV.1 Necessity for the existence of a dynamical barrier

In the extrapush model, an extra energy above the contact configuration was sometimes necessary to reach the conditional saddle point and to get fusion. In other dynamical models for heavy ion collisions the notion of extra energy also exists but has a different meaning. It is illustrated in fig. 9 for a head-on and for a non-central collision. The potential energy surface defines a static barrier. Because some kinetic energy in the relative motion can be lost, due to friction, before the system reaches this barrier, we see that some extra energy above the static fusion threshold is necessary to perform fusion. In other words the dynamical barrier is different from the static one. If ΔE is the energy difference between the dynamical and the static fusion barriers, it is possible to show that[5]) :

$$\sigma_F = \pi R^2_{f\ell_{cr}} \left(1 - \frac{V(R_{f\ell_{cr}}) + \Delta E}{E} \right) \tag{6}$$

where $R_{f\ell_{cr}}$ is the position of the static fusion barrier corresponding to $\ell = f\ell_{cr}$. ℓ_{cr} is the critical angular momentum and f the amount of orbital angular momentum remaining in the relative motion. $V(R_{f\ell_{cr}})$ is the value of the nuclear plus Coulomb potentials at distance $R_{f\ell_{cr}}$. The extra energy, ΔE, is zero for light systems as well as for small angular momentum. It can be parametrized by the following expression :

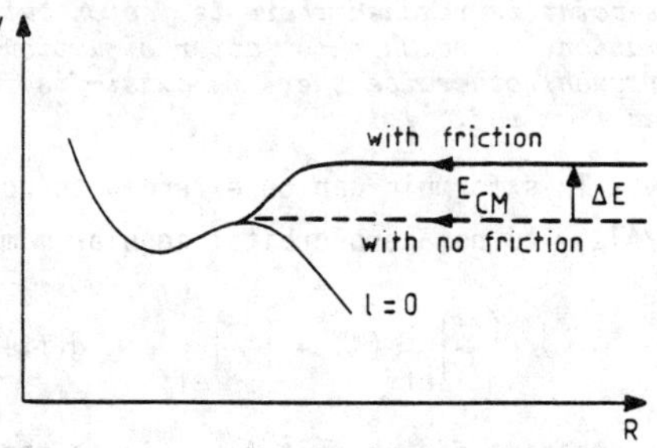

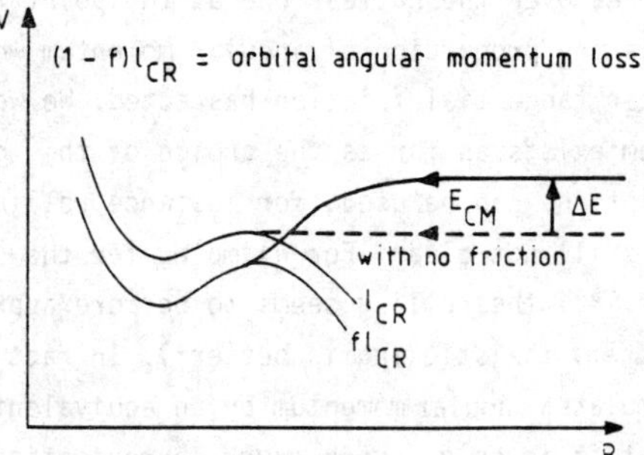

Fig. 9 – Schematic illustration of the fact that some extra kinetic energy is needed to overcome the static fusion barrier : on top is the case of a head-on collision. In the bottom is the case when the orbital angular momentum is equal to ℓ_{cr}. There the total interaction potential including centrifugal energy changes due to angular momentum loss.

$$\Delta E = 0 \qquad \text{for} \quad X < 0.68$$
$$\Delta E \approx 2000\,(X_{eff} - 0.68)^2\ \text{MeV} \qquad \text{for} \quad X > 0.68 \tag{7}$$

where

$$X_{eff} = \frac{1}{12.54}\left(\frac{Z_1 Z_2 e^2}{C_1 C_2 (C_1 + C_2)} + f^2 \ell^2 \hbar^2 \, \frac{A_1 + A_2}{A_1 A_2} \, \frac{1}{C_1 C_2 (C_1 + C_2)^2}\right) \tag{8}$$

where C_1 and C_2 are the central radii of the nuclei. Again we are faced with the choice of the f factor telling us how much angular momentum is dissipated.

The above considerations explain why very heavy systems, like those investigated in ref.[15]), have a reduction of fusion cross section : it is just because the dynamical barrier becomes larger than the static one. For a given system, at large bombarding energies, it also explains the fusion cross section defect because of the same reason.

Since static fusion models based only on potential energy considerations reproduce quite well the fusion data of not too heavy systems and at not too high bombarding energies, it means that the dynamical and static barriers are the same. This is likely to indicate that friction does not extend too far outside and probably extend similarly, for all systems, from the point where the nuclear force is maximum. As $Z_1 Z_2$ increases, or as ℓ increases, the position of the fusion barrier becomes more and more compact and friction can act making a difference between the dynamical and the static barriers.

IV.2 Simple dynamical model for fusion

To check the above ideas, a simple model has been developed in refs.[26,27]) to describe fusion. It uses two macroscopic variables : the distance, R, between the two nuclei and the polar angle θ. Friction forces proportional to the velocities are introduced in the radial and tangential motion with the following form factor :

$$g(R) = \frac{1}{1 + \exp\left(\frac{s - 0.75}{0.2}\right)} \tag{9}$$

where

$$s = R - C_1 - C_2 \tag{10}$$

and C_1, C_2 are the central radii of each nuclei. The friction force in the radial motion is : $- C_R\, g(R)\, \dot{R}$ with $C_R = 31\,000$ MeV.fm^{-2} $\times 10^{-23}$s. The tangential motion is responsible of orbital angular momentum damping which is introduced in the following way :

$$\frac{d\ell}{dt} = -\frac{C_t}{\mu}(\ell - \ell_{st}) \tag{11}$$

where ℓ_{st} is the angular momentum corresponding to sticking and μ the reduced mass. According to the one body picture[23]), $C_t = C_R/2$. The special form (eq.(11)) of the orbital angular momentum loss is chosen to ensure that the sticking limit is not overcome, as expected in the collision of two rigid bodies.

In ref.[27]) the excitation functions of a large number of systems have been calculated and compared with experiment. The agreement turns out to be surprisingly good despite the extreme simplicity of the model. This is illustrated in fig. 10.

This simple macroscopic model gives an overall understanding of the fusion process. The unexpected results quoted in section I are understood in terms of dynamical fusion barriers which become, in some cases, larger than the static one.

V. CONCLUSION

During the last few years a great progress has been done in understanding fusion. New concepts like fast fission, quasifission, extrapush and dynamical barriers have been proposed and seem to be quite successful. This big progress has only be done by a simultaneous attack of the problem by experimentalists and theoreticians showing the efficiency of such a close connection between them. If fusion seems now qualitatively understood, there still remains open problems for a good quantitative description of the data. Especially the role of orbital angular momentum is still not clear and the approximations used to treat it have to be removed in the near future.

ACKNOWLEDGMENTS

I would like to use the opportunity of this manuscript to thank my colleagues from Caen, Heidelberg, Orsay and Saclay for the good collaborations we had in the fast fission experiments. A large part of the materials presented in these lectures have been obtained together with Christian Grégoire, Renée Lucas, Hélène Ngô, Bernard Rémaud, Tiina Suomijarvi and Eglé Tomasi. I would like to thank them a lot. Finally, it is a pleasure to acknowledge Mmes F. Lepage and E. Thureau, as well as M.J. Matuszek who have carefully ensured the material preparation of the manuscript.

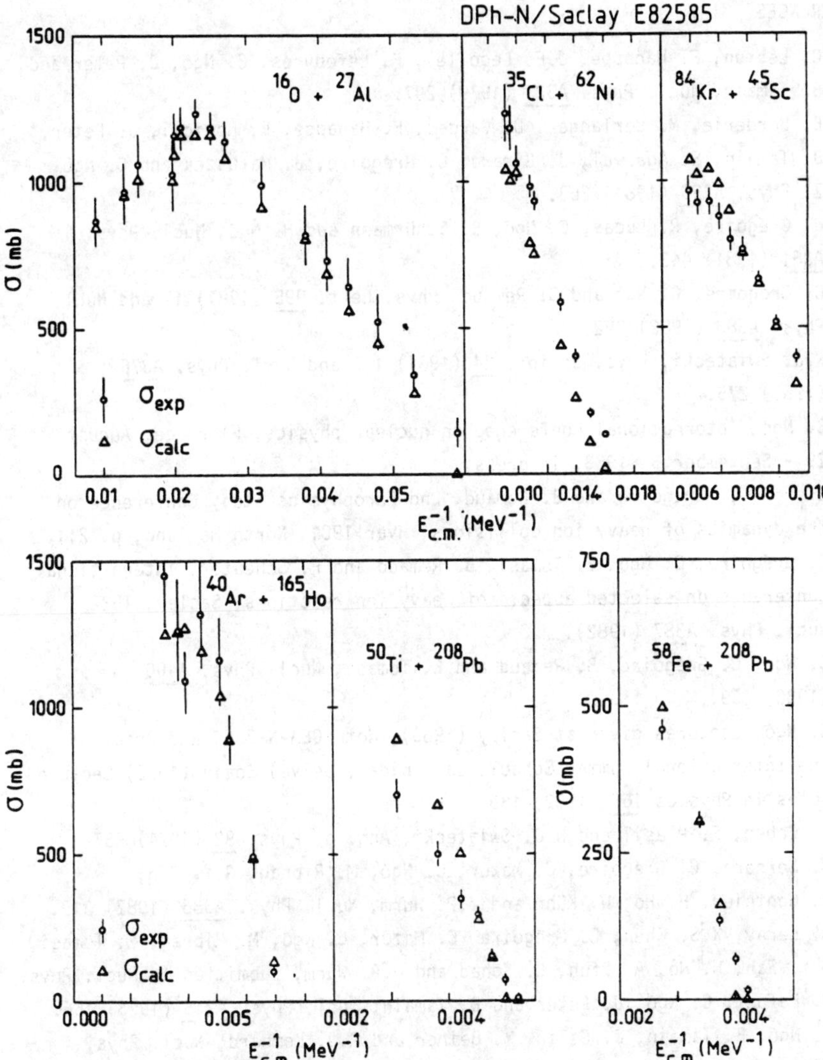

Fig. 10 - Comparison of the experimental cross sections with those computed using the dynamical model of ref.[27]). Experimental points have been taken from the compilation of ref.[28]). This figure has been extracted from ref.[26]).

REFERENCES

1) C. Lebrun, F. Hanappe, J.F. Lecolley, F. Lefebvres, C. Ngô, J. Peter and B. Tamain, Nucl. Phys. A321 (1979) 207.
B. Borderie, M. Berlanger, D. Gardes, F. Hanappe, L. Nowicki, J. Peter, B. Tamain, S. Agarwal, J. Girard, C. Grégoire, J. Matuszek and C. Ngô, Z. Phys. A299 (1981) 263.
2) C. Grégoire, R. Lucas, C. Ngô, B. Schürmann and H. Ngô, Nucl. Phys. A361 (1981) 443.
3) C. Grégoire, C. Ngô and B. Remaud, Phys. Lett. 99B (1981) 17 and Nucl. Phys. A383 (1982) 392.
4) W.J. Swiatecki, Phys. Script. 24 (1981) 113 and Nucl. Phys. A376 (1982) 275.
5) C. Ngô, International Conference on nuclear physics, Florence, August 29 - September 3, 1983, in press.
6) C. Ngô, C. Grégoire and B. Remaud, 2nd Europhysics Study Conference on the dynamics of heavy ion collisions, Hvar 1981, North Holland, p. 211.
7) C. Grégoire, C. Ngô, E. Tomasi, B. Rémaud and F. Scheuter, International Conference on selected aspects of heavy ion reactions, Saclay, 1982 ; Nucl. Phys. A387 (1982).
8) C. Ngô, C. Grégoire, B. Rémaud and E. Tomasi, Nucl. Phys. A400 (1983) 259.
9) C. Ngô, Lectures given at Saclay (1983), Note CEA-N-2354 and Proc. of the International Summer School, La Rabida (Huelva) Spain (1982) Lecture Notes in Physics 168 (1982) 185.
10) S. Cohen, F. Plasil and W.J. Swiatecki, Ann. of Phys. 82 (1974) 557.
11) V. Bernard, C. Grégoire, C. Mazur, C. Ngô, M. Ribrag, G.Y. Fan, P. Gonthier, H. Ho, W. Kühn and J.P. Wurm, Nucl. Phys. A385 (1982) 319.
12) S. Leray, X.S. Chen, C. Grégoire, C. Mazur, C. Ngô, M. Ribrag, E. Tomasi, G.Y. Fan, H. Ho, A. Pfoh, L. Schad and J.P. Wurm, submitted to Nucl. Phys.
13) M. Lefort, C. Ngô, J. Peter and B. Tamain, Nucl. Phys. A216 (1973) 166.
14) C. Ngô, B. Tamain, J. Galin, M. Beiner and R.J. Lombard, Nucl. Phys. A240 (1975) 353.
15) R. Bock, Y.T. Chu, M. Dakowski, A. Gobbi, E. Grosse, A. Olmi, H. Sann, D. Schwalm, U. Lynen, W. Müller, S. Bjornholm, H. Esbensen, W. Wölfli and E. Morenzoni, Nucl. Phys. A388 (1982) 334.
16) R. Bass, Nucl. Phys. A231 (1974) 45.
17) J. Galin, D. Guerreau, M. Lefort and X. Tarrago, Phys. Rev. C9 (1974) 1018.

18) C. Ngô, B. Tamain, M. Beiner, R.J. Lombard, D. Mas and H.H. Deubler, Nucl. Phys. A252 (1975) 237.
H. Ngô and C. Ngô, Nucl. Phys. A348 (1980) 140.
19) T. Ledergerber and H.C. Pauli, Nucl. Phys. A207 (1973) 1.
20) H. Hofmann and P.J. Siemens, Nucl. Phys. A275 (1977) 467.
21) H. Ho, private communication.
22) C. Grégoire and F. Scheuter, Z. Phys. A303 (1981) 337.
23) J. Blocki, Y. Boneh, J.R. Nix, J. Randrup, M. Robel, A.J. Sierk and W.J. Swiatecki, Ann. Phys. 113 (1978) 330.
24) H. Kramers, Physica VII (1940) 284.
25) S. Bjornholm and W.J. Swiatecki, Nucl. Phys. A391 (1982) 471.
26) R. Lucas, T. Suomijarvi, C. Ngô, E. Tomasi and O. Granier, XXII International Winter Meeting on nuclear physics, Bormio (1984), preprint DPh-N/Saclay n°2122.
27) T. Suomijarvi, R. Lucas, C. Ngô, E. Tomasi, D. Dalili and J. Matuszek, to be published.
28) J.R. Birkelund and J.R. Huizenga, Ann. Rev. Nucl. Sc. 33 (1983) 265.

THE AVERAGE NUCLEON-NUCLEUS POTENTIAL

C. MAHAUX

Institut de Physique B5, Université de Liège,
B-4000 Liège 1, Belgium

Abstract. The main empirical properties of the average nucleon-nucleus potential are reviewed. Their theoretical interpretation is outlined. Emphasis is put on theoretical calculations which are based on realistic nucleon-nucleon interactions, and on the existence of a unified picture which encompasses negative as well as positive nucleon energies.

I. INTRODUCTION

The interpretation of many nuclear phenomena involves the description of the motion of a nucleon inside a nucleus. Two obvious examples are the analysis of nucleon-nucleon scattering on the one hand, and the nuclear shell model of the other hand. In the present lectures, we survey some empirical characteristics of the average nucleon-nucleus potential, and progress recently accomplished in their theoretical understanding.

The first lecture, contained in sect. II, is devoted to the main features of the optical model. It provides a connection with the basic formalism described by L. Fonda in previous lectures at this College.

Section III deals with the theoretical investigation of the real and imaginary parts of the optical-model potential, using as input "realistic" nucleon-nucleon interactions i.e. two-body potentials fitted to the nucleon-nucleon scattering data. Two main approaches are briefly outlined and are confronted with recent experimental data on nucleon-nucleus scattering.

In sect. IV we turn to the case of negative energy nucleons i.e. to quasibound single-particle states. Emphasis is put on the main properties of the corrections to the Hartree-Fock approximation. These corrections lead to a distribution of the single-particle strength. This distribution is quite different in the case of a weakly bound single-particle state on the one hand, and of a deeply bound single-particle state on the other hand. It appears that it is only in the latter case that the analog of an optical model is convenient for the description of bound single-particle states.

II. THE EMPIRICAL AVERAGE NUCLEON-NUCLEUS POTENTIAL

The average potential between a target nucleus and a bombarding nucleon is called the optical-model potential. The foundation of the optical model and the empirical properties of the optical-model potential are described in textbooks and monographs, see e.g. refs. [1,2].

II.a. Analogy with optics

The propagation of an electromagnetic wave in a material is governed by the index of refraction n which is related to the permittivity ε by

$$n = \sqrt{\varepsilon} \quad . \tag{II.1}$$

For a wave with frequency ω , one has

$$\varepsilon = \varepsilon_d + i\sigma/\omega \tag{II.2}$$

where ε_d is the dielectric constant and σ the conductivity of the medium. In a conductor, ε and n are therefore complex, and so is the wave-number k defined by

$$k = \omega c/n = k_R + i\, k_I \quad . \tag{II.3}$$

The energy density of a plane electromagnetic wave sent on a medium is damped when the wave propagates through the medium. The associated penetration depth is given by

$$\Delta L = (2\ k_I)^{-1} \quad . \tag{II.4}$$

II.b. The optical model

Concepts similar to those outlined in sect. II.a apply to the propagation through a target of the wave associated with a beam of projectiles. In that case the energy-momentum relation reads

$$E = \frac{\hbar^2 k^2}{2m} + \mathcal{M} \quad , \tag{II.5}$$

where $\mathcal{M}$ is the mean potential energy. Since the projectiles exchange energy with the particles contained in the medium, k must be complex. Therefore the potential energy $\mathcal{M}$ is also complex :

$$\mathcal{M} = V + i\,W \quad . \tag{II.6}$$

In a finite medium, the potential $\mathcal{M}$ is an operator, called the optical-model potential. The present lectures are devoted to the real part V and to the imaginary part W of the optical-model potential for protons and neutrons.

II.c. Empirical properties

The optical theorem states that the total cross section for neutrons is proportional to the imaginary part of the forward scattering amplitude. If the target nucleus would be completely opaque to the incoming beam, the total cross section would be a monotonically decreasing function of the neutron energy. Figure 1 shows that this model disagrees with the experimental data : the measured total cross sections present oscillations when plotted versus energy. These oscillations can be interpreted as arising from the interference between the part of the incoming wave which passes through the nucleus and the part which goes around it (Ramsauer effect), see fig. 2. They provide a direct

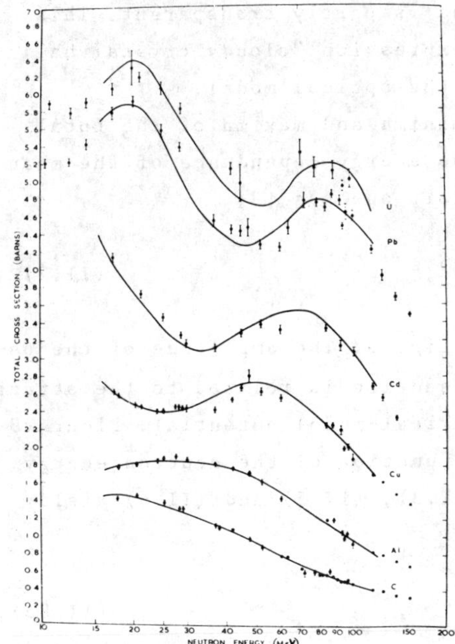

Fig. 1. *Taken from ref.* [3]. *Dependence upon energy of the total cross section for neutrons on C , Al , Cu , Cd , Pb and U .*

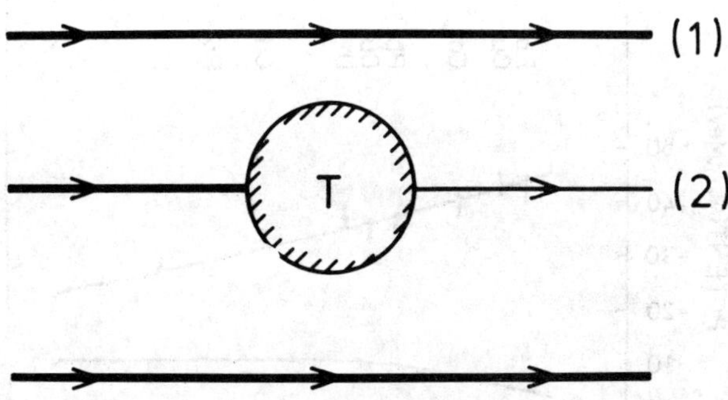

Fig. 2. The optical theorem relates the total cross section to the forward scattering amplitude. Minima and maxima occur in the total cross section when the phase difference between the part (2) of the beam which goes through the nucleus and that (1) which goes around it is a even multiple of π *(Ramsauer effect).*

evidence that the target nucleus is partly transparent. This property is the origin of the expression "cloudy crystal ball model" which has been used for the optical model.

From the location of the minima and maxima of the total cross section one can derive the energy dependence of the mean potential energy. In heavy nuclei, one has [3]

$$V \approx (-50 + 0.3\,E)\ \text{MeV} \quad , \qquad \text{(II.7)}$$

for $20\ \text{MeV} < E < 80\ \text{MeV}$, see fig. 3. The amplitude of the oscillations of the total cross section is related to the strength of the imaginary part of the optical-model potential. Figure 3 shows that it is an increasing function of the neutron energy.

Since $W_o \ll E - V$, eqs. (II.3), (II.5) and (II.6) yield ($V < 0$, $W < 0$)

$$\hbar k_R = mv = [2m(E-V)]^{\frac{1}{2}} \quad , \qquad \text{(II.8a)}$$

$$\hbar k_I = -W/v \quad , \qquad \text{(II.8b)}$$

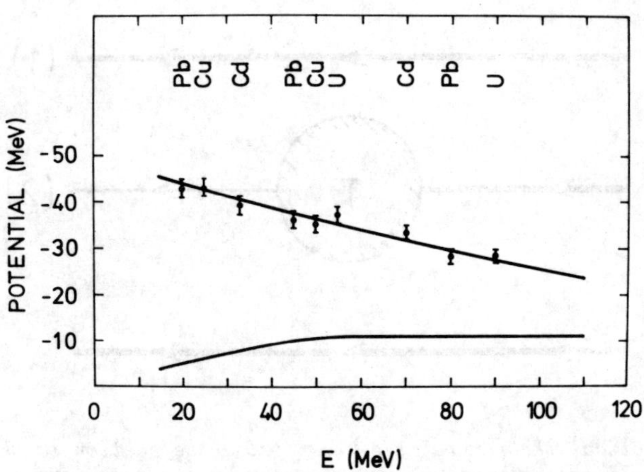

Fig. 3. Adapted from ref. [3]. Dependence upon neutron energy of the strength of the real (upper line) and imaginary (lower curve) parts of the optical-model potential, as derived from the analysis of the total cross sections shown in fig. 1.

where v is the velocity that a nucleon with momentum k_R would have in free space. The mean free path ΔL of a nucleon in the medium is given by eq. (II.4)

$$\Delta L = -\hbar v / 2W \quad . \tag{II.9a}$$

The mean lifetime $\Delta\tau$, i.e. the time which elapses before a collision takes place, is equal to

$$\Delta\tau = \Delta L / v \quad . \tag{II.9b}$$

The uncertainty relation $\Delta E . \Delta\tau = \hbar$ yields the corresponding energy spread :

$$\Delta E = -2W \quad . \tag{II.9c}$$

This quantity ΔE essentially measures the energy range in which the single-particle strength of the incoming nucleon remains located.

II.d. Dispersion

Figure 3 shows that the nuclear medium is dispersive, i.e. that the average potential depends upon the nucleon energy. According to eq. (II.7), the mean potential energy is approximately a linear function of energy in the energy interval 20 MeV < E < 80 MeV . In that domain one can write

$$V(E) \approx V^0 + \left(1 - \frac{m^*}{m}\right) E \quad , \tag{II.10a}$$

where $V^0 \approx -50$ MeV while

$$m^*/m \approx 0.7 \quad . \tag{II.10b}$$

The quantity m^* is called the "effective mass".

In a dispersive medium, the velocity of the projectile should be identified with the group velocity v_g which reads

$$v_g = \hbar^{-1} \frac{dE}{dk} = \frac{\hbar k}{m^*} \quad ; \qquad \text{(II.11)}$$

equations (II.5) and (II.11) give

$$\frac{m^*}{m} = 1 - \frac{dV(E)}{dE} \quad . \qquad \text{(II.12)}$$

This relation can be considered as a more general definition of the effective mass m^* ; it is valid even when m^* depends upon the nucleon energy, i.e. when $V(E)$ is not a linear function of E .

II.e. Main sources of absorption

As in the case of optics, the imaginary part W of the mean field accounts for the fact that a fraction of the incoming beam excites the target, and therefore disappears from the elastic channel. In the nuclear case, the nucleon-nucleon collisions lead to the formation of a compound nucleus on the one hand, and to direct reactions on the other hand. Let us illustrate this with a two-channel model. The entrance (elastic) channel is denoted by p and the inelastic channel by q . The coupled equations read

$$(E - H_{pp})\chi_p = V_{pq}\,\chi_q \quad , \qquad \text{(II.13a)}$$

$$(E - \varepsilon_q - H_{qq})\chi_q = V_{qp}\,\chi_p \quad . \qquad \text{(II.13b)}$$

Here, H_{pp} is the Hamiltonian in the elastic channel p in the absence of coupling, ε_q is the threshold energy of channel q and V_{qp} is the coupling between the two channels. Solving eq. (II.13b) for χ_q and substituting the result in eq. (II.13a), we obtain $(V_{pq} = V_{qp})$

$$(E - H_{eff})\chi_p = 0 \quad , \qquad \text{(II.14a)}$$

where the effective Hamiltonian $H_{eff}^{(p)}$ in channel p reads [4]

$$H_{eff}^{(p)} = H_{pp} + V_{pq} \frac{1}{E - \varepsilon_q - H_{qq} + i\eta} V_{qp} \quad . \qquad \text{(II.14b)}$$

The (+ iη) in the energy denominator on the right-hand side of eq. (II.14b) derives from the property that in channel q the wave is purely outgoing.

In eq. (II.14b) one can write the unit operator in the form

$$1 = \int_0^\infty dE' \, |\varphi_{E'}^{(q)}><\varphi_{E'}^{(q)}| + \sum_n |\varphi_n^{(q)}><\varphi_n^{(q)}| \quad , \qquad \text{(II.15)}$$

where one has introduced the complete set $\{\varphi^{(q)}\}$ of eigenstates of H_{qq} with eigenvalues λ_n et E', respectively. The right-hand side of eq. (II.14b) then reads

$$H_{eff}^{(p)} = H_{pp} + \int_0^\infty dE' \frac{|\varphi_{E'}^{(q)}><\varphi_{E'}^{(q)}|}{E - \varepsilon_q - E' + i\eta} + \sum_n \frac{|\varphi_n^{(q)}><\varphi_n^{(q)}|}{E - \varepsilon_q - \lambda_n} \quad . \qquad \text{(II.16)}$$

The operator $H_{eff}^{(p)}$ is a many-body operator. It depends upon the (A+1) coordinates of the projectile and of the target nucleons. For a qualitative discussion, one can assume that the mean potential M felt by the projectile is obtained by taking the expectation value of $H_{eff}^{(p)} - T_{pp}$ in the target ground state $|\phi_o)$. Here T_{pp} denotes the sum of the kinetic energy operators. One gets then

$$M = (\phi_o|V_{pp}|\phi_o) + \int_0^\infty dE' \frac{\rho_{E'}^{(q)}(\vec{r}') \, \rho_{E'}^{(q)}(\vec{r})}{E - \varepsilon_q - E' + i\eta} + \sum_n \frac{\rho_n^{(q)}(\vec{r}') \, \rho_n^{(q)}(\vec{r})}{E - \varepsilon_q - \lambda_n} \, , \qquad \text{(II.17)}$$

where

$$\rho_n^{(q)}(\vec{r}) = (\phi_o|\varphi_n^{(q)}> \qquad \text{(II.18a)}$$

$$\rho_{E'}^{(q)}(\vec{r}) = (\phi_o|\varphi_{E'}^{(q)}> \qquad \text{(II.18b)}$$

are form factors.

The first term on the right-hand side of eq. (II.17) can be identified with the Hartree-Fock approximation. The second term is due to direct inelastic scattering; it is complex and depends smoothly upon energy. The third term corresponds to the feeding of resonances, i.e. of quasi-bound states $\varphi_n^{(q)}$. It is real and is a wild function of energy : it has poles at $E = \lambda_n + \varepsilon_q$, i.e. for real values of the energy.

The optical-model potential $M(E)$ is defined by requiring that $M(E)$ yields the energy-average of the scattering function $S(E)$. It is obtained by replacing on the right-hand side of eq. (II.17) E by $E + iI$, where I is the energy averaging interval. The imaginary part of the optical-model potential is then given by

$$W(E) \approx -\pi \rho_{E-\varepsilon_q}^{(q)}(\vec{r}') \rho_{E-\varepsilon_q}^{(q)}(\vec{r}) - I \sum_n \frac{\rho_n^{(q)}(\vec{r}') \rho_n^{(q)}(\vec{r})}{(E-\varepsilon_q-\lambda_n)^2 + I^2} \qquad \text{(II.19)}$$

This explicitly exhibits the two main sources of absorption, i.e. the direct feeding of the inelastic channel on the one hand and the formation of compound nuclear resonances on the other hand. The model can be extended to an arbitrary number of inelastic channels.

The coupled-channel description encounters difficulties in properly taking the antisymmetrization into account. The latter is best handled by using the quantum field approach to the many-body problem. Then one finds that the right-hand side of eq. (II.17) should include an additional term, sometimes called the "correlation correction" to the Hartree-Fock approximation. This correction is real for positive values of E but is complex for negative values of E . This is opposite to the properties of the so-called "polarization correction" which is given by the sum of the second and third terms on the right-hand side of eq. (II.17). Indeed, this polarization correction is complex for positive values of E (see eq. (II.19)). The polarization correction accounts for the existence of a spreading width ΔE (see eq. (II.9c)) for nucleons with positive energy. The correlation correction accounts for the spreading width for hole states, i.e. for the nucleons in the target [5].

III. CALCULATIONS OF THE OPTICAL-MODEL POTENTIAL

The optical-model potential plays a basic role in the analysis of most nuclear reactions. Much effort has therefore been devoted to its theoretical investigation. Many calculations are based on effective nucleon-nucleon interactions analogous to those used in the Hartree-Fock approximation. Recent examples can be found in refs. [6-8]. Below, we only survey calculations based on so-called "realistic" nucleon-nucleon potentials, i.e. on potentials which reproduce the nucleon-nucleon scattering phase shifts between zero and several hundreds MeV. These calculations have the merit of reducing the number of adjustable parameters. They face the difficulty that the realistic nucleon-nucleon potentials have a strong short range repulsion at small internucleon distances. The correlated basis function approach has recently been used for handling this difficulty [9]. Most calculations adopt another method which is based on an idea due to Brueckner and that we survey below.

III.a. Nuclear matter

Nuclear matter is a hypothetical uniform medium which has an equal number of neutrons and of protons and where no Coulomb interaction exists. Nuclear matter is characterized by its density ρ , i.e. the number of nucleons per unit volume.

In the independent particle model of nuclear matter, each nucleon is described by a plane wave $\langle\vec{r}|\vec{k}\rangle = \exp(i\,\vec{k}.\vec{r})$ (we henceforth set $\hbar = 1$) . In the ground state all momenta $j = |\vec{j}|$ are occupied up to a maximum value k_F called the "Fermi momentum" and related to the density by

$$\rho = \frac{2}{3\pi^2} k_F^{\,3} \quad . \tag{III.1}$$

III.b. Impulse approximation

Let v_{NN} denote a realistic nucleon-nucleon potential. The transition matrix which describes the scattering between a nucleon with momentum $\vec{k}$ and a nucleon with momentum $\vec{j}$ is the solution of the integral equation

$$t[w] = v_{NN} + v_{NN} \frac{1}{w - h_o + i\eta} t[w] \quad , \tag{III.2}$$

where

$$w = (k^2 + j^2)/(2m) \tag{III.3}$$

is the "starting energy" while h_o is the kinetic energy operator

$$h_o|\vec{a},\vec{b}\rangle = [(a^2 + b^2)/(2m)]|\vec{a},\vec{b}\rangle \quad . \tag{III.4}$$

Let us put a nucleon with momentum k on top of the nuclear matter ground state. At intermediate energy, i.e. for $k \gg k_F$ the mean potential felt by this nucleon is given by the following "impulse approximation"

$$M^{(IA)} = \sum_{j<k_F} \langle\vec{k},\vec{j}|t[w]|\vec{k},\vec{j} - \vec{j},\vec{k}\rangle \quad . \tag{III.5}$$

In the impulse approximation, the bombarding energy E can be identified with $k^2/2m$. Since $k \gg k_F$, one can neglect the

Fermi motion inside the target, i.e. set $j = 0$ in the matrix element on the right-hand side of eq. (III.5). Then one obtains

$$M^{(IA)} = \frac{\pi}{2m} \rho (3 f_1 + f_0) , \tag{III.6}$$

where f_T is the forward scattering amplitude in the nucleon-nucleon channel with isospin T. One has

$$f_T = (2ik)^{-1} \sum_{L,J} (2T+1) [\exp (2i\delta_{LJT}) - 1] , \tag{III.7}$$

where δ_{LJT} is the nucleon-nucleon scattering phase shift with standard notation.

The Hartree-Fock approximation reads

$$M^{(HF)} = \sum_{j<k_F} \langle \vec{k},\vec{j} | v_{eff} | \vec{j},\vec{k} - \vec{k},\vec{j} \rangle , \tag{III.8}$$

where v_{eff} is an effective nucleon-nucleon interaction (e.g. of the Skyrme type). The formal analogy between the Hartree-Fock approximation (III.8) and the impulse approximation (III.5) leads one to identify the effective interaction with the nucleon-nucleon transition matrix. This is only valid at medium energy and cannot be extrapolated to low and a fortiori to negative energies. In order to exhibit the main physical effects which are omitted in the impulse approximation, let us solve the integral equation (III.2) in second order in the strength of the nucleon-nucleon interaction v_{NN} and let us omit antisymmetrization. The impulse approximation then yields

$$M^{(IA)} \approx \sum_{j<k_F} \langle \vec{k},\vec{j} | \{ v_{NN} + 2m\, v_{NN} \sum_{\vec{a},\vec{b}} \frac{|\vec{a},\vec{b}\rangle\langle\vec{a},\vec{b}|}{k^2 + j^2 - (a^2+b^2)} v_{NN} \} | \vec{k},\vec{j} \rangle . \tag{III.9}$$

The first term contained inside the curly brackets in eq. (III.9) yields a contribution analogous to the Hartree-Fock approximation (III.8), except that the realistic nucleon-nucleon interaction v_{NN} takes the place of the effective interaction v_{eff}. The contribution of this first term is certainly very different from the full impulse approximation since v_{NN} is singular at short distance in contrast to the transition matrix. The second term contained inside the curly brackets should therefore at least partly cancel this singular behaviour. This second term takes into account the fact that the two nucleons with momenta $\vec{k}$ and $\vec{j}$ can collide, and thereby excite a two particle $(\vec{a},\vec{b})$ -one hole $(\vec{j})$ configuration. This is depicted in fig. 4. Note, however, that in eq. (III.9) the sum over $\vec{a}$ and $\vec{b}$ is unrestricted. This omits the role of the Pauli exclusion principle which requests that $\vec{a}$ and $\vec{b}$ should both be unoccupied momentum states.

III.c. Brueckner-Hartree-Fock approximation

In order to take the Pauli exclusion principle into account, Brueckner introduced an operator $g[w]$ called Brúeckner's reaction matrix and defined by the following integral equation

$$g[w] = v_{NN} + v_{NN} \frac{P}{w - h_o + i\eta} g[w] \quad , \qquad \text{(III.10)}$$

where the projection operator

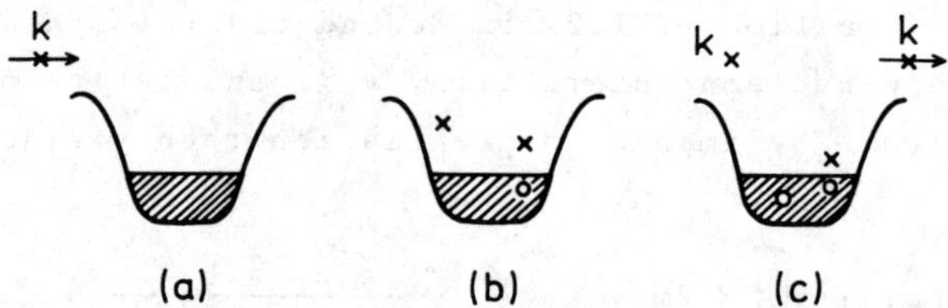

Fig. 4. In the initial state, a nucleon with momentum k is put on top of the uncorrelated ground state (a) . The nucleon-nucleon interaction leads to the excitation of two particle-one hole configurations (b) and subsequently of more complicated configurations. In reality the ground state is correlated and contains two particle-two hole admixtures. Configurations like the one depicted in (c) are forbidden; this is taken into account by the "correlation correction" to the Hartree-Fock approximation.

$$P = \sum_{a,b>k_F} |\vec{a},\vec{b}><\vec{a},\vec{b}| \qquad (III.11)$$

takes the Pauli exclusion principle into account.

The *Brueckner-Hartree-Fock approximation* to the single-particle field is given by

$$M^{(BHF)}(k;E) = \sum_{j<k_F} <\vec{k},\vec{j}|g[E+e(j)]|\vec{k},\vec{j} - \vec{j},\vec{k}> \quad . \qquad (III.12)$$

Note that $M^{(BHF)}(k;E)$ is a function of two variables, namely k and E. However, these variables are related by

$$E = k^2/(2m) + V^{(BHF)}(k;E) \qquad (III.13)$$

inasmuch as the optical model is valid. Here, $V^{(BHF)}$ is the real part of $M^{(BHF)}$. The solution of eq. (III.13) is denoted by $e(k)$ and appears on the right-hand side of eq. (III.12); this is Brueckner's self-consistency requirement.

In most works on the optical-model potential, one also chooses the operator h_o in a self-consistent way, i.e. one replaces eq. (III.4) by

$$h_o|\vec{a},\vec{b}> = [e(a) + e(b)]|\vec{a},\vec{b}> \quad . \qquad (III.14)$$

The physical origin of the self-consistent choice is that it takes into account binding corrections, i.e. the fact that the (bound or unbound) nucleons feel the average potential [10,11].

The imaginary part of the transition matrix (III.2) differs from zero for all values of k. In contrast, the imaginary part of $M^{(BHF)}$ differs from zero only for $E > \varepsilon_F$, where ε_F is the Fermi energy

$$\varepsilon_F = k_F^2/(2m) + V^{(BHF)}(k_F,\varepsilon_F) \quad . \qquad (III.15)$$

The Fermi energy is the binding energy of the least bound single-particle state in the independent particle model of the target.

III.d. Reaction matrix and transition matrix

The reaction matrix and the transition matrix are defined by the integral equations (III.10) and (III.2), respectively. The operator $g[w]$ is more complicated than $t[w]$ mainly because it depends upon one additional variable, namely k_F or, equivalently, the density ρ of the medium. This density-dependence arises from the Pauli operator P and from the binding corrections. Note that $g[w]$ becomes identical to $t[w]$ in the low density limit i.e. for $k_F \to 0$.

The operators $g[w]$ and $t[w]$ can be used as effective interactions for calculating the optical-model potential. They can also be used as effective interactions in the distorted wave Born approximation for evaluating the inelastic scattering cross section. It is therefore useful to construct parametric expressions for these operators. This can be performed in the space or in the momentum representation.

In the space representation, one usually approximates an interaction between nucleons 1 and 2 by an expression of the form

$$v(1,2) \approx v_{st}(1,2) + v_{LS}\,\vec{L}_{12}\cdot\vec{S}_{12} + v_{tens}\,T_{12} \quad , \qquad \text{(III.16)}$$

where $\vec{L}$ and $\vec{S}$ are the orbital and total spin operators and T_{12} is the tensor operator. Even the static part $v_{st}(1,2)$ of the interaction is a complicated object since one can write

$$v_{st}(1,2) = v_c(r_{12}) + v_\sigma(r_{12})\,\vec{\sigma}_1\cdot\vec{\sigma}_2 + v_\tau\,\vec{\tau}_1\cdot\vec{\tau}_2$$

$$+ v_{\sigma\tau}(r_{12})\,(\vec{\sigma}_1\cdot\vec{\sigma}_2)\,(\vec{\tau}_1\cdot\vec{\tau}_2) \quad . \qquad \text{(III.17)}$$

Let t_c denote the central part of the nucleon-nucleon transition matrix, the matrix element

$$t_c^0 = \langle\vec{k},0|t_c[E]|\vec{a},\vec{b}\rangle \qquad \text{(III.18a)}$$

only depends upon the modulus of the momentum transfer

$$\vec{Q} = (\vec{a} - \vec{b}) - \vec{k} \qquad \text{(III.18b)}$$

and upon $E = k^2/2m$. The full curve in fig. 5 shows the dependence of $|t_c^0|$ upon Q for $E = 140$ MeV . The dashes in fig. 5 represent the dependence upon Q of the modulus of the reaction matrix element

$$g_c^0 = \langle\vec{k},0|g_c[E]|\vec{a},\vec{b}\rangle \qquad \text{(III.19)}$$

as calculated from two different realistic potentials, for the Fermi momentum $k_F = 1.4\ \text{fm}^{-1}$ which approximately corresponds to the density at the centre of a heavy nucleus. Figure 5 exhibits the quantitative importance of the density dependence of the reaction matrix. Note that g_c^0 depends upon the nucleon-nucleon interaction v_{NN} , while this is not the case for t_c^0 .

One can conclude from fig. 5 that the medium corrections to the calculated effective interaction are important and depend upon the realistic nucleon-nucleon potential on which they

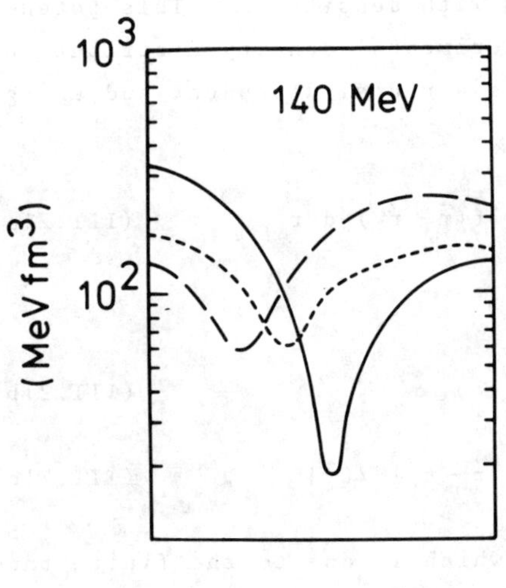

Fig. 5. Dependence upon the momentum Q of the modulus of the central part of the effective interaction at 140 MeV . The full curve corresponds to the free nucleon-nucleon transition matrix [12]. The long and the short dashes both correspond to a Brueckner reaction matrix for the Fermi momentum $k_F = 1.4\ fm^{-1}$ which corresponds to the centre of a heavy nucleus. The long dashes are associated with the Hamada-Johnston potential [13] and the short dashes with the Paris potential [14].

are based. Hence, comparisons [14,15] between specific experimental and calculated cross sections should be judged with caution. It is likely that reliable experimental evidence for medium corrections will emerge only from systematic comparisons between experiment and theory.

III.e. The optical-model potential in nuclei

Two main approaches have been used for calculating the optical-model potential in nuclei from calculations performed in nuclear matter. They correspond to two versions of the local density approximation.

(i) Local density approximation for the optical-model potential

The simplest form of the local density approximation has been proposed and used by the Liège group [16]. It starts from the optical-model potential (see eq. (III.13))

$$\mathcal{M}_\rho(k;E(k)) = M_\rho(E) = V_\rho(E) + W_\rho(E) \qquad \text{(III.20)}$$

calculated in an infinite medium with density ρ. This potential is then folded with the experimental density distribution $\rho(r)$ of the target. This yields a radial-dependent and energy-dependent complex potential :

$$M(r;E) = \int \rho(r_t)\,\frac{M_{\rho(R)}(E)}{\rho(R)}\,f(\vec{r}-\vec{r}_t)\,d^3r \qquad \text{(III.21a)}$$

with

$$\vec{R} = \frac{1}{2}(\vec{r}+\vec{r}_t) \quad , \qquad \text{(III.21b)}$$

$$f(\vec{r}-\vec{r}_t) = (b\sqrt{\pi})^{-3}\exp[-|\vec{r}-\vec{r}_t|^2/t^2] \quad . \qquad \text{(III.21c)}$$

Here, $b \approx 1$ fm is a parameter which is due to the finite range of the effective nucleon-nucleon interaction. The overall strength of the effective interaction is measured by the ratio

$$v_\rho(E) = M_\rho(E)/\rho \quad . \qquad \text{(III.22)}$$

The dependence of $v_\rho(E)$ upon ρ and E has been parametrized in ref. [16], see also ref. [17].

(ii) Local density approximation for the effective interaction

This version of the local density approximation has been proposed by Brieva and Rook [18] and further developed by Von Geramb [13]. In this approach, the reaction matrix g is written in the form (III.16), (III.17) and parametrizations of the functions $v_c(r_{12})$, $v_\sigma(r_{12})$, ... are provided. This interaction is then used as a complex effective interaction in a Hartree-Fock code for computing the projectile-target average potential. In practice, the exchange (Fock) contribution is rendered local by means of a local momentum approximation.

The versions (i) and (ii) of the local density approximation have recently been confronted with the experimental elastic scattering cross sections of protons and neutrons from ^{54}Fe and ^{56}Fe [19]. In order to estimate the accuracy of the theoretical predictions, two adjustable strength parameters λ_V and λ_W have been introduced. The theoretical cross sections have thus been computed from the optical-model potential

$$M_{th}(r;E) = \lambda_V V_{LDA}(r;E) + \lambda_W W_{LDA}r;E) \quad , \qquad \text{(III.23)}$$

where V_{LDA} and W_{LDA} are obtained from the two versions of the local density approximation that have been outlined above. Figures 6a,6b correspond to the local density approximation (i) for the potential and figs. 7a and 7b to the local density approximation (ii). The agreement between the experimental data and the theoretical cross sections is good in both cases. Figures 6b and 7b show that the values of λ_W are closer to unity in the version (i) of the local density approximation. This, however, should not be viewed as an argument for favouring the version (i) over the version (ii). Both of these lead to a surprizingly good agreement with the experimental data.

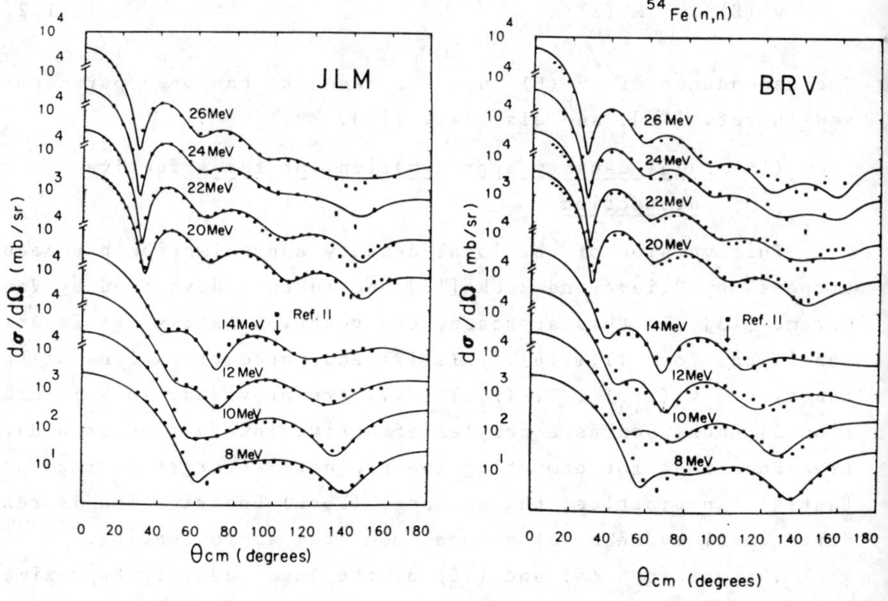

Fig. 6a. Taken from ref. [19]. Comparison between the measured $^{54}Fe(n,n)$ differential cross sections (dots) and theoretical curves calculated from the local density approximation (i) of sect. III.e. The real and the imaginary parts of the potential have been renormalized by the quantities λ_V and λ_W shown in fig. 6b.

Fig. 7a. Taken from ref. [19]. Same as fig. 6a, except that the theoretical curves have been obtained from the local density approximation labelled (ii) in sect. III.e.

Fig. 6b (shown next page). Taken from ref. [19]. Renormalization coefficients λ_V and λ_W associated with the calculated differential cross sections shown in fig. 6a, see eq. (III.23).

Fig. 7b (shown next page). Taken from ref. [19]. Renormalization coefficients λ_V and λ_W associated with the calculated differential cross sections shown in fig. 7a, see eq. (III.23).

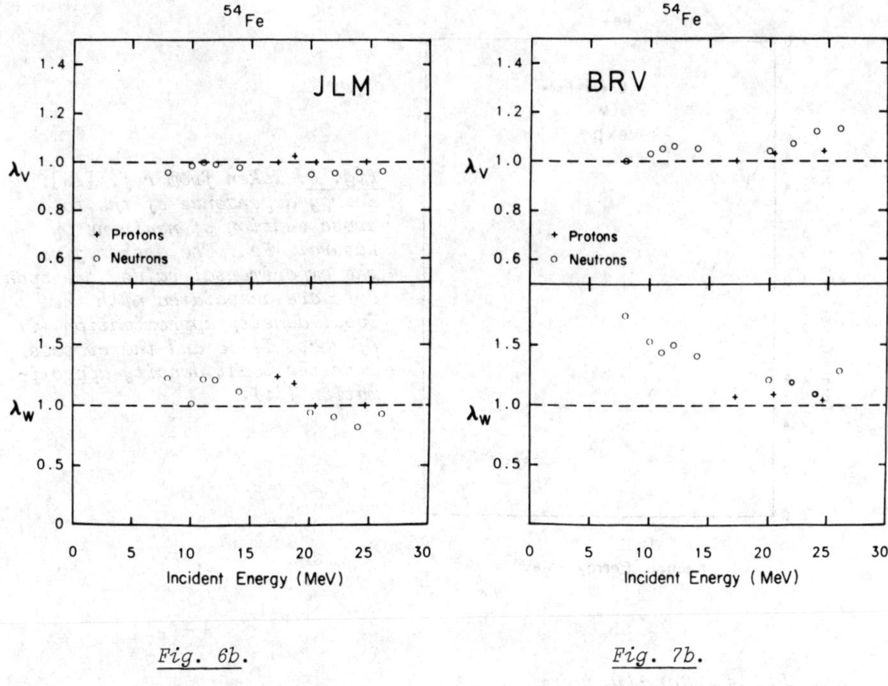

Fig. 6b.

Fig. 7b.

Figure 8 shows a comparison between the empirical cross section of neutrons on natural Fe and the theoretical values computed from the versions (i) (open circles) and (ii) (crosses) of the local density approximation. The fact that version (i) is in better agreement with the measured total cross section is in keeping with the property that it gives a better fit to the elastic differential cross section in the forward direction. Indeed, these two quantities are related by the optical theorem.

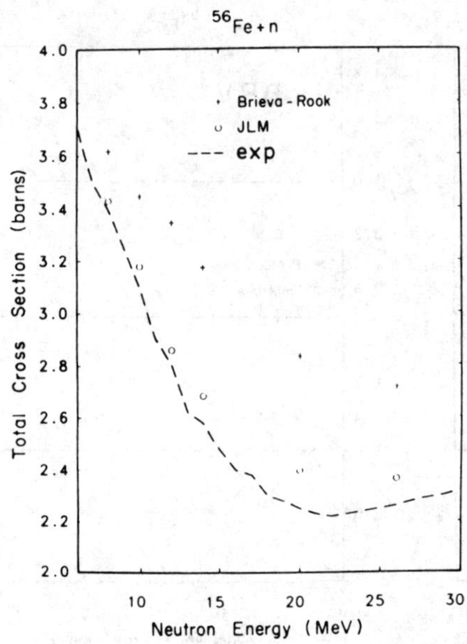

Fig. 8. Taken from ref. [19]. Energy dependence of the total cross section of neutrons by natural Fe . The dashes give the experimental value, the open dots are associated with the local density approximation (i) of sect. III.e and the crosses with the local density approximation (ii).

III.f. Conclusions

From the examples given above one can conclude that rather satisfactory methods have been constructed for calculating the optical-model potential from a realistic nucleon-nucleon interaction. Two main versions of the local density approximation exist. The first one, labelled (i) in sect. III.e, is easy to use and nevertheless yields a satisfactory agreement with empirical optical-model potentials and with elastic and total cross sections. The second one, labelled (ii) in sect. III.e, is more complicated to apply. It does not appear to yield more satisfactory results than method (i) as far as elastic scattering data are concerned. However, it presents the significant advantage of providing a complex effective interaction which can be used for evaluating inelastic scattering cross sections. Much work remains to be accomplished on the theoretical method as well as on the systematic comparison between calculated and measured quantities.

The two versions of the local density approximation make use of empirical information on the density distribution $\rho(r)$ in nuclei. In their present form they do not attempt to calculate $\rho(r)$ from the nucleon-nucleon interaction.

The theoretical approaches that we outlined both yield better agreement with the empirical optical-model potential than microscopic calculations in which nuclear excitations are explicitly taken into account [6,7]. It is conceivable that this success is somewhat accidental [20].

IV. THE OPTICAL-MODEL POTENTIAL AT NEGATIVE ENERGY

In sects. II and III we dealt with empirical and theoretical properties of the average nucleon-nucleus potential for nucleons which are scattered by a target and have thus a positive energy. We now turn to the case of negative energy nucleons, i.e. to the interpretation of the bound single-particle states of the average nucleon-nucleus potential.

No physical reason exists for treating on a different footing the average nucleon-nucleus potential for scattering nucleons on the one hand and for bound nucleons on the other hand [21]. In the first case the average field describes the main properties of the system obtained by adding a nucleon to the system. In the second case the average field describes the main properties of the system obtained by taking one nucleon out from the system. As illustrated by fig. 9, the transition between these two domains is provided by the weakly bound single-particle states which belong to the valence shells which are located on each side of the Fermi energy.

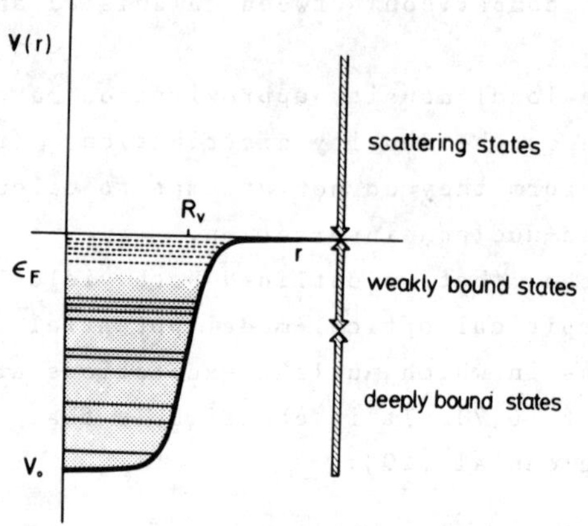

Fig. 9. *Taken from ref.* [22]. *Schematic representation of the three main energy domains for single-particle states. They correspond to the scattering states, the weakly bound states and the deeply bound states.*

IV.a. Weakly bound single-particle states

The single-particle states which belong to the main valence shell located above the Fermi energy of ^{208}Pb are strongly excited in stripping reactions, e.g. ^{208}Pb(d,p) or ^{208}Pb(d,n). Similarly, the single-particle states which belong to the main valence shell located below the Fermi energy are strongly excited in pick-up reactions, e.g. (p,d) or (t,α) .

The resulting "experimental" values of the single-neutron energies in ^{208}Pb are plotted in the left-hand column in fig. 10. There, the middle and the right columns show single-particle energies derived from typical Hartree-Fock potentials. These energies approximately correspond to those which should be computed from a Woods-Saxon potential whose depth depends linearly upon energy according to eq. (III.10a), with $m^*/m = 0.75$ (SIII) and 0.5 (SV) , respectively.

The most conspicuous failure of the Hartree-Fock approximation is that the average energy separation between the two valence shells is two or three MeV larger than the experimental value. This points to the existence of sizeable corrections to the Hartree-Fock approximation in the vicinity of the Fermi surface.

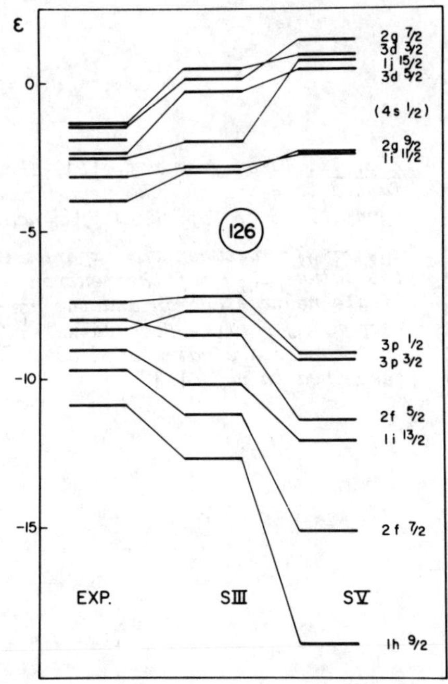

Fig. 10. Taken from ref. [23]. *The experimental values of the single-neutron energies in* ^{208}Pb *are plotted on the left-hand side. The Hartree-Fock energies shown in the middle and in the right-hand side columns have been computed from the Skyrme III and from the Skyrme V effective interactions, respectively.*

Let us identify the Fermi energy ε_F with the energy located half-way between that of the last occupied subshell and that of the first unoccupied subshell. In practice, the strength of the effective interaction used in a Hartree-Fock calculation is adjusted in such a way that the experimental value of ε_F is reproduced. This is apparent in fig. 10. A closer look at fig. 10 also shows that the difference between a calculated single-particle energy $\varepsilon^{(o)}_{n\ell j}$ and the corresponding experimental value $\varepsilon_{n\ell j}$ tends to increase when the distance of $\varepsilon_{n\ell j}$ from the Fermi energy increases. This is exhibited in fig. 11.

One global quantity which characterizes a potential well $V(r;E)$ is the volume integral per nucleon $J_V(E)$:

$$J_V(E) = A^{-1} \int V(r;E) \, d^3r \quad . \qquad \text{(IV.1)}$$

This quantity is represented by the full dots in fig. 12.

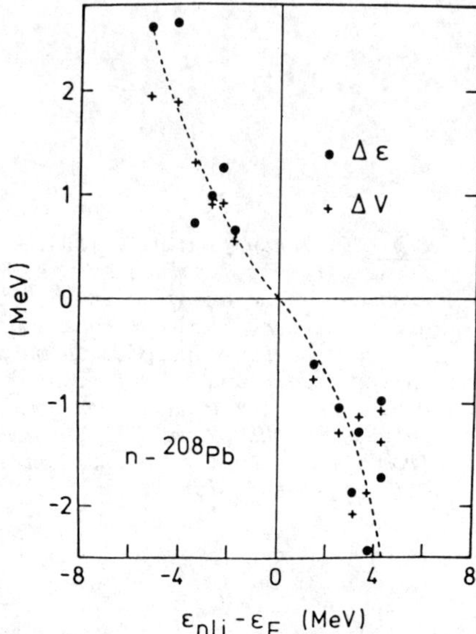

Fig. 11. Taken from ref. [5]. The full dots represent the dependence upon $\varepsilon_{n\ell j} - \varepsilon_F$ of the difference $\varepsilon_{n\ell j} - \varepsilon^{(o)}_{n\ell j}$ between the experimental value $\varepsilon_{n\ell j}$ of the neutron single-neutron energy and the corresponding Hartree-Fock value $\varepsilon^{(o)}_{n\ell j}$. The crosses are calculated as described in ref. [5].

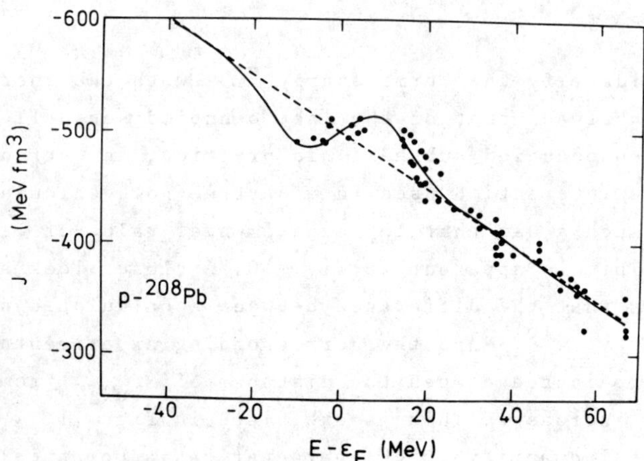

Fig. 12. Adapted from ref. [24]. The dots show the energy dependence of the volume integral per nucleon of the real part of a potential well which reproduces the elastic scattering cross section $(E - \varepsilon_F > 10$ MeV$)$ and the experimental single-particle energies $(E - \varepsilon_F < 10$ MeV$)$ in the case of protons in ^{208}Pb. The full curve is drawn by eye and the dashed straight line would be obtained from a Hartree-Fock approximation.

In the Hartree-Fock approximation, the energy dependence of $J_V(E)$ would be linear, see the dashes in fig. 12. This confirms that there exist significant corrections to the Hartree-Fock approximation in the vicinity of the Fermi energy. In recent years, much effort has been devoted to the evaluation of these corrections. An example is shown in fig. 13.

At positive energy, the existence of corrections to the (real) Hartree-Fock approximation is made evident by the necessity of introducing an imaginary part in the optical-model potential. Equation (II.9c) indicates that this imaginary part is related to the size of the energy domain in which the single-particle configuration can be found. Similar concepts remain meaningful at negative energy. Indeed, the existence of a residual interaction implies that the whole single-particle strength is not contained in a single state of the ^{207}Pb nucleus. At negative energy, the calculation of the corrections to the Hartree-Fock approximation therefore amounts to evaluating the distribution of the single-particle strength. One example is shown in fig. 14.

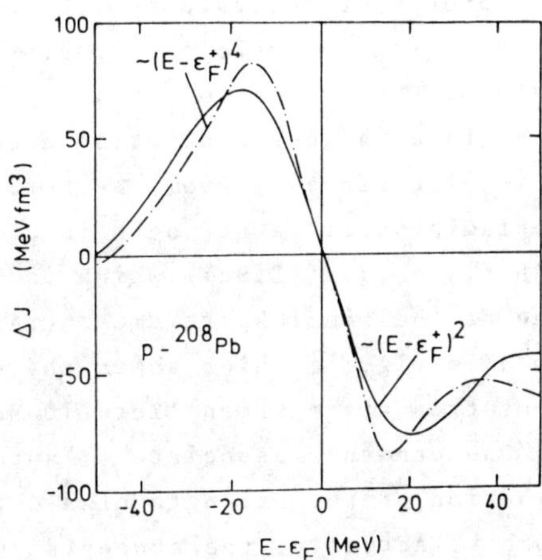

Fig. 13. Taken from ref. [5]. Dependence upon $E-\varepsilon_F$ of the volume integral per nucleon of the calculated corrections to the Hartree-Fock approximation as evaluated from two versions of the dispersion relation approach.

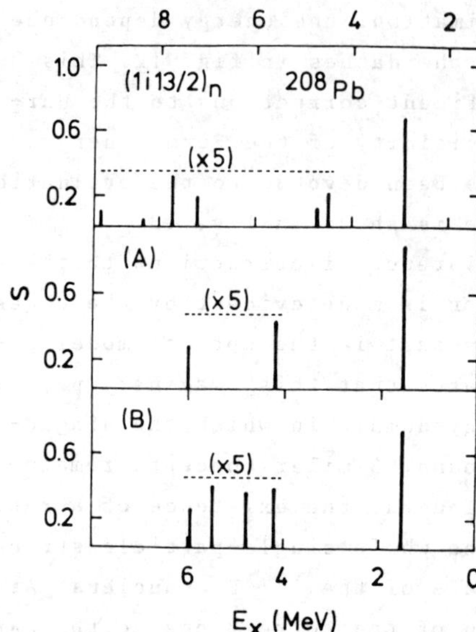

Fig. 14. Taken from ref. [25]. Calculated (top) and measured (middle [26] and bottom [27]) distribution of the 1i13/2 single-particle strength in ^{207}Pb. Note that the small spectroscopic factors have all been multiplied by five. The experimental single-particle energy plotted in fig. 10 corresponds to the location of the state which has the large spectroscopic factor.

IV.b. Deeply bound state

The typical property of the valence subshells is that a large fraction of the single-particle strength is located in one low-lying state of the ^{207}Pb or of the ^{209}Pb nucleus; the remaining bits lie at higher excitation energies. In the case of shells which lie far away from the Fermi energy one expects that the single-particle strength is distributed in a domain of size ΔE (see eq. (II.9c)) which is centered on the Hartree-Fock value of the single-particle energy. This expectation is confirmed in fig. 15 which shows the example of the (1h11/2) single-neutron state which lies 10 MeV away from the Fermi energy. One can thus associate a "spreading width" ΔE to the distribution of the strength of a deeply bound single-particle orbit. Actually, the concepts outlined in sect. II.c apply to this situation as well as to the case of a positive energy nucleon.

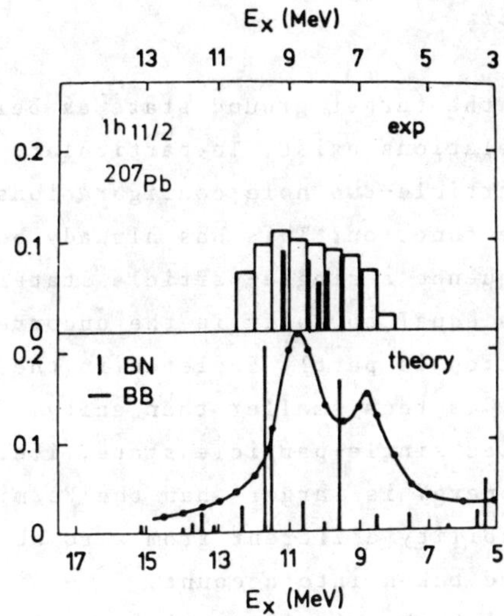

Fig. 15. Taken from ref. [28]. Measured (top, refs. [26,29]) and calculated (bottom, refs. [25,30]) distribution of the (1h11/2) single-neutron strength in ^{207}Pb .

Figure 16 exhibits the striking property that the strength W of the imaginary part of the average nucleon-nucleus potential takes approximately symmetric values on each side of the Fermi energy.

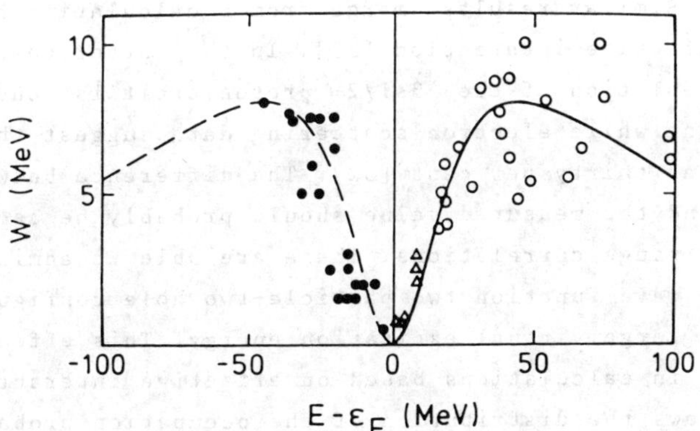

Fig. 16. Taken from ref. [31]. The dots give the strength of the imaginary part of the average nucleon-nucleus potential as compiled from data for nuclei whose mass number is contained in the range $12 \leqslant A \leqslant 60$.

IV.c. Occupation probability

Until now we have treated the target ground state as being uncorrelated. In reality, correlations exist. In particular, these correlations admix two particle-two hole configurations in the actual ground state wave function. This has already been depicted in fig. 4. As a consequence a single-particle state whose occupation probability is equal to unity in the uncorrelated Hartree-Fock parametrization is partly depleted in the actual nucleus. Its probability is thus smaller than unity. Conversely, a normally unoccupied single-particle state, i.e. a single-particle state whose energy is larger than the Fermi energy, has an occupation probability different from zero when nucleon-nucleon correlations are taken into account.

The occupation probability of the single-particle orbits have drawn recent attention because of the claim that some of these can be measured experimentally [32]. The results of two calculations are shown in figs. 17 and 18, respectively. In fig. 17 the dots show the occupation probability which is calculated from a particle-phonon coupling model [33]. The calculated deviation from the uncorrelated distribution is less than 10 per cent. Similar results emerge from a calculation based on the Gogny effective interaction [35]. In the latter case, the calculated depletion of the $3s1/2$ proton orbit is found equal to 11 per cent while electron scattering data suggest that it is as large as thirty per cent [32]. The difference between the calculated and the measured value should probably be ascribed to the short range correlations. These are able to admix in the ground state wave function two particle-two hole configurations which have a large virtual excitation energy. This effect is not included in calculations based on effective interactions. Figure 18 shows the distribution of the occupation probability which emerges from a model which includes short range correlations in a schematic way [5]. The corresponding deviation from the uncorrelated distribution is much larger than the one computed from effective interactions. It is not yet known how this effect can be implemented in nuclear models.

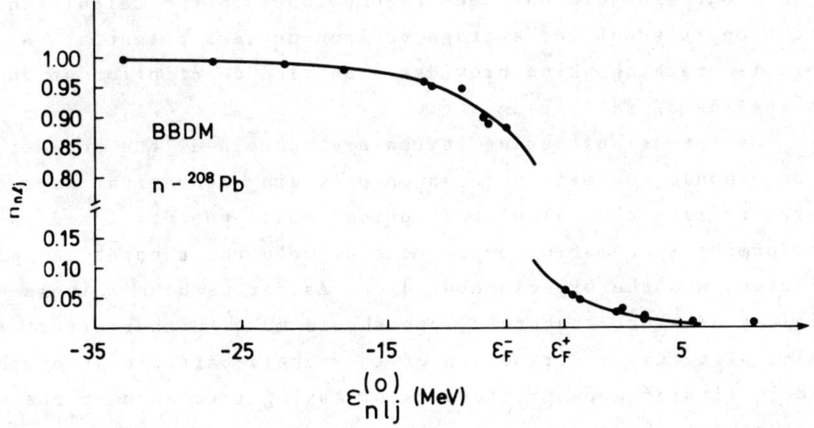

Fig. 17. Taken from ref. [33]. The full dots show the occupation probabilities for neutron orbits in ^{208}Pb as calculated in a particle-vibration model.

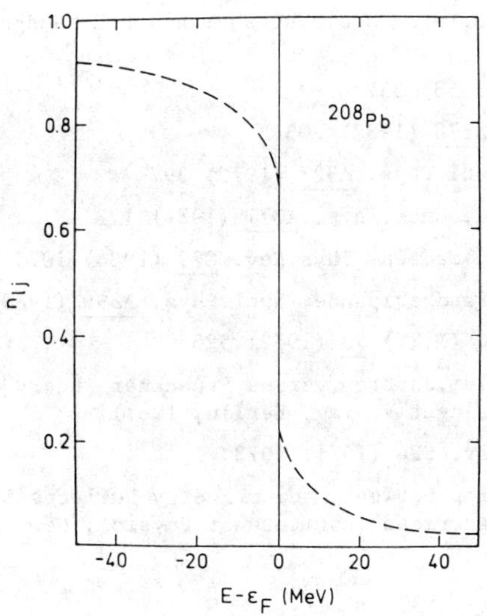

Fig. 18. Taken from ref. [34]. Distribution of the occupation probabilities in a model which schematically takes the short range correlations into account [5].

V. CONCLUSIONS

Great progress has been accomplished in the calculation of the properties of the average nucleon-nucleus potential. A picture is emerging which provides a unified description of bound as well as of scattering states.

As far as scattering states are concerned, the investigations should now mainly focus on providing convenient ways of parametrizing the calculated optical-model potential and on performing systematic comparisons between the theoretical predictions and the experimental data. As far as bound single-particle states are concerned, one should make further effort in calculating the distribution of the single-particle strength and in finding appropriate ways of taking into account the effect of the short range correlations.

REFERENCES

[1] P.E. Hodgson, Nuclear Reactions and Nuclear Structure (Clarendon Press, Oxford, 1971)

[2] G.R. Satchler, Direct Nuclear Reactions (Clarendon Press, Oxford, 1983)

[3] P.H. Bowen, J.P. Scanlon, G.H. Stafford, J.J. Thresher and P.E. Hodgson, Nucl.Phys. 22 (1961) 640

[4] H. Feshbach, Ann.Phys. (N.Y.) 5 (1958) 357

[5] C. Mahaux and H. Ngô, Nucl.Phys. A378 (1982) 205

[6] V. Bernard and Nguyen Van Giai, Nucl.Phys. A327 (1979) 397

[7] A. Bouyssy, H. Ngô and N. Vinh Mau, Nucl.Phys. A371 (1981) 173

[8] H. Dermawan, F. Osterfeld and V.A. Madsen, Phys.Rev. C29 (1984) 1075

[9] S. Fantoni, B.L. Friman and V.R. Pandharipande, Nucl.Phys. A399 (1983) 51

[10] J. Hüfner and C. Mahaux, Ann.Phys. (N.Y.) 73 (1972) 525

[11] C. Mahaux, in "The Many-Body Problem.Jastrow versus Brueckner Theory", R. Guardiola and J. Ros, eds. (Springer Verlag, Berlin, 1981)

[12] W.G. Love and M.A. Franey, Phys.Rev. C24 (1981) 1073

[13] H.V. Von Geramb, in "The Interaction between Medium Energy Nucleons in Nuclei - 1982", H.O. Meyer, ed. (American Institute of Physics, New York, 1983) p. 44

[14] M. Hugi, W. Bauhoff and H.O. Meyer, Phys.Rev. C (in press)

[15] J. Kelly, W. Bertozzi, J.N. Buti, F.W. Hersman, C. Hyde, M.V. Hynes, B. Norum, F.N. Rad, A.D. Backer, G.T. Emery, C.C. Foster, W.P. Jones, D.W. Miller, B.L. Berman, W.G. Love and F. Petrovich, Phys.Rev.Lett. 45 (1980) 2012

[16] J.P. Jeukenne, A. Lejeune and C. Mahaux, Phys.Rev. C16 (1977) 80

[17] J.P. Jeukenne and C. Mahaux, Z.Phys. A302 (1981) 233

[18] F.A. Brieva and J.R. Rook, Nucl.Phys. A291 (1977) 299, 317; ibid. A297 (1978) 206); ibid. A307 (1978) 493

[19] S. Mellema, R.W. Finlay, F.S. Dietrich and F. Petrovich, Phys.Rev. C (in press)

[20] C. Mahaux, in "Microscopic Optical Potentials", H.V. Von Geramb, ed. (Springer Verlag, Berlin, 1979) p. 1

[21] C. Mahaux, in "Common Problems in Low- and Medium-Energy Nuclear Physics", B. Castel, B. Goulard and F.C. Khanna, eds. (Plenum Press Corp., 1979) p. 265

[22] C. Mahaux, P.F. Bortignon, R.A. Broglia and C.H. Dasso, Phys.Reports (submitted)

[23] P. Quentin and H. Flocard, Ann.Rev.Nucl.Sci. 28 (1978) 523

[24] C. Mahaux and H. Ngô, Phys.Lett. 126B (1983) 1

[25] Nguyen Van Giai, in "Proceedings of the 1980 Symposium of Highly Excited States in Nuclear Reactions", H. Ikegami and M. Muraoka, eds. (Osake University, 1980) p. 682

[26] J. Guillot, J. Van de Wiele, H. Langevin-Joliot, E. Gerlic, J.P. Didelez, G. Duhamel, G. Perrin, M. Buenerd and J. Chauvin, Phys.Rev. C21 (1980) 879

[27] H. Langevin-Joliot, E. Gerlic, J. Guillot, M. Sakai, J. Van de Wiele, A. Devaux, P. Force and G. Landaud, Phys.Lett. 114B (1982) 103

[28] S. Galès, in "Proceedings of the 1980 Masurian Summer School" (Mikolajki, Poland)

[29] S. Galès, G.M. Crawley, D. Weber and B. Zwieglinski, Phys.Rev. C18 (1978) 2475

[30] P.F. Bortignon and R.A. Broglia, Nucl.Phys. A371 (1981) 405

[31] C. Mahaux and H. Ngô, Phys.Lett. 100B (1981) 285

[32] B. Frois, J.M. Cavedon, D. Goutte, M. Huet, Ph. Leconte, C.N. Papanicolas, H.-H. Phan, S.K. Platchkov and S.E. Williamson, Nucl.Phys. A396 (1983) 409c

[33] P.F. Bortignon, private communication

[34] C. Mahaux, in "Perspectives in Nuclear Physics at Intermediate Energies" (World Scientific Publ. Comp., Singapore) in press

[35] J. Dechargé and L. Sips, Nucl. Phys. A407 (1983) 1; J. Dechargé and D. Gogny, private communication.

MICROSCOPICAL DESCRIPTION OF COLLISIONS OF LIGHT COMPLEX NUCLEI BASED ON VARIATIONAL METHODS

M.V. Mihailovic

J. Stefan Institute, University "E. Kardelj", Jamova 39,
61000 Ljubljana, Yugoslavia

1. INTRODUCTION

The motivation to develop a microscopic theory for collisions between complex nuclei is the need in some reactions to take properly into account the Pauli principle and details of the structure of colliding nuclei. In this way the exchange of a nucleon or a cluster between the colliding nuclei can be included properly into description. At least this property of the microscopic theory is of great value in the description of collisions of lightest nuclei in the astrophysical situations. Difficulties in describing collisions of ions with a phenomenological projectile – target potential indicates that such a theory would be useful.

The motivation to base the theory of collisions on a variational principle is the belief that variational methods offer significant advantage from the calculational point of view. The variational methods proved extremely effective in studies of bound states of many body systems. However, variational approaches for the description of collisions sometimes provide false predictions, eliminating or controlling these is another motivation for the study.

This lecture contains the description of a microscopic theory of collisions between complex nuclei based on the variational method for calculating parameters of reaction. Scattering ware functions are expressed in the generator coordinate (GC) representation, where the generator coordinate is the distance S between the centres of two shell model potentials, the single particle functions of those are used to describe colliding nuclei separately. The only parameters that appear in the theory are those which describe the effective two – body interaction between nucleons.

In this paper it is to shown:
How the formalism can be used to describe in details collisions of (light) nuclei[1]. The combination of existing variational methods can offer the possibility of avoiding false predictions in most cases.

It will be shown also how the GC basis can be made to include also the scattering boundary conditions which are different than that in the Coulomb scattering.

A recent result[2] is obtained on the source of the false predictions of the whole class of approaches which are most often used and a proposal is given for a new constrained variational principle which eliminates false predictions in the elastic scattering. The open problem of extending this result to many channel collisions will be stated.

Contains: In Ch. 2 the formalism is explained. The GC basis is described in Ch. 3. The emphasis is on the calculation of the GC scattering amplitude in a wide (almost whole) interval of values of the distance between colliding nuclei. Variational principle for scattering are discussed in Ch. 4. Chapter 5 contains an example, the elastic scattering (^{3}H, ^{4}He). The last Chapter 6 contains a discussion on related problems.

2. THE FORMALISM

We consider reactions of the type

$$A_i + B_i \to A_j + B_j, \qquad j \in 1,2,\ldots,N_c \tag{1}$$

where N_c is the number of open channels. The channel index j is specified by the total intrinsic angular momenta s_1 and s_2 of the fragments A_j and B_j, respectively, their z–components ν_1 and ν_2, the relative angular momentum l and its z–component m and the wave number k of the relative motion, and all the other quantum numbers α needed to specify the channel uniquely: $j = (\alpha k l m s_1 \nu_1 s_2 \nu_2)$.

If there is an incoming wave in the j-th channel, the solution of the Schrödinger equation, corresponding to the system (1), has to satisfy the following boundary conditions:

$$|r_j|^{-l+1}\Psi_j \to 0, \qquad |r_j| \to 0, \tag{2}$$

$$\Psi_j \to \mathscr{F}_j + \sum_{j'=1}^{N_c} K_{jj'}\mathscr{G}_{j'}, \qquad |r_j| \to \infty, \tag{3}$$

if the channel j is open, or

$$\Psi_j \to 0, \qquad |r_j| \to \infty,$$

if the channel j is closed. Here $r_j = X^{A_j}_{c.m} - X^{B_j}_{c.m}$ is the relative separation of the centres of mass of the fragments A_j and B_j. The N–body functions $\mathscr{F}$ and $\mathscr{G}$, which are antisymmetric in all the coordinates $x_1, x_2, \ldots, x_{A+B}$ are given by

$$\begin{Bmatrix} \mathscr{F}_j(x) \\ \mathscr{G}_j(x) \end{Bmatrix} = \mathscr{A}\left\{\begin{pmatrix} F_l(k_j|r_j|) \\ G_l(k_j|r_j|) \end{pmatrix} Y_{lm}(\hat{r}_j) i^l (k_j|r_j|)^{-1} \phi_{A_j s_1 \nu_1}(\xi_{A_j}, \sigma_{A_j}) \phi_{B_j s_2 \nu_2}(\xi_{B_j}, \sigma_{B_j})\right\}, \tag{4}$$

where $x = (x_1,\ldots,x_{A+B})$ and F_l and G_l are regular and irregular Coulomb wave functions, respectively. Functions $\phi_{A_j}(\xi_{A_j}, \sigma_{A_j})$, $\phi_{B_j}(\xi_{B_j}, \sigma_{B_j})$ antisymmetric in all their arguments, where

$$\xi_{A_j} = (x_1 - X^{A_j}_{c.m.}, x_2 - X^{A_j}_{c.m.}, \ldots, x_{A_j-1} - X^{A_j}_{c.m.}),$$

$$\xi_{B_j} = (x_{A_j+1} - X^{B_j}_{c.m.}, \ldots, x_{A_j+B_j-1} - X^{B_j}_{c.m.}),$$

describe the states of the fragments A_j and B_j, respectively. The coefficients K_{jj} are the elements of the K–matrix. The spin–isospin variables are denoted by σ_{A_j} and σ_{B_j}.

The many – body wave function Ψ_j is very complicated object as it has to describe well the separated fragments and the composite system. In Ch. 3 a wave function of the GC type is described in detail which offer a basis which is appropriate to many applications with light nuclei.

In our calculations we use the microscopic hamiltonian of the form

$$H = -\frac{\hbar^2}{2m}\sum_{i=1}^{A+B}\nabla_i^2 + \sum_{i<j=1}^{A+B}(V_{ij}^{\mathrm{N}} + V_{ij}^{\mathrm{C}}). \tag{5}$$

Here V_{ij}^N stands for the effective nucleon–nucleon interaction. We usually choose a central potential of the form

$$V_{ij}^{\mathrm{N}} = \{\sum_k V_k \exp[-(r_{ij}/\rho_k)^2]\}(1 - m + mP_{ij}^r), \tag{6}$$

where P_{ij}^r is the space exchange operator. The parametrization of the radial part of the potential in terms of gaussian functions is very convenient, because it allows us to perform analytically the major part of the calculation of matrix elements. For the same reason the Coulomb potential

$$V_{ij}^{\mathrm{C}} = \tfrac{1}{4}(1 + \tau_3(i))(1 + \tau_3(j))\frac{e^2}{4\pi\varepsilon_0|\boldsymbol{r}_i - \boldsymbol{r}_j|} \tag{7}$$

is represented by

$$V_{ij}^{\mathrm{C}} = \tfrac{1}{4}(1 + \tau_3(i))(1 + \tau_3(j))e^2\pi^{-\frac{3}{2}}(2\varepsilon_0 b)^{-1}\int_0^{\infty}\exp[-a^2(\boldsymbol{r}_i - \boldsymbol{r}_j)^2/b^2]\,da.$$

The effective nucleon–nucleon tensor potential is usually chosen of the form

$$V_{ij}^{NT} = \sum_{k=1}^{2}(1 + M^T P^M)V_{ok}^T r_{ij}^2 \exp(-\gamma_k^T/2\, r_{ij}^2)[3\cdot\sigma_i(r_i - r_j)\cdot\sigma_j(r_i - r_j))/r_{ij}^2 - (\sigma_i\,\sigma_j)],$$

where σ_i and σ_j are the spins of the nucleons.

The central problem of microscopic theories of collisions is the feasibility of calculations of reaction parameters. For this reason in all variational calculations the functional and its domain are chosen to lead to linear variational equations for variational parameters. The chosen trial function is of the form

$$\Psi_j^t = \sum_{p=1}^{N_p} c_{jp}\phi_p + \mathscr{F}_j + \sum_{j'=1}^{N_c} K_{jj'}^t \mathscr{G}_{j'},$$

where ϕ_p is a totally antisymmetric and square integrable function of the coordinates $x_1, \ldots, x_{A+B}$. The coefficients c_{jp} and $K_{jj'}^t$, are variational parameters. Functions $\mathscr{F}_j$ and $\mathscr{G}_j$ are given by eq. (4), except that $G_l(k_j|r_j|)$ is regularized at the origin.

The functionals of the Kohn type

$$K[\Psi]_{ij} = K_j^t - \frac{1}{w_{ij}} < \Psi_i | H - E | \Psi_j > , \tag{9}$$

is of the desired form as it contains the squares of the trial functions and their derivatives. In the original Kohn functional, where there is no other variational parameters except c's and K_{ij} (see Ch. 4) the stationary solution is

$$K_{ij} = -\frac{1}{w_{ij}}\{(\Delta_{\mathscr{F}\mathscr{F}})_{ij} - [(\Delta^{T}_{\mathscr{G}\mathscr{F}})(\Delta_{\mathscr{G}\mathscr{G}})^{-1}(\Delta_{\mathscr{G}\mathscr{F}})]_{ij}\}, \tag{10}$$

where w_{ij} is the Wronskian of the functions $\mathscr{F}_i$ and $\mathscr{G}_i$. The operator Δ is defined by

$$\Delta = (H-E)-(H-E)\sum_{p}\sum_{p'}|\phi_p\rangle\langle\phi_p|(H-E)^{-1}|\phi_{p'}\rangle\langle\phi_{p'}|(H-E), \tag{11}$$

and Δ^T is its transpose. The matrix elements of these operators are denoted by

$$(\Delta_{\mathscr{F}\mathscr{F}})_{ij} = \langle\mathscr{F}_i|\Delta|\mathscr{F}_j\rangle, \quad (\Delta_{\mathscr{F}\mathscr{G}})_{ij} = \langle\mathscr{F}_i|\Delta|\mathscr{G}_j\rangle, \quad (\Delta_{\mathscr{G}\mathscr{F}})_{ij} = \langle\mathscr{G}_i|\Delta|\mathscr{F}_j\rangle,$$
$$(\Delta_{\mathscr{G}\mathscr{G}})_{ij} = \langle\mathscr{G}_i|\Delta|\mathscr{G}_j\rangle. \tag{12}$$

The K–matrix is explicitely symmetric and the corresponding S–matrix is unitary.

The result (10) shows that the calculation of reaction parameters reduces to the calculation of the integrals (12). A very convenient feature is that the asymptotic functions $\mathscr{F}_i$, $\mathscr{G}_i$ corresponding different channels are not coupled together.

Some conclusion is valid when that functionals of the Kohn type are used.

Same conveniencies are valid when other functionals of the Kohn type (see Ch. 4) are used.

3. THE GENERATOR COORDINATE FORMALISM

3.1. The two–centre generator coordinate basis

During the collision of two complex nuclei most nucleons remain in their fragments. For this reason it seems useful to work with generator coordinate two–centre wave functions. They are constructed in the following way:

$$\phi_{A_jB_j}(x,S) = \mathscr{A}\{\phi_{A_j}(x_1,\ldots,x_{A_j},S_1)\phi_{B_j}(x_{A_j+1},\ldots,x_{A_j+B_j},S_2)\}$$
$$S = S_1 - S_2, \qquad S_1 = \frac{B_j}{A+B}S, \qquad S_2 = \frac{-A_j}{A+B}S, \tag{13}$$

where $\phi_{A_j}(x_1,\ldots,x_{A_j},S_1)$ and $\phi_{B_j}(x_{A_j+1},\ldots,x_{A_j+B_j},S_2)$ are linear combinations of Slater determinants; the corresponding single–particle functions are eigenfunctions of the harmonic oscillators with centres at S_1, S_2, respectively, and with the oscillator parameter β. In order to avoid the problem of spurious c.m. motion, we restrict ourselves to the shell–model configurations with no holes in the closed shells. The generator coordinate two–centre wave functions (13) can be written in the following form:

$$\phi_{A_jB_j}(x,S) = \mathscr{N}_j(S)\phi_{\text{c.m.}}(R)\mathscr{A}\{\exp[-\tfrac{1}{2}\beta'_j(r_j-S)^2]\phi_{A_j}(\xi_{A_j},\sigma_{A_j})\phi_{B_j}(\xi_{B_j},\sigma_{B_j})\}.$$

Here $\phi_{c.m.}(R)$ is the wave function describing the motion of the c.m.

$$R = \frac{1}{A+B}\sum_{i=1}^{A+B} x_i, \qquad \beta'_j = \frac{A_j B_j}{A+B}\beta, \tag{14}$$

and $N_j(S)$ is a normalization factor.

We then define the two–centre generator coordinate representation of the trial functions by

$$\Psi^t_j = \sum_{p=1}^{N_p} c_{jp}\phi_j(x, S_p) + \int d^3S f_j(S)\phi_j(x, S) + \sum_{j'=1}^{N_p} K_{jj'}\int d^3S g_{j'}(S)\phi_{j'}(x, S), \tag{15}$$

where S_p are members of a discrete set of N_p values of S. Since the sum $\Sigma_{p=1}^{N_p} c_{jp}\phi_j(x, S_p)$ is supposed to approximate the true wave function Ψ_j in the region of the overlap of the nuclei, we restrict S_p to within a sphere of finite radius, which is of the magnitude of the sum of radii of the fragments. The generator coordinate amplitudes $f_j(S)$ and $g_j(S)$ are defined by the following equation:

$$\begin{Bmatrix} F_l(k_j|r_j|) \\ G_l(k_j|r_j|) \end{Bmatrix} Y_{lm}(\hat{r}_j)i^l/|r_j| = \int d^3S \exp\left[-\tfrac{1}{2}\beta'_j(r_j - S)^2\right] \begin{Bmatrix} f_{lm}(S) \\ g_{lm}(S) \end{Bmatrix}, \tag{16}$$

so that the trial wave function (15) satisfies the correct boundary conditions (2) and (3).

With the generator coordinate trial functions (15) we can write the matrix elements (12) [from which the K–matrix (10) is expressed directly] in the following form:

$$(\Delta_{\mathcal{F}\mathcal{F}})_{jj'} = \int d^3S \int d^3S' f^*_{lm}(S)\Delta_{jj'}(S, S')f_{l'm'}(S'), \tag{17}$$

where

$$\Delta_{jj'}(S, S') = \langle\phi_j(S)|\Delta|\phi_{j'}(S')\rangle. \tag{18}$$

Using the generator coordinate wave functions (15) we get for the form factor (18):

$$\Delta_{jj'}(S, S') = \langle\phi_j(S)|H - E|\phi_{j'}(S')\rangle \tag{19}$$
$$- \sum_n (E_n - E)^{-1}\sum_{j''p} b_{j''p}\langle\phi_j(S)|H - E|\phi_{j''}(S_p)\rangle \sum_{j'''p'} b_{j'''p'}\langle\phi_{j'''}(S_{p'})|H - E|\phi_{j'}(S')\rangle.$$

Here we have expressed the inverse of the operator $(H - E)$ in the subspace spanned by $\phi_j(S_p)$ by the eigenvalues E_n of the hamiltonian:

$$H|n\rangle = E_n|n\rangle, \tag{20}$$

where

$$|n\rangle = \sum_{j''p} b_{j''p}|\phi_{j''}(S_p)\rangle. \tag{21}$$

The generator coordinate kernel. – In order to calculate the form factor (19) one has to calculate the generator coordinate kernels: the hamiltonian kernel, $<\phi_j(S)|H|\phi_{j'}(S')>$, and the normalization kernel, $<\phi_j(S)|\phi_{j'}(S')>$. First one evaluates the unprojected kernels, $<\phi_{A_jB_j}(S)|H|\phi_{A_j'B_j'}(S')>$ and $<\phi_{A_jB_j}(S)|\phi_{A_j'B_j'}(S')>$, using the GC functions (13). We calculate these kernels by the algebraic programming systems REDUCE 2 or SCHOONSCHIP[3]). Both kernels have the following form

$$\sum_N \exp[-(\alpha_N S+\beta_N S')^2] \sum_{\substack{n_1n_2n_3\\ n_1'n_2'n_3'}} \mathcal{O}^{N,jj'}_{n_1n_2n_3n_1'n_2'n_3'} S_x^{n_1}S_y^{n_2}S_z^{n_3}(S_x')^{n_1'}(S_y')^{n_2'}(S_z')^{n_3'}. \tag{22}$$

For calculation of the transition amplitudes (12), where the channels i and j are partially characterized by angular momenta, we have to convert the kernels into a spherical representation. In expression (22) we group together the polynomials $S_x^{n1}S_y^{n2}S_z^{n3}$ and $(S_x')^{n'1}(S_y')^{n'2}(S_z')^{n'3}$ which have the same value of $\mathcal{L} = n_1 + n_2 + n_3$ and $\mathcal{L}' = n_1' + n_2' + n_3'$, respectively. We can express them in the form

$$S_x^{n_1}S_y^{n_2}S_z^{n_3} = |S|^{\mathcal{L}} \sum_{L=0}^{\mathcal{L}} \sum_{M=-(n_2+n_1)}^{n_2+n_1} D^{LM}_{n_1n_2n_3\mathcal{L}} Y_{LM}(\hat{S}), \tag{22'}$$

where L runs over even or odd integers, depending on whether $\mathcal{L}$ is even or odd. The sum over M goes in steps of 2. Expanding in addition the gaussian functions $\exp[(\alpha_N S + \beta_N S')^2]$ in partial waves,

$$\exp[-(\alpha_N S+\beta_N S')^2] = \sum_{L_0M_0} Y_{L_0M_0}(\hat{S}) a^N_{L_0}(S,S') Y_{L_0M_0}(\hat{S}'), \tag{22''}$$

where the expansion coefficients are given in terms of modified spherical Bessel functions

$$a^N_{L_0}(S,S') = 4\pi \exp(-\alpha_N^2 S^2 - \beta_N^2 S'^2) i_{L_0}(-2\alpha_N\beta_N SS'),$$

we obtain the generator coordinate kernel in the following form

$$\langle\phi_j(S)| \begin{Bmatrix} H \\ I \end{Bmatrix} |\phi_{j'}(S')\rangle = \sum_N \sum_{\substack{\Lambda\Lambda'\\ M_\Lambda M'_\Lambda}} \sum_{\mathcal{L}\mathcal{L}'} \sum_{\substack{LM\\ L'M'}} \sum_{L_0M_0} (-1)^{M_0} a^N_{L_0}(S,S')$$

$$\times S^{\mathcal{L}}(S')^{\mathcal{L}'} \begin{Bmatrix} \mathcal{H}^{Njj'}_{\mathcal{L}LM,\mathcal{L}'L'M'} \\ \mathcal{I}^{Njj'}_{\mathcal{L}LM,\mathcal{L}'L'M'} \end{Bmatrix} \langle LML_0M_0|\Lambda M_\Lambda\rangle\langle L'M'L_0-M_0|\Lambda'M'_\Lambda\rangle \tag{23}$$

$$\times \langle L0L_00|\Lambda 0\rangle\langle L'0L_00|\Lambda'0\rangle \left[\frac{(2L+1)(2L'+1)}{(2\Lambda+1)(2\Lambda'+1)}\right]^{\frac{1}{2}} \frac{2L_0+1}{4\pi} Y_{\Lambda M_\Lambda}(\hat{S}) Y_{\Lambda' M'_\Lambda}(\hat{S}'),$$

where

$$\mathcal{H}^{Njj'}_{\mathcal{L}LM,\mathcal{L}'L'M'} = \sum_{n_1n_2n_3n_1'n_2'n_3'} h^{Njj'}_{n_1n_2n_3n_1'n_2'n_3'} D^{LM}_{n_1n_2n_3\mathcal{L}} D^{L'M'}_{n_1'n_2'n_3'\mathcal{L}'}, \tag{24}$$

and similarly for the normalization kernels. Here we sum over those values of n_1, n_2, n_3 and n'_1, n'_2, n'_3 whose sum is equal to $\mathcal{L}$ and $\mathcal{L}'$ respectively. The sum over Λ, Λ', M_Λ and M'_Λ performs couplings of L's and spins of fragments in channels to the total angular momenta of channels j and j'.

3.2. The generator coordinate amplitude for scattering

In the generator coordinate method for scattering the proper boundary conditions is accomplished by requiring the GC amplitude to satisfy the integral equation of the first bind (15). There were several attempts[4] to solve this integral equation for the case of the Coulomb interaction. Very often solutions were not good enough in the whole region of the separation $|S|$ as needed for an efficient calculation. We refer here results published in a recent paper[4] in which is derived an improved approximation for the GC scattering amplitude for Coulomb case as well as for a wide class of non Coulomb cases. For the validity of results and details of derivation see ref. 4.

3.2.1. GC amplitude in the assymptotic region

The Eq. (15), is written with dimensionalness quantities

$$\int_{R^3} K(x,y) f_L(z,y) \mathrm{d}^3 y = g_L(z,x), \qquad K(x,y) = \exp[-(x-y)^2], \tag{25}$$

for $|y| \to \infty$. Eq. (25) results from Eq. (16) if all distances are measured in the scale unit $\sqrt{2/\beta'}$.

$$x = \sqrt{\tfrac{1}{2}\beta'}\, r, \qquad y = \sqrt{\tfrac{1}{2}\beta'}\, S,$$

$$z = \sqrt{\frac{2}{\beta'}}\, k, \quad g_L(z,x) = F_L\left(\sqrt{\frac{\beta'}{2}}\, z, \sqrt{\frac{2}{\beta'}}\, x\right), \quad f_L(z,y) = f_L\left(\beta', \sqrt{\frac{\beta'}{2}}\, z, \sqrt{\frac{2}{\beta'}}\, y\right) \tag{26}$$

In this section we shall assume that g_L is such that it gives rise to a unique solution in some space of ordinary functions.

From the Fourier transform of Eq. (25) it follows that

$$\tilde{f}_L(q) = \pi^{-\frac{3}{2}} e^{q^2/4} \tilde{g}_L(q).$$

We are interested in $\tilde{f}_L(q)$ for small values of $|q|$. Although it is difficult to define $e^{q^2/4}$ by its Taylor series $\Sigma_{n=0}^{\infty} (1/n!)(\frac{1}{2}q)^{2n}$, for the domain of q^{2n} becomes narrower for increasing n since q^2 is an unbounded operator, we shall nevertheless use it in the hope that it will suffice for our modest purpose to construct $f_L(y)$ only for large $|y|$.

We define the n-th order approximation to the solution of Eq. (25) by the coordinate representation of the Taylor series, truncated after the n-th term:

$$f_L^{(0)}(z, x) = \pi^{-\frac{3}{2}} g_L(z, x),$$
$$f_L^{(1)}(z, x) = \pi^{-\frac{3}{2}}(1+\tfrac{1}{4}\Delta_x) g_L(z, x),$$
$$f_L^{(2)}(z, x) = \pi^{-\frac{3}{2}}(1+\tfrac{1}{4}\Delta_x+\tfrac{1}{32}\Delta_x^2) g_L(z, x), \ldots .$$

Since $g_L(z, x)$ is supposed to be specified in polar coordinates

$$g_L(z, x) = i^{-L} P_L(\cos\hat{\omega}) \frac{1}{zx} \phi_L(\eta, zx), \ \cos\hat{\omega} = \frac{z \cdot x}{zx},$$

where ϕ_L is a solution of the Coulomb equation in the form

$$\left[\frac{1}{\rho^2}\frac{d}{d\rho}\left(\rho^2 \frac{d}{d\rho}\right) + 1 - V(\eta, \rho) + \frac{L(L+1)}{\rho^2}\right] \frac{\phi_L(\eta, \rho)}{\rho} = 0.$$

we express $f^{(i)}$'s in these coordinates:

$$f_L^{(1)}(z, x) = \pi^{-\frac{3}{2}} i^{-L} P_L(\cos\hat{\omega})\{1+(\tfrac{1}{2}z)^2[V(\eta, zx)-1]\} \frac{\phi_L(\eta, zx)}{zx},$$

$$f_L^{(2)}(z, x) = \pi^{-\frac{3}{2}} i^{-L} P_L(\cos\hat{\omega}) \left\{[1+(\tfrac{1}{2}z)^2(V-1)+\tfrac{1}{2}(\tfrac{1}{2}z)^4(V-1)^2] \frac{\phi_L(\eta, zx)}{zx} + \tfrac{1}{2}(\tfrac{1}{2}z)^4 \left(V'' \frac{\phi_L}{zx} + 2V' \frac{\phi_L'}{zx}\right)\right\}, \tag{27}$$

where prime denotes the derivative with respect to the second argument of functions:

$$V'(\eta, \rho) = dV(\eta, \rho)/d\rho, \quad \phi_L'(\eta, \rho) = d\phi_L(\eta, \rho)/d\rho.$$

The first three terms on the r.h.s. of Eq. (27) are the first three terms of the Taylor series of $i^{-L} P_L(\cos\omega) \exp[(\frac{1}{2}z)^4(V-1)] \phi_L(\eta, zx)/zx$, so that we expect that $f_L(z,y)$ contains a term of such an exponential form. Such an approximation can be obtained by embedding the original problem (25) into a generalized problem

$$f_L(\lambda, z, x) = \pi^{-\frac{3}{2}} e^{-(\lambda/4)\Delta_x} g_L(z, x), \qquad \lambda \in \mathbf{R} \tag{28}$$

from the solution of which one can recover the solution of Eq. (25) by finding $\lambda = 1$.

For $\lambda = 0$ the solution of Eq. (28) is

$$f_L(0, z, x) = \pi^{-\frac{3}{2}} g_L(z, x). \tag{29}$$

In order to get an approximation to $f_L(\lambda, z, x)$ for $\lambda \neq 0$, we differentiate Eq. (28) with respect to λ and get

$$\begin{aligned} \frac{\partial}{\partial\lambda} f_L(\lambda, z, x) &= -\pi^{-\frac{3}{2}}(\tfrac{1}{2}z)^2 e^{-(\lambda/4)\Delta_x}(V-1) g_L(z, x) \\ &= -(\tfrac{1}{2}z)^2 e^{-(\lambda/4)\Delta_x}(V-1) e^{+(\lambda/4)\Delta_x} f_L(\lambda, z, x). \end{aligned} \tag{30}$$

In Eq. (30) we then use again the Taylor series of the exponential functions:

$$e^{-(\lambda/4)\Delta_x}(V-1)e^{(\lambda/4)\Delta_x} = (V-1)-\tfrac{1}{4}\lambda[\Delta_x,(V-1)] + \tfrac{1}{32}\lambda^2[\Delta_x,[\Delta_x,(V-1)]]\pm\ldots \tag{31}$$

If the series (31) is terminated already after the first term, the corresponding approximant $f_L^{(0)}(\lambda,z,x)$ satisfies the equation

$$\frac{\partial}{\partial\lambda} f_L^{(0)}(\lambda,z,\boldsymbol{x}) = -(\tfrac{1}{2}z)^2[V(\eta,z\cdot\boldsymbol{x})-1]f_L^{(0)}(\lambda,z,\boldsymbol{x})$$

whose solution, obedient to the initial condition (29), is

$$f_L^{(0)}(\lambda,z,\boldsymbol{x}) = \pi^{-\frac{3}{2}}g_L(z,\boldsymbol{x})\exp\{(\tfrac{1}{2}z)^2[1-V(\eta,z\cdot\boldsymbol{x})]\lambda\}.$$

This approximation to $f_L(z,x)$ can be obtained also by semiclassical arguments[5].

If the series (31) is terminated after the second term and the ansatz

$$f_L^{(n)}(\lambda,z,\boldsymbol{x}) = i^{-L}P_L(\cos\hat{\omega})\,\frac{\varphi_L^{(n)}(\lambda,zx)}{zx}$$

is made for the corresponding approximant to $f_L(\lambda,z,x)$ then $\varphi_L^{(1)}(\lambda,zx)$ obeys the equation

$$\frac{\partial}{\partial\lambda}\varphi_L^{(1)}(\lambda,\rho) = -(\tfrac{1}{2}z)^2\left\{[V(\eta,\rho)-1-\lambda V''(\eta,\rho)]\varphi_L^{(1)}(\lambda,\rho) - 2V'(\eta,\rho)\lambda\frac{\partial}{\partial\rho}\varphi_L^{(1)}(\lambda,\rho)\right\}.$$

This homogeneous partial differential equation of first order can be readily solved for many potentials by the method of characteristics[4].

If we are interested in $\phi_L^{(1)}(1,\rho)$ for $\rho\to\infty$. The result written in powers of $1/\rho$ is

$$\varphi_L^{(1)}(1,\rho) = \pi^{-\frac{1}{2}}\left\{\phi_L(\eta,\rho) - \frac{1}{\rho^2}ab\phi_L'(\eta,\rho) - \frac{1}{\rho^3}a^2b\phi_L(\eta,\rho) - \frac{1}{\rho^4}\frac{a^2b^2}{2}\phi_L(\eta,\rho)\right\}\exp\left[a\left(1-\frac{2\eta}{\rho}-\frac{1}{3}\frac{b^2}{\rho^4}\right)\right] + O(\rho^{-5}),$$

with $a = \frac{1}{2}\frac{b^2}{\beta'}$ and $b = \eta\frac{b^2}{\beta'}$.

One can pursue the programme to find the asymptotic form of $\varphi_L^{(1)}(1,\rho)$ for a rich class of potentials $V(\eta,\rho)$, for example for

$$V(\eta,\rho) = H(\rho_0-\rho)\sum_{n=0}^{N}a_n\rho^n + H(\rho-\rho_0)\sum_{m=0}^{M}b_m\rho^{-m},$$

where ρ_0 is some finite number, corresponding to the distance below which V starts to deviate from its long–range behaviour and $H(\rho_0-\rho)$ is the step function of the interval $[0,\rho_0]$.

With the manoeuvre as exemplified in obtaining the asymptotic form of $\varphi_L^{(1)}(1,\rho)$ we can attack also approximations of higher order in λ. The approximant $\varphi^{(2)}(\lambda,\rho)$, defined by truncating the r.h.s. of eq. (28) after the third term ψ).

How closely the functions $\varphi^{(n)}(1,\rho)$, n = 1,2, ... satisfy the integral eq.(25) is measured by the smallness of the difference

$$\Delta^n(x) = |\int_0^\infty xy\, e^{-(x^2+y^2)}\, i_L(2xy)\, \varphi_L^{(n)}(1,zy)\, dy - \phi_L(\eta,zx)| \tag{32}$$

An example is shown in Fig. 1.

3.2.2. The GC amplitude at small distances

In this section we discuss the ill–posed nature of the defining eq. of the radial part of the GC amplitude for scattering and apply the method of statistical regularization to define an approximation to the amplitude inside a finite interval around the origin of the generator coordinate. The method does not rely directly on the potential $V(\eta,\rho)$ like the method presented in the previous section, but only indirectly through $\phi_L(\eta,\rho)$.

From the computational standpoint the most troublesome feature of the integral equation

$$\int_0^\infty K(x,y)\varphi_L(zy)\mathrm{d}y = \phi_L(\eta,zx), \tag{25}$$

$$K(x,y) = xye^{-(x^2+y^2)}i_L(2xy),$$

is that it is incorrectly posed, by which we mean that the solution $\varphi_L(zy)$, if it exists, does not depend continuously on the function $\phi_L(\eta,zx)$; a very small perturbation on ϕ_L can give rise to an arbitrary large perturbation in φ_L [4)].

In previous sect. 3. we described a way to a series of successively better, in the uniform convergence sense, approximations $\varphi_L^{(n)}$ to φ_L in the asymptotic region, $y \to \infty$. Since the kernel K(x,y) for a given value of x drops off very fast as a function of y it is possible for any given number $\epsilon > 0$ to find such a value y_n of y that

$$|\delta\phi_L^{(n)}(x)| \equiv \left|\int_{y_n}^\infty K(x,y)[\varphi_L^{(n)}(1,zy) - \varphi_L(zy)]\mathrm{d}y\right| < \varepsilon.$$

We now rewrite eq. (22) in the form

$$\int_0^{y_n} K(x,y)\varphi_L(zy)\mathrm{d}y = \phi_L(\eta,zx) - \int_{y_n}^\infty K(x,y)\varphi_L(zy)\mathrm{d}y \tag{33}$$

$$= \phi_L(\eta,zx) - \int_{y_n}^\infty K(x,y)\varphi_L^{(n)}(1,zy)\mathrm{d}y + \delta\phi_L^{(n)}(x)$$

$$\equiv \phi_L^{(n)}(\eta,zx) + \delta\phi_L^{(n)}(x).$$

Due to the localized nature of K(x,y) there exists an x_n such that

$$\left|\left(\int_{y_n}^{\infty}-\int_{0}^{\infty}\right)K(x,y)\varphi_L^{(n)}(1,zy)\mathrm{d}y\right|<\varepsilon\quad\text{for } x\geqq x_n.$$

If also expression (32) is less than ϵ for $x \geqslant x_n$, then the function $\phi_L^{(n)}(\eta,zx)$, defined in eq. (33), is below 3ϵ in magnitude on the interval (x_n,∞).

Then we treat $\phi_L^{(n)}(\eta,zx)$ as if it had a node at $x = x_n$, make the Fourier–Bessel expansion of it, and express $\varphi_L(zy)$ in the form of a Dini series on the interval $0 \leqslant y < y_n$:

$$\phi_L^{(n)}(\eta,zx)=\sum_{j=1}^{p}b_j j_L(\beta_j zx), \tag{34}$$

where $\{\beta_j\}$ is the set of numbers determined by the equation $j_l(\beta_i z x_n) = 0$, and

$$\varphi_L(zy)=\sum_{j=1}^{m}a_j j_L(\alpha_j zy), \tag{35}$$

where $\{\alpha_j\}$ are zeroes of the expression $[Aj_L(\alpha_j\rho)+B\mathrm{d}j_L(\alpha_j\rho)/\mathrm{d}\rho]_{\rho=zy_n}$ with

$$\frac{A}{B}=-\left(\frac{\frac{\mathrm{d}}{\mathrm{d}\rho}\varphi_L^{(n)}(1,\rho)}{\varphi_L^{(n)}(1,\rho)}\right)_{\rho=zy_n}$$

In this way the integral equation (33) is transformed into a system of linear algebraic equations

$$\sum_{i=1}^{p}K_{ji}a_i=b_j+\delta b_j;\qquad j=1,\ldots,m. \tag{36}$$

The fact that the integral eq. (25) is incorrectly posed is reflected in the ill-conditioning of the system of equations (36). By increasing the number p of expansion functions we can get as ill–conditioned a system as we want[4,6].

We have attacked the problem by the method of statistical regularization[7]. We assume that the errors δb form a random vector δ in the space R^m and we denote by $P_\delta(\delta)$ the probability density of this vector. Quantities b_j are random as well and they depend on α and $\underline{b}$. The density of conditional probability of the vector b is

$$P(\boldsymbol{b}|\boldsymbol{a})=P_\delta(\boldsymbol{b}-K\boldsymbol{a}). \tag{37}$$

We are then confronted with the following problem: given a vector $\underline{b}$ what is the probability density P(a|b) of the vector $\underline{a}$. In order to answer the question we have to introduce some a priori information about the solution a by means of a probability denstiy $P(\underline{a})$. Then the a posteriori density is given by the theorem of Bayes

$$P(\boldsymbol{a}|\boldsymbol{b})=P(\boldsymbol{a})P(\boldsymbol{b}|\boldsymbol{a})\Big/\int P(\boldsymbol{a})P(\boldsymbol{b}|\boldsymbol{a})\mathrm{d}\boldsymbol{a}. \tag{38}$$

Let us now specify the probability density of the vector $\underline{\delta}$. We assume that the errors δb_j for different j are not correlated among themselves and that the distribution function of the random vector $\underline{\delta}$ is normal with vanishing mathematical expectation. Let the root mean square of the error δb_j be s_j. Then the density of conditional probability (37) is

$$P(\boldsymbol{b}|\boldsymbol{a}) = (2\pi s^2)^{-m/2} \exp(-(\boldsymbol{c}-L\boldsymbol{a})^2/2s^2), \tag{39}$$

where

$$s^m = \prod_{j=1}^{m} s_j, \qquad c_j = \frac{s}{s_j} b_j, \qquad L_{ji} = \frac{s}{s_j} K_{ji}.$$

Now, let us assume that the unknown function φ_L is bounded:

$$\Omega[\varphi_L(zx)] \equiv \int_0^{y_n} \varphi_L^2(zx) z\mathrm{d}x \leqslant \omega < \infty,$$

or

$$(\boldsymbol{a}, \Omega\boldsymbol{a}) = \sum_{i=1}^{p} a_i^2 \leqslant \omega. \tag{40}$$

Then the a priori density distribution of probability $P(\underline{a})$ has to satisfy the condition

$$\int (\boldsymbol{a}, \Omega\boldsymbol{a}) P(\boldsymbol{a}) \mathrm{d}\boldsymbol{a} = \omega.$$

Among the probability densities $P(\underline{a})$, which satisfy this condition, we choose the one with minimal information

$$P(\boldsymbol{a}) = \text{const} \exp(-\tfrac{1}{2}\alpha(\boldsymbol{a}, \Omega\boldsymbol{a})), \tag{41}$$

where $\alpha = p/\omega$. From eqs. (39) and (41) we then get according to the theorem of Bayes the following a posteriori density distribution:

$$P(\boldsymbol{a}|\boldsymbol{b}) = \text{const} \exp\left\{-\frac{1}{2s^2}[(\boldsymbol{a}, (L^{\mathrm{T}}L + s^2\alpha\Omega)\boldsymbol{a}) - 2(L^{\mathrm{T}}\boldsymbol{c}, \boldsymbol{a})]\right\}. \tag{42}$$

We then define the regularized solution of the system of algebraic equations (36) by the average over the ensemble corresponding to the probability density (42):

$$a_i^0 = \int a_i P(\boldsymbol{a}|\boldsymbol{b}) \mathrm{d}\boldsymbol{a}, \qquad i = 1, 2, \ldots, p. \tag{43}$$

From eqs. (42) and (43) we get the result that the regularized solution a^0 satisfies the following equation:

$$(L^T L + s^2 \alpha \Omega) a^0 = L^T c \tag{44}$$

and that its mean square error is

$$\sigma_i^2 = s^2[(L^T L + s^2\alpha\Omega)^{-1}]_{ii}. \tag{45}$$

One can show that, if the solution of the integral equation (25) exists and if it is unique, then the regularized solution converges in the mean to the true solution if and only if $\lim_{\alpha\to 0} s/\alpha = 0$.

By the method of statistical regularization we are also able to test the hypothesis (40). Let us first notice that the mean square error s^2 is the average of $(\underline{c} - L\underline{a})^2/n$ over the complete ensemble, characterized by the probability density (38) with $P(\underline{a})$ = = const., i.e. no a priori constraints on the unknown a. On the other hand we can calculate the mean square error s'^2 of the same quantity over the a posteriori ensemble (42):

$$s'^2 = \frac{1}{n}\int (c - La)^2 P(a|b)\mathrm{d}a = \frac{s^2}{n}\sum_{i=1}^{n} [(L^T L + s^2\alpha\Omega)^{-1}]_{ii}.$$

We have obtained the a posteriori ensemble by choosing from the complete ensemble only those vectors $\underline{a}$ which belong also to the a priori ensemble. If the choice of the a priori ensemble is correct, then their members on the average are better solutions than the members of the complete ensemble, and $s' \leqslant s$. But if the choice of the a priori ensemble, is a bad one, then $s' > s$ and the hypothesis is not correct.

We applied the method of statistical regularization to find an approximate solution of eq. (25) in the case of S-wave scattering of two point α–particles whose c.m. energy is 15 MeV. We borrowed the asymptotic part of the approximate solution from sect. 3.2.1. We chose y_n to be the left edge of the interval on which $\Delta^n(x) \leqslant \epsilon = 10^{-(n+1)}$. Then we determined the regularized solution on the interval $[0, y_n]$ by the method described in this section with only a few ($5 \leqslant p, m \leqslant 10$) terms in the Dini and Fourier–Bessel expansions (34), (35). In fig. 1 we show the absolute value of the difference $\Delta^{(n)}_{\mathrm{reg}}(x)$ between $\phi_L(\eta, zx)$ and the result of folding of the regularized amplitude thus obtained with kernel (25). Similar results were obtained at energies 1 MeV and 50 MeV. For given ϵ the solution does not depend only on the order n of the asymptotic approximant $\varphi_L^{(n)}(1, zy)$ but also on the value of the regularization parameter α introduced in eq. (41).

The way of choosing this variational parameter is described in ref. 2.

Let us mention that the method described in this section enables one to calculate better approximations to the amplitude if needed by going to higher orders in λ.

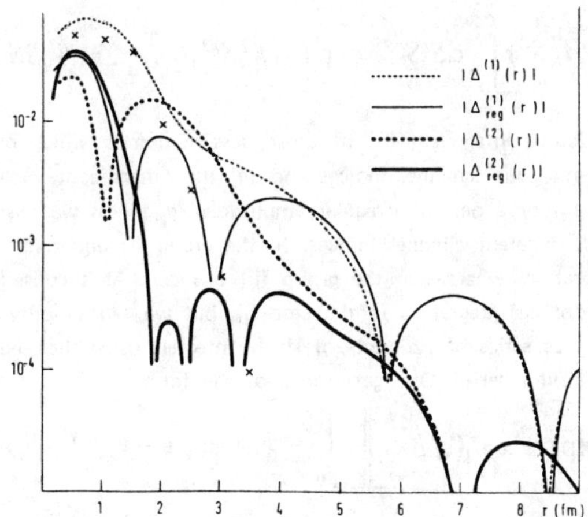

Fig. 1. The absolute value of the difference between the Coulomb wave function and the folded GC amplitude for the case of S-wave scattering of two α-particles whose c.m. energy is 15 MeV: $\Delta^{(n)}(r)$ for asymptotic GC amplitudes, $\Delta^{(n)}_{reg}(r)$ for GC amplitudes regularized on the intervals [0.35 fm] and [0.47 fm] for n = 2 and n = 1, respectively.

3.3. Calculation of matrix elements of the operator Δ

The matrix elements (17) of the operator Δ are the building blocks of the proposed reaction theory. It follows from the expression for the $\Delta_{ij'}(S,S')$, eq. (20), that these elements can be obtained from integrals of the type

$$\int d^3S \int d^3S' \begin{Bmatrix} f_{lm}(S) \\ g_{lm}(S) \end{Bmatrix} \langle \phi_j(S)|H-E|\phi_{j'}(S')\rangle \begin{Bmatrix} f_{l'm'}(S') \\ g_{l'm'}(S') \end{Bmatrix} , \tag{46}$$

and those of the type

$$\int d^3S \begin{Bmatrix} f_{lm}(S) \\ g_{lm}(S) \end{Bmatrix} \langle \phi_j(S)|H-E|\phi_{j'}(S_p)\rangle. \tag{47}$$

After expanding the integrand in partial waves the integrals (47) reduce to a sum of single radial integrals. They are of the form (see eqs. (21') and (21")).

$$\int_0^\infty SdS \begin{Bmatrix} f_l(S) \\ g_l(S) \end{Bmatrix} S^{\mathscr{L}} \exp(-\alpha_N^2 S^2) i_{L_0}(-2\alpha_N \beta_N S_p S).$$

The integrals (46) reduce to the sum of double integrals and their evaluation is the most time consuming part of the calculation of scattering parameters. These double integrals are of the form

$$\int_0^\infty \mathrm{d}S \begin{Bmatrix} f_l(S) \\ g_l(S) \end{Bmatrix} S^{\mathscr{L}} \exp(-\alpha_N^2 S^2) \int_0^\infty \mathrm{d}S' S'^{\mathscr{L}'} \exp(-\beta_N^2 S'^2) i_{L_0}(-2\alpha_N \beta_N SS') \begin{Bmatrix} f_{l'}(S') \\ g_{l'}(S') \end{Bmatrix}. \tag{48}$$

There are two kinds of double integrals (48): (i) those, less numerous, which contain the amplitudes f and g with equal channel indices and (ii) the others containing a pair of regular amplitudes (f_l, $f_{l'}$) or a pair of irregular amplitudes (g_l, $g_{l'}$) as well as those containing f_l and $g_{l'}$ with different channel indices. In the group (i) one cannot exchange the order of integration, whereas in the group (ii) one can. As a consequence we have to perform the double integration in the case (i), but we can simplify those of (ii) by expressing them as series of products of single integrals using the expansion of $i_{L_0}(-2\alpha_N \beta_N SS')$ in Taylor series. One gets series of the form

$$\sum_n A_n \left[\int_0^\infty x^{\alpha_n} \mathrm{d}x \exp(-x) f_l(\sqrt{\beta x}) \right] \left[\int_0^\infty y^{\alpha'_n} \mathrm{d}y \exp(-y) f_{l'}(\sqrt{\beta' y}) \right].$$

This converges satisfactorily (it gives an accuracy of better than 10^{-6} by taking 20 terms).

In the calculation of double integrals (i), one can use the following "simplification":

Let us denote by $\bar{H}$ the matrix of the operator $H - E$ in the basis $\phi_j(x, S_p)$ defined by the type of clustering C_j and the value of the generator coordinate, $\gamma_p \equiv (C_j, S_p)$, $\gamma \equiv (C_j, S)$. Let us call $\bar{H}^{-1}$ the inverse of $\bar{H}$ in the same basis. The elements of the type (18) can be written then as

$$(\Delta_{\mathscr{F}\mathscr{G}})_{\gamma\gamma'} = \int_0^\infty \mathrm{d}S S f_{C_j l}(S) \int_0^\infty \mathrm{d}S' S' g_{C_{j'} l'}(S') [\bar{H}_{\gamma\gamma'} - \sum_{\gamma_p \gamma'_p} \bar{H}_{\gamma\gamma_p} \bar{H}^{-1}_{\gamma_p \gamma'_p} \bar{H}_{\gamma'_p \gamma'}]. \tag{49}$$

When calculating these integrals numerically, we get

$$\int_0^\infty \mathrm{d}S S f_{C_j l}(S) \sum_n w_n g_{C_{j'} l'}(S_n) [\bar{H}_{\gamma\gamma_n} - \sum_{\gamma_p \gamma'_p} \bar{H}_{\gamma\gamma_p} \bar{H}^{-1}_{\gamma_p \gamma'_p} \bar{H}_{\gamma'_p \gamma_n}].$$

If one chooses the integration points S_n, so that they coincide with the points S'_p of γ'_p, then the summation over γ'_p provides $\delta_{\gamma_p \gamma'_p}$ and the summation over the γ_p then makes the integrand vanishing over this region.

If additional integration points are chosen, one has to sum over n running only over these additional points. A numerical test on the system $^3H + {}^4He$ showed that the phase shifts are affected by less than 2 %.

The reason that one can use the described "simplification" is a consequence of the fact that the contribution to the integral in eq. (48) from a region where both S and S' are small is negligible provided the sum $\Sigma_p C_{jp} \phi_j(x, S_p)$ in formula (15) describes well the composite system. Well means that the value of the integral (49) will not change to the desired accuracy if one exchanges $f_{c_j l}(S)$ and $g_{c_j l}(S)$ by their projections out of the space of square integrable functions $\phi_j(x, S_p)$.

4. VARIATIONAL PRINCIPLE FOR SCATTERING

1. There is no straightforward way to obtain a variational formulation of the quantum mechanics of scattering which leads to a calculable procedure and provides no false predictions. Most of the proposed formulations up to now were based on solving variational equations without introducing addition constraining equations. The simplest type of the functional used in solving the inverse variational problem contains only independent variables, the trial functions and their first derivatives. Douglas[8] showed how to derive equations for such functionals with two trial functions. Also there are two realizations[9] for scattering, but both leads to spurious resonances.

Useful existing variational formulations were based on (local improper) functionals, which contains independent variables, trail function, its first and second derivatives [10]. Some of these methods like the Kohn, the inverse-Kohn, the Hulten methods provides false predictions (false resonances and complex phase shifts or both) which are proportional to the number of correlation functions ϕ_p, eq. (14). Some of these methods [11] contain in the trial function an additional parameter, the value of which one determines after variation by requiring that the approximate trial solution satisfies one or more relations which are valid for the exact solution of the Schroedinger equation. These methods with the projection after the variation (as tested in potential elastic scattering) provided less false resonances, but it is proved[2] that are not immune from false predictions. The worse part of the evaluation is that in all cases, untrivial testing is necessary.

An important step toward resolving this annoying situation is made recently[2] by showing that false resonances can be eliminated by formulating for elastic scattering a constrained variational approach which guaranties nonviolation of the principle of causality. This constrain requires that the norm of the solution for free dynamics is equal to the norm of the solution for interacting dynamics, as well as that the trial solution satisfies the dispersion relation valid for the exact solution. These conditions can be realised by treating the reaction parameters as distributions. The equation for variational scattering parameters are however nonlinear, but they can be put in a form which allows an iterative method of solving. It is convenient, that the input requires same elements as those necessary in unconstrained methods. There is no obvious way, however, how this constrained method can be extended to non elastic scattering including the transfer of nucleons. Dispersion relations for multichannel scattering are quite complicated and till now we have not obtained useful relations for conditions on the variations.

2. In practical calculations of reaction parameters in collision of light nuclei we were using together the Kohn and the inverse-Kohn variational methods. The latter one corresponds to same type of the functional(9), but the trial function is of the form

$$\Psi_j^t = \sum_{p=1}^{N_p} c_{jp}\,\phi_p + \mathcal{G}_j + \sum_{j'=1}^{N_c} (K^{-1})_{jj'}\,\mathcal{F}_j$$

The Kohn — Kato variational expression for the elements of the (K^{-1}) — matrix are given by

$$(K^{-1})_{ij} = \frac{1}{w_{ij}} \{(\Delta_{\mathscr{G}\mathscr{G}})_{ij} - [(\Delta^{T}_{\mathscr{F}\mathscr{G}})(\Delta_{\mathscr{F}\mathscr{F}})^{-1}(\Delta_{\mathscr{F}\mathscr{G}})]_{ij}\}, \tag{51}$$

As mention before both variational principles can lead to imprecise results to the K and (K^{-1}) matrix due to spurious resonances. However, these spurious resonances are on different places for the two variational methods. So the Kohn method has spurious resonances for

$$<\mathscr{G}_o, (H-E)\mathscr{G}_o> - \sum_{n=1}^{N_p} |<\mathscr{G}_o, (H-E)\phi_n>|^2/(E_n - E) = 0$$

and the inverse – Kohn method for

$$<\mathscr{F}_o, (H-E)\mathscr{F}_o> - \sum_{n=1}^{N_p} |<\mathscr{F}_o, (H-E)/\phi_n>|^2/(E_n - E) = 0$$

In order to be sure that spurious resonances do not influence the results we compare the K matrix and the matrix $(K^{-1})^{-1}$ and we accept the result if these differ for less then a reasonable value which is consistent with numerical errors. If, however, the difference between K and $(K^{-1})^{-1}$ is large we repeat the calculation of both K and K^{-1} in an enlarged (or changed) basis of the square integrable functions φ_p . This usually helps, as the false resonances are positioned between the eigenstates of the Hamiltonian in the basis $\{\varphi_p\}$, different eigenvalues are obtained in a changed basis.

Instead the Kohn and the inverse Kohn method one can use any two of method based on the local improper functionals. It is a convenient feature of all these methods that same matrix elements entree into expression for reaction parameters. However, the untrivial and costly testing of the precision of results is unavoidable.

5. AN EXAMPLE: THE ELASTIC SCATTERING ^{4}He (^{3}H, ^{3}H)^{4}He

This section contains short report on the calculation of the elastic scattering of ^{4}He on ^{3}H. The aim of the work on this example was two fold: (1) to test the feasibility of the calculation on a rather realistic example and to test the reliability of the procedures proposed in previous sections and (2) to see what the predictions of the method are when a simple effective interaction is used and a restricted basis of functions is chosen to span the subspace of the scattering space. (For more details see ref. 1).

(1) We use the following model in this calculation:

The basis is generated out of the square integrable functions ϕ_p in eq. (14) which are two–centre functions corresponding to α–t clustering

$$\phi_p \equiv \phi(x, S_p) = P^{JM\pi}\{\mathscr{A}[\prod_{i=1}^{4} \varphi_i(x_i - \tfrac{3}{7}S_p) \prod_{j=5}^{7} \varphi_j(x_j + \tfrac{4}{7}S_p)]\}. \tag{52}$$

The functions φ_i are the s–wage single–particle functions of the harmonic oscillator and $P^{JM\pi}$ is the operator which projects the antisymmetrized product onto the eigenstates of angular momentum labelled by quantum numbers J, M and parity π.

The channel wave function (15) has the channel index j = $(lm\frac{1}{2}\pm\frac{1}{2}00)$. Each function ϕ_{A_j} and ϕ_{B_j} is constructed out of single determinants of the same single–particle wave functions as in (50). We choose everywhere $\beta = 0.5$ fm^{-2}.

As the effective nucleon–nucleon potential (6) we used the expressions V1 and V2 proposed by Volkov[12], each consisting of a sum of two gaussian functions.

In order to eliminate the influence of spurious resonances the method of testing the reliability of results described in Ch. 4 is used. We found in this concrete case that if we include more than five functions ϕ_p the relative difference between the two K–matrices is smaller than 4×10^{-4}.

The relative difference between the K–matrix calculated with five square integrable functions ϕ_p and that calculated with ten ϕ_p is less than 10^{-2} for the most sensitive case $l = 3$ at the energy near the resonance.

(2) Although our model for scattering of ^{4}He nuclei on ^{3}H is too simple to allow for a quantitative agreement with experiment its general features provide some insight into the scattering process.

(i) The calculated phase shifts for all relative angular momenta except $l = 3$ show the same trend as the experimental data, except that there is no structure in the calculated ones, (fig. 2).

The calculated phase shift for $l = 3$ shows a resonance at the relative kinetic energy 4.906 MeV (4.05 MeV) for the effective interaction V1 (V2). The experiment[13] revealed two resonances in this energy region, one with angular momentum and parity $\frac{5^-}{2}$ at 2.2 MeV and another $\frac{7^-}{2}$ at 4.25 MeV.

The calculation would give rise to two resonances if we used a non–central effective potential instead of the central one (6).

The difference between the calculated and measured postion of the resonance can also be accounted for in the following way. In this energy region the two colliding nuclei scatter through a compound state. Therefore it is more appropriate to measure the energy of the colliding system with respect to the ground state of the compound system than with respect to the threshold. The difference between the calculated position of the resonance and the calculated energy position of the ground state is 5.787 MeV (5.605 MeV) for the potential V1 (V2) whereas experimentally this energy difference is 4.67 MeV for the resonance $\frac{5^-}{2}$ and 6.27 MeV for $\frac{7^-}{2}$. From this standpoint the calculated energy of the resonance lies between the measured energies of the resonance $\frac{5^-}{2}$ and $\frac{7^-}{2}$, whose average $\frac{3}{7}E^{exp}_{7^-/2} + \frac{4}{7}E^{exp}_{5^-/2} = 5.56$ MeV.

The experimental data are usually given as functions of the relative kinetic energy of the colliding nuclei. In view of the remark made in the preceeding paragraph we compare the measured differential cross section for the elastic scattering at a given relative kinetic energy close to the measured resonant energy with the cross section calculated at the energy which is greater than the measured relative kinetic energy by

$E^{cal}_{resonance} - (\frac{4}{7} E^{exp}_{7^-/2} + \frac{3}{7} E^{exp}_{5/2})$. These two differential cross sections are shown in fig. 3.

The cross sections at energy away from the energy of the resonance is shown in fig. 4.

(ii) The calculation gives also some insight into the reaction mechanism:

The probability amplitudes of finding the compound system in one of the quasi–bound states ϕ_n which are eigenfunctions of the hamiltonian in the basis of square integrable functions (52) are given by

$$A_{ln}(E, E_{ln}) = \sum_{pp'} C_{lp}(E) b_{ln}(S'_{p'}) \langle \phi_l(S_p) | \phi_l(S'_{p'}) \rangle .$$

The amplitudes $b_{in}(S_p)$ are the expansion coefficients of the functions ϕ_n in the basis (52), eq. (21). The amplitudes $A_{ln}(E, E_{ln})$(l = 0, ... 4, n = 1, ... 6) are shown in fig. 4 for the bombarding energies in the c.m. system E = 1.217, 4.906 and 10.0 MeV.

The amplitudes $A_{ln}(E, E_{ln})$ as functions of E_{ln} for l = 0, 2 and 4 do not change with the bombarding energy E. For fixed E and different l (= 0, 2 and 4) the distri–bution of absolute values of $A_{ln}(E, E_{ln})$ are very similar having a broad Lorentzian–like shape.

For odd l and fixed E the distribution of absolute values of A_{ln} over E_{ln} has a completely different shape peaked around the lowest two values of E_{ln}: for l = 3 the lowest value of E_{ln} is the resonance energy. At the energy E equal to the resonance energy 4.906 MeV, the amplitude $A_{ln}(E, E_{ln})$ at each E_{ln} has a sign opposite to the sign at E = 1.217 MeV and at E = 10.0 MeV.

The behaviour of the amplitudes $A_{ln}(E, E_{ln})$ indicates that for l = 0, 2, 4 several quasibound states are needed to describe the elastic scattering whereas for l = 1, 3 few lowest states are needed. This can be understood as if the reaction in the channels l = 0, 2,4 is of a direct type and in l = 1, 3 it is of a compound type. This shows that the described method of calculation enables one to get information on the reaction mechanism by analysing the scattering wave functions in terms of wave functions of quasibound states.

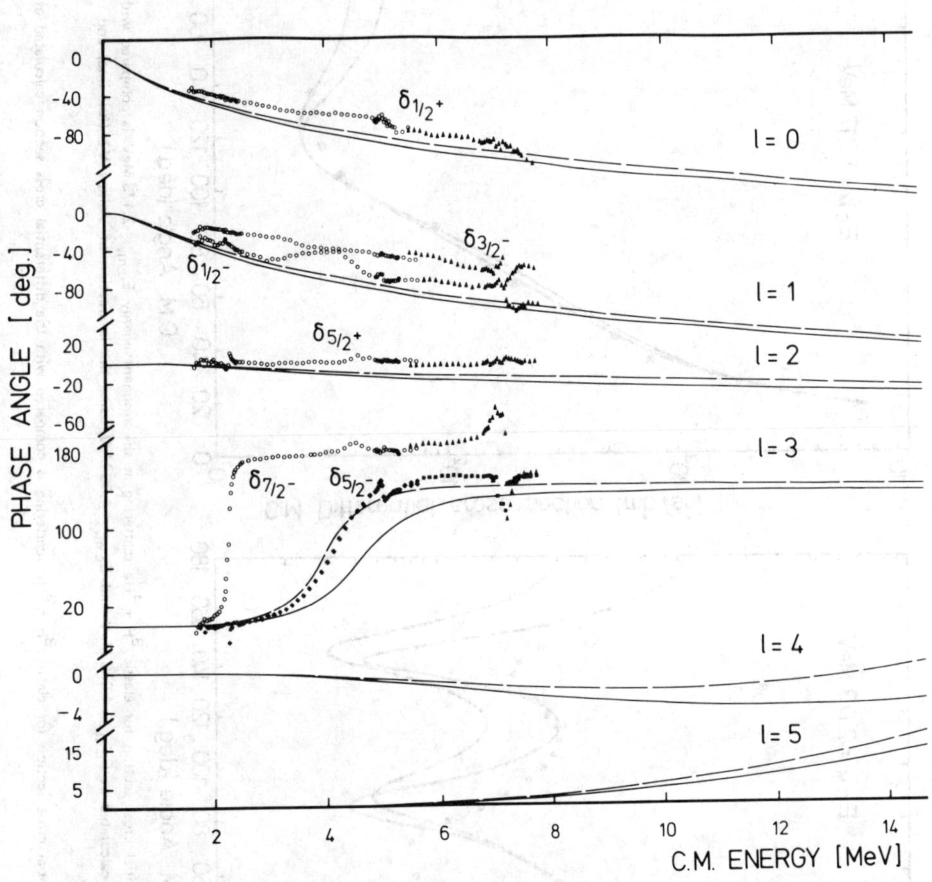

Fig. 2. Phase shifts for elastic $^3H + {}^4He$ scattering extracted from experimental data are compared with those calculated by using the interaction of Volkov V1 (full line) and V2 (dashed line).

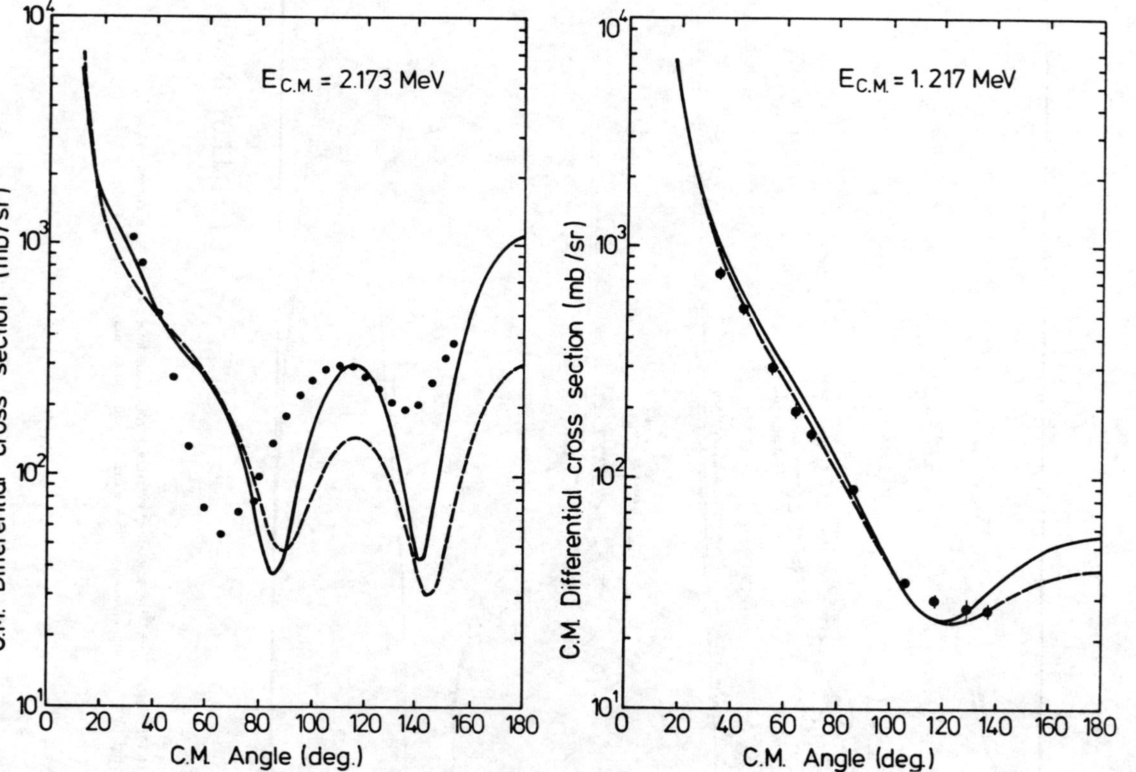

Fig. 3. The measured differential cross section for elastic $^3H + ^4He$ scattering at the resonant energy $E_{c.m.}$ = 2.173 MeV is compared with the differential cross section calculated at $E_{c.m.}$ = 4.906 MeV and $E_{c.m.}$ = 4.05 MeV using the interaction of Volkov V1 (full line) and V2 (dashed line), respectively.

Fig. 4. The measured differential cross section for elastic $^3H + ^4He$ scattering is compared with the differential cross section calculated at $E_{c.m.}$ = 1.217 MeV .

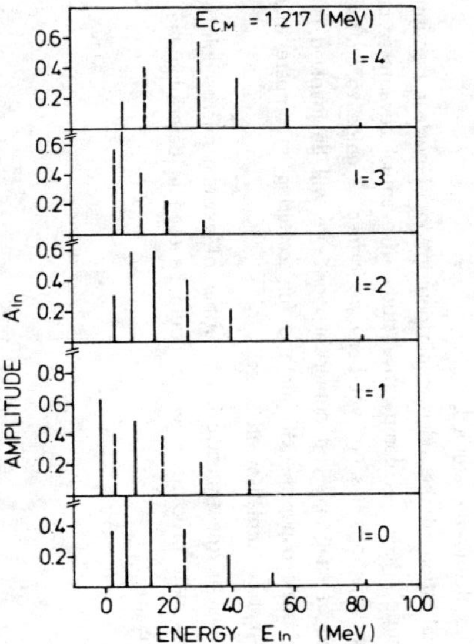

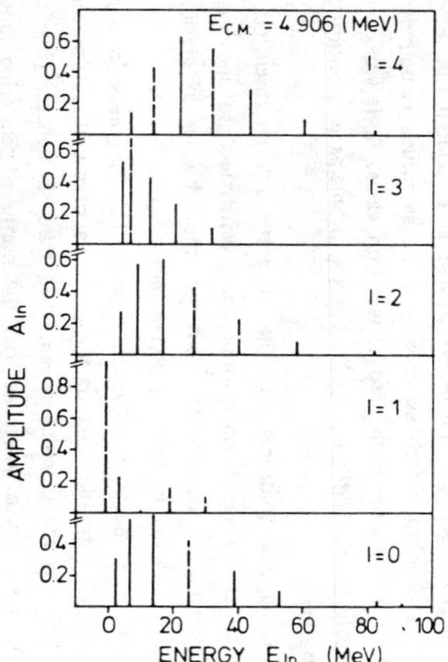

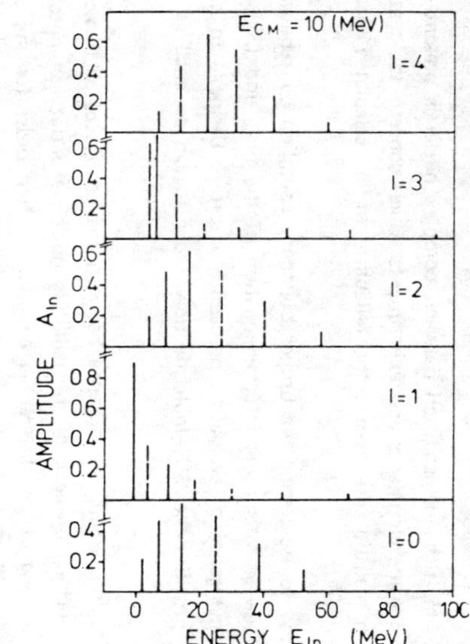

Fig. 5. The probability amplitudes of finding the compound system in one of the quasibound states which are eigenfunctions of the hamiltonian in the basis of square integrable functions corresponding to the energies E_{In} are shown for the bombarding energies $E_{c.m.}$ = 1.217, 4.906 and 10 MeV. Full lines denote the amplitudes with positive sign and dashed lines those with negative sign.

6. CONCLUSION

A formalism for the multichannel collision between complex nuclei is presented. The emphasis is put on new developments and experiments which concern the feasibility of the calculation with realistic systems and the reliability of the calculated reaction parameters.

(i) In the GC method for scattering the proper boundary condition by requiring the GC scattering amplitude to satisfy the integral equation of the first kind (22). In order to optimize the convergence of evaluation procedure at proves desirable to have an approximation which provides the GC amplitude down to distances as small as possible.

In the assymptotic region, where the distance between the centers of shell model wells is larger than the touching distance of the colliding nuclei, a better assymptotic solution is obtained by deriving partial differential equation of first order for the terms of an assymptotic series in 1/E, E is the scattering energy. The method of statistical regularization is used to obtain a useful and controlable expression in the form of a Dini series, for the GC amplitude for small distances. (Extracting these informations for small distances is an ill—posed problem). The GC is treated as a random function with a"a priory" distribution of probability which assumes that the amplitude is bounded and that the errors in the input are random with zero expectation value. Both improved approximations are applicable to a wider class of scattering problems in addition to the long range Coulomb scattering.

(ii) The problem of using the variational principle for scattering was discussed. It is reported how a combination of the variational formulations, which lead to linear variational equations for parameters, can be used to avoid that the influence of spurious resonances to no influence the evaluated reaction parameters.

Also it is reported also on a recent work which showed that the cause for the appearance of false resonances is in the violation of the principle of causality. For the ellastic scattering a formulation is found which provides no false resonances. This is realized by introducing constrains which guarantee that the norms of scattering functions are conserved and that the trial function satisfies dispersion relations.

(iii) In an example, the elastic scattering $^4He(^3H,{}^3H)^4He$, the feasibility and the usefullness of the proposed theory are studied.
There is an alternative realization of the microscopic description of nuclear reactions, the resonating group method (RGM). A comparative study of eventual advantages or drawbacks of either of the methods has not yet been performed, a guide to an eventual new user as to what kind of analytical and numerical work he will be involved is written in ref. 1. In the same paper the GC method for scattering is compare to other macroscopic methods (R-matrix formalism etc.).

Finally, let us note that in the framework of neither of existing microscopic is used till now to develop approximate procedures, which could be used in describing the scattering of heavier nuclei.

REFERENCES:

1. R. Beck, M.V. Mihailović and M. Poljšak, Nucl. Phys. A351 (1981), 295
2. M. Poljšak, Ph. D. Thesis, University E. Kardelj, Ljubljana, 1982
3. A.C. Hearn, REDUCE 2, A system and language for algebraic manipulations, User's manual, UCP–19 (1973),

   H. Strubbe, Comp. Phys. Comm. 8 (1974) 1
4. M.V. Mihailović and M. Poljšak, Nucl. Phys. A388 (1982) 317; A rather complete list of references on solving the GC amplitude is in this reference.
5. R. Beck, J. Borysowicz, D.M. Brink and M.V. Mihailovic, Nucl. Phys. A244 (1975) 45, 58
6. D.K. Faddeev and V.N. Faddeeva, J. Numer. Math. and Math. Phys. 1 (1961) 412
7. V.F. Turchin, V.P. Kozlov and M.S. Malkerich, Usp. Fiz. Nauk 102 (1970) 345
8. J. Douglas, Trans. Am. Math. Soc. 50 (1941) 71
9. P.M. Morse and H. Feshbach, Methods of Theoretical Physics (Mc Graw–Hill, New York, 1953), str. 1127

   N.E. Tamm, Ž.E.T.F. 18 (1948) 337
10. W. Kohn, Phys. Rev. 74 (1948) 1763;

    L. Hulten, Arkir. Mat. Astron. Fys. 35A (1948) 25; L. Hulten and S. Skarlem, Phys. Rev. 87 (1952) 297
11. R.K. Nesbet and R.S. Oberoi, Phys. Rev. A6 (1972) 1855;

    F.E. Harris and H.H. Michels, Phys. Rev. Lett. 22 (1969) 2036;

    K. Takatsuka and T. Fueno, Phys. Rev. A19 (1979) 1011
12. A.B. Volkov, Nucl. Phys. 74 (1964) 33;
13. R.J. Spieger and T.A. Tombrello, Phys. Rev. 163 (1967) 964

    T.A. Tombrello, Nucl. Phys. A195 (1972) 250.

FUSION OF DEFORMED HEAVY IONS

N. Rowley
SERC Daresbury Laboratory, Warrington WA4 4AD, U.K.

1. INTRODUCTION

Simple phenomenological expressions have proved rather successful in describing heavy-ion fusion cross sections at energies not too high above the Coulomb barrier. This is a strong reflection of the insensitivity of this process to the detailed intrinsic structure of the target and projectile. The fusion cross section may be written

$$\sigma_f = \frac{\pi}{k^2} \sum_{\ell=0}^{\infty} (2\ell + 1)\, T_\ell(E) \tag{1.1}$$

where $T_\ell(E)$ is the probability that a given partial wave ℓ will fuse at an incident energy $E = 2mk^2$, with m the reduced mass of the system. In terms of the S-matrix elements for the problem we may write

$$T_\ell = 1 - \sum_i \left|S_\ell{}^i\right|^2 \tag{1.2}$$

where the sum runs over the elastic and all direct reaction channels so that some dependence on the structure should be expected if certain states are strongly coupled.

An extremely useful approximation to σ_f may, however, be obtained by assuming that all the flux which penetrates the Coulomb plus centrifugal barrier fuses, whereas that which is reflected goes into direct channels. By approximating the barrier top with an inverted oscillator potential (frequency ω) we may use an analytic approximation for the $T_\ell(E)$. Then replacing the sum over ℓ in Eq.(1.1) by an integral we obtain the formula of Wong[1]

$$\sigma_f = \frac{\hbar\omega R_B^2}{2E} \ln\{1 + \exp[2\pi(E - V_B)/\hbar\omega]\} \tag{1.3}$$

where R_B amd V_B are the position and height of the s-wave barrier (see Fig.1). It may be shown[2] that typical values of $\hbar\omega$ are around 3 MeV, so for energies $E - V_B \gg \hbar\omega/2\pi \approx 0.5$ MeV Eq.(1.3) reduces to the simpler result

$$\sigma_f = \pi R_B{}^2(1 - V_B/E) \tag{1.4}$$

which one would have obtained on the assumption that all partial waves up to the grazing angular momentum ℓ_g fuse (see Fig.1). The great advantage of Eq.(1.3), however, is that it is good close to the barrier and in particular <u>below</u> the Coulomb barrier it is capable of describing the exponentially decreasing cross section coming from "fusion by penetration". For systems comprising a spherical target and projectile this expression works rather well with values of $\hbar\omega$ consistent with those quoted above. However, for heavy deformed systems one finds a large enhancement of σ_f below the barrier which can be fitted only with an artificially large value of $\hbar\omega$ (very narrow barrier) in this model.

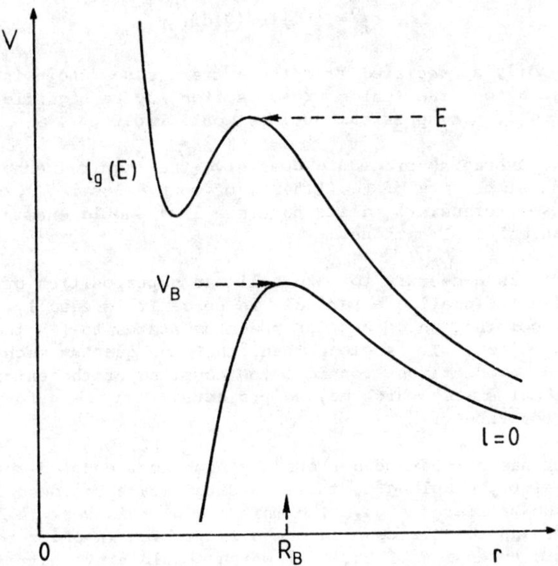

Fig.1. A typical nucleus-nucleus potential including nuclear, Coulomb and centrifugal terms. The s-wave barrier has position and height R_B and V_B. For the incident energy E we indicate the total potential for the grazing partial wave ℓ_g.

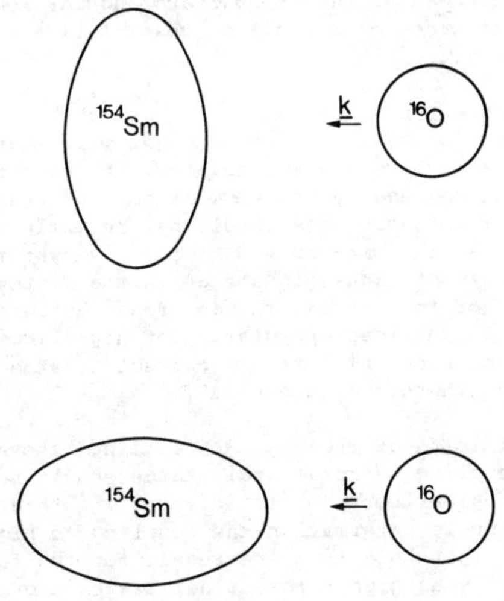

Fig.2. The two extreme orientations for which fusion is most inhibited (top) and most enhanced (bottom).

Semi-classically the reason for this enhancement is rather clear — a deformed even-even nucleus in its 0^+ ground state should be thought of as having an equal probability of being oriented in any direction. The fusion cross section should, therefore, be evaluated for each orientation and then averaged[3] i.e.

$$\sigma_f = \frac{1}{4\pi} \int \sigma_f(\underline{\Omega})d\underline{\Omega}. \tag{1.5}$$

It will be readily appreciated from the extreme cases shown in Fig.2 that for some orientations the fusion cross section may be significantly larger than scattering from a spherical nucleus would imply.

The above averaging procedure does give the required enhancement but the analytical simplicity of Eq.(1.3) is of course lost. In addition this procedure is very classical in its nature and we should examine closely its quantum mechanical implications.

Just as it is necessary to take a linear superposition of momentum states in order to localise a particle in space it is equally necessary to take a linear combination of angular momentum states to fix the orientation of a deformed object. It is clear then, that the quantum mechanical approach to the problem must consider the coupling of the entrance channel to the rotational states which may be projected from the deformed intrinsic structure of the target.

Indeed it has already been shown[4] by numerical coupled-channels calculations that simply coupling to the 2^+ excited state can describe the major features of sub-barrier fusion. The main aim of this work is to use an adiabatic approach to this coupled-channels problem in order to derive a simple analytic extension of Eq.(1.3) which should prove extremely useful in fitting experimental data.

Other simple adiabatic models (e.g. reference[5]) have been used to explain the qualitative features of this problem. The major difference between these models and the work presented in this talk is that we shall give a rigorous treatment of the appropriate angular momentum algebra in order to obtain an expression capable of quantitative comparison with full coupled-channels results.

One might, however, have thought that for sub-barrier fusion the s-wave would dominate since the barrier height will increase further with increasing ℓ. For a massive system, however, the barrier increases rather slowly compared with the energy scale $\hbar\omega/2\pi$ so that even below the barrier many partial waves may fuse. This result has recently been elegantly verified experimentally at the Daresbury NSF[6] for the system ^{80}Se + ^{80}Se. By observing the γ-decays of high-spin states in the evaporation residues a qualitative picture of the partial fusion cross sections may be built up. This demonstrates a significant population of high-ℓ compound nucleus states even at sub-barrier energies. A correct treatment of the angular momentum algebra is, therefore, essential.

A possible criticism of the approach outlined above is that for large deformations the coupling to rotational states should be strong and it may be inadequate to consider coupling to only the 2^+ state. However, for nuclei with a quadrupole deformation the coupling to higher states will be sequential and one might hope that the result for the fusion cross section would converge rapidly as higher rotational states are included. We shall demonstrate that this appears to be the case by explicitly including coupling to the 4^+ state in a case where the nuclear deformation is large. We shall see that this causes very little change in the theoretical fusion cross section.

In the case of just a 2^+ coupling we shall show that the system of 4 coupled equations reduces drastically to two uncoupled elastic scattering problems[7] and that the fusion cross section may be expressed as a simple weighted average of its values for these two channels. A further advantage of our method is that we shall also obtain an expression for the total inelastic cross section which is numerically easier to evaluate than the DWBA and which gives, for large deformations, a more accurate result than this approximation.

2. ILLUSTRATIVE EXAMPLE

Consider the scattering of two spin-zero nuclei one of which has a 0^+ excited state at a low excitation energy ε which is strongly coupled to the ground state. (Although this is a rather unrealistic physical problem it illustrates rather well the technique used in the case of coupling to a low-lying 2^+ state.) For the partial wave ℓ the coupled-channels equations for the problem may be written (in an obvious notation)

$$\begin{aligned}(T_\ell - V_{00} + E)\phi_0 &= V_{01}\ \phi_1 \\ (T_\ell - V_{11} + E - \varepsilon)\phi_1 &= V_{10}\ \phi_0\end{aligned} \tag{2.1}$$

with

$$T_\ell = \frac{\hbar^2}{2m}\left(\frac{d^2}{dr^2} - \frac{\ell(\ell+1)}{r^2}\right).$$

If we take the (possibly complex) central potentials to be identical i.e. $V_{00} = V_{11} = V$, take $V_{10} = V_{01} = F$ and let $\varepsilon \to 0$ we obtain the coupled system of equations

$$H_D\phi = (T_\ell - V + E)\begin{pmatrix}1 & 0\\ 0 & 1\end{pmatrix}\begin{pmatrix}\phi_0\\ \phi_1\end{pmatrix} = F(r)\begin{pmatrix}0 & 1\\ 1 & 0\end{pmatrix}\begin{pmatrix}\phi_0\\ \phi_1\end{pmatrix}. \tag{2.2}$$

Since the coupling form factor is multiplied by a constant matrix it is possible to diagonalise these equations with a unitary transformation which is <u>independent of r</u> i.e.

$$U\ V_{CC'}\ U^+ = \begin{pmatrix}\lambda_0 & 0\\ 0 & \lambda_1\end{pmatrix} F(r) = \begin{pmatrix}1 & 0\\ 0 & -1\end{pmatrix} F(r) \tag{2.3}$$

with

$$U = \frac{1}{\sqrt{2}}\begin{pmatrix}1 & 1\\ -1 & 1\end{pmatrix}. \tag{2.4}$$

In other words our coupled equations reduce to[8]

$$\begin{aligned}\left[T_\ell - (V + F) + E\right]\tilde{\phi}_0 &= 0 \\ \left[T_\ell - (V - F) + E\right]\tilde{\phi}_1 &= 0\end{aligned} \tag{2.5}$$

with

$$\begin{aligned}\tilde{\phi} &= U\ \phi \\ \phi &= U^+\ \tilde{\phi}\ .\end{aligned} \tag{2.6}$$

Thus all the physical information may be obtained from the solution of two <u>elastic scattering</u> problems using a standard optical code.

One must, however, be careful to apply the correct boundary conditions to the uncoupled wavefunctions. Since the physical problem will have unit incident flux in the elastic channel the initial conditions for the uncoupled wavefunctions are given by

$$\begin{pmatrix}\tilde{\Phi}_0 \\ \tilde{\Phi}_1\end{pmatrix}_{\text{incident}} = U\begin{pmatrix}1\\0\end{pmatrix}. \tag{2.7}$$

This may clearly be written as

$$\begin{pmatrix}\tilde{\Phi}_0 \\ \tilde{\Phi}_1\end{pmatrix}_{\text{incident}} = \begin{pmatrix}U_{00} & 0 \\ 0 & U_{10}\end{pmatrix}\begin{pmatrix}1\\1\end{pmatrix} = \Lambda\begin{pmatrix}1\\1\end{pmatrix} \tag{2.8}$$

i.e. $\Lambda_{ij} = \delta_{ij}\, U_{jo}$. Thus we may take unit incident flux in each of the unphysical channels so long as we map back to the physical wavefunctions with the <u>non-unitary</u> transformation

$$\begin{pmatrix}\psi_0 \\ \psi_1\end{pmatrix} = U^+ \Lambda\begin{pmatrix}\tilde{\Phi}_0 \\ \tilde{\Phi}_1\end{pmatrix}. \tag{2.9}$$

Writing this explicitly for the S-matrix elements we have

$$\begin{pmatrix}S_{e\ell} \\ S_{in}\end{pmatrix} = \frac{1}{2}\begin{pmatrix}1 & 1 \\ 1 & -1\end{pmatrix}\begin{pmatrix}S_0 \\ S_1\end{pmatrix} \tag{2.10}$$

The fusion cross section is now

$$\begin{aligned}\sigma_f &= \frac{\pi}{k^2}\sum_\ell (2\ell+1)\left(1 - |S_{e\ell}|^2 - |S_{in}|^2\right) \\ &= \frac{\pi}{k^2}\sum_\ell (2\ell+1)\left(1 - \frac{1}{2}|S_0|^2 - \frac{1}{2}|S_1|^2\right) \\ &= \frac{1}{2}\left(\sigma_f^{(0)} + \sigma_f^{(1)}\right)\end{aligned} \tag{2.11}$$

i.e. the fusion cross section is just the average of that evaluated for the two potentials $V(r) + F(r)$ and $V(r) - F(r)$.

Eq.(1.3) may now be readily extended to this problem by using it in Eq.(2.11) with the values of R_B, V_B and $\hbar\omega$ appropriate to the two channels. A good approximation would generally be obtained by using the values of R_B and $\hbar\omega$ appropriate to the potential $V(r)$ but by taking two different barrier heights $V_B \pm F(R_B)$. It is clear from Eq.(1.3) that although one barrier is increased and the other decreased, the coupling will greatly enhance σ_f at energies below the barrier if $|F(R_B)| \gg \hbar\omega/2\pi$ irrespective of the sign of F. It is, however, also clear from Eq.(1.4) that at energies high above the barrier the effect of coupling will be almost entirely negligible.

A simple equation for the total inelastic cross section may be obtained from Eq.(2.10) i.e.

$$\sigma_{inel} = \frac{\pi}{k^2} \sum_{\ell} (2\ell + 1) \left|S_{in}\right|^2 = \frac{\pi}{4k^2} \sum_{\ell} (2\ell + 1) \left|S_0 - S_1\right|^2. \qquad (2.12)$$

Clearly no simple phenomenological expression can be given for this quantity since we require the relative phases of the S-matrix elements in the two uncoupled channels. Note, however, that the evaluation of σ_{inel} is actually simpler in this model than in the DWBA. In this latter approximation the elastic scattering wavefunctions in the potential V(r) must be folded with the coupling form factor F(r). In our method, however, we need only calculate elastic S-matrix elements in the potentials $V(r) \pm F(r)$.

3. COUPLING TO A LOW-LYING 2^+ STATE

When coupling to a state of spin 2 is considered the angular momentum in the final inelastic channels may be $\ell' = J, J \pm 2$ where J is the total angular momentum. Thus in addition to having a different energy from the entrance channel we will also have different angular momentum barriers. For a system with large reduced mass however we may ignore these differences and take the barrier to be that for angular momentum J throughout. We thus again obtain an equation similar to Eq.(2.2) i.e.

$$(T_J - V + E) \begin{pmatrix} \psi_0 \\ \psi_1 \\ \psi_2 \\ \psi_3 \end{pmatrix} = V^J_{CC'} \begin{pmatrix} \psi_0 \\ \psi_1 \\ \psi_2 \\ \psi_3 \end{pmatrix} \qquad (3.1)$$

where 0 is the elastic channel and 1,2,3 are the three inelastic channels $\ell' = J - 2,\ J,\ J + 2$. In the rotational model the coupling may be written[9]

$$V^J_{CC'} = V^J_{I'\ell' I\ell} = 5^{-1/2}(-1)^J\, i^{\ell-\ell'+I-I'}\, \hat{\ell}\hat{\ell}'\hat{I}\hat{I}' \times \langle \ell 0 \ell' 0 | 20\rangle W(\ell I \ell' I'; J2) \langle I0I'0 | 20\rangle F(r) \qquad (3.2)$$

where $\hat{I} = (2I + 1)^{1/2}$ etc. Since this is again a constant matrix times the form factor F(r) it is clear that we may apply the same techniques as in the previous section to write the physical S-matrix elements in terms of the S-matrix elements for four uncoupled elastic channels (though the transformation now depends explicitly on J) i.e.

$$\begin{pmatrix} S_{e\ell} \\ S^{J-2}_{in} \\ S^{J}_{in} \\ S^{J+2}_{in} \end{pmatrix} = (U^+ \Lambda)^J \begin{pmatrix} S_0 \\ S_1 \\ S_2 \\ S_3 \end{pmatrix} \qquad (3.3)$$

where the S_i are evaluated in the potentials $V(r) + \lambda^J_i F(r)$ with λ^J_i the eigenvalues of the constant part of the coupling matrix in Eq.(3.2). The diagonal matrix Λ^J is again given by $\Lambda^J_{ij} = \delta_{ij}\, U^J_{jo}$ and is again necessary to preserve the correct boundary conditions.

The transformation $\tilde{\phi} = U^J \phi$ which diagonalises $V^J_{CC'}$ may readily be shown to transform our original basis functions $|(\ell I)J\rangle$ into the strong coupling basis[10] so that each of the decoupled channels may be labelled by a quantum number K which is the projection of the nuclear spin I along the nuclear symmetry axis. Since I = 0,2 we obtain two K = 0 states, one K = 1 and one K = 2. Since the entrance channel has I = 0 it must be pure K = 0 and thus only two of the decoupled channels are needed to reproduce all four physical S-matrix elements i.e.

$$S_{physical} = (U^{+} \Lambda)^{J} \begin{pmatrix} S_0 \\ 0 \\ 0 \\ S_3 \end{pmatrix} . \tag{3.4}$$

(Note that our approximation of setting all centrifugal barriers equal essentially ignores Coriolis coupling which would have mixed the different K states.) In addition to the above simplification the eigenvalues λ_i^J may be shown to be independent of J. The relevant ones are $\lambda_0 = 0.73$ and $\lambda_3 = -1.37$. So our problem once again reduces to two elastic scattering problems with different potentials which are both independent of J.

Although the non-unitary transform in Eq.(3.4) does depend on J a little further angular momentum algebra yields the additional simplification

$$\begin{aligned} \sigma_f &= \frac{\pi}{k^2} \sum_{J,\ell'} (2J+1)\left(1 - |S_{el}|^2 - |S_{in}^{\ell'}|^2\right) \\ &= \frac{\pi}{k^2} \sum_J (2J+1)\left(1 - 0.652\, |S[V + \lambda_0 F]|^2 - 0.348\, |S[V + \lambda_3 F]|^2\right) \\ &= 0.652\, \sigma_f(V + 0.73F) + 0.348\, \sigma_f(V - 1.37F), \end{aligned} \tag{3.5}$$

with no explicit J-dependence in the coefficients. Again this gives no change in the fusion cross section at high energies since $0.652\lambda_0 + 0.348\lambda_3 = 0$. At sub-barrier energies, however, it is clear that a substantial enhancement will occur if $|F(R_B)| \gg \hbar\omega/2\pi$ irrespective of the sign of F and thus of whether the nuclear deformation is prolate or oblate.

The expression analogous to Eq.(2.12) for the total inelastic cross section is

$$\sigma_{inel} = 0.227 \frac{\pi}{k^2} \sum_J (2J+1)\, |S_0 - S_3|^2 . \tag{3.6}$$

In ref.[7] the case of coupling to a vibrational 2^+ state has been treated. The coupling matrix is the same as in the rotational model except that the re-orientation terms connecting the three inelastic channels ($\ell' = J, J \pm 2$) to one another vanish. The result for σ_f reduces very simply to Eq.(2.11) as for coupling to a 0^+ state.

4. COMPARISON WITH EXACT CALCULATIONS

a) Fusion cross section

In order to test our expression for the fusion cross section we now make a comparison with some full coupled-channels calculations for the $^{16}O + {}^{154}Sm$ system. This case has been studied in detail by Stokstad and Gross[4] and we use here the same potential as these authors. Here we must be clear about what we interpret as the fusion cross section. In a calculation without coupling we shall take the total reaction cross section σ_r

to be σ_f. In the case where we have coupling we define $\sigma_f = \sigma_r - \sigma_{inel}$ (see Eq.(2.1)). In other words we assume that all absorption corresponds to compound nucleus formation except for the channels explicitly coupled in.

The dashed line in Fig.3 shows the fusion cross section for the $^{16}O + ^{154}Sm$ reaction if coupling to the 2^+ state is ignored. The solid line shows the fusion cross section when the 2^+ state is included in the exact coupled-channels problem and displays a large enhancement relative to the no-coupling result shown by the dashed line. The crosses in this figure show our adiabatic approximation to the full curve and we see that on the scale of this figure the results are indistinguishable. The differences are in any case significantly less than the experimental error bars for this system[11] (see Fig.6). The dotted line in Fig.3 shows how the fusion cross section behaves if coupling to the 4^+ state is also included (see below).

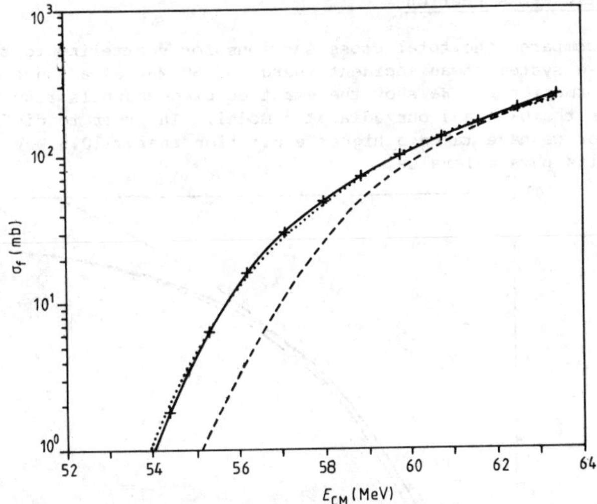

Fig.3. The fusion cross section for $^{16}O + ^{154}Sm$ as calculated a) without coupling (- - -), b) with the exact coupled-channels code and coupling to the 2^+ state only (———), c) our adiabatic approximation to this (+ + +), and d) full coupled-channels including the 4^+ state (•••••). The calculations use the potential of Stokstad and Gross[4]. Figs.3-7 are reproduced from ref.[7] courtesy of the Institute of Physics.

The remarkable quality of the adiabatic model fit to the exact calculation may be understood as follows. The energy scale for the sub-barrier fusion reaction is essentially $\hbar\omega/2\pi \approx 0.5$ MeV and clearly a large enhancement will be obtained if $F(R_B)$ is significantly larger than this. In the present case $F(R_B) \approx 2$ MeV (including contributions from the nuclear and Coulomb fields) and enhancements of the order of a factor 5 are obtained. We have, however, ignored the excitation energy of the 2^+ state and differences in the centrifugal barriers and it is clear that these must be <u>small</u> compared with $\hbar\omega/2\pi$. For this case $\varepsilon(2^+) \approx 80$ keV and the difference is the centrifugal potentials at the barrier may be written $\Delta V_\ell \approx 60\ell$ keV and the necessary inequalities are fulfilled except for high angular momenta which should not contribute too much at energies below the barrier.

In a collective model the nuclear part of the coupling potential may be written in terms of the derivative of the central nuclear potential[9] thus

$$F(R_B) \propto \beta R_B \frac{dV_N}{dr}\bigg|_{R_B}, \tag{4.1}$$

where β is the nuclear deformation. However the barrier occurs at a point where $dV_N/dr = - dV_c/dr$ with $V_c = Z_1 Z_2 e^2/r$. It is clear, therefore, that in such a model $F(R_B) \propto \beta Z_1 Z_2$ and thus we should only expect significant enhancement of σ_f for heavy deformed systems. The Coulomb contribution to $F(R_B)$ is also clearly proportional to $\beta Z_1 Z_2$ though smaller and of the opposite sign[7].

b) Inelastic cross section

Fig.4 compares the total cross sections for scattering to the 2^+ state for the above system at an incident energy of 58 MeV as a function of the nuclear deformation β. We show the exact coupled-channels results as well as those for the DWBA and our adiabatic model. In order to disfavour our model however we have taken a higher excitation energy (0.5 MeV) for the 2^+ state than its physical value.

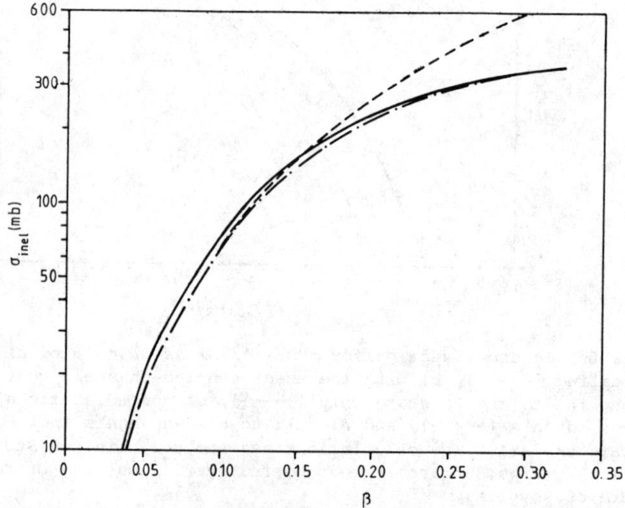

Fig.4. The total inelastic cross section for scattering to a 2^+ state at 0.5 MeV and an incident energy of 58 MeV as a function of the deformation β. The figure shows a) full coupled channels results (-•-•), b) our adiabatic model (———) and c) the DWBA. The potentials are again those of ref.[5].

Since the DWBA takes account of this energy loss and of changes in the centrifugal potential it fits rather well the exact results at small deformation. At larger deformations though this first-order theory breaks down.

Here our adiabatic model is vastly superior since it takes into account all orders in the coupling which are now more important than the neglected ε and ΔV_ℓ.

5. EXTENDING THE PHENOMENOLOGICAL MODEL

In the Wong approximation we have seen that we may write

$$\sigma_f = 0.652\ \sigma_w\left[V_B + 0.73\ F_B\right] + 0.348\ \sigma_w\left[V_B - 1.37\ F_B\right] \qquad (5.1)$$

where σ_w is the expression of Eq.(1.3) and $F_B = F(R_B)$. This expression assumes that no absorption takes place outside the Coulomb barrier and so we compare it in Fig.5 with a full coupled-channels calculation where the strong imaginary potential was confined to the interior of the Coulomb barrier. The comparison is seen to be excellent except for a long way below the Coulomb barrier where the Wong assumption of parabolic barrier transmission coefficients is poor.

Fig.6 shows experimental data[11] for the $^{16}O + {}^{154}Sm$ system and a fit to these data using a) the proximity potential of Blocki et al.[12] to obtain $\hbar\omega$ = 4.3 MeV and b) the coupling strength of Stockstad and Gross[4] in Eq.(5.1). We see that Eq.(5.1) is capable of an excellent fit to experiment and it could prove extremely useful in this respect since it only contains one additional parameter and is simple to apply.

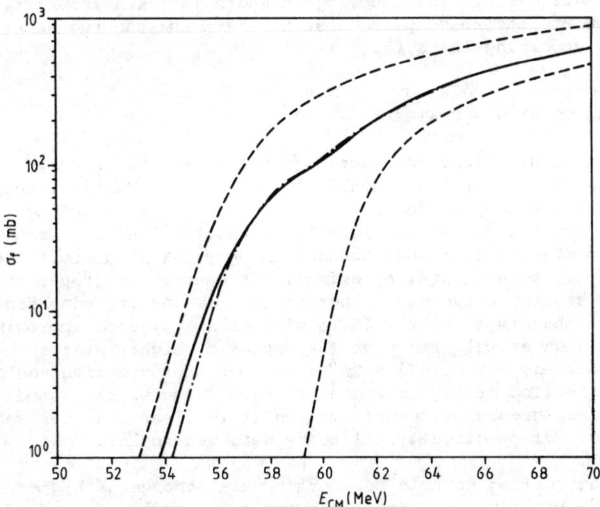

Fig.5. The fusion cross section of the extended Wong formula (Eq.(5.1)) (———) compared with the full coupled-channels result (-•-•). The dashed lines show σ_f for the two separate channels which make up the Wong result.

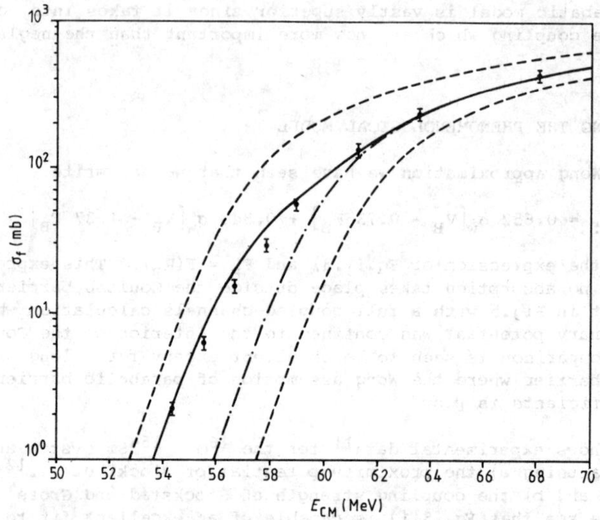

Fig.6. The extended Wong fit to the data of ref.[11] (———) using $\hbar\omega$ = 4.3 MeV[12]. The coupling strength is taken from ref.[4]. Also shown are the no-coupling result (-•-•) and the two separate channels appearing in Eq.(5.1) (- - - -).

6. COUPLING TO HIGHER STATES

At large deformations the couplings become strong and maybe higher states in the rotational band should be coupled in. We have, therefore, performed an exact coupled-channels calculation of σ_f including the 4^+ state. The result is shown in Fig.3 by the dotted line and is seen to be very little different from that found with only the 2^+ state. Inclusion of the 6^+ state has an even smaller effect. It seems, therefore, that the fusion cross section converges rather rapidly as one includes higher states and thus that the simple result (5.1) will be a very good approximation to σ_f. This is just as well since as the energy of higher states increases as $I(I + 1)$ we already have $\varepsilon(6^+) \approx \hbar\omega/2\pi$ and our approximation would be poor for the contribution of higher states to σ_f. Since we have studied a system with a large ground state deformation it is clear that our result should be good for practically all heavy deformed nuclei.

It is interesting to note how our formulae change as higher states in the rotational band are included (ignoring their excitation energies and changes in centrifugal barriers). Each additional state gives one extra K = 0 channel in the uncoupled system. This leads to one extra elastic channel in which the fusion cross section must be calculated and averaged over e.g. for inclusion of the 4^+ state we find

$$\sigma_f = 0.17\ \sigma_f(V - 1.8\ F) + 0.36\ \sigma_f(V - 0.35\ F) + 0.47\ \sigma_f(V + 0.93\ F). \quad (6.1)$$

In the limit of including <u>all</u> states the expression should finally become equivalent to averaging over <u>all</u> orientations where the coupling term would be the appropriate quadrupole field for the given orientation [13]. We have, however, shown that an extremely good analytic approximation to this situation (which in any case may not provide a more accurate result since higher states may not be considered as degenerate) may be obtained by considering a simple average over just <u>two</u> different potentials!

REFERENCES

1. C.Y. Wong, Phys. Rev. Lett. 31 (1973) 766.

2. N. Poffé, N. Rowley and R. Lindsay, Nucl. Phys. A410 (1983) 498.

3. W. Scobel et al., Phys. Rev. C11 (1975) 1701.

4. R.G. Stokstad and E.E. Gross, Phys. Rev. C23 (1981) 281.

5. C.H. Dasso, S. Landowne and A.Winther, Nucl. Phys. A405 (1983) 381; Nucl. Phys. A407 (1983) 221.

6. P.J. Nolan et al., private communication.

7. R. Lindsay and N. Rowley, J. Phys. G10 (1984) 805.

8. N.F. Mott and H.S.W. Massey, Theory of Atomic Collisions 3rd edn. (Oxford: Clarendon, 1965) 349.

9. B. Buck, A.P. Stamp and P.E. Hodgson, Phil. Mag. 8 (1963) 1805.

10. A. Bohr and B.R. Mottelson, Nuclear Structure vol.II (Reading, Mass.: Benjamin, 1975).

11. R.G. Stokstad et al., Phys. Rev. C21 (1980) 2427.

12. J. Blocki et al., Ann. Phys., NY 105 (1977) 427.

13. P.M. Jacobs and U. Smilansky, Phys. Lett. 127B (1983) 313.